GEO-PLATINUM 87

The Proceedings of the Symposium Geo-Platinum 87 held on 22–23 April 1987 at the Open University, Milton Keynes, UK

GEO-PLATINUM 87

Edited by

H. M. PRICHARD, P. J. POTTS,

Department of Earth Sciences, The Open University, Milton Keynes, UK

J. F. W. BOWLES

and

S. J. CRIBB

Mineral Industry Research Organisation
Lichfield, UK

ELSEVIER APPLIED SCIENCE
LONDON and NEW YORK

ELSEVIER SCIENCE PUBLISHERS LTD
Crown House, Linton Road, Barking, Essex IG11 8JU, England

Sole Distributor in the USA and Canada
ELSEVIER SCIENCE PUBLISHING CO., INC.
52 Vanderbilt Avenue, New York, NY 10017, USA

WITH 57 TABLES AND 124 ILLUSTRATIONS

© 1988 ELSEVIER SCIENCE PUBLISHERS LTD
Softcover reprint of the hardcover 1st edition 1988

British Library Cataloguing in Publication Data

Geo-Platinum 87.
1. Platinum metals. Geochemical aspects
I. Prichard, H. M.
553.4'95

Library of Congress Cataloging-in-Publication Data

Geo-platinum 87 (1987 : Open University)
Geo-platinum 87.

'Proceedings of the symposium Geo-Platinum 87,
held on 22–23 April 1987, at the Open University,
Milton Keynes, UK.'—Half t.p.
Bibliography: p.
Includes index.
1. Platinum-group—Congresses. I. Prichard, H. M.
II. Open University.
TN490.P7G42 1987   622'.3424   88-3910

ISBN-13: 978-94-010-7102-4   e-ISBN-13: 978-94-009-1353-0
DOI: 10.1007/978-94-009-1353-0

Phototypesetting by Keyset Composition, Colchester, Essex

PREFACE

The Geo-Platinum 87 Symposium, held at the Open University during April 1987, was designed as a forum for presentation of new research results on the occurrence, genesis, geochemistry, mineralogy and analysis of the platinum-group elements (PGE). With the support of the Open University and the Mineral Industry Research Organisation, the symposium was attended by 115 representatives of university departments, research institutions and members of the mining and mineral exploration industries.

An introduction to the symposium was provided by two invited papers from C. J. Morrissey (Riofinex North) and C. R. N. Clark (Johnson Matthey) which were designed to give perspective to the goals of PGE research work. The first of these papers gave a provocative insight into the aims and objectives of an exploration manager, examining the influence of supply, demand and perceived world reserves on exploration strategy. The second invited paper gave a valuable view of the industrial uses, market trends and predicted changes in the commercial value of the platinum-group elements from the standpoint of a refining company and supplier. These invited papers are reproduced in this volume and are followed by twenty-four full papers and twenty abstracts that reflect the wide range of research topics presented at the symposium.

Developments in analytical methods and experimental and theoretical aspects of the PGE are described as well as papers evaluating the distribution and processes concentrating the PGE in a variety of geological settings. Indeed a special feature of both the symposium and this volume is the diversity of occurrences of the platinum-group metals. Papers presented here report platinum in areas such as Norway, Finland, Scotland, Italy, Spain, Morocco, Canada, California, Hawaii, Brazil, Sierra Leone, Zimbabwe, Namibia, the Philippines, Indonesia as well as the more familiar Bushveld and Stillwater deposits.

A common theme through these papers is that the processes of magmatic concentration often produce a mineralogical association of Os, Ru, Ir with chromite and of Pt, Pd and Rh with sulphides. Several papers discuss the role of fluid-rich magmas as collectors of the PGE. Furthermore, a substantial advance in an understanding of the low-temperature chemistry of the platinum-group elements is reported in this volume together with developments in the analysis of these elements by bulk techniques and autoradiography. If this volume has a single aim, it is to give the reader a taste of contemporary research activities which are now being pursued outside of the traditional field areas and lithologies and to further an understanding of concentration processes that lead to the development of potential deposits.

H. M. PRICHARD
P. J. POTTS
J. F. W. BOWLES
S. J. CRIBB

CONTENTS

Theoretical and Experimental Studies

Europe and North Africa

North and South America

Equatorial and South Africa

CO-ORDINATING COMMITTEE

H. M. Prichard	Department of Earth Sciences, Open University
P. J. Potts	Department of Earth Sciences, Open University
J. F. W. Bowles	Department of Earth Sciences, Open University
S. J. Cribb	Mineral Industry Research Organisation

ADVISERS

N. J. Roberts	Mineral Industry Research Organisation
I. G. Gass	Department of Earth Sciences, Open University
G. C. Brown	Department of Earth Sciences, Open University

ORGANISING BODIES

The Open University
The Mineral Industry Research Organisation

CHAIR and REFEREES

S.-J. Barnes	Université du Québec à Chicoutimi
G. C. Brown	Open University
L. J. Cabri	Canadian Centre for Mineral and Energy Technology
J. H. Crocket	McMaster University, Ontario
E. A. Mathez	American Museum of Natural History
A. J. Naldrett	University of Toronto
R. W. Talkington	Stockton College, New Jersey

ADDITIONAL REFEREES

G. R. Gilmore	Universities Research Reactor Centre, Risley
A. G. Gunn	British Geological Survey, Keyworth
C. J. Hawkesworth	Open University
G. T. Ibbs	Caleb Brett Ltd
R. A. Ixer	University of Aston
R. A. Lord	Open University
J. W. Lydon	Geological Survey of Canada
J. W. Morgan	US Geological Survey
C. R. Neary	Natural Environment Research Council
B. Orberger	RWTH, Aachen
S. J. Parry	Imperial College Reactor Centre
S. Roberts	Southampton University
N. W. Rogers	Open University
E. F. Stumpfl	Mining University, Leoben, Austria

M. T. Styles	British Geological Survey, Keyworth
M. Tredoux	University of the Witwatersrand
D. H. Watkinson	Carleton University, Ottawa
P. Whitehead	Johnson Matthey Technology Centre, Reading
P. A. Williams	University College, Cardiff
A. H. Wilson	University of Natal
S. A. Wood	McGill University, Montreal

1

Exploration for Platinum: a Contemporary Viewpoint

C. J. MORRISSEY
Managing Director, Riofinex North Ltd, PO Box 50, Castlemead, Lower Castle Street, Bristol BS99 7YR, UK

There are two ways of looking at the recent emergence of platinum as a prime exploration target. One view is that it simply illustrates the abiding fatuousness of exploration geologists—their herd instinct, their inability to concentrate on more than one thing at a time and their slender grasp on economic realities. From the other point of view it is a classic case of market forces doing their thing—investment rushing to meet opportunity and plug anticipated gaps in the supply of essential materials.

The first view is easier to support with statistics than the second, and indeed it is hard to deny that exploration companies are creatures of fashion and do get fixated on particular metals. (This is especially true of junior companies, many of which have to trim their sails to the fickle winds of the Stock Market.) These characteristics of the exploration business are well illustrated by what has happened with gold—the supreme exploration bandwaggon of this decade (Fig. 1).

The statistics for gold exploration are all the more astonishing when you consider that:

1. Non-gold expenditures cover all the ferrous and ferroalloy ore metals, all the base metals and also uranium.
2. Much of the non-gold expenditure is made by companies that feel compelled to protect their market share in other metals.
3. Gold accounts for less than 20% of the value of the world trade in metals, and is one of a dozen metals in which international trade is worth more than a billion dollars a year.

Could platinum go the same way as gold and become the universal favourite of the exploration industry? I believe the straight answer to that is no, but I do think we are likely to see a lot more exploration for platinum in the coming years, especially among junior companies.

Comparisons with gold are inescapable when discussing platinum exploration and in my view we could be entering a phase of exploration activity that in the case of gold began 15 years ago when the fixed dollar price for gold—then $35—was abandoned. Since then its price has gone as high as $850 an ounce, and over the past couple of years has averaged about ten times what it was in 1970.

C. J. Morrissey

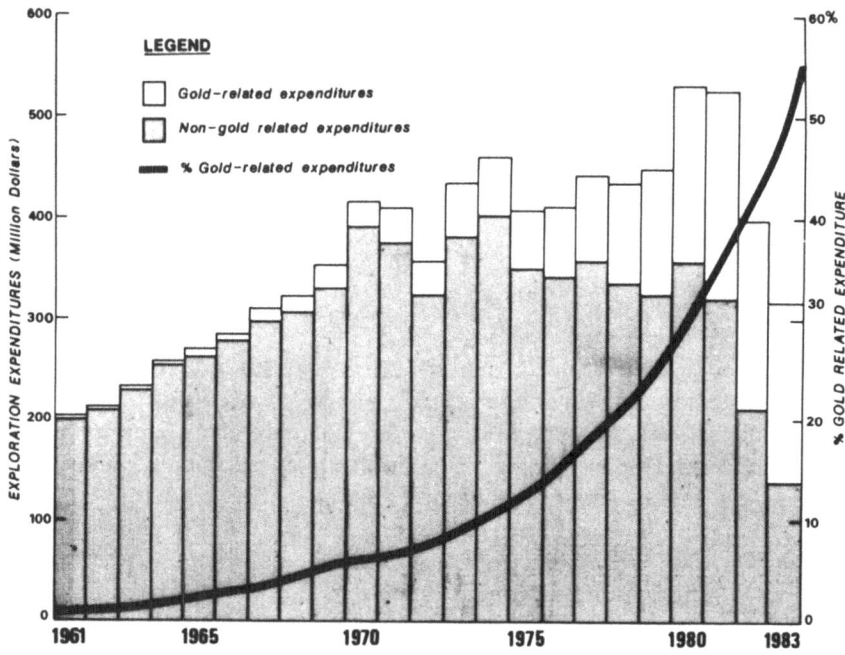

FIG. 1. US exploration expenditures (30 companies) (expressed in 1983 $). (Source: Schrieber & Emerson, *Engineering and Mining Journal*, October 1984.)

With this stimulus there have been tremendous advances in the understanding of the geology of gold, and these have been paralleled and promoted by advances in the technology of gold exploration and in the processing of gold ores.

Taking these one by one, it is not all that long ago that the geology of gold could be largely encapsulated in three words—Witwatersrand, placers and lodes. Now we have numerous different models to use in exploration—the hotspring model, the Carlin model, the porphyry model, the Hemlo model and so forth. Where exploration technology is concerned, there is now much greater interest in geochemical and isotopic fingerprinting of auriferous systems, and gold itself can be determined down to low ppb levels for a few dollars a sample.

This means that large areas can be screened for signs of gold mineralisation for little more than it used to cost to scout for base metals. Where process technology is concerned, activated carbon and ion exchange technology have allowed gold to be recovered economically from materials that were far too lean to count as ore, and the advent of improved analytical procedures has allowed much better control of processing operations.

The stimulus for these advances was a big increase in the price of gold, so before discussing where similar advances could lead us with the platinum-group metals (PGM) I want to highlight some similarities and differences between platinum and gold that are likely to have a bearing on the future price of platinum.

Table 1 gives some basic statistics about the volume and spread of primary production of platinum and gold in the non-communist world. One fundamental point of difference between the two metals is that production of gold in the year concerned was fifteen times greater than that of platinum and came from hundreds of mines spread around dozens of countries. By contrast, about 92% of the entire Western World's platinum supply in 1985 came from three mines, all of them in South Africa. (I include in that term the self-governing territory of Bophutatswana, and apologise to anyone who may regard that as glossing over an important distinction.) A fourth major producer is due to start up in South Africa in 1991, and the other three mines are all in the process of expanding their production.

South Africa's dominance of primary production is, therefore, much greater with platinum than it is with gold, for which the corresponding figure is about 55%. It also rests on a much more solid reserve base in that South African reserves down to depths well within the reach of existing mining technology would last for several hundred years at current rates of production and account for about 80% of Western World platinoid reserves overall (Table 2).

Outside South Africa, platinum production in the Western World is almost entirely as a by-product, notably from nickel ores in Canada and Australia. Again there is a parallel with gold, a small proportion of which comes as a by-product of mining for other metals, especially copper. The implication is that some platinum

TABLE 1

| | Western world production (1985) (tonnes) | % from S. Africa | Countries responsible for western world production | | | Number of primary producers |
			>10%	>5%	>1%	
Gold	1 212	55	1	4	12	Hundreds
Platinum	80	92	1	1	1	Very few

TABLE 2
Reserves of platinum group metals

	Tonnes	% of total
South Africa	24 570	80
Soviet Union	5 900	19
Canada	250	—
USA	30	—
Colombia	<1	—
Zimbabwe	<1	—
Total	30 750	100%

Source: Phillip Crowson, *Minerals Handbook 1986–7*.

could be driven out of the market by a slump in the price of other metals, but the amount involved is only a few per cent of total Western World production and in practice platinum tends to subsidise the production of the other metals.

As with gold, imports of metal from the USSR meet a proportion of the Western World demand for platinum. The figure in 1985 was about 8%, which is a good deal less than the corresponding figure for palladium (about 53%). As things stand, exports from the USSR are a major force in the palladium market but not nearly as influential in the platinum market, and to build up platinum exports the Soviet Union would probably have to produce a lot more palladium than it could sell. How much platinum the Soviet Union will want to sell to the Western World in coming years is very hard to say since it depends to a large extent on extraneous factors that affect their need for foreign exchange such as the price of oil, the size of their harvest and the pace of expansion and modernisation of their industry.

Their scope for dumping large quantities of platinum on Western markets may be fairly limited, but even so I would expect Soviet exports to have a restraining influence on the price of metal for quite some time.

Turning now to the nature of platinum as a market commodity, there are three points that I want to make. The first is that industrial demand for platinum-group metals has grown rapidly since the last war, though less rapidly in the last decade than in the ones before. I am happy to take it on the word of Johnson Matthey (Clark, this volume) that there is definite potential for strong growth in consumption over the next 10 years, and won't waste words on reiterating their pronouncement on this subject.

The second point is that the main area of growth over the past 5 years has been in investment demand. One can invest in platinum metal (as opposed to bits of paper and electronic signals that represent platinum) in two ways: by buying coins such as the Isle of Man noble, which people tend to hold on to, and by buying ingots, which tend to be traded on a more short-term basis. As with anything else, people will buy platinum if they think the price is going to go up or regard it as a safer store of wealth than others that are on offer. In this context platinum is much like gold: investment demand follows similar patterns and the price behaviour is similar (Fig. 2). Investment demand is likely to drive up the price of platinum, and follow it up some distance, if enough people believe one of three things:

1. That supplies from South Africa are likely to be seriously curtailed by the political problems of that country and their impact on its mining industry.
2. That demand for platinum will outpace its availability from the usual sources—new mine production, producer stocks and scrap, and exports from the USSR.
3. That a lot of other people believe one of the things I have just mentioned.

Until recently the trade in new-mined platinum was largely under long-term contracts based on producer prices, but since the precious metals boom of 1979/80 there has been a gradual change to free market pricing that has been brought about to some extent by greater interest in platinum as a medium for investment and speculation. The market in platinum is nothing like as free as that in gold, and I

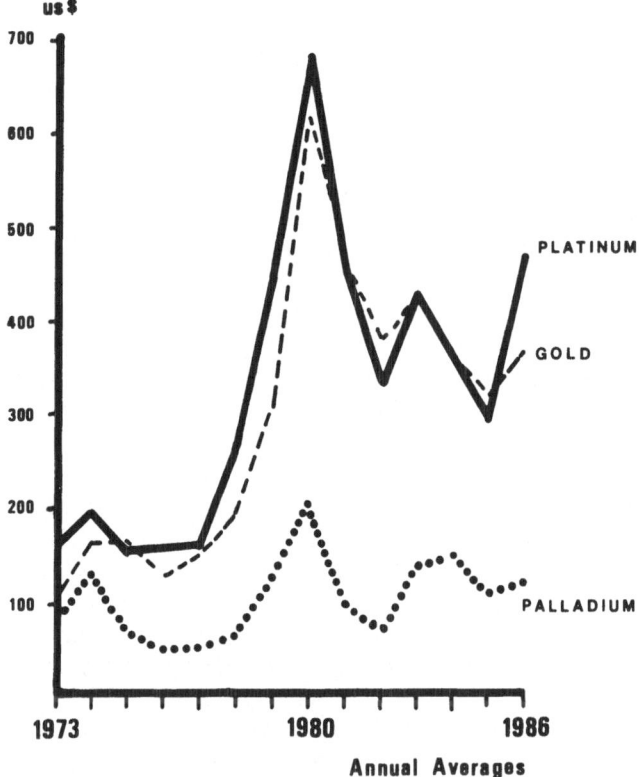

Fɪɢ. 2. Free market prices of platinum, palladium and gold.

wouldn't expect it to attain the same degree of freedom in the near future. In practical terms this means that any company that aspires to become a major platinum producer must put a lot more effort into securing markets for its output than it would have to with gold, and may stand a greater risk of being squeezed out of the market by existing producers. Against that, consumers might be particularly willing to enter into long-term contracts with aspiring producers that had large orebodies outside South Africa and weren't demanding a large premium for political security.

It is against this background, especially the statistics given in Table 2 and the knowledge that platinum mines in South Africa are technically quite capable of meeting increasing demands for platinum over many decades, that exploration companies in other parts of the world have to consider their stategy towards this metal. In doing so, wise exploration companies are guided by one of the most important principles of their business. This is that at any time and in almost any place the best type of orebody to look for is one that, by virtue of its geological quality and its location, can get its products to the market at a relatively low price and over a long period of time. Whether platinum orebodies of that sort are likely to be found using current concepts and current techniques is the next subject of my paper.

Unfortunately it is impossible to offer you a comprehensive review of the target concepts that are being used in platinum exploration at present. Exploration companies are secretive, especially when they are trying to break into a market that is dominated by a few producers, and they do not copy competitors in the perceptions and theories that underpin their efforts. Even so, it is possible to make some general comments about the types of target that seem to be receiving most attention these days, and also about other types that may warrant more attention than they are getting.

A fundamental point is that most exploration strategies for platinum attempt to exploit a strong association between platinum group metals and igneous rocks of mafic to ultramafic composition. Empirically, this association is so clear-cut that many people regard it as exclusive. Where, after all, does one find platinum mineralisation of economic interest in granites, andesites or dolomites? I am going to leave that as a rhetorical question, but I would point out that the laws of nature do not restrict platinum ores to mafic and ultramafic rocks. It is always a good idea in exploration to pay some attention to mineral occurrences that do not obey the known rules, and that suggests to me that platinum may still have a few tricks up its sleeve when it comes to forming ores.

Most exploration is based on analogy, and the most powerful analogue in this context is the Bushveld Complex. Any mafic mass that resembles the Bushveld has a head start as an exploration target, and some highly imaginative claims about the platinum potential of other complexes have been made largely on that basis. So, what is the Bushveld Complex, and what are the key features that make it so important as a host of platinum ores?

There is no hard-and-fast answer to either part of that question, but certain aspects of the Bushveld may be crucial to its value as a model of what to look for elsewhere. They are its size, its geological age, its geological setting (in which I include the nature of host rocks), and its internal make-up. Put baldly, it is a very large, early Proterozoic, intra-cratonic mafic body, strongly differentiated and, it seems, not compositionally independent of its host rocks. There is evidence of multiple magma injections and of hydrothermal activity during some stages of its evolution.

There is, however, a lack of consensus about the origin of the Bushveld platinum ores that leaves them in an equivocal position as an exploration target model.

What really matters about the Bushveld—its size, its age, the provenance of its magmas, their cooling history, sulphur inputs from country rocks or what?

One thing that nobody can argue about is the exceptionally large size of the complex, and I find it difficult to avoid the conclusion that that has a direct bearing on its fertility for platinum ores. Table 3 lists the published dimensions of the Bushveld and Stillwater Complexes, and of some other layered mafic complexes that are known to be platiniferous. I have not quoted published grades for current exploration targets on the list because not all of the available figures can be taken as

TABLE 3
Published outcrop dimensions of some platiniferous layered mafic intrusives

	Surface area/strike extent	Thickness (km)
Bushveld Complex	240 × 400 km	7–9
Stillwater, Montana	47 × 5 km	?
Great Dyke, Zimbabwe	530 × 12 km	>3·5
Freetown Complex, Sierra Leone (onshore)	48 × 14 km	6
Muskox, N.W.T., Canada	120 × 11 km	>2
Bird River Sill, Manitoba, Canada	>32 km	0·7
Fox River Sill, Manitoba, Canada (discontinuous)	250 km	2–2·5
Lac des Iles, Ontario	11 × 3 km	
La Perouse, Alaska	20 × 10 km	6
Munni Munni, Australia	200 km²	?
Jimberlana, Australia	180 km	5·5
Penikat, Finland	23 km	1·5–3·5

Source: various.

accurate, let alone representative. The only purpose of the table is to show what I mean by 'large' in this context.

Where age is concerned, believers in closely time-bracketed metallogenesis will no doubt find it significant that the Bushveld Complex is very similar in age to the Sudbury Complex (the main source of by-product platinum in Canada), and also reportedly to the platiniferous sulphide deposits in the Ungava province and the Thomson Nickel Belt in Canada. Another age coincidence occurs in the mid-Phanerozoic, exemplified by platiniferous sulphide deposits in the Noril'sk region of the USSR and the Insizwa Complex in South Africa. However, in view of the reputedly Archaean age of the Stillwater Complex and recent reports of ore grade PGM values in certain Tertiary complexes this is not a point that I feel confident in labouring. Likewise, there is nothing very hard-and-fast one can say about similarities in the geotectonic setting of platiniferous complexes, the provenance and original composition of their magmas, the role of contamination by sulphide-bearing country rocks or the importance of hydrothermal activity in ore formation. So it seems to me that the field is wide open where layered mafic complexes of sill-like character are concerned.

That seems to have been the experience of the exploration industry in recent years. The Stillwater Complex fulfils most of the main criteria I have mentioned in relation to the Bushveld, and the same can be said with varying degrees of certainty about several other large mafic complexes that seem to be giving promising exploration results.

A dark horse in the race may be ophiolite complexes. Platinoids are commonly reported in so-called alpine-type intrusions that are an integral part of such complexes, and they also occur in discordant, concentrically zoned intrusions in the ophiolitic terrains of Alaska, the Urals and some other parts of the world. Outside the USSR, where they may see things differently, ophiolitic complexes have a bad reputation as potential platinum sources of economic importance, the conventional wisdom being that:

1. Platinum itself is generally only a minor component of the PGM suite in alpine-type intrusions;
2. PGM only attain interesting concentrations in chromite-rich rocks, which generally do not amount to much in tonnage terms and are very difficult to deal with analytically and metallurgically.

All this is true, but I believe it is also true that a long-standing preoccupation with chromite-associated PGM in ophiolites has left other associations and modes of occurrence under-researched, with the result that the bad reputation that I referred to may be in the nature of a self-fulfilling prophecy.

The fallibility of conventional wisdom is reason enough for exploration geologists to take heed of some peculiar platinum occurrences that could have important metallogenic implications. One such is the well-documented occurrence of high PGM and gold values in a thin thucholitic shale horizon in the Zechstein of Poland (Kucha, 1982). The platiniferous unit is very organic, and heavy metals are thought to have been concentrated in it by the action of algae.

On the same lines, I was intrigued to read recently that trial recovery tests on oil shales in Colorado had yielded almost 70 g/t of gold, 186 g/t of silver and 'significant platinum values', with one sample yielding 130 g/t of platinum. This story may have got garbled in the telling, so I am waiting for further information before trying to build platiniferous oil shales into my view of the geological world.

The Polish example, in any case, underlines the poor state of existing knowledge about the geochemical behaviour of platinum-group metals in many types of natural environment. Ignorance seems to be particularly profound about the behaviour of platinoids in sea-floor and surface weathering environments.

I referred earlier to the way in which sustained interest in gold as an exploration objective has given rise to a range of new target concepts. Some of them hinge on the recognition that, under certain conditions of weathering, large quantities of gold can be taken into solution by meteoric waters and reprecipitated either as an enhancement of primary values or in deposits that are essentially supergene in origin. Fortuitously and deliberately, large deposits of supergene gold have been discovered in recent years, but because the platinum-group metals are still generally regarded as being very immobile in weathering systems I doubt whether anyone is looking seriously for, say, a PGM equivalent of the Boddington auriferous bauxite deposit in Western Australia. They may be quite right not to do so while there are still so many ultramafic sills, layered mafic complexes and ophiolite suites to test, but I for one do not believe that supergene platinum deposits are pure moonshine.

One straw in the wind, I believe, is that gold in weathered hematite schists in the Iron Quadrangle of Brazil, which is almost certainly supergene in origin, is alloyed with 4% palladium and also with traces of platinum and other metals of the group.

Another is that studies of alluvial platinoids have shown that the grains tend to be much larger than those in primary mineralisation from presumed bedrock sources (Cousins & Kinloch, 1976), which seems to lend some support to the fascinating proposal that unabraded euhedral aggregates of platinum-group minerals in drainages of the Freetown Complex in Sierra Leone have grown *in situ* through processes that must be common in laterite terrains (Bowles, 1986).

The author of the last paper referred to was quite explicit about the economic significance of his proposal—'that near-surface pockets high in platinum-group minerals may be found in appropriate horizons in laterites overlying basic or ultrabasic intrusions'.

EXPLORATION METHODS

Speculating about new types of exploration targets is an intellectual exercise and sometimes leads on to worthwhile field programmes, but the fact remains that exploration for platinum is still largely focused on conventional types of target. I suspect that most of the targets under current exploration were found simply by homing-in on large mafic and ultramafic intrusives shown on existing geological maps, or by hunting down chromite and copper–nickel occurrences in areas known to contain such intrusives. A few may have been found by aero-magnetic surveys or by regional geochemical surveys that are likely to have used copper, nickel and chromium as pathfinder elements. Where large mafic complexes are concerned, finding a prospective environment is probably the easiest part of the job.

Where real expertise is required is in checking the target out for platinum concentrations of sufficient grade, continuity and treatability to make ore, and in coming up with figures that will not look too limp and imprecise when subjected to the stern gaze of potential investors. That demands expertise in sampling and in assaying for PGM, and obviously it helps to know where the samples should be taken.

I suppose it would be possible to locate something like the Merensky Reef by sampling, metre by metre, the entire thickness of the layered mafic complex. However, sampling in that fashion would be tantamount to saying that there is no way of knowing where PGM concentrations are most likely to occur in such a complex, which is well wide of the truth. Analogy does offer us some guidelines for selective sampling of layered mafic complexes, but they are no more than guidelines, and other papers in this volume call into question some of the conventional wisdom on this subject by drawing attention to PGM concentrations in the upper levels of such complexes and in association with magnetite rather than chromite.

How should one go about sampling other types of geological target? Here again a geological concept can be very helpful—if it is right. If it is wrong, it can throw you off the scent entirely. I have to say that with many target types the suck-it-and-see approach may be the best—admit ignorance and sample everything.

I have been talking as if rock sampling was the only way to find platinum mineralisation. Of course, it is not: where geochemistry is concerned, almost everything one can do to find gold can in theory now be done with PGM, or at least some of them.

Panning—visual recognition of geochemical anomalies—is certainly applicable to PGMs and I am assured by professional colleagues that they have been able to make effective use of soils, stream sediments and certain types of vegetation as sampling media in geochemical exploration for platinum group metals. Where soils are concerned, that suggests to me that they are using laboratories capable of detecting platinum down to a few ppb and of achieving good precision in the concentration range below 100 ppb.

Table 4 indicates the amount of geochemical contrast that exists between ore and unmineralised rocks where platinum, palladium and gold are concerned. The indicative ore grades for platinum and palladium are based on figures for the Bushveld and Stillwater, while the range given for gold covers all but the richest and leanest of the hardrock mines in operation at present. The figures on the right of the table are disputable, but they do give some idea of what might constitute a geochemical anomaly in these rock types and in residual soils derived from them.

In many natural situations the particulate nature of gold would seem to weigh the dice against being able to find contourable gold anomalies in soils, but in my experience careful sampling, sample preparation and analysis can show up repeatable anomalies that can be contoured with confidence at levels as low as 15 ppb.

With gold, significant geochemical contrast in the low ppb range can now be quite easily detected at an analytical cost of about £4 or US$6 per sample. (This does not include the cost of sample preparation or of getting samples to the laboratory.) For the same cost either platinum or palladium can be measured down to 10 ppb—the cost of the two elements together being about £5 per sample. So far so good, and that's about as far as many exploration managers will want to go on the budgets they have these days. When they start to want assays for all the platinum-group metals they have to look carefully at their budgets because the cost per sample shoots up to anything between £50 and £200 depending on where and how the job is done. Even at those prices they could get rubbish results if they chose the wrong laboratory.

Quite apart from the problem of nugget effects, getting good precision for PGM

TABLE 4
Geochemical contrast between ores and unmineralised rocks
(values in ppb)

		Average abundance in unmineralised rocks			
	Indicative ore grade	Ultramafic	Mafic	Granitic	Sediments
Platinum	3 700– 4 800	32	30	8	?
Palladium	2 000–15 000	13	21	2	?
Gold	1 000–15 000	3–5	3–5	2–4	1–60

Sources: Govett, G. J. S. (ed.) (1983). *Handbook of Exploration Geochemistry*, Vol. 3, Elsevier, Amsterdam. Rose, A. W., Hawkes, H. E. & Webb, J. S. (1979) *Geochemistry in Mineral Exploration*, 2nd edn, Academic Press, London.

determinations on natural materials such as soils is not easy. Many things can go wrong, and they often do. For a start there are well over a hundred known mineral hosts of platinum-group metals, many more than there are for gold, so it is not always possible to be sure that the methods of attack and collection you have chosen will be 100% efficient. Then there is the question of matrix: some of the common matrices of platinum-group minerals are very refractory so again there can be problems with attack and collection. The next problem that sometimes crops up is extraneous platinum in the collector used in fire assay, which is commonly nickel sulphide.

After fire assay you can use atomic absorption, neutron activation or inductively coupled plasma to measure the PGM content of your prill. All are prone to inter-element interferences, between the platinum-group elements themselves and between them and other metals that get into the prill. There are ways of getting around this, but they do add to the cost.

Laboratories are very reluctant to admit to analytical errors—they have a number of let-outs that are very difficult to disprove—so the best advice I can give is to stick to laboratories with an excellent track record in PGM analyses and make constant checks on the reliability of their results by inserting standards and blanks into the sample stream and sending a proportion of all samples/pulps to a different laboratory.

I have laboured the analytical side of platinum exploration because, along with skill in target selection and in the design of sampling programmes, it is absolutely crucial to success. I should also mention the importance of mineralogical studies and preliminary metallurgical testwork in this connection. With noble metals in general, and platinum-group metals in particular, whole-rock assays are often not enough to show whether one has found potential ore: it is also necessary to show that the ore metals occur in a way that allows them to be concentrated economically by some chemical or physical process. With other metals that type of information is commonly not sought at an early stage in the exploration process, but the platinum-group metals have such an extensive mineralogy and, in certain types of deposit, such a tendency to form very fine-grained minerals enclosed in refractory silicates and spinels that assays alone can be very misleading about the economic worth of platiniferous mineralisation. To avoid wasting money on mineralisation that is going to be impossible to treat at a reasonable cost I advocate very early attention to the recoverability of platinum-group metals, involving both detailed studies of their mineralogy and distribution in the rock and bench-scale testing of their response to flotation or alternative methods of concentration.

There is very little I want to say about the role of geophysics in platinum exploration. The techniques that are currently used—magnetics for locating and defining mafic masses and getting a sense of their internal organisation, electro-magnetics for finding and tracing relatively sulphide-rich units, and so forth—are quite standard so their capabilities and limitations in this respect will be well known to experienced exploration geologists. Similarly, I don't think there is anything I can say about diamond drilling that hasn't already been said at dozens of symposia on gold exploration. As with gold, diamond drilling for grade information will be a

C. J. Morrissey

futile exercise if there is poor recovery of crucial sections of core, if the cores are sampled in a slovenly or unintelligent way, or if the samples are sent to a laboratory that switches them, contaminates them, or contrives in some other way to come up with rubbish numbers.

So much, so obvious, but even if it is done well where is all the platinum exploration that is going on now likely to leave us 10–20 years hence? Is there really likely to be a rash of major new platinum discoveries in places like Canada and Australia, Brazil and Unst, that will eventually topple South Africa from its perch? After more than 10 years of feverish gold exploration, Western world gold production has only increased by about 5% per annum in recent years, and South Africa is still responsible for over 50% of it. Why should things turn out any differently with platinum?

Where platinum consumption is concerned, Johnson Matthey are projecting that it will grow by 30% over the next 10 years, which is certainly some incentive to explore for new sources of supply. However, they are also projecting that the increased growth in demand will be largely matched by increased production in South Africa: they are expecting no significant change in the production of Canada over the next 10 years, and no new producers in Australia. Their projection recognises that it usually takes 5–10 years from the completion of a successful exploration programme to the start of commercial production whereas it only takes a year or two to expand the production of an existing mine. Against that the profit motive has lately been very effective in bringing new gold mines into production well within 10 years from discovery. In North America it is now taking as little as 2 years from discovery to production for gold deposits that are scheduled to produce 25 000–50 000 ounces per year (as many as the ounces of platinum that Johnson Matthey expect to come from Stillwater each year), and in Brazil production at an initial annual rate of 100 000 ounces will start later this year (1987) from a deposit that my own company did not begin to explore until 1981.

So, if platinum remains in fashion as an exploration target, which it will do if the price does not slump, it may turn out that Johnson Matthey have been too pessimistic about the chances of new discoveries leading to important new production over the next 10 years. Current exploration activities seem unlikely to lead to a spate of important ore discoveries outside South Africa (which would probably take the shine off platinum as an exploration objective), but they could bring at least one within hailing distance of production within the next 5 years. I hope that this symposium will help to make that possible.

REFERENCES

Bowles, J. F. W. (1986). The development of platinum-group minerals in laterites. *Econ. Geol.*, **81**, 1278–85.

Cousins, C. A. & Kinloch, E. D. (1976). Some observations on textures and inclusions in alluvial platinoids. *Econ. Geol.*, **71**, 1377–98.

Kucha, H. (1982). Platinum-group metals in the Zechstein copper deposits, Poland. *Econ. Geol.*, **77**, 1578–91.

2

The Platinum Market: Recent History and Future Developments

C. R. N. CLARK

Director of Platinum Marketing, Johnson Matthey plc, 43 Hatton Garden, London EC1N 8EE, UK

INTRODUCTION

Platinum-group metals are of the greatest strategic importance to the industrialised world. There is a vital need for platinum in many of the processes upon which we depend for our standard of living. These processes range from the production of basic fuels, chemicals and materials, through to high-technology applications in the protection of the environment and the treatment of pernicious disease. As developing countries emerge they too find a growing need for platinum's unique combination of intrinsic properties.

A radical change has occurred in demand patterns for platinum over the last 30 years stimulated by both the factors listed above and the public's growing awareness of them (e.g through driving cars with platinum catalysts or having a cancer cured by a platinum drug). Furthermore, special situations have arisen to highlight the unique applications of platinum metals. These factors have brought about a basic change in the structure of the market over the past decade.

I would like to start by reviewing the history of the platinum market, highlighting the major changes that have occurred in demand and supply since the last war and then to outline some thoughts on what might happen in the future, in particular over the next 5 years.

THE HISTORY OF THE MARKET

For many of its early years platinum was something of a nuisance. It contaminated the gold being sought by the Spaniards in Colombia, its intractable nature making its removal difficult. Some people even used platinum as a counterfeit for silver! Even after techniques were developed to separate and work platinum, its uses were rather limited, consisting largely of jewellery and coinage in Russia. These early metal workers probably had no idea of the future importance of this metal. In fact it was not until after 1945 that rapid growth in technology (stimulated by the war) gave rise to the need for more sophisticated materials to meet the complex requirements of newly developing technology.

Platinum has a unique blend of physical, electrical and chemical properties that makes this metal the sole answer to a host of problems created by the needs of modern industrial processes. Platinum melts at a higher temperature than most metals (1769°C), exhibiting strength at high temperatures. It can be worked easily into sheet or drawn into wire and, being a metal, is a good conductor of electricity. Furthermore, this noble metal is extremely inert. Like gold it will not easily corrode or react with other elements and remains bright and tarnish free in the most hostile of environments. Finally, platinum has a unique ability to catalyse many chemical reactions, a property that leads to its largest current use.

These factors have meant that *demand for this metal has doubled in each of the three decades between 1946 and 1976* and has increased by a further 40% in the last 10 years. This increase is more than three times the growth in demand for nickel over the same period. Overall demand has risen by a factor of ten in the last 40 years, an increase which, by any measure, is spectacular. But along with this growth, there have been significant changes in the pattern of demand.

The three major events, which accounted for most of this increase, each led to the doubling of demand in its decade. The 1950s saw the use of platinum as a catalyst by the oil industry, the Japanese jewellery market blossomed in the 1960s and a decade later catalytic converters to control exhaust pollution from cars were introduced in the USA and Japan. It is interesting to examine the impact of these developments on consumption of platinum over the years (Table 1). These data include figures for both 1986 and 1991 (based on a personal estimate) and predict little variation in the proportion of platinum used in different applications but with the 1991 consumption totalling some 17% higher than that for 1986.

I believe that these predictions are realistic and may even err on the side of caution. I now propose to examine the development of some of the sectors individually.

TABLE 1
Consumption of platinum in '000 oz troy
for 1946–1986 and the predicted consumption for 1991

	1946	1956	1966	1976	1986	1991
Autocatalyst	—	—	—	490	1 015	1 150
Jewellery	180	60	150	950	880	1 050
Investment	—	—	—	—	450	400
Chemical	40	200	300	280	195	200
Electrical	50	80	200	240	180	220
Glass	—	80	180	95	90	140
Petroleum	—	220	370	90	25	40
Fuel cells	—	—	—	—	—	10
Other	30	40	100	285	45	190
Total	300	680	1 300	2 430	2 880	3 400

Notes: Net consumption.
'Other' includes sales to Comecon/China.

FUTURE TRENDS IN THE USE OF PLATINUM

Automobile Catalysts

The imposition of stringent exhaust emission controls is spreading throughout the world from its origins in North America and Japan. Emission control legislation is already in place in Australia; Korea intends to introduce controls in 1988 and Taiwan has indicated the early 1990s for its programme. Recently Mexico has stated that it too would like to see controls imposed. However, it is Europe and Scandinavia which are, at the present time, the most important areas of growth as far as the use of this catalyst is concerned. Bearing in mind that the size of the market in terms of the number of vehicles is as large as, if not larger than, North America, a few comments on how the situation is developing are appropriate.

On the legislative front there is still considerable progress to be made within the EEC. Unfortunately, the historical perspective shows that in negotiating any consensus policy, each country, whatever its final objective, endeavours to secure as much benefit for itself (usually in completely disparate areas) before finally committing its support to the proposals. This is certainly the case with exhaust emission negotiations. Suffice to say that however protracted the exhaust control negotiations turn out to be, European car companies, especially those that already sell cars in the USA, seem to be proceeding along a voluntary course of compliance in advance of legislation. For example Mercedes are fitting catalysts to all their cars manufactured for the German market, BMW sell 'clean' cars as an option and Volvo from Holland sell catalyst-equipped small cars at no extra cost.

By 1991, it is certain that most of the legislation will be in place in Europe and that many new cars will be fitted with emission catalysts. These developments would stimulate an increase in demand for new platinum of 200 000 oz. However, it is possible that recovery from spent autocatalysts, especially in the USA and Japan, may also grow. I think that it is prudent, therefore, to discount this estimate of an increase of 200 000 oz and reduce the forecast *increase* for 1991 to 135 000 oz.

Jewellery

As can be seen from data in Table 1, the proportion of platinum used by the jewellery industry during the 1940s (after the war) was substantial, being concentrated in the USA. The reason would seem to be that during the 'Art-deco' period of the 1920s and 1930s, platinum became popular, since it matched the style of the times. When hostilities ceased people picked up where they left off. So why did its use fade away? For two reasons, I would suggest.

The first may well have been that the new generations growing up in the 1950s wanted something different from what had been stylish for their parents.

The second was that platinum was not readily available. The producers had control over supply and wished to foster and expand supply to new industrial uses. The producer price was available to industry but not always to the jewellers in the quantity they wanted. Through the late 1940s and 1950s platinum was often priced at three or so times more expensive than gold—which was itself fixed at $35 per oz. In consequence it was at this time that white gold became popular as a substitute for platinum. It is interesting to speculate on what the US platinum jewellery demand might have been today had more supplies been available in earlier years at competitive prices.

Present-day demand is primarily from Japan where significant use of platinum in jewellery began in the 1960s, grew steadily and peaked in 1975. At that time, restrictions on the imports of gold into Japan were lifted and the two metals competed on a more equal footing for the jewellery business. Subsequently escalation in prices curtailed demand even though the Japanese liking for platinum had not been displaced by gold.

Over the past 2 or 3 years we have seen demand increasing again in Japan and some new offtake in other parts of the world, particularly West Germany and Italy. On this basis, the figure predicted for 1991 in Table 1 is by no means unachievable. If earlier US taste for platinum were to return and the Japanese reverted to 1975 levels, the figure could be substantially higher.

Investment

Investment represents a new sector of demand which, from Table 2, can be seen to have grown over the last 5 years from virtually nothing to well over 450 000 oz per annum. In this context we define investments as coins, medallions and bars

TABLE 2
Platinum investment demand (troy oz)
1982–1986

1982	1983	1984	1985	1986
45 000	90 000	170 000	260 000	450 000

Note: Johnson Matthey defines investment as the net acquisition by private individuals of platinum investment products (small bars, medallions and bullion coins) weighing up to 10 oz each.

weighing 10 oz or less. This is of course very arbitrary but we need some means of distinguishing investment from speculative buying and selling, which is essentially of a short-term nature.

It was the gyration of the market in the late 1970s and 1980 and the accompanying media comment that drew the public's attention to platinum. The vital uses of the metal began to be publicised, not least because of its presence in most automobiles. Platinum began to move out of the exclusive domain of the rich speculator. The general public wanted to become involved in platinum investment and began to approach dealers and fabricators to purchase small amounts. This development was motivated by the desire to invest in something that was both tradeable and in which there was financial confidence. The platinum industry reacted to this new demand by producing a range of products to meet the varying requirements. These are now available in one form or another virtually everywhere in the Western world.

Between 1982 and 1985 the climate for precious metals as an investment vehicle was inhospitable. Prices were under pressure, inflation was falling, real interest rates were extremely high and stock markets were booming. In spite of this, demand in 1985 reached around 260 000 oz.

The price rises in 1986, falling interest rates and worries about currencies produced a more attractive atmosphere and demand in that year expanded to 450 000 oz.

If such a demand can be established during a period when circumstances are unfavourable, I believe it is likely to be sustained in the future. It is not realistic to predict with complete confidence a level of demand 5 years ahead for this type of activity. The figure for 1991 of 400 000 oz listed in Table 1 assumes that growth in the world economy over intervening years will be small. This is, I suggest, a reasonable approach. As investment demand tends to be more buoyant when the price is on the move, this figure could prove to be embarrassingly short of reality if market conditions become volatile before 1991.

Chemical, Electrical, Glass

These applications of platinum are the more mature uses that have become established solidly over the years. That is not to say there are no new developments in these areas. More economical ways of using the precious metals are constantly being found.

Demand in these areas tends to be related more to the overall level of industrial activity; increase in demand in particular is expected where that activity has been growing over a period and has culminated in an expansion of capacity.

While there has been steady economic growth over the last few years, its pace has been slow and production requirements have been met more by higher utilisation of existing capacity rather than by expansion. Industry's confidence to expand is still not as great as it was before the oil crises.

The view for 1991 is that growth, sporadic or otherwise, will have taken up all capacity by that date and that plant expansion will have become necessary. The figures used in Table 1 are certainly not extravagant—no higher than have already been seen at times in the past.

Fuel Cells

Substantial use of platinum as a catalyst in fuel cells is, of necessity, a long-term development. I do not expect significant offtake until the 1990s. For 1991 I have included a figure of 10 000 oz of platinum in Table 1.

However, phosphoric acid fuel cell (PAFC) technology is no longer at the experimental stage. In the USA, research and development work in this area no longer receives government support in the belief that fundamental research has now advanced beyond the stage where this is necessary. The problem that is being addressed at the moment in the US is how to overcome the initial high capital cost of building PAFC units in small numbers, until the economics of large-scale production reduce the price to a level where they are fully competitive with alternative sources of energy.

The situation in Japan is rather different. The government is still providing massive assistance, including financial support for the establishment of a fuel cell manufacturing capability. A growing number of cells are being built to prove their efficacy. It is quite likely that the first commercial PAFCs will come from Japan in spite of the huge technological lead gained by the USA a few years ago.

Oil
Demand for platinum in the oil industry has been affected greatly by technology. Improvements in catalysts and processes have perhaps halved the platinum content of the catalyst and, in some cases, tripled the catalyst's life. It is not easy to gauge the latter because many catalysts are now constantly being regenerated on site.

As demand for oil products grew in the 1960s and 1970s much of the expansion was met by surpluses created by technical improvements. The massive oil price increases imposed by OPEC in the mid-1970s stopped in its tracks the growth in oil demand. Everyone looked for ways of reducing their use of oil and sought alternative energy sources. The recent fall in oil prices will probably lead to renewed growth in demand. Surpluses have mostly been sold off and any expansion is likely to lead to new platinum buying. The 1991 forecast on Table 1 is modest and allows for only a small rise in demand.

SUPPLY

Having examined, in detail, the demand for platinum—what of supply? Data in Table 3 list the historical demand and supply for platinum since 1979.

History shows, not surprisingly, that growth in demand has been generally met from South African sources, where platinum is mined and extracted as a primary product. For a while Soviet sales to the West expanded to reach a peak of about 700 000 oz in 1976. Since then they have steadily declined. No one positively knows the reason for the reduction in their sales, but it is most likely to be due to growing domestic consumption, either in the Soviet Union itself or in other Eastern bloc countries. Soviet palladium deliveries to the West have not been reduced and in fact have increased. It is unlikely that the USSR would have preferentially stockpiled

TABLE 3
Platinum demand and supply ('000 troy oz)
1979–1986

	1979	1980	1981	1982	1983	1984	1985	1986
Demand	2 800	2 360	2 460	2 350	2 200	2 660	2 840	2 880
Supply	2 800	2 820	2 330	2 490	2 480	2 720	2 740	2 820
Movement in stocks	(80)	460	(130)	140	280	60	(100)	(60)

Note: Demand includes Western sales in China.

platinum for strategic reasons, not the least because there must be pressure in that country to maximise foreign currency earnings to compensate for falling oil revenues. But this is only an opinion.

If my forecast of a demand in 1991 of 3·4 million oz materialises, where is the extra 600 000 oz going to come from?

At present, although there is considerable activity in the exploration for new resources all over the world, only the Stillwater venture in the USA is progressing down the expensive path towards production. The developers estimate that in 5 years' time production may be up to 50 000 oz per year of platinum. If any other venture were to decide to develop a new mine today, a realistic time scale to achieve commercial mining and refining would be 5 years. We can therefore discount any other new mining ventures from the formula for 1991.

Existing output in the production of platinum in Canada may if anything decline slightly over the next 5 years. This leaves the two major producers, the Soviet Union and South Africa, to satisfy an increase in excess of 550 000 oz in new production.

Nobody outside the Soviet bloc can accurately predict the USSR sales levels to the West in 5 years' time. A credible estimate is that the level will increase by 50 000 oz to 300 000 oz. This leaves a balance of 2 880 000 oz that will have to be supplied by South Africa; an increase of 500 000 oz per year over current output. Western Platinum, South Africa's third largest supplier after Rustenburg Platinum Mines and Impala Platinum, have recently announced plans to double output adding around 100 000 oz per year to their production of platinum in 5 years' time. It must now be down to the top two producers to announce work on increasing their capacity. The investment required for this kind of expansion might be of the order of US$1·0 billion and one may imagine the degree of difficulty associated with a decision to invest that amount of money in South Africa under the present conditions of uncertainty.

CONCLUSION

By way of a conclusion perhaps I can pull together the threads of my argument and make a comment on prices.

Apart from the greater general awareness of platinum, for the variety of reasons I have suggested, the metal has attracted considerable comment over the past year because of what has been perceived as a dramatic price increase. Of course the price behaviour owes a great deal to the metal's higher profile.

The very probability of a growth in demand set against the massive investment required for expansion that is only viable in South Africa or Russia leads me to conclude that the price will be substantially underpinned in the medium to long term.

I have reviewed the recent past, the present and the future for platinum. I acknowledge my bias as a 25-year JM man but I believe this paper presents an exciting scenario for the rapid development of an element which, until perhaps 100 years ago, was a little-known curiosity. Now, a large part of the world's industry is dependent upon platinum and its exploitation is a world industry in itself. The fact that many of the metal's uses are in the forefront of modern-day technology truly makes it a metal of the future.

May I leave you with the following thought? Something like 70% of the world's platinum is used in just two countries, Japan and the USA, with populations totalling perhaps 350 million. Excluding the Eastern bloc, the rest of the world's population is about ten times larger. Most countries are striving to raise their living standards and will gradually succeed. What will be the prospects for platinum and the platinum-group metals as they progress? What sources will satisfy these future demands?

3

Evaluation of the Nickel Sulphide Bead Method of Fire-assay for the Platinum-Group Elements using Neutron Activation Analysis

S. J. PARRY, I. W. SINCLAIR & M. ASIF

Imperial College Reactor Centre, Silwood Park, Ascot, Berkshire, SL5 7PY, UK

ABSTRACT

Fire-assay has been implemented as a preconcentration step before neutron activation analysis to determine the platinum-group elements (PGE) and gold in representative samples (50 g). The procedure is based on the established nickel sulphide bead method, with lithium tetraborate in the flux and a fusion temperature of 1000°C. When the analytical technique was tested by measuring the PGE in reference materials SARM 7 and PTC-1, the data agreed with recommended values.

The method was modified for chromitites both by the addition of sodium hydroxide to the flux and by raising the temperature to 1150°C. The accuracy of the modified technique was checked with a MINTEK 'in house' chromite standard, 2/77, and the concentrations of the PGE measured agreed with the preferred values. Replicates of two chromites, UG2 and LG6, were analysed to measure the analytical reproducibility. Results confirmed that the variations were acceptable. Concentrations of the PGE in the blanks were insignificant compared to those in the samples but high gold blanks may be the source of some variation in the results. A second source of error could be losses during dissolution.

It is concluded that, even though the method appears to give satisfactory results for high concentrations of the PGE, there may be significant problems due to losses and reagent blanks when it is applied to samples with lower values. There is an urgent need for geological reference materials, particularly chromites, containing 0·1–1·0 ppm total PGE, to check analytical procedures. Meanwhile, in the absence of such reference samples, it is proposed that the use of radiochemical techniques may be the only way to ensure that accurate data are obtained for samples containing low concentrations of the PGE.

INTRODUCTION

Fire-assay, with nickel sulphide bead collection, is used to concentrate the PGE and gold from large rock samples, prior to neutron activation analysis. The technique is very sensitive and therefore it is widely used in geochemical studies of the PGE.

21

The procedure was developed by Hoffman *et al.* (1978) based on the original fire-assay method of Robert *et al.* (1971). The rock powder is fused with a flux at 1000°C and the PGE and gold are collected in the nickel sulphide button which separates from the slag. The button is crushed and the nickel sulphide matrix dissolved in concentrated hydrochloric acid. The insoluble residue containing PGE and gold is filtered off and analysed by neutron activation.

This procedure was implemented at the Reactor Centre to enable large and representative samples to be analysed for the PGE. The modification, proposed by Borthick & Naldrett (1984), replacing sodium tetraborate in the flux with lithium tetraborate, was adopted. In addition, sodium hydroxide was added to reduce the viscosity of the melt. Temperature of the fusion melt was increased to 1150°C, to ensure complete dissolution of chromite grains.

This paper describes in detail the methods used at Imperial College Reactor Centre for the determination of the PGE and gold in rocks. The methods are evaluated by the analysis of standard reference materials and the reproducibility of the technique for chromite is demonstrated by the repeated analysis of 'in house' standards.

METHOD

Reagents
Analytical-reagent grade chemicals from BDH Chemicals (Poole, England) were used unless otherwise specified.

Di-lithium tetraborate, anhydrous (Spectrosol, BDH Chemicals)
Lithium carbonate
Sodium carbonate, anhydrous
Sodium hydroxide pellets
Sand, purified by acid (GPR, BDH Chemicals)
Sulphur, sublimed (Laboratory reagent, BDH Chemicals)
Nickel powder (Inco Metals Co.)

Standards
The multi-element chemical standard used in this work was prepared by dissolution and dilution of complexes of individual PGE (Johnson Matthey Chemicals). The solutions contained $29\,\mu$g Rh, $225\,\mu$g Pd, $199\,\mu$g Pt, $21\cdot4\,\mu$g Ir, $29\,\mu$g Os and $6\cdot66\,\mu$g Ru per $0\cdot1$ ml. A gold standard was prepared by diluting a $1000\,\mu$g/ml solution available commercially as an atomic absorption standard (Ventron GmbH). The dilution contained $10\,\mu$g Au per $0\cdot1$ ml of solution. Aliquots of each solution ($0\cdot1$ ml) were pipetted onto filter paper and dried to reproduce the geometry of the preconcentrated samples for activation analysis.

Blanks
All reagent-blank determinations were carried out by replacing the rock sample with sand, purified by acid.

Reference Material Description

SARM 7:   A composite of samples taken from the Merensky Reef in the Bushveld Complex, South Africa. It consists mainly of fels-phathic pyroxenite. The principal platinum minerals are ferro-platinum, cooperite, sperrylite, braggite and moncheite. It was prepared by the National Institute for Metallurgy (South Africa) and distributed by SA Bureau of Standards.

PTC-1:   A sulphide flotation concentrate from Sudbury, Ontario. The major constituents are copper (5·16%), nickel (9·42%), sulphur (23·5%) and iron (26·9%). It was prepared and distributed by the Mines Branch, Department of Energy, Mines and Resources, Ottawa, Canada.

Mintek 2/77:   An 'in house' standard chromitite with preferred values for PGE. Provided to the authors to assist with the evaluation of the fire-assay method for chromites.

Fusion Mixtures

For the preconcentration of silicates and sulphides, mix di-lithium tetraborate (100 g), sodium carbonate (50 g) and nickel powder (20 g). Add sulphur (14 g) except in the case of sulphides where excess sulphur (above 14 g) will cause brittleness in the nickel sulphide bead. In addition, sand (10 g) is included with sulphide material.

Chromites are mixed with di-lithium tetraborate (100 g), lithium carbonate (50 g), sodium hydroxide (50 g), nickel (20 g), sulphur (14 g) and sand (10 g).

Fire-assay Preconcentration

The fire-assay fusions were carried out in fire-clay crucibles (Beecroft and Partners, Sheffield, England) heated in a muffle furnace (Gallenkamp, Loughborough, England). The nickel sulphide beads were crushed in a steel hammer-cutter mill (Glen Creston, Stamford, England).

Procedure

Add 50 g of powdered rock sample to the fusion mixture, in a polythene bag to ensure adequate mixing of the reagents. Fuse in a fire-clay crucible, in a muffle furnace, for 1 h at 1000°C (silicates and sulphides) or 1150°C (chromites). Pour the melt into an iron mould and when cool, separate the button from the slag. Crush the button in a micro hammer-cutter mill (Glen Creston) and heat with 500 ml of concentrated hydrochloric acid until it has dissolved to form a clear green solution.

Filter off the remaining insoluble PGE and gold on Whatman 42 filter paper (4 cm diam.), wash with deionised water followed by ethanol and leave to dry. Fold the filter paper into four and seal in polythene sheet, using a heat-sealing device.

Neutron Activation Analysis

PGE residues were irradiated in the Imperial College 100 kW 'Consort' reactor. The 5 min irradiations were made in the epithermal neutron flux (10^{10}ncm^{-2}s^{-1},

$R_{CdAu} = 2 \cdot 4$) of the cadmium-lined pneumatic transfer system. Gamma-ray analysis was performed with a lithium-drifted germanium detector (Princeton Gamma-Tech, FWHM = $1 \cdot 69$ keV, peak/Compton ratio : $27 \cdot 7$ at $1 \cdot 33$ MeV, efficiency : $4 \cdot 9\%$). The $7 \cdot 5$ h irradiations were made in either the thermal neutron flux (10^{12}ncm^{-2}s^{-1}) of the core-tube facility or in the epithermal neutron flux (10^{10}ncm^{-2}s^{-1}, $R_{CdAu} = 2 \cdot 4$) of the cadmium-lined core-tube facility. All subsequent gamma-ray spectrometry was carried out with a lithium-drifted germanium detector (Princeton Gamma-Tech, FWHM : $1 \cdot 8$ keV, peak/Compton : $36 \cdot 3$, at $1 \cdot 33$ MeV, efficiency : 11%). Both gamma-ray detectors were connected to a ND6700 multichannel analyser with software for neutron activation analysis.

Procedure

Pack the unknowns and standards into individual polythene capsules (15 mm $\times 31$ mm) and irradiate for 5 min in the epithermal neutron flux of the pneumatic transfer system. Let the sample decay for 5 min and then count on the gamma-ray detector for 5 min. Measure rhodium and palladium with the 51 and 188 keV gamma-ray energies of ^{104m}Rh ($T_{1/2} = 4 \cdot 41$ min) and ^{109m}Pd ($T_{1/2} = 4 \cdot 69$ min), respectively.

Transfer all the unknowns and standards into a single capsule (21 mm $\times 76$ mm) and irradiate for $7 \cdot 5$ h in the epithermal neutron flux of the core-tube facility. After a 4 d decay, count for 10 min to measure platinum and gold with the 158 and 411 keV gamma-ray energies of ^{199}Au ($T_{1/2} = 3 \cdot 14$ d) and ^{198}Au ($T_{1/2} = 2 \cdot 7$ d), respectively.

Re-irradiate the samples for $7 \cdot 5$ h in the thermal neutron flux of the core-tube facility and count after 10 d decay, for 1–2 h. Measure osmium, ruthenium and iridium from the 129, 497 and 468 keV gamma-ray energies of ^{191}Os($T_{1/2} = 15 \cdot 4$ d), ^{103}Ru($T_{1/2} = 39 \cdot 4$ d) and ^{192}Ir($T_{1/2} = 73 \cdot 8$ d), respectively.

Calculate the concentrations of all the elements relative to the filter paper standards using the neutron activation analysis package of the ND6700, with corrections for dead-time, pulse pile-up and neutron flux gradients.

RESULTS

In the absence of any chromite certified as a reference material, the accuracy of the method was tested with SARM 7 (a silicate) and PTC-1 (a sulphide), both of which have certified or recommended values for the PGE and gold. Table 1 shows that the technique gives good agreement with the recommended values except for rhodium. The rhodium values were low due to inaccurate evaluation of the 51 keV gamma-ray peak, which suffers background interference due to overlapping X-rays. Once the error had been located, all further analyses for rhodium, i.e. in the chromitite samples subsequently analysed, were corrected.

Since the purpose of establishing the method was to analyse chromitite, the reproducibility of the technique was checked with two unknown samples of chromitite from the Bushveld complex: LG6 and UG2, and the Mintek 'in house'

TABLE 1

Analyses of reference materials
(concentrations in ppm)

Element	SARM 7	PTC-1
Ru	0·40 ±0·13	0·40 ± 0·12
	(0·43 ±0·06)	—
Os	0·070 ± 0·028	0·28 ± 0·04
	(0·063 ± 0·007)	—
Rh	0·16 ±0·02	0·36 ± 0·06
	(0·24 ±0·01)	(0·62 ± 0·07)
Ir	0·077 ± 0·019	0·16 ± 0·02
	(0·074 ± 0·012)	—
Pd	1·58 ± 0·10	10·6 ± 1·4
	(1·53 ± 0·03)	(12·7 ± 0·7)
Pt	3·4 ± 0·4	2·6 ± 0·7
	(3·74 ± 0·05)	(3·0 ± 0·2)
Au	0·31 ± 0·08	0·5 ± 0·5
	(0·31 ± 0·02)	(0·65 ± 0·07)

Mean concentrations ±1 sigma errors are calculated from the results of four replicate analyses. Certified (SARM 7), and recommended ($\overline{\text{PTC}}$-1) values are given in parentheses.

TABLE 2

Reproducibility of chromitite analyses
(concentrations in ppm)

Sample	Ru	Os	Rh	Ir	Pd	Pt	Au
UG2	1·20	—	0·516	0·272	1·95	2·76	0·014
	0·99	—	0·536	0·253	1·88	1·81	0·012
	1·22	0·171	0·523	0·238	1·76	2·54	0·024
	1·12	0·205	0·522	0·250	1·72	2·42	0·013
Mean	1·13	0·188	0·524	0·253	1·83	2·38	0·016
(error)	(0·10)	(0·024)	(0·008)	(0·014)	(0·11)	(0·41)	(0·006)
LG6	0·104	0·015	0·099	0·043	0·175	0·189	0·0016
	0·197	0·021	0·088	0·031	0·118	0·289	0·0015
	0·084	0·024	0·082	0·040	0·120	0·316	0·0011
Mean	0·128	0·020	0·089	0·038	0·138	0·265	0·0014
(error)	(0·060)	(0·005)	(0·009)	(0·006)	(0·032)	(0·067)	(0·0003)
2/77	0·48	—	0·33	0·16	1·36	3·1	0·023
	0·64	—	0·52	0·20	1·52	3·6	0·012
	0·44	0·082	0·52	0·15	1·63	2·6	0·011
	1·22	0·13	0·65	0·21	1·88	4·6	0·107
Mean	0·70	0·11	0·51	0·18	1·60	3·5	0·04
(error)	(0·36)	(0·03)	(0·13)	(0·03)	(0·22)	(0·9)	(0·05)
Preferred value	0·93	—	0·61	<0·2	1·67	3·11	—

S. J. Parry, I. W. Sinclair, M. Asif

TABLE 3
Reagent blanks (concentrations in ppb)

Ru	Os	Rh	Ir	Pd	Pt	Au
<32	<2	1·3	<0·5	<19	<15	1·3
<16	4	1·1	0·25	<14	<12	18·8
<17	8	0·6	0·29	<16	<9	1·2
<56	4	<0·4	<0·3	<12	<21	0·99

Values determined using the proposed procedure with acid washed sand substituted for the sample.

standard chromitite 2/77. The results of the individual analyses of replicates are given in Table 2, with the mean values and one sigma error. The results for 2/77 are in reasonable agreement with the Mintek 'preferred values'. The reproducibility of values for UG2 is much better than that for the other samples, which would suggest that the lack of reproducibility may be due to heterogeneities of the PGE and particularly gold in the samples. It is also possible that poor reproducibility for LG6 and UG2 is due to inadequate sampling of the rock powder, despite the use of manual 'coning and quartering' techniques.

In general, the data for Ru, Os, Pt and Au in all the samples show poor reproducibility, often with errors above 25%. Robert *et al.* (1971) suggested that incomplete collection of platinum and gold was possible and a recent radiotracer study carried out by Parry *et al.* (1987) has indicated that losses of Ru and Au may occur at the dissolution stage. On the other hand, the reproducibility of the data for Rh, Pd and Ir is good, with almost all errors below 20%, showing that the technique can provide reliable data for these elements.

The reagents used during the preconcentration stage may contain PGE and gold, causing errors in the analyses. The results of repeated analyses with acid-washed sand as the blank, in Table 3, show that the concentrations of PGE in the reagents are well below the values measured in the samples. However, there is one high value for gold and the source has not been established, but it may be from the di-lithium tetraborate, as suggested by Borthick & Naldrett (1984).

CONCLUSIONS

The techniques evaluated in this paper provide accurate analyses of silicates and sulphides with reasonable reproducibility. In the absence of a suitable standard it is difficult to assess the accuracy of the method for chromitite but acceptable data are obtained for Mintek 2/77, where preferred values are available. Again, it is possible to obtain reasonable reproducibility but, in all the samples analysed, the significant variation in data for Ru, Os, Pt and Au is of concern. This problem may be caused by reagent blanks in the case of Au but not for the PGE. Losses during dissolution may be the reason for high errors in Au and Ru.

These two problems, reagent blanks and losses during dissolution, are likely to become even more serious when the analytical technique is applied to ppb concentrations of the PGE. Radiotracer studies have already highlighted the losses of Ru and Au during dissolution at high concentrations (Parry *et al.*, 1987) and further work is required to study losses at lower levels. It may be necessary to use radiochemical fire-assay techniques, thus enabling carriers to be used for yield determinations. A further advantage would be that reagent blanks are eliminated if the rock is activated before separation.

Finally, although the analytical technique provides acceptable data, it could be further improved using radiochemical fire-assay techniques, removing the problem of reagent blanks and giving a yield determination during the preconcentration stage. Such a technique could be used to standardise reference materials including a chromitite and other rock types containing 0·1–1 ppm total PGE. Once suitable reference materials are available, they can be used to check the accuracy of the other analytical procedures used to determine PGE and gold.

ACKNOWLEDGEMENTS

The authors would like to thank Dr C. A. Lee for providing the material used to prepare their 'in house' standards. Rock-crushing facilities were kindly provided by the Applied Geochemistry Section of the Department of Geology, Imperial College of Science and Technology.

REFERENCES

Borthick, A. A. & Naldrett, A. J. (1984). Neutron activation analysis for platinum-group elements and gold in chromitites. *Anal. Letters*, **17**, 265.

Hoffman, E. L., Naldrett, A. J. & Van Loon, J. C. (1978). The determination of all the platinum-group elements and gold in rocks and ore by neutron activation analysis after preconcentration by a nickel sulphide fire-assay technique on large samples. *Anal. Chim. Acta*, **109**, 157–66.

Parry, S. J., Asif, M. & Sinclair, I. W. (1987). Radiochemical fire-assay for determination of the platinum-group elements. Presented at *Methods and Applications in Radioanalytical Chemistry, ANS Topical Conference*, 6–10 April 1987, Hawaii (to be published in *J. Radioanal. Chem.*).

Robert, R. V. D., Van Wyk, E. & Palmer, R. (1971). Concentration of the noble metals by a fire-assay technique using nickel sulphide as a collector. National Institute for Metallurgy (S. Africa), Report No. 1371.

4

Preconcentration of Precious Metals by Tellurium Sulphide Fire-assay Followed by Instrumental Neutron Activation Analysis

I. Shazali, L. Van't dack & R. Gijbels*

Department of Chemistry, University of Antwerp (UIA), B-2610 Wilrijk-Antwerp, Belgium

ABSTRACT

A fire-assay method is described to separate the platinum-group elements, silver and gold from rocks and ores using tellurium sulphide as the collector.

The residue remaining after dissolution of the tellurium sulphide button in hydrochloric acid is filtered off and irradiated with reactor neutrons, and analysed by gamma-ray spectrometry. By adding stannous chloride to the filtrate, tellurium is reduced and the precious metals that have passed into solution are recovered by coprecipitation. This improves the yields of gold and especially silver (respectively 20 and 90% recovered).

The proposed method is compared with the more conventional nickel sulphide fire-assay. Using the experimental conditions adopted in this laboratory, the latter method is still superior, as evaluated from the analysis of standard platinum ore SARM 7.

Data are also reported on the blank precious metal content of various types of nickel and tellurium used in the fire-assay methods.

INTRODUCTION

The platinum-group metals, silver and gold are often present in rocks and ores at very low concentrations and their distribution in nature can be inhomogeneous. These two factors favour analysis by techniques based on the fire-assay approach, where a large sample size may be used and where the precious metals are concentrated into relatively small buttons or beads. Many fire-assay procedures have been proposed for the collection of precious metals, the most common ones using

---

*To whom correspondence should be addressed.

I. Shazali, L. Van't dack, R. Gijbels

lead, tin, copper, copper-nickel-iron or nickel sulphide as a collector. For a detailed account (practical procedures, advantages and disadvantages of various collectors) reference may be made to the literature (Beamish, 1966; Robert *et al.*, 1971, 1972; Wall & Chow, 1974; Sulcek *et al.*, 1976; Kallmann & Maul, 1983).

The realization that (i) precious metals can be associated in nature with tellurium, (ii) precious metals can be beneficially collected from aqueous solution by coprecipitation with tellurium and (iii) the incomplete collection of gold by the nickel sulphide fire-assay used in this laboratory (Shazali *et al.*, 1987), motivated us to investigate a Te-precious metals-sulphide fire-assay procedure, for which there appears to be no analytical literature. Tellurium has been added to the tin fire-assay charge as a carrier or fixing agent (Moloughney & Faye, 1976). The separation of precious metals from complex matrices and solutions by coprecipitation with tellurium has been described in the literature (Sandell & Neumayer, 1951; Elson & Chatt, 1983; Stockman, 1983). Since selenium and tellurium are geochemically similar elements the system Se-precious metals–sulphide was also investigated.

Selenium and tellurium are chalcophile elements, their widespread geochemical analogue is sulphur. Selenium is geochemically associated with gold and to a lesser extent with silver, lead, mercury, bismuth, cobalt and nickel. Tellurium is associated with gold and bismuth (Smirnov *et al.*, 1983). These associations are responsible for the large number of naturally occurring selenide and telluride

TABLE 1
Selenide and telluride intermetallic precious metal minerals

Formula	Name	Reference
Au_2Te_3	Montbrayite	1
$AuTe_2$	Calaverite	2
$(Au, Ag)Te_2$	Krennerite	1
$(Au, Ag)Te$	Muthmannite	1
$(Au, Ag)Te_4$	Sylvanite	1
$Pb_5Au_{0.7}Sb_{1.1}(Te_{2.1}S_{5.4})_{7.5}$	Nagyagite	1
$AuAg_3Te_2$	Petzite	1
Ag_2Te	Hessite	1
$Ag_{(5-8)}Te_3$	Stützite	1
$AgTe_3$	Empressite	1
Ag_4SeS	Aguilarite	3
Ag_2Se	Naumannite	3
Ag_3AuSe_2	Fischesserite	3
$RhSe_2$		3
Rh_3Se_8		3
$IrSe_2$		3
$Pt(SnTe)$	Niggliite	4
$Pd(Te, Bi)_{1-2}$	Kotalskite	4
$(Pt, Pd)(Te, Bi)_2$	Moncheite	4

1. Zemann & Leutwein (1969).
2. Smirnov *et al.* (1983).
3. Fisher & Leutwein (1969).
4. Mertie (1969).

minerals (Table 1). There is a surprisingly large number of gold–tellurium minerals as well as those of silver. From a mineralogical and economic point of view, the most interesting tellurium deposits are sulphide bearing gold veins. Selenium forms a homogeneous series of mixed crystals with sulphur. It should be noted that selenium gold-bearing minerals do not exist, even though selenides often coexist with gold and gold minerals (Fisher & Leutwein, 1969). Selenides and tellurides contain much higher platinum metal contents than sulphides (Smirnov *et al.*, 1983). The simultaneous occurrence of high selenium, tellurium and platinum metal contents in some copper–nickel sulphide ore deposits and the occurrence of platinum-metal tellurides indicate a strong geochemical association of these elements with platinum-group metals (Genkin, 1968).

In this study it was decided to investigate the use of TeO_2 or SeO_2 plus sulphur in place of a mixture of nickel powder plus sulphur, as a collector for precious metals during the fusion process in fire-assaying. The behaviour of precious metals during parting with HCl was determined by the coprecipitation technique (Shazali *et al.*, 1987). The results were compared to those obtained by conventional nickel sulphide fire-assay.

EXPERIMENTAL PROCEDURE

Fire-assay

Fire-assay crucibles (Epteroder 'Tonschmelztiegel' No. 4, from Fletcher) were cut to 9 cm height, to give a volume of $230\,cm^3$. An electrical furnace with the capacity for three fusion crucibles was used at a temperature of 1000°C for the nickel sulphide fire-assay. For the tellurium sulphide fire-assay, a temperature of 800–1000°C was used. Several blank fire-assay charges were analysed. All reagents were crushed and sieved through a 400 mesh sieve before use.

Nickel sulphide fire-assay buttons were prepared from 20–50 g of ore samples following the fusion procedure of Robert *et al.* (1971) with some modifications (Shazali *et al.*, 1987).

Tellurium sulphide fire-assay buttons were prepared from 25–50 g samples by fusing with 60 g boric acid or $Li_2B_4O_7$, 60 g of Na_2CO_3, 15 g sulphur, 5 g TeO_2 and silica at a temperature of 800 and 1000°C for 60 and 80 min respectively. In some experiments, TeO_2 was replaced by SeO_2.

After dissolving the nickel or tellurium sulphide buttons with hydrochloric acid, the solutions were filtered through Whatman No. 542 (5·5 cm diameter) filter paper in Gelman funnels. The filter papers with the precious metals residues were dried, folded and pressed into pellets prior to irradiation.

Coprecipitation with Tellurium

The precious metals which pass into solution during dissolution of the nickel and tellurium sulphide buttons were recovered by coprecipitation with tellurium. The filtrates were collected in 600 ml beakers, covered with a watch glass and slowly evaporated to dryness overnight. The residues were dissolved with 100 ml 1N HCl and heated to boiling; 1 ml of a tellurium solution ($1\,mg\,ml^{-1}$ Te, prepared as described by Sandell & Neumayer, 1951) was added in the case of nickel sulphide fire-assay only. While stirring, 10–12 ml of a freshly prepared stannous chloride solution was added. The solutions were heated to boiling and kept near boiling for 30 min, or until the tellurium precipitate was well coagulated; sometimes a second portion of reducing agent was necessary to ensure complete reduction. The precipitate was filtered, washed with hot water, dried in a desiccator, folded and pressed into pellets for neutron irradiation.

Calibration Standards
Standards were prepared by pipetting 100–250 μl of a standard solution of each precious metal onto filter papers, together with 0·1–0·2 g of nickel or tellurium sulphide residue prepared from a blank fire-assay, and 0·1–0·2 g residue of a granite rock powder fire-assay. In this way, the effects of neutron and gamma self-absorption in standards were matched with those of the samples.

Neutron Activation and Gamma-ray Spectrometry
Nuclear data for radioisotopes used in this work are listed in Tables 2 and 3. The procedure followed was similar to that of Hoffman *et al.* (1978).

All the neutron irradiations were carried out in the research reactor Thetis of the State University of Ghent. The pneumatic rabbits system of channel 9 allows for rapid transfer of samples into the reactor and back within 2·4 s. This channel has a thermal flux of $2·6 \times 10^{12}$ ncm^{-2} s^{-1} and was chosen for the determination of rhodium. Pellets were irradiated for 5 min; after transfer to the detector (about 40 s), measurement of the gamma spectrum was carried out for 1200 s with the sample placed at a distance of 10 cm from the top of the low energy photon detector (LEPD), i.e. a planar high purity germanium detector. A nickel flux monitor was used to correct for differences in neutron dose from one irradiation to the next (^{65}Ni $T_{1/2} = 2·52$ h, $E_\gamma = 366·5, 1115·5$ and $1481·7$ keV).

The remaining precious metals were determined after a long (7 h) irradiation in channel 5 or 14 using a neutron flux of $1·0 \times 10^{12}$ ncm^{-2} s^{-1}. Seven samples or standards could be irradiated in the same rabbit. To correct for vertical gradients in the neutron flux within the rabbit, copper monitors were co-irradiated. Copper wire was mounted around individual sample containers, being held securely in position in a groove. After irradiation, the wires were counted and after the copper activity had decayed, they were weighed.

Samples were allowed to decay for 16 h before measuring the palladium content (3000 s count time) with the LEPD. Osmium, platinum, iridium and gold were also measured on the LEPD using a 3000 s count time but after a 7 d cooling time. Ruthenium and silver were determined by counting for 10 000 s, after a 4–5 weeks decay period, using a large Ge(Li) detector.

RESULTS AND DISCUSSION

A detailed discussion of nuclear considerations and interfering reactions when determining the precious metals by neutron activation analysis has been given by

TABLE 2
Nuclear data of the radioisotopes used for determining the precious metals by neutron activation analysis

Element	Reactor neutron induced (n, γ) reaction	Half-life of nuclide produced	Gamma-ray energy used (keV)	Detector used	Decay time
Rhodium	^{103}Rh (n, γ) ^{104m}Rh	4·35 min	51	LEPD	40–50 s
Platinum	^{198}Pt (n, γ) ^{199}Pt (β^-) ^{199}Au	3·15 d	158; 208	LEPD	7 d
Palladium	^{108}Pd (n, γ) ^{109}Pd	13·47 h	88	LEPD	16 h
Iridium	^{191}Ir (n, γ) ^{192}Ir	74·02 d	316	LEPD	7 d
Osmium	^{190}Os (n, γ) ^{191}Os	14·6 d	129	LEPD	7 d
Ruthenium	^{102}Ru (n, γ) ^{103}Ru	38·9 d	497	Ge(Li)	4–5 weeks
Gold	^{197}Au (n, γ) ^{198}Au	2·69 d	412	LEPD	7 d
Silver	^{109}Ag (n, γ) ^{110m}Ag	249·9 d	658	Ge(Li)	4–5 weeks

TABLE 3

Nuclear data of radionuclides formed by thermal neutron irradiation of tellurium

Nuclear reaction	Abundance of stable nuclide (%)	Thermal neutron capture cross-section (barn)	Half-life of produced radionuclide	Main gamma-ray energies (keV) and intensities (%)
^{120}Te(n, γ) ^{121m}Te	0·096	0·3	154 d	212·3 (82%) + Sb X-rays
^{120}Te(n, γ) ^{121}Te	0·096	2	16·8 d	507·4 (18%) 572·9 (80%) + Te and Sb X-rays
^{122}Te(n, γ) ^{123m}Te	2·60	38	117 d	158·8 (84%) + Te X-rays
^{126}Te(n, γ) ^{127m}Te	18·95	0·13	109 d	59 (0·19%) 88·7 (0·08%) + Te X-rays + daughter rad. from ^{127}Te
^{126}Te(n, γ) ^{127}Te	18·95	0·9	9·5 h	361·0 (0·05%) 417·4 (0·3%) + I X-rays
^{128}Te(n, γ) ^{129m}Te	31·69	0·015	33·4 d	695·8 (6%) + Te X-rays + daughter rad. from ^{129}Te
^{128}Te(n, γ) ^{129}Te	31·69	0·20	69·5 m	287·5 (1·7%) 459·5 (15%) + I X-rays
^{130}Te(n, γ) ^{131m}Te	33·80	0·02	30 h	200·6 (8%), 240·9 (8%) 334·3 (9%), 782·5 (60%) 852·2 (31%), 1 125·5 (13%) 1 206·6 (11%), +Te and I X-rays, +daughter rad. from ^{131}Te, ^{131}I
^{130}Te(n, γ) ^{131}Te	33·80	0·22	25·0 m	149·7 (68%), 452·4 (16%) 492·7 (5%), 602·1 (4%) 654·4 (1·5%), 997·2 (4%) 1 147·8 (6%) + I X-rays
^{130}Te(n, γ) ^{131m}Te (β^-) ^{131}I	33·80	—	8·04 d	80·2 (2·6%), 284·3 (5·4%) 364·5 (82%), 636·4 (6·8%) 722·1 (1·6%) + Xe X-rays

Gijbels (1971). For results and discussion on the nickel sulphide fire-assay reference is made to Shazali *et al.* (1987). The present paper is mainly limited to an evaluation of the tellurium sulphide fire-assay technique. The isotope ^{199}Au is normally used to determine platinum via its 158 keV γ-ray. However, this peak suffers strong interference from ^{123m}Te. The ^{199}Au γ-ray at 208 keV, although less sensitive, was therefore used in this work to determine platinum. The above interference could be corrected by spectrum stripping (that is by determining the actual tellurium content and then ratioing the peak in a pure tellurium spectrum). However, this generally complicates the determination of platinum and represents an additional source of error. Heating samples to reduce the tellurium content by evaporation is not recommended, since osmium and ruthenium can also be lost by volatilization as their tetra-oxides. In general, the gross activity induced in tellurium posed no problem in the γ-ray spectrometry of the other precious metals.

The tellurium sulphide fire-assay buttons weigh approximately 1·6–3·8 g by the proposed method as compared with 25–35 g for the nickel sulphide method. A fusion temperature of 1000°C was found to be satisfactory. At 800°C a second light phase of tellurium, containing traces of precious metals, formed on top of the slag. This phase has not been identified. In any case, at this lower temperature (800°C), decomposition of the samples was not complete, resulting in poor recovery of precious metals.

No button was formed at either fusion temperature when selenium was substituted as a collector. Addition of excess SeO_2 (3–20 g) to the charge did not alter this observation. Flux to sample ratio, flux composition, and inappropriate fusion conditions are all believed to have caused incomplete dissolution of samples, but no work was carried out to investigate these factors further.

It was difficult to achieve satisfactory fusion conditions for chromite samples or other samples rich in chromium. The tellurium sulphide button did not separate, remaining as droplets in the slag no matter how fluid the melt. In nickel sulphide fire-assay this problem was resolved by substituting $Li_2B_4O_7$ for $Na_2B_4O_7$. In tellurium fire-assay this change in flux had no effect. It appears that for such samples pretreatment is necessary. The effect of various matrix elements on the recovery of precious metals by the proposed method was not studied.

A significant proportion of the tellurium collected was found in the slag and caused staining of the crucible wall. No traces of precious metals were, however, detected on 0·5 g samples of the slag, although the sample size used may have been too small and not fully representative. It should be noted, however, that, in the nickel sulphide fire-assay procedure, traces of gold were found in the slag. The residue after HCl dissolution of tellurium sulphide fire-assay buttons contained traces of the collector as well as a considerable ^{131}I activity ($T_{1/2} = 8·1$ d) derived from the β^- decay of ^{133m}Te. Furthermore, activities due to small amounts of chromium, iron, antimony and cobalt as well as precious metals were detected in residues from either tellurium sulphide or nickel sulphide fire-assay. These components can readily be identified from the respective gamma spectra (Figs 1 and 2).

The results obtained by the proposed method for standard platinum ore SARM 7 (Steele *et al.*, 1975) are listed in Table 4. The collection of ruthenium and osmium after fire-assay fusion and parting with hydrochloric acid is essentially complete. However, there are contradictory statements in the literature about the mechanism for losses of osmium and ruthenium during the heating of HCl solutions in the absence of oxidizing agents. Van Loon (1984) stated that osmium, and to some extent ruthenium, may be lost by volatilization during the dissolution step, osmium being particularly susceptible. Van Loon & Beamish (1965) found that losses amounting to 90–100% osmium occurred on evaporating chloro-osmate solutions

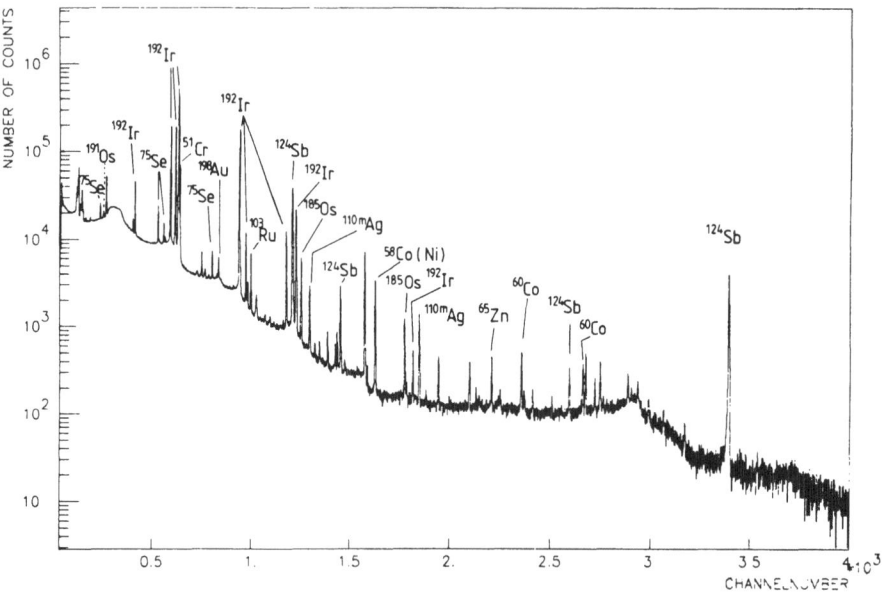

F<small>IG</small>. 1. Gamma-ray spectrum of noble metal sulphide residue from standard platinum ore SARM 7 after nickel sulphide fire-assay, recorded with a conventional Ge(Li) detector, 4 weeks after neutron irradiation.

containing 5 g copper and iron. These losses were reduced to 5–27% in solutions containing 5 g of copper alone. Gijbels *et al.* (1971) added Fe^{2+} to osmium monitor solutions to create a sufficiently reducing medium and prevent losses of osmium by volatilization during evaporation of monitor solutions on silica powder. Van Loon (1984) claimed that addition of small amounts of NaCl minimized the volatilization of osmium when evaporating solutions. Faye (1972) reported that at low acid concentration, hydrolysis of chloro-osmate can result in large losses due to solid material (probably hydrated osmium oxide) adhering to the walls of the beaker. By maintaining the concentration of HCl solution at 0·5 M or higher allowed the solution to be boiled vigorously in an open beaker without losing osmium by volatilization.

Silver is nearly completely collected by tellurium sulphide fire-assay but it passes into solution as the $HAgCl_2$ complex when the button is dissolved. Similar results were found in the nickel sulphide fire-assay (Shazali *et al.*, 1987). More than 90% of the silver sulphide is reported to be decomposed by treatment with concentrated HCl (Kuznetsov *et al.*, 1974; Steele *et al.*, 1975). Kallmann & Maul (1983) indicated that 20–100 mg of silver may remain in solution. These values are rather ambiguous without stating the parting procedure and temperature of dissolution. However, in the proposed method, dissolved silver is recovered by the subsequent tellurium-coprecipitation treatment (see Table 4). In this procedure, it is not necessary to add

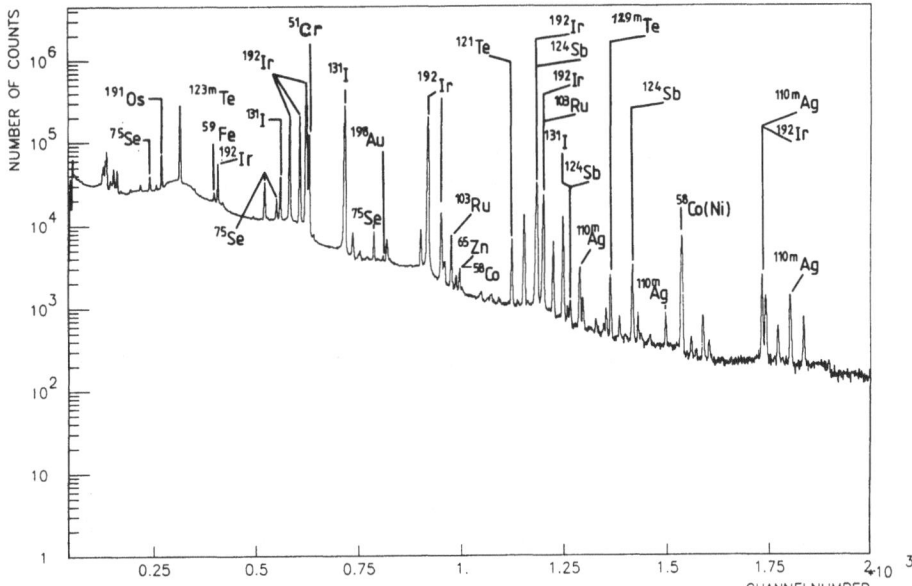

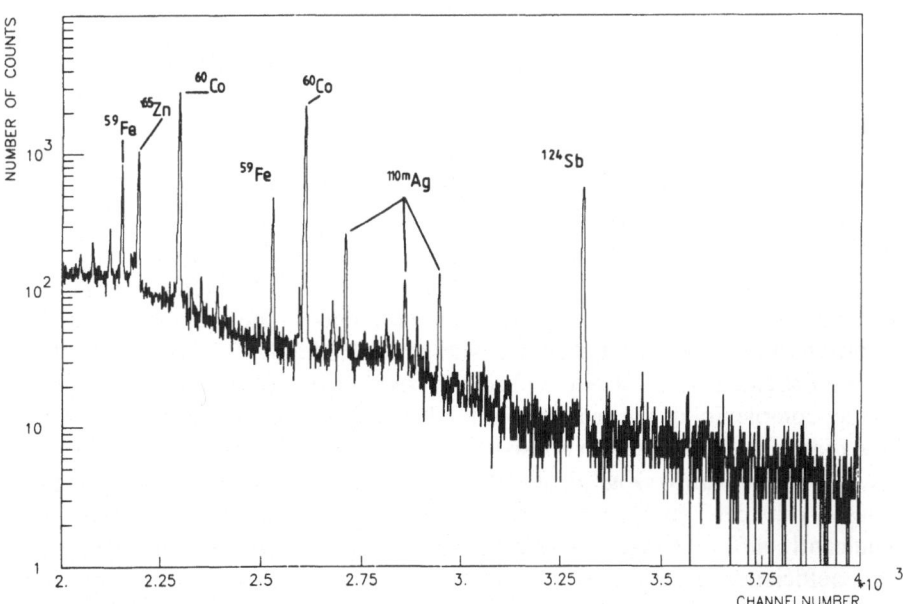

Fig. 2. Gamma-ray spectrum of noble metal sulphide residue from standard platinum ore SARM 7 after tellurium sulphide fire-assay, recorded with a conventional Ge(Li) detector, 4 weeks after neutron irradiation.

TABLE 4

Results for individual precious metals determined in standard platinum ore SARM7 by tellurium sulphide fire-assay followed by coprecipitation on tellurium. All values in ppm (average of 8 determinations)

	Rh	Pd	Os	Pt	Ir	Au	Ru	Ag
TeS residue (a)	0·171	1·23	0·063	3·13	0·066	0·164	0·415	0·041
Coprecipitation treatment (b)	0·001	N.D.	0·001	N.D.	0·002	0·042	N.D.	0·425
Total a + b	0·17	1·23	0·064	3·13	0·068	0·21	0·42	0·47
% Recovery by coprecipitation treatment	0·6	—	1·6	—	2·9	20	—	91
Recommended value	0·24	1·53	0·063	3·74	0·074	0·31	0·43	0·42

tellurium solution to the filtrate during the coprecipitation step, since a substantial amount of tellurium was already present after dissolution of the button by HCl.

The losses of other precious metals during parting of both the nickel and tellurium sulphide buttons were found insignificant except for gold (referred to above). Losses of the precious metals in solution by adsorption to the container walls were eliminated by high normality of HCl and immediate evaporation of the filtrate to dryness.

Moloughney & Faye (1976) reported that a substantial fraction of silver and rhodium dissolved during parting of a tin button even when tellurium was present. It is possible that a tin alloy with rhodium promotes dissolution of rhodium. The effect of base metals (the presence of which may be detected in Figs 1 and 2) causing losses of precious metals into solution has been reported by Aggarwal & Beamish (1964), Beamish (1966), Snodgrass (1972), Robert & Van Wyk (1975), Szaploncczay (1983) and Tarapcik & Mikulaj (1983).

In Table 5 the results obtained by the proposed method are compared with those of the more conventional nickel sulphide fire-assay. Student's t-test was applied to determine whether there was any significant difference in the average obtained by the different procedures for standard platinum ore SARM 7 (Steele et al., 1975). The high coefficients of variation (Table 5) could be due to inhomogeneity of the sample, complexity of the reactions in the fusion step and difficulty in controlling these reactions. This variation is also in agreement with the observations of Robert et al. (1977) and Van Wyk & Dixon (1983). These authors stated that the agreement between duplicate analyses in a run is usually good, but repeat analyses give significantly different values indicating that day-to-day variations are significant. The coefficients of variation of the results of the proposed method (tellurium sulphide) tend to be higher than that of the nickel sulphide fire-assay. However the total average results of the precious metals (always including the fractions recovered by tellurium coprecipitation) are not statistically different at the 95% confidence level, except for palladium.

The recovery of gold in both treatments was found to be incomplete by com-

TABLE 5

Average concentrations (ppm) of precious metals found in standard platinum ore SARM 7, by nickel sulphide fire-assay and by tellurium sulphide fire-assay, both followed by a tellurium coprecipitation step

	Rh	Pd	Os	Pt	Ir	Au	Ru	Ag
Preferred values[a]	0·24	1·53	0·063	3·74	0·074	0·31	0·43	0·42
Nickel sulphide fire-assay + coprecipitation on Te[b]	0·22	1·53	0·054	3·47	0·071	0·23	0·41	0·41
Coeff. of variation (%)	20	11	28	11	17	9	15	12
Tellurium sulphide fire-assay + coprecipitation on Te (this work)	0·17	1·23	0·064	3·13	0·068	0·21	0·42	0·47
Coeff. of variation (%)	30	10	25	30	18	18	24	14

[a] Steele *et al.*, 1975.
[b] Shazali *et al.*, 1987.

parison with the preferred values. The reasons for the losses of gold have not been established unambiguously but could be due to the following factors:

1. Incorrect composition of the fusion mixture and incorrect fusion conditions.
2. Presence of gold in a metallic form in the sample which does not readily form sulphide (Robert *et al.*, 1971).
3. Gold is largely reduced by sulphur to the free metal during the fusion process, passing partly into the slag and partially being retained by the wall of the fire-assay crucible (Kallmann & Maul, 1983).
4. Poor reproducibility. Hoffman & Ernst (1982) reported poor reproducibility of gold using lead fire-assay, even when a 50- g sample was used. This was attributed to the heterogeneous distribution of gold-bearing minerals in the sample and would apply to other precious metals which show a high coefficient of variation.

Further investigations are required to clarify the above points.

All reagents must be tested to ensure that they are free of noble metals. Some commercially available nickel products contain substantial amounts of noble metals, as appears from Table 6. These are of course not suitable for use in nickel sulphide fire-assay, but even the purer qualities yielded occasionally a few tens or even a few hundreds ppb wt (ng g^{-1}) for some of the precious metals (values in parentheses). The apparent concentration which would be calculated for a sample is about three times lower, if 16 g of nickel is used with 50 g of sample (Shazali *et al.*, 1987).

The tellurium dioxide used in this work was obtained from Fluka (pract., > 95%). Its noble metal content appeared to be sufficiently low to explore the possibilities of the described tellurium sulphide fire-assay procedure, and the recovery of noble metals by coprecipitation on tellurium in the case of a relatively

I. Shazali, L. Van't dack, R. Gijbels

TABLE 6
Average blank values of precious metals in nickel and tellurium (ppb-wt)

	Ru	Rh	Pd	Ag	Os	Ir	Pt	Au
Nickel powder 99·99% Aldrich Gold label[a]	≤3	<1	<8	<3	<3	<1	<15 (90)	<1
Carbonyl nickel powder Inco (types 123, 255 and 287 averaged)[b]	<3	<1	<8	<3 (20;45)	<3 (30)	≤4	<15 (25–515)	<2
Nickel powder Janssen Chimica[a]	900	750	<20	<10	90	250	350	≥60
Tellurium 'high purity' Metallurgie Hoboken Overpelt[c]	<4	<1	<4	<2	<8	<8	<10	<6

[a] Average of 4 determinations.
[b] Average of 12 determinations.
[c] Obtained by spark source mass spectrometry (Swenters, K., Verlinden, J., Adriaenssens, E. & Gijbels, R., to be published).

rich ore, such as SARM 7; no representative blank values were determined. It is of interest to note that high-purity tellurium is now being produced for the semi-conductor industry, which would be suitable for tellurium sulphide fire-assay of rock samples with a precious metal content at the ng g^{-1} level, especially if one considers that only 5 g of TeO_2 (or Te) are used for 25–50 g of sample.

CONCLUSION

From the results obtained by the two fire-assay procedures, it appears that nickel sulphide is a superior collector. The total recovery of rhodium, palladium, platinum, iridium and gold with tellurium sulphide is 74, 81, 84, 92 and 92% respectively of that by nickel sulphide fire-assay. The results for osmium, ruthenium and silver are similar for both types of fire-assay. It appears that the applicability of the proposed tellurium sulphide method for samples of different composition is doubtful without sample pretreatment. The parameters affecting the collection of precious metals during the fusion procedure of the proposed method are still not completely understood.

REFERENCES

Aggarwal, K. C. & Beamish, F. E. (1964). Studies of the fire assay for platinum metals by lead collection. *Talanta*, **11**, 1449–57.

Beamish, F. E. (1966). *The Analytical Chemistry of the Noble Metals.* Pergamon Press, Oxford.

Elson, C. & Chatt, A. (1983). Determination of gold in silicate rocks and ores by co-precipitation with tellurium and neutron activation γ-spectrometry. *Anal. Chim. Acta,* **155**, 305–10.

Faye, G. H. (1972). Review of methods for the determination of osmium and ruthenium. *J. South African Chem. Inst.,* **25**, 311–19.

Fisher, R. & Leutwein, F. (1969). In *Handbook of Geochemistry, Vol. II-3,* Chapter 34, *Selenium,* ed. K. H. Wedepohl. Springer-Verlag, Berlin.

Genkin, A. D. (1968). Mineralogy of the platinum metals and their association in the copper-nickel ores of the Noril'sk deposits. *Akad. Nauk. SSSR, Inst. Geol. Rud. Mostorokh. Petrogr. Mineral. Geokhim.,* 106–20.

Gijbels, R. (1971). Determination of noble metals by neutron activation analysis. *Talanta,* **18**, 587–601.

Gijbels, R., Millard, H. T., Desborough, G. A. & Bartel, A. J. (1971). Neutron activation analysis for osmium, ruthenium and iridium in some silicate rocks and rock-forming minerals. In *Activation Analysis in Geochemistry and Cosmochemistry,* ed. A. O. Brunfelt & E. Steinnes. Universitetsforlaget, Oslo, pp. 359–69.

Hoffman, E. L. & Ernst, P. C. (1982). Analytical geochemistry advanced by neutron activation. *J. Radioanal. Chem.,* **71**, 447–62.

Hoffman, E. L., Naldrett, A. J. & Van Loon, J. C. (1978). The determination of all the platinum group elements and gold in rocks and ores by neutron activation analysis after preconcentration by a nickel sulphide fire-assay technique on large samples. *Anal. Chim. Acta,* **102**, 157–66.

Kallmann, S. & Maul, C. (1983). Referee analysis of precious metal sweeps and related materials. *Talanta,* **30**, 21–39.

Kuznetsov, A. P., Kikushkin, Yu. N. & Makarov, D. (1974). Use of nickel matte as a collector for the noble metals during the analysis of lean materials. *Zhur. Anal. Khim.,* **29**, 2155–60.

Mertie, J. B. (1969). Economic geology of the platinum metals. US Geological Survey, Professional paper 630.

Moloughney, P. E. & Faye, G. H. (1976). A rapid fire-assay/atomic-absorption method for the determination of platinum, palladium and gold in ores and concentrates; a modification of the tin-collection scheme. *Talanta,* **23**, 377–86.

Robert, R. V. D. & Van Wyk, E. (1975). The effects of various matrix elements on the efficiency of the fire-assay procedure using nickel sulphide as the collector. National Institute for Metallurgy (S. Afr.), Report No. 1705.

Robert, R. V. D., Van Wyk, E. & Palmer, R. (1971). Concentration of the noble metals by a fire-assay technique using nickel sulphide as the collector. National Institute for Metallurgy (S. Afr.), Report No. 1371.

Robert, R. V. D., Van Wyk, E. & Palmer, R. (1972). The development of a fire-assay procedure using nickel sulphide as the collector for the noble metals. *J. South African Chem. Inst.,* **25**, 179–89.

Robert, R. V. D., Van Wyk, E. & Ellis, P. J. (1977). An examination of fire-assay techniques as applied to chromite-bearing materials. National Institute for Metallurgy (S. Afr.), Report No. 1905.

Sandell, E. B. & Neumayer, J. J. (1951). Photometric determination of traces of silver. *Anal. Chem.,* **23**, 1863–5.

Shazali, I., Van't dack, L. & Gijbels, R. (1987). Determination of precious metals in ores and rocks by thermal neutron activation/γ-spectrometry after preconcentration by nickel sulphide fire-assay and coprecipitation with tellurium. *Anal. Chim. Acta,* **196**, 49–58.

Smirnov, V. I., Ginzburg, A. I., Grigoriev, V. M. & Yakovlev, G. F. (1983). *Studies of Mineral Deposits* (English translation). Mir Publishers, Moscow.

Snodgrass, B. A. (1972). The determination of platinum-group metals in gold-bearing ores. *J. South African Chem. Inst.,* **25**, 268–74.

Steele, T. W., Levin, J. & Copelowitz, I. (1975). The preparation and certification of a reference sample of a precious metal ore. National Institute for Metallurgy (S. Afr.), Report No. 1696.

Stockman, H. W. (1983). Neutron activation determination of noble metals in rocks, a rapid radiochemical separation based on tellurium coprecipitation. *J. Radioanal. Chem.*, **78**, 307–17.

Sulcek, Z., Povondra, P. & Dolezal, J. (1976). Decomposition procedures in inorganic analysis. *Crit. Rev. Anal. Chem.*, **6**, 255–323.

Szaplonczay, A. M. (1983). Effect of copper on the solubility of palladium in nitric acid solutions in analysis of electroplated palladium layers. *Anal. Chem.*, **55**, 2202–4.

Tarapcik, P. & Mikulaj, V. (1983). Separation of [103]Pd from cyclotron irradiated rhodium targets. *Radiochem. Radioanal. Letters*, **47**, 15–19.

Van Loon, J. C. (1984). Accurate determination of noble metals in rocks, ores and alloys. *Tr. Anal. Chem.*, **3**, 272–9.

Van Loon, J. C. & Beamish, F. E. (1965). Inclusion of osmium in an assay method for six platinum metals by iron–copper–nickel collection. *Anal. Chem.*, **37**, 113–16.

Van Wyk, E. & Dixon, K. (1983). The recovery of platinum-group metals and gold by the lead-collection step of fire assay procedure. MINTEK (S. Afr.) Report M88.

Wall, G. & Chow, A. (1974). The determination of losses in the fire assay of gold. Part I: Cupellation and parting losses. *Anal. Chim. Acta*, **69**, 438–50.

Zemann, J. & Leutwein, F. (1969). In *Handbook of Geochemistry, Vol. II-4*, Chapter 52, Tellurium, ed. K. H. Wedepohl. Springer-Verlag, Berlin.

5

The Potential of Fire-assay and ICP-MS for the Determination of Platinum-Group Elements in Geological Materials

The late ALAN R. DATE, ALAN E. DAVIS & YUK YING CHEUNG

British Geological Survey, 64 Gray's Inn Road, London WC1X 8NG, UK

The analytical chemistry of the platinum-group elements may be said to reflect their occurrence in nature as rare, discrete, inhomogeneously distributed, mineral species, and their very high financial value. These factors have resulted in the need for an analytical method offering *complete dissolution* of a *representative sample* allied with an *accurate measurement* technique. For many years these criteria were satisfied by fusion of large sample aliquots followed by gravimetric measurement of the various components after 'partition', and the classical lead assay reigned supreme. The more recent requirement for accurate determination of individual platinum-group elements at low concentration, following developments in geo-chemical research methods, has resulted in a reappraisal of platinum-group element analytical chemistry. A useful two-part review has recently appeared (Van Loon, 1984, 1985).

In general, the classical lead assay provides pre-concentration for platinum, palladium, rhodium (and gold), although their efficient collection on a silver cupellation bead is critically dependent on 'flux' composition and assay conditions. The neo-classical fire assay procedure with nickel sulphide collection (Robert *et al.*, 1971) is gaining popularity because it offers efficient collection for the complete platinum element group, although special techniques are necessary to avoid the loss of volatile osmium compounds at the button dissolution stage.

The most commonly used final measurement techniques include atomic absorption spectrometry, instrumental neutron activation analysis and inductively coupled plasma-atomic emission spectrometry. The most widely used, atomic absorption spectrometry, is basically a single-element technique, but offers high sensitivity when used with graphite furnace techniques. The comparable high sensitivity of nuclear methods is off-set by the requirement for sophisticated hardware (a reactor), while the multi-element capability of ICP emission is accompanied by poor sensitivity for heavy elements. The advent of ICP-MS, with its multi-element capability and high sensitivity for heavy elements, offers a valuable alternative approach for the determination of the platinum-group elements. Recent development work in the field of fire-assay/ICP-MS at the British Geological Survey will be discussed.

43

REFERENCES

Robert, R. V. D. van Wyk, E. & Palmer, R. (1971). Concentration of the noble metals by a fire assay technique using nickel sulphide as the collector. Report Nat. Inst. Metall. (S. Afr.) No. 1371, 14 pp.

Van Loon, J. C. (1984). Accurate determination of the noble metals. I. Sample decomposition and methods of separation. *Trends in Analytical Chemistry*, **3**, 272–5.

Van Loon, J. C. (1985). Accurate determination of the noble metals. II. Determination methods. *Trends in Analytical Chemistry*, **4**, 24–9.

6

ICP-AES Determination of Precious Metals in Difficult Materials

PAUL WHITEHEAD

Johnson Matthey Technology Centre, Blount's Court, Sonning Common, Reading RG4 9NH, UK

Precious metals occur both during mineral treatment and in many of their industrial applications at low concentrations in a wide variety of complex matrices. Their accurate determination is of considerable economic and technical significance. Plasma emission spectrometry is now the most widely used technique for this type of analysis due to its advantages of speed and relative freedom from interferences. However, in many cases there are still significant effects from the matrix constituents and for accurate analysis the alternatives are to separate the precious metals prior to measurement or to compensate for matrix effects by a combination of matched standards and/or internal standards and/or mathematical correction. This presentation is concerned primarily with the value and limitations in the use of internal standards in this field.

Widely used internal standards such as Sc, Y and In have been compared with a range of other possible elements with diverse atomic properties. Their effectiveness in overcoming matrix effects is described. Whereas many internal standards will adequately compensate for errors arising from changes in sample introduction characteristics, those due to changes in the 'power' available in the plasma pose far more problems. The non-linearity of these matrix effects means that they may often be a significant source of error at relatively low interferant concentrations and the inter-line variation imposes severe limitations on the scope of single-element internal standards. In addition to the use of individual internal standards the application of a modified version of the PRISM method of Ramsey & Thompson (1985) has been investigated.

REFERENCE

Ramsey, M. H. & Thompson, M. (1985). Correlated variances in simultaneous inductively coupled plasma-atomic emission spectrometry: its causes and correction by a parameter related internal standard method. *Analyst*, **110**, 519–30.

7

New Detection Techniques for Locating Precious Metal Minerals by Beta Autoradiography: Preliminary Results for Rhodium and Silver Grains

PHILIP J. POTTS

Department of Earth Sciences, Open University, Walton Hall, Milton Keynes MK7 6AA, UK

ABSTRACT

The effectiveness of current beta autoradiography techniques in locating small discrete minerals containing elements such as Au, Ir, Pt, Os, Sb, REE, Ta, etc., is enhanced by substituting a screen-type X-ray film (Fuji RX) for the nuclear research plates (Ilford K2) used in earlier studies. This X-ray film offers equivalent resolution to the nuclear research plates but is less easily damaged during handling and both reduces the cost and increases the efficiency since autoradiography measurements of more than 30 thin sections may be recorded simultaneously on a single 180 × 240 mm sheet of film.

Tests using a calcium tungstate intensifying screen confirm that the beta sensitivity of the X-ray film can be significantly enhanced, as has been shown in biological applications, although the corresponding degradation in resolution detracts from routine use with geological samples.

Data are also presented to demonstrate that the scope of beta autoradiography may be significantly extended by recording autoradiographs of short-lived isotopes within about 1 min of irradiation. Minerals containing rhodium, silver, indium and rare earth elements should readily be distinguished from matrix activity (due to $^{28}Al, ^{52}V$ and ^{56}Mn) under these conditions. Preliminary measurements confirm that it is feasible to record autoradiography data under these circumstances. However, it is desirable to substitute perspex (5 mm thick) for conventional glass as the section mounting material in order to avoid the high background activity that would otherwise be induced in ^{28}Al.

INTRODUCTION

Studies of gold and platinum-group element (PGE) mineralisation often rely on an interpretation of bulk analytical data. However, valuable complementary information can be obtained if such specimens are examined to locate and analyse discrete

47

noble metal grains. Unfortunately, there are practical difficulties involved in undertaking such an analysis: not the least being the problem of finding statistically significant numbers of PGE grains, which often range from 1 to 15 μm in diameter and possess optical properties similar to other common opaque minerals.

Beta autoradiography is a simple, but effective, technique for locating systematically some minerals of interest (Potts, 1984, 1986). Samples are prepared as polished thin sections and irradiated in a nuclear reactor. After a suitable decay period (optimised to the half-life of the isotope of interest) the distribution of those elements that form isotopes which decay by beta emission following neutron activation is detected on a nuclear track detector exposed in contact with the surface of the sample. Since the technique responds to a wide range of elements, autoradiographs require careful interpretation. However, highest sensitivity is shown by a range of elements including gold and iridium that are often present in discrete grains at overall bulk concentrations in the ppm or even ppb range (Potts, 1984). Indeed, the technique has been shown to be highly effective in locating discrete grains of Au, Ag, Ir, Sb, Pt, Os, REE, Ta, etc. (Potts & Prichard, 1986; Prichard *et al.*, 1986).

In this paper, the results of new developments in this technique are described involving:

(1) The use of modern 'screen'-type X-ray film detectors that permit simultaneous exposure of large numbers of samples.
(2) Tests on intensifier screens that enhance the sensitivity of X-ray film.
(3) Preliminary results of experiments designed to evaluate the detection of rhodium and silver grains from very short-lived isotopes ($T_{1/2} < 5$ min) using samples mounted on perspex rather than glass slides.

DETECTION USING SCREEN-TYPE X-RAY FILM

The traditional beta-sensitive nuclear track detector (Ilford K2 Nuclear Research Plates) used in earlier work consists of a gelatin-based silver halide emulsion, 25 μm thick, mounted on a glass supporting plate. Although these plates have been successfully used in radio-tracer studies applied to experimental petrology (Mysen & Seitz, 1974; Mysen, 1978), some disadvantages became apparent when used for beta autoradiography measurements; particularly the sensitivity of the gelatin film to bruising when handled and the relatively high cost. Beta autoradiography involving the isotopes ^{3}H, ^{14}C, ^{32}P and ^{125}I is routinely used in studies of biological materials and a number of investigations have been undertaken to evaluate the most sensitive film type in this application (Laskey & Mills, 1975, 1977; Swanstrom & Shank, 1978). In a review of this work, Laskey (1984) identified a screen-type (medical) X-ray film (Fuji RX) as offering the advantages of high sensitivity, long shelf-life and ease of handling in the detection of autoradiographs from biological samples.

This screen-type film was evaluated for neutron activation-induced beta autoradiography of geological samples by preparing and irradiating polished thin sections as described in previous studies (Potts, 1984; Potts & Prichard, 1986).

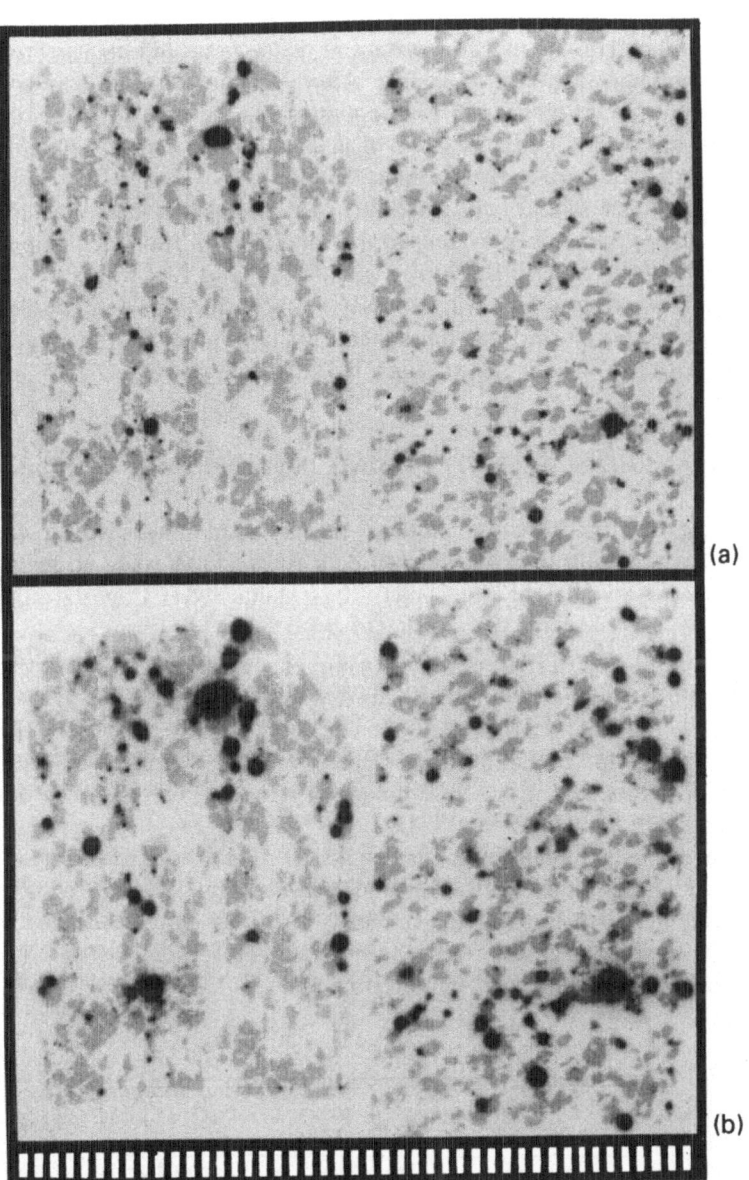

FIG. 1. Beta autoradiographs of two chromite thin sections from Cliff (Unst, Shetland Isles) recorded on Fuji RX film 6 weeks after irradiation. (a) 24-h exposure. (b) 20·5 h exposure with an intensifying screen placed behind the film. Each scale interval represents 1 mm.

After a decay period of 12 and 40 days, autoradiographs were recorded by mounting irradiated thin sections in a commercially available exposure cassette (Genetic Research Instrumentation, Felstead, Essex). A sheet of Fuji RX film was placed in contact with the polished surface of the sections and the cassette closed and sealed in a light-tight polythene bag. After exposure, the film was developed using Fuji Photosol developer CD-18 (for about 5 min at 20°C) and CF-40 fixer (for about 10–15 min at 20°C). Manipulation of all undeveloped film must be carried out in a dark room under dim red safety lights.

Typical results of an autoradiograph recorded on this film are shown in Fig. 1. Image resolution appears equivalent to that obtained on Ilford K2 plates. However, the sensitivity of Fuji RX film is greater, as is its robustness to handling. Furthermore, a considerable saving arises not only in the cost of the Fuji film, but also from the facility of exposing simultaneously a large number of samples on each sheet of film (at least 30 thin sections on a 180×240 mm sheet).

ENHANCED SENSITIVITY USING INTENSIFYING SCREENS

Biological applications have also led to the development of intensifying screens to enhance the sensitivity of X-ray film to beta detection (Laskey & Mills, 1977; Swanstrom & Shank, 1978; Laskey, 1980, 1984; Hahn, 1983). Commercially available intensifying screens consist of a sheet loaded with a scintillant such as calcium tungstate. During exposure of irradiated samples, this screen is placed behind the X-ray film (Fig. 2). The film is then exposed both by direct interaction of beta particles and by secondary photons generated when energetic beta particles penetrate through the film and cause scintillation of the intensifying screen. This arrangement is reported to increase the sensitivity to ^{32}P detection in biological studies by a factor of 10 (Laskey, 1984).

The effectiveness of measurements using a CAWO 409 F4 intensifying screen may be assessed by comparing the autoradiographs of two chromite samples from Unst, Shetland shown in Fig. 1(a) (intensifying screen not used) and Fig. 1(b) (recorded with an intensifying screen). These autoradiographs were recorded one after the other using similar exposure times. Although beta active minor phases

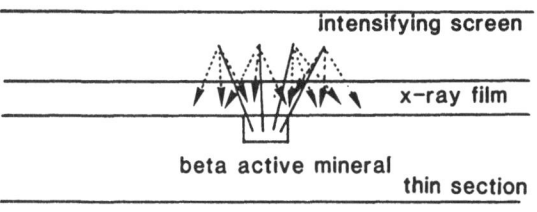

FIG. 2. Schematic diagram showing the generation of secondary photons (dotted lines) that result from placing an intensifying screen behind the X-ray film to enhance the sensitivity of the beta autoradiograph image.

containing iridium and antimony show significantly enhanced detection sensitivity in Fig. 1(b), this is associated with some degradation in image resolution, similar to that reported in biological applications (Laskey, 1984). Since the principal application of beta autoradiography is in locating very small mineral grains, where resolution is at a premium, it is likely that no advantage will normally arise from the use of intensifying screens for conventional measurements on geological samples.

AUTORADIOGRAPHY STUDIES FOR RHODIUM AND SILVER USING VERY SHORT HALF-LIFE ISOTOPES

Measurements to date have been restricted to the detection of beta-emitting isotopes having half-lives longer than 1–2 days. One reason for this restriction is the high activity induced in ^{24}Na ($T_{1/2} = 15$ h) during the irradiation of samples mounted on silica glass slides. It is necessary to leave such samples for at least 7 days to permit this activity to decay to safe levels for handling.

If this experimental restriction can be overcome, calculations (Table 1) show that several elements of mineralogical interest form isotopes after neutron irradiation that decay with half-lives in the range 0·5–5 min. Of particular interest is the high sensitivity shown by isotopes of rhodium, silver, indium and selected rare earth elements. Since successful autoradiography measurements can only be made if sufficient contrast in beta activity is observed between these isotopes and the matrix, account must also be made of data in Table 1 for ^{28}Al, ^{52}V and ^{56}Mn. These elements can occur in appreciable concentrations in matrix phases—particularly in the study of chromites.

The relative decay rates of the isotopes listed in Table 1 are demonstrated by data

TABLE 1
Isotopes that decay with high beta activity after thermal neutron irradiation

Isotope	^{28}Al	^{52}V	^{56}Mn	^{66}Cu	^{70}Ga	^{80}Br	^{104m}Rh	^{104}Rh
$T_{1/2}$ (min)	2·3	3·7	155	5·1	21·1	17·6	4·34	0·705
A	14 100	98 200	6 650	8 260	3 210	16 200[a]	260[a]	6 000 000

Isotope	^{108}Ag	^{110}Ag	^{116m}In	^{128}I	^{152m}Eu	^{165m}Dy	^{176m}Lu	^{233}Th	^{239}U
$T_{1/2}$ (min)	2·4	0·41	54·2	25·0	558	1·3	221	22·3	23·5
A	255 000[a]	2 440 000	125 000	11 150[a]	61 500[a]	98 000[a]	2 870	7 060	5 990

A: beta activity in decays per second per μg after a 1 min irradiation in a thermal neutron flux of 10^{13} neutrons cm^{-2} s^{-1} (Erdtmann, 1976).

[a]Total activity data corrected to account for beta component only using data from Wilson (1966) as follows: ^{80}Br (95%), ^{104m}Rh (0·13%), ^{108}Ag (96%), ^{128}I (93·7%), ^{152m}Eu (78%), ^{165m}Dy (1·3% of total decays occur by beta emission).

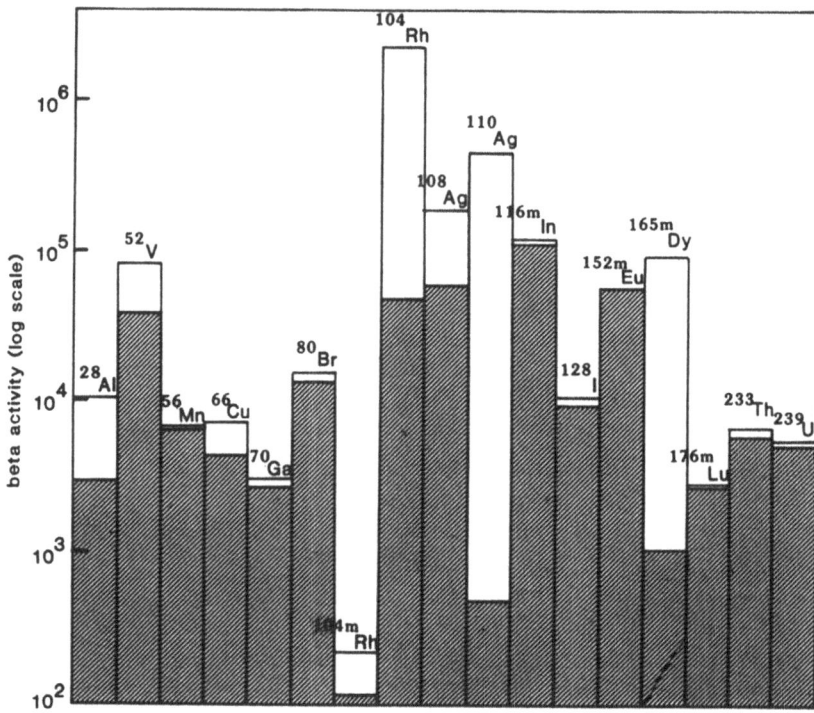

FIG. 3. Relative beta activity of short-lived isotopes that show high specific activity (a) 1 min (total column height) and (b) 5 min (hatched area) after irradiation. Activity data were calculated by applying the appropriate exponential decay factor to the values listed in Table 1.

plotted in Fig. 3, which shows the relative beta activity of each isotope after (a) 1 min and (b) 5 min decay. These results show that rhodium and silver are likely to show high sensitivity 1 min after irradiation but that the beta contrast with aluminium, vanadium and manganese (matrix elements) is much less marked after a decay of 5 min. These data suggest an initial experimental strategy involving a 1 min irradiation, followed by a 0·5–1 min decay, followed by a 1 min auto-radiograph exposure.

A strategy such as this imposes significant experimental constraints, not the least being the high activity induced in ^{28}Al $(T_{1/2} = 2\cdot3\,\text{min})$ if conventional glass slides are used as the thin section mount. Some preliminary measurements were made, therefore, to evaluate the use of perspex as a substitute for glass, since the former is both transparent and rigid in sheet form. These measurements (Table 2) showed a substantial reduction in the activity induced in perspex compared with glass after a 1-min irradiation confirming its suitability as mounting material in this application. To investigate the feasibility of beta autoradiography for short-lived isotopes, a chromite sample was prepared as a thin section, mounted on a perspex slide 45 mm × 25 mm made from relatively thick sheet (5 mm) to prevent flexing during

TABLE 2

Dose rate from test samples after a 1 min irradiation in a neutron flux of 2×10^{12} neutrons $cm^{-2} s^{-1}$

Sample	Surface dose rate (mGy/h)		Dominant activity
	After irradiation	*After 5 min decay*	
Glass slide (0·6 g)	1·0	0·40	^{28}Al
Perspex (0·86 g)	0·026	~0	^{19}O (+ Na contamination)
Chromite thin section[a]	3·0	1·0	$^{28}Al, {}^{52}V, {}^{56}Mn$

Data by courtesy of Susan Parry and Ian Sinclair, Imperial College Reactor Centre, Ascot.
[a]Mounted on a perspex slide (0·5 g).

polishing. Dose rate measurements on this thin section after a 1 min irradiation (Table 2) showed that relatively high dose rates were again observed; presumably due to substantial ^{28}Al, ^{52}V and ^{56}Mn activity induced in the chromite matrix, despite the fact that only about 60 mg of chromite remained after polishing.

Autoradiograph measurements were made as follows: the chromite section was sealed in a polythene irradiation can and irradiated at the Imperial College Reactor Centre, Ascot using the 'In Core Irradiation System' (ICIS) to transfer the sample pneumatically into and out of the reactor. The sample received a 1 min irradiation in a flux of about 2×10^{12} neutrons $cm^{-2}s^{-1}$ giving an integrated flux of $1·2 \times 10^{14}$ neutrons cm^{-2}. After ejection from the reactor, the shortest time expended in unloading the section from its irradiation can and transferring it to a temporary dark room adjacent to the reactor was 30 s. Thereafter, a series of autoradiographs were recorded to optimise exposure on Fuji RX film.

An example of one of these autoradiographs (exposed for 10 s) 1 min after irradiation is shown in Fig. 4(b). The thin section selected for this investigation (Fig. 4(a)) was prepared from a chromite from Cliff, Unst (Shetland Isles) chosen because abundant platinum-group element grains have been described in this sample (Prichard *et al.*, 1986) including rare rhodium phases (Tarkian & Prichard, 1987). However, no discrete phases containing elements with short-lived beta activity are apparent in Fig. 4(b). The autoradiograph image matches the distribution of chromite (^{28}Al, ^{52}V and ^{56}Mn activity) with some indication that the intensity of the image varies with thickness of the sample. The fact that this sample is not completely barren of discrete phases is demonstrated in Fig. 4(c). This second autoradiograph was recorded for 5·5 days commencing 5 days after irradiation and delineates the distribution of longer half-life isotopes. The discrete grains revealed in this autoradiograph are expected to be iridium and antimony phases known to occur in this sample (Tarkian & Prichard, 1987).

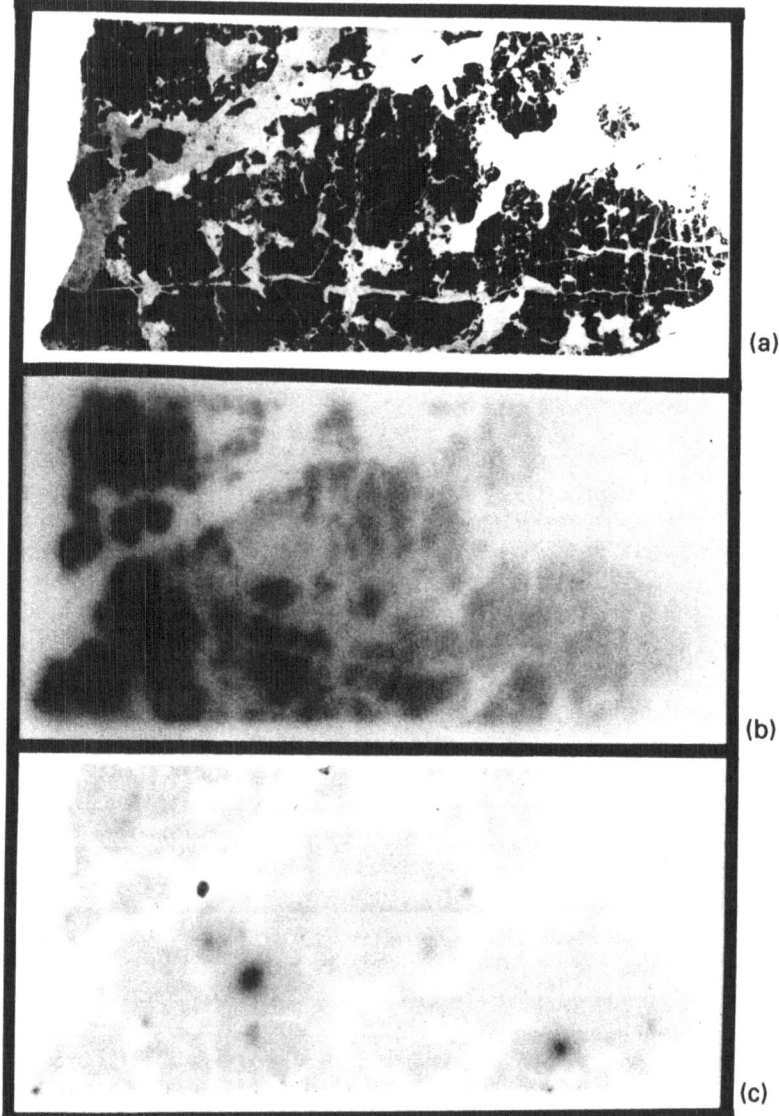

FIG. 4. Beta autoradiographs of short-lived isotopes. (a) Chromite thin section (Unst, Shetland Isles) mounted on a perspex slide. (b) Beta autoradiograph recorded for 10 s after a decay of 1 min following a 1 min irradiation showing short half-life activity induced in the chromite matrix (^{28}Al, ^{52}V, ^{56}Mn). (c) Beta autoradiograph of the same sample exposed for 5·5 days starting 5 days after irradiation showing the presence of some discrete grains containing longer half-life isotopes (^{192}Ir, ^{124}Sb).

CONCLUSIONS

The beta autoradiography technique may be improved and extended in several ways. The substitution of Fuji RX screen-type film offers greater sensitivity and durability at lower unit costs than Ilford K2 nuclear research plates. Sensitivity may be further enhanced by the use of calcium tungstate intensifier screens although with some loss in image resolution. For autoradiography measurements made within 5 min of irradiation, perspex has been shown to be a suitable mounting material to avoid the high activity that would otherwise be induced in ^{28}Al if conventional glass slides are used to prepare thin sections. The feasibility of recording autoradiographs for short-lived isotopes after a decay of 1 min has been established although the potential application in the sensitive location of rhodium, silver, indium and rare earth minerals must await further work on better-characterised samples.

ACKNOWLEDGEMENTS

The results presented in this paper would not have been possible without the assistance and advice of Susan Parry and Ian Sinclair and the facilities of the Imperial College Reactor Centre, as well as the help of Ian Chaplin (preparation of thin sections) and Pam Owen and John Taylor (preparation of manuscript).

REFERENCES

Erdtmann, G. (1976). *Neutron Activation Tables*. Verlag Chemie, Weinheim, 146 pp.

Hahn, E. J. (1983). Autoradiography—a review of basic principles. *Am. Lab.*, July, 64–71.

Laskey, R. A. (1980). The use of intensifying screens or organic scintillators for visualizing radioactive molecules resolved by gel electrophoresis. *Meth. Enzymol.*, **65**, 363–71.

Laskey, R. A. (1984). Radioisotope detection by fluorography and intensifying screens. Review 23, Amersham International, Amersham, 39 pp.

Laskey, R. A. & Mills, A. D. (1975). Quantitative film detection of ^{3}H and ^{14}C in polyacrylamide gels by fluorography. *Europ. J. Biochem.*, **56**, 335–41.

Laskey, R. A. & Mills, A. D. (1977). Enhanced autoradiographic detection of ^{32}P and ^{125}I using intensifying screens and hypersensitized film. *FEBS Lett.*, **82**, 314–16.

Mysen, B. O. (1978). Experimental determination of rare earth element partitioning between hydrous silicate melt, amphibole and garnet peridotite minerals at upper mantle pressures and temperatures. *Geochim. Cosmochim. Acta*, **42**, 1253–63.

Mysen, B. O. & Seitz, M. G. (1974). Trace element partitioning determined by beta track mapping: an experimental study using carbon and samarium as examples. *J. Geophys. Res.*, **80**, 2627–35.

Potts, P. J. (1984). Neutron activation-induced beta autoradiography as a technique for locating minor phases in thin sections: application to rare earth element and platinum group element mineral analysis. *Econ. Geol.*, **79**, 738–47.

Potts, P. J. (1986). Neutron activation-induced beta autoradiography as a technique for locating minor phases in thin sections: emulsion response characteristics. *Mater. Sci. Forum*, **7**, 35–44.

Potts, P. J. & Prichard, H. M. (1986). Mineralogical applications of neutron activation-induced beta autoradiography: the search for gold mineralisation in thin section. In *Metallogeny of Basic and Ultrabasic Rocks*, ed. M. J. Gallagher, R. A. Ixer, C. R. Neary & H. M. Prichard. Institution of Mining and Metallurgy, London, pp. 455–65.

Prichard, H. M., Neary, C. R. & Potts, P. J. (1986). Platinum group minerals in the Shetland ophiolite. In *Metallogeny of Basic and Ultrabasic Rocks*, ed. M. J. Gallagher, R. A. Ixer, C. R. Neary & H. M. Prichard. Institution of Mining and Metallurgy, London, pp. 394–414.

Swanstrom, R. & Shank, P. R. (1978). X-ray intensifying screens greatly enhance the detection by autoradiography of the radioactive isotopes ^{32}P and ^{125}I. *Anal. Biochem.*, **86**, 184–92.

Tarkian, M. & Prichard, H. M. (1987). Irarsite-hollingworthite solid-solution series and other associated Ru, Os, Rh bearing PGM's from the Shetland ophiolite complex. *Mineral Deposita*, **22**, 178–84.

Wilson, B. J. (ed.) (1966). *The Radiochemical Manual*, 2nd edn. The Radiochemical Centre, Amersham, 327 pp.

8

Solubility and Transport of Platinum-Group Elements in Hydrothermal Solutions: Thermodynamic and Physical Chemical Constraints

Bruce W. Mountain & Scott A. Wood

Department of Geological Sciences, McGill University, 3450 University Street, Montreal, PQ, Canada H3A 2A7

ABSTRACT

Thermodynamic calculations and physical chemical considerations suggest that at least some of the platinum-group elements (PGE) may be mobile in geological fluids such as chloride, hydroxide, bisulphide, polysulphide, thiosulphate or ammonia complexes, depending on pH, fO_2, temperature and lignad concentration.

In saline fluids (i.e. $\Sigma Cl^- > 1 \cdot 0$ m) at 25°C, Pt and Pd are soluble as $PtCl_4^{2-}$ and $PdCl_4^{2-}$, at a level of 10 ppb or greater, only in very acidic and oxidizing fluids. The complexes $OsCl_6^{3-}$, $IrCl_6^{3-}$, $RuCl_5^{2-}$ and $RhCl_5^{2-}$ and their hydration and hydrolysis products may also be responsible for solubilities of the order of 10 ppb under more restricted conditions. The thermodynamically predicted order of solubility as chlorides at 25°C is $Pd > Pt > Os > Ir$, Au. It was possible to extrapolate the data for Pt and Pd chloride complexes to 300°C where significant Pt and Pd solubility are attained only in highly acidic and oxidizing solutions.

Under mildly oxidizing, near neutral to basic conditions, hydroxy species such as $Pt(OH)_4^{2-}$ or $Pd(OH)_4^{2-}$ appear to give solubilities as high as 10 ppb. For the other PGE, oxyanionic species such as OsO_4^{2-} and RuO_4^{2-} can become important in very basic and oxidizing solutions. At low temperatures, thiosulphate and polysulphide species may be important. For example, at 25°C and $\Sigma S = 0 \cdot 1$ m, 10 ppb of Pt and Pd can be attained as $Pt(S_2O_3)_4^{6-}$ and $Pd(S_2O_3)_4^{6-}$ at near neutral to basic pH near the $HS^-/S_2O_3^{2-}$ boundary. Although no stability constant data are available, it is expected that the soft (in the Pearson sense) PGE metal ions would form strong complexes with HS^-. It was possible to estimate the stability constants of $Pt(HS)_4^{2-}$ and $Pd(HS)_4^{2-}$ at 300°C. These result in calculated solubilities of 1–10 ppt (parts per trillion) in the region near the pyrite–pyrrhotite–magnetite triple point. Higher solubilities may be expected if additional bisulphide species or mixed bisulphide–hydroxide species exist. Preliminary experimental data suggest that this is the case.

Ammonia may also be effective in mobilizing the PGE, however, only under near neutral to basic, oxidizing conditions. The hydroxide complexes may dominate over the ammonia complexes in this region. The effect of As, Se and Te appears to be to

decrease the solubility of the PGE although the possibility of complexing by ligands
containing these elements cannot be completely ignored.

Calculations also indicate that at temperatures greater than 1200K, 10 ppt Pd and
0·1 ppt Ru may be transported as volatile halide species in an aqueous vapour phase
at $\log f_{HCl} = 100$ bars and $\log fO_2$ of the quart–fayalite–magnetite buffer.

INTRODUCTION

There is accumulating evidence that hydrothermal fluids may play a role in the
redistribution of platinum-group elements (PGE) in a variety of environments.
Several Pt–Pd deposits have been described which appear to be largely hydrothermal in origin such as those of the Waterberg District, South Africa (Wagner,
1929); Messina, South Africa (Sohnge, 1945; Mihalik *et al.*, 1974); New Rambler,
Wyoming (McCallum *et al.*, 1976); Rathbun Lake, Ontario (Rowell & Edgar,
1986). Furthermore, many workers have proposed the involvement of hydrothermal fluids in the mobilization of PGE during or after the formation of such
'magmatic' PGE deposits as those in the Bushveld Complex, South Africa; the
Stillwater Complex, Montana; the Donaldson West Deposit, Quebec, etc. (Razin,
1968; Cousins, 1973; Stumpfl, 1974; Volborth & Housley, 1984; Ballhaus &
Stumpfl, 1985*a,b*; Boudreau & McCallum, 1986; Boudreau *et al.*, 1986*a,b*;
Watkinson *et al.*, 1986; Dillon-Leitch *et al.*, 1986).

Despite the geological evidence suggesting the possible mobility of the PGE in
aqueous fluids, few attempts have been made (e.g. Fuchs & Rose, 1974; Bowles,
1986) to quantify the solubility and transport of the PGE either experimentally or
theoretically. This paper summarizes our recent attempts to theoretically determine the efficacy of hydrothermal fluids in dissolving and transporting PGE,
particularly Pt and Pd. Many of the calculations performed involve extrapolation or
estimation of the necessary thermodynamic data and represent only a first
approximation. The approach will be considerably refined as more experimental
data become available. However, it is hoped that this work will both stimulate and
act as a framework for additional experimental and field studies involving the
mobilization of PGE in hydrothermal fluids.

GENERAL AQUEOUS CHEMISTRY OF THE PGE

Platinum and palladium may occur in either the +2 or the +4 valence state in
aqueous solution. The divalent state greatly predominates over the tetravalent
state at 25°C (Hartley, 1973; Westland, 1981; Table 1) except under extremely
oxidizing conditions and it is expected that only the divalent state will be important
in most hydrothermal solutions. The calculations described in Table 1 indicate that
neither Pt nor Pd can be transported as a simple aqueous ion at 25°C except in
extremely oxidizing and acidic solutions. The equilibrium constants for the
dissolution of Pd as Pd^{2+} and Pt as Pt^{2+} decrease with temperature; consequently,

TABLE 1

Activities of Pt^{2+} and Pd^{2+} in equilibrium with Pt and Pd metal respectively at 25°C

$log\ fO_2$	pH	$log\ a_{Pt2+}$	$log\ a_{Pt4+}$	$log\ a_{Pd2+}$
−68·6 (hematite-magnetite)	4	−43·3	−156	−32
	7	−49·3	−168	−38
−0·7 (surface conditions)	4	−9·3	−88	[a]2·3
	7	−15·3	−100	−3·7

Calculated from the following equations:

$Pt + 2H^+ + 1/2O_2 = Pt^{2+} + H_2O(l)$ $log\ K = -1·0$

$Pt + 4H^+ + O_2 = Pt^{4+} + 2H_2O(l)$ $log\ K = -71$

$Pd + 2H^+ + 1/2O_2 = Pd^{2+} + H_2O(l)$ $log\ K = 10·6$

Log K's calculated using data in Barner & Scheuerman (1978).

[a]Pd phases such as $Pd(OH)_2(s)$ become stable before such a high Pd^{2+} activity is reached.

the activities of the simple Pt and Pd ions are expected to be small under most hydrothermal conditions and it is thus necessary to invoke complexing to account for aqueous transport of these metals.

The possible oxidation states of the remaining PGE (Ru, Rh, Os and Ir) are more varied (cf. Hartley, 1973; Livingstone, 1973; Westland, 1981). Osmium and iridium do not exist in aqueous solutions as simple ions but only as complexes and hydrolysed species. The most common oxidation states for Os in aqueous solution appear to be +3, +4, +6 and +8 but under geologically reasonable conditions it is likely that the +3 species is of most importance. Iridium exhibits similar oxidation states. The hydrated Ru^{3+} and Rh^{3+} ions are known to exist and this oxidation state appears to be most important for these metals in complexes under geologically reasonable conditions as well.

Both Pt^{2+} and Pd^{2+} are relatively large ions with high electronegativity and high polarizability. The stability constants of their complexes with the halide ions increase in the order $Cl^- < Br^- < I^-$ (see Table 2) at 25°C. The Pt^{2+} and Pd^{2+} ions also form extremely stable complexes with ligands such as cyanide, sulphite, and thiosulphate at 25°C (Table 2). In fact, the stability constant for the $Pt(CN)_4^{2-}$ complex is estimated to be as high as 10^{78} (Hancock et al., 1977). These properties indicate that Pt^{2+} and Pd^{2+} can be considered soft acids according to the Pearson (1963) classification; they prefer to complex with ligands capable of predominantly covalent interactions, i.e. soft ligands. Such behaviour is also observed for Au^+, Ag^+ and Hg^{2+} (Crerar et al., 1985). The Cu^+ ion also demonstrates some affinity towards soft ligands which may help explain the common association of Cu with Pt and Pd in many of the hydrothermal deposits mentioned above.

Quantitative data for the relative stabilities of a wide variety of Os, Ir, Ru and Rh complexes are almost completely lacking. Qualitatively, the lower oxidation states of these metals should tend to prefer soft ligands such as CN^-, CO, etc., while the

TABLE 2

Cumulative stability constants for the complexes of platinum and palladium at 25°C

Ligand n =	Log β_n (Pt^{2+})				Ref.	Log β_n (Pd^{2+})				Ref.
	1	2	3	4		1	2	3	4	
Cl⁻	4·97	8·97	11·89	13·99	1	4·47	7·76	10·17	11·54	2
Br⁻	4·9	9·0	12·6	15·4	3	5·17	9·42	12·7	14·9	3
I⁻	—	24·4	27·9	29·6	4	10·0	—	—	24·9	4
OH⁻	(15·8)	(31·4)	(46·8)		5			(38·4)	(50·8)	5
OH⁻				62	6	13·0	25·8			7
NH₃	—	—	—	35·3	4	9·6	18·5	26·0	32·6	7
SO₃²⁻	—	—	—	37·9	8	—	—	—	29·1	8
S₂O₃²⁻	—	—	—	43·7	8	—	—	—	35·0	8
NO₂	—	—	—	24·1	8	—	—	—	21·0	8
CN⁻	—	—	—	(78)	8	—	—	—	63	9
SCN⁻	—	—	—	33·6	8	—	16·9	—	25·6	3
HS⁻	—	—	—	(51)	5	—	—	—	(41)	5
SO₄²⁻	—	—	—	—		—	3·16	—	—	3

1. Elding (1978); 2. Elding (1972); 3. Högfeldt (1982); 4. Sillen & Martell (1971); 5. This study (see text for derivation); 6. Calculated from data in Peshchevitskiy *et al.* (1962) and Elding (1978); 7. Smith & Martell (1976); 8. Hancock *et al.* (1977); 9. Hancock & Evers (1976).
Parentheses denote estimated values. Dashes indicate datum not available.

higher oxidation states ($+6$ to $+8$) should prefer to bind with oxygen-containing ligands and perhaps fluoride.

Because of the soft nature of Pt^{2+} and Pd^{2+}, their complexes with hard anions, such as carbonate, bicarbonate, sulphate, phosphate and fluoride, will have low stability. Indeed, the only available stability constant for such complexes, that of $Pd(SO_4)_2^{2-}$ at 25°C, is quite small (log $\beta_2 = 3·16$; Högfeldt, 1982). Although quantitative data are lacking, probably the same can be said about the stability of Ru, Rh, Os and Ir complexes with hard ligands, except under highly oxidizing conditions where the higher valence states ($+6$, $+8$) are stable. For example, although complexes such as RuF_6^{3-}, RuF_6^{2-}, RhF_6^{3-} and RhF_6^{2-} are known to occur in solid compounds, they are described as being insoluble in water or susceptible to complete hydrolysis (Livingstone, 1973). Simple fluoride complexes are apparently unknown for Pt, Pd, Os and Ir. In any case, the harder alkali metal cations and the hydrogen ion are expected to successfully compete with the PGE for the hard ligands, especially at higher temperature (Barnes, 1979).

In spite of the relatively high stability constants of Pt and Pd complexes with Br⁻ and I⁻ (Table 2), very little metal is expected to be carried in this form because of the low abundance of Br⁻ and I⁻ relative to Cl⁻ in natural fluids (Seward, 1984). Similar remarks probably apply to the other PGE as well. Platinum, palladium and perhaps the other PGE as well form strong complexes with cyanide (CN^-), nitrite (NO_2^-), and thiocyanate (SCN^-). However, the concentrations of these in most common hydrothermal fluids are likely to be small and their thermal stability would be somewhat in question. Cyanide could possibly be important on a limited basis at low temperatures due to either anthropogenic or biogenic processes.

After having eliminated several possibilities, the ligands considered most likely to contribute to PGE transport and which warrant more detailed discussion include chloride, hydroxide, bisulphide and other sulphur ligands and ammonia.

Chloride

High chlorinity fluids appear to be associated with several types of PGE deposits. For example, Ballhaus & Stumpfl (1985b) have found highly saline fluid inclusions (halite-saturated) in quartz in the Merensky reef. Biotites and apatites from the Merensky Reef (Boudreau & McCallum, 1985) and the J-M Reef in the Stillwater complex (Boudreau & McCallum, 1985; Boudreau et $al.$, 1986b) are unusually chloride-rich (up to 100% Cl in the volatile site). These observations have led to the suggestion that chloride complexing may play a role during remobilization of the PGE in these deposits. It therefore seems desirable to quantitatively evaluate the role of chloride complexing in the transport of the PGE.

At 25°C, complete thermodynamic stability data is available only for the Pt^{2+} and Pd^{2+} chloride complexes. These data allow the calculation of the distribution of Pt^{2+} and Pt^{2+} chloride species as a function of chloride concentration in aqueous solution. This distribution is illustrated for 25°C in Fig. 1. It is immediately apparent that the $PtCl_4^{2-}$ and $PdCl_4^{2-}$ complexes predominate over all others at free chloride concentrations as low as 0·01 and 0·1 molal respectively. Thus, if the chloride concentrations to be considered are restricted to be greater than these values, total Pt and Pd solubilities as chloride complexes in moderately oxidizing solutions can be simply represented by the divalent tetrachloro complexes and in extremely oxidizing solutions by the tetravalent hexachloro complexes.

Platinum and palladium solubility as chloride complexes is thus depicted for 25°C in the form of Eh–pH diagrams in Figs 2(a) and 2(b). Platinum solubility greater than 10 ppb as $PtCl_4^{2-}$ at 25°C is restricted to quite oxidizing and acidic conditions. Palladium solubility as $PdCl_4^{2-}$ is slightly higher. The tetravalent hexachloro complexes are restricted to even higher Eh conditions than the divalent tetrachlorides (Fig. 2).

The lack of complete stability constant data precludes a comprehensive calculation of the distribution of Ru, Rh, Os and Ir among their various chloride complexes. However, the data available allow the construction of Figs 3 and 4, which may be regarded as preliminary representations of the distribution of Ir^{3+} and Rh^{3+} among their chloride complexes. It appears that over a wide range of geologically reasonable chloride activities $RhCl_5^{2-}$ is the dominant chloride complex of Rh^{3+}. Figure 4 reveals the same conclusion for Ir^{3+}. There are also some data to suggest that $OsCl_5^-$ is predominant over $OsCl_6^{2-}$ below a chloride activity of approximately $5\,m$ at 80°C (Miano & Garner, 1965). However, in as much as the stability data for the pentachloro complexes are not all available at room temperature and the same ionic strength, and because of the uncertainty in the data, we have chosen to represent the solubility of Os and Ir in chloride solutions as

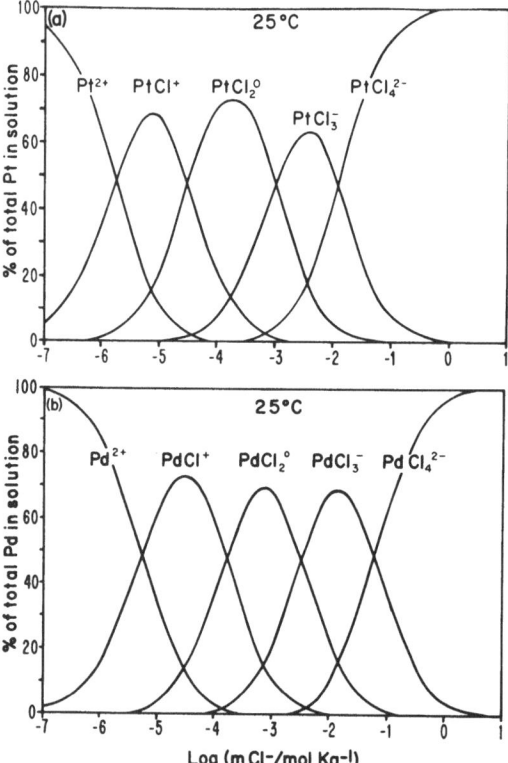

FIG. 1.   Percentage distribution of metal chloride complexes as a function of free chloride concentration for: (a) Pt at 25°C; (b) Pd at 25°C. Data to construct diagram was taken from Elding (1972, 1978).

the hexachloro complexes, both because the thermodynamic data for these seem to be better known and also in order to maintain some consistency (in terms of temperature) amongst the metals. The chloride complexes of Os and Ir may actually hydrate and hydrolyse under the chosen conditions; however, the solubilities of Ir and Os calculated using these complexes are probably a good first approximation to the true solubility. If other complexes are important, the actual solubilities of these metals may be somewhat higher than calculated. Figures 5 and 6 demonstrate that the solubility (10 ppb) of Os and Ir as chloride complexes is quite restricted as both require relatively high *Eh* and low pH. Complete and reliable thermodynamic data for Ru and Rh chloride complexing are not available and therefore meaningful *Eh*–pH diagrams cannot be drawn for these metals.

That kinetics plays an important part in the chemistry of the PGE is certain. For example, according to Llopis & Tordesillas (1976), bulk Rh is not noticeably attacked by highly corrosive aqua regia solutions but finely divided Rh powder can be dissolved in either hot, concentrated sulphuric acid or aqua regia. Even homogeneous equilibria among the chloride complexes in solution are slow at 25°C (Chang & Garner, 1965; Mestre *et al.*, 1982). However, neglecting kinetics, the

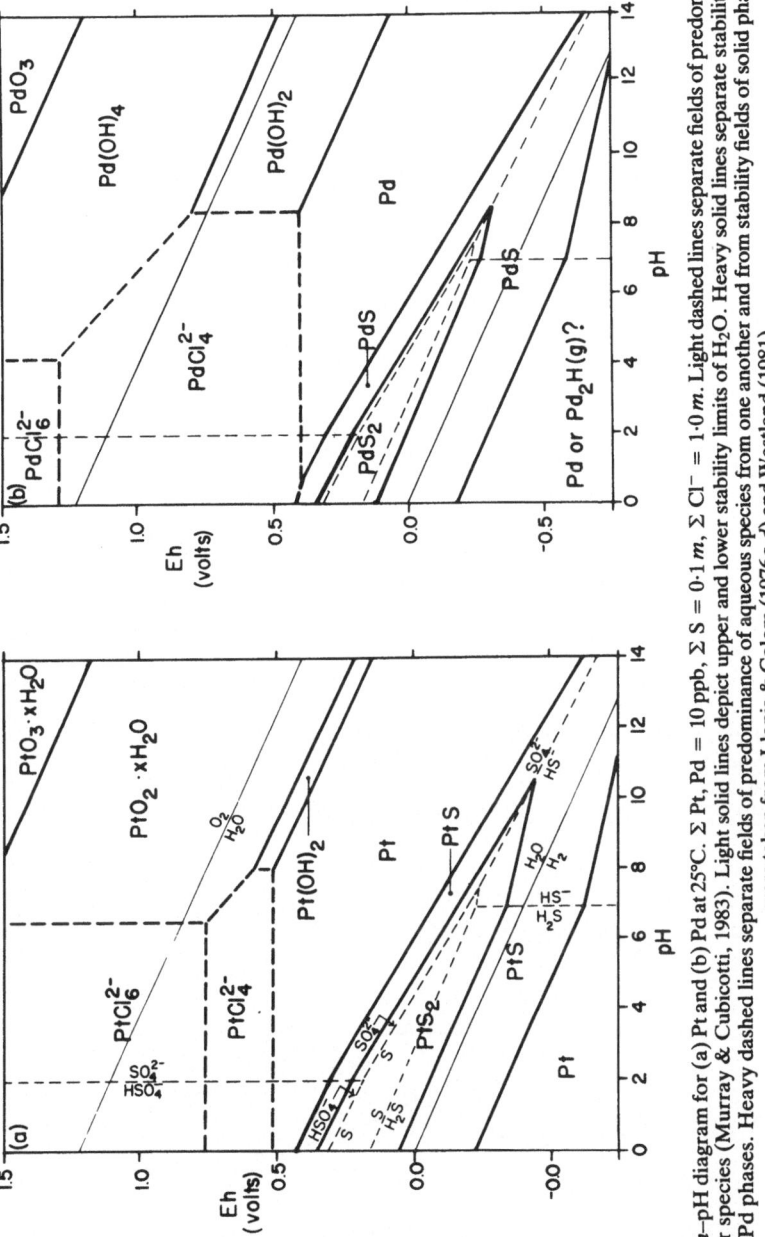

FIG. 2. *Eh–*pH diagram for (a) Pt and (b) Pd at 25°C. Σ Pt, Pd = 10 ppb, Σ S = 0·1 m, Σ Cl⁻ = 1·0 m. Light dashed lines separate fields of predominance of the sulphur species (Murray & Cubicotti, 1983). Light solid lines depict upper and lower stability limits of H_2O. Heavy solid lines separate stability fields of solid Pt or Pd phases. Heavy dashed lines separate fields of predominance of aqueous species from one another and from stability fields of solid phases. Data were taken from Llopis & Colom (1976a,d) and Westland (1981).

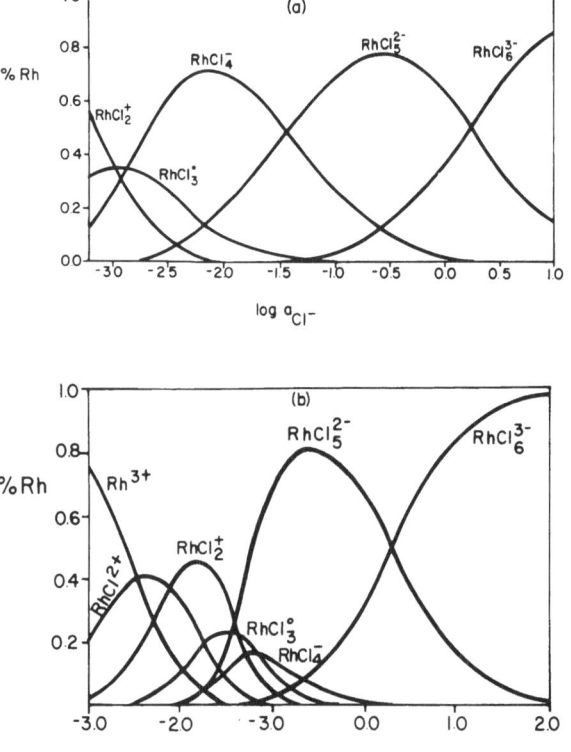

FIG. 3. Percentage distribution of Rh^{3+} chloride complexes as a function of free chloride activity. (a) At 120°C and a total ionic strength of 6·0 M. Calculated from data given by Wolsey *et al.* (1965). (b) At 25°C and an ionic strength equal to total chloride activity at each point. After Cozzi & Pantani (1958). Note that all concentrations in this diagram are given on the molar (M) rather than the molal (*m*) scale.

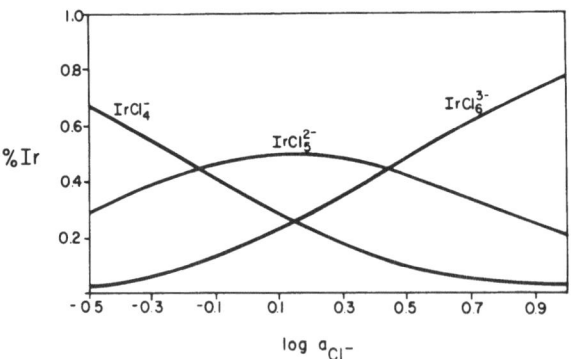

FIG. 4. Percentage distribution of Ir^{3+} chloride complexes as a function of free chloride activity at 50°C and an ionic strength of 3·7 M. Data from Poulsen & Garner (1962) and Chang & Garner (1965). Note that all concentrations in this diagram are given on the molar (M) rather than the molal (*m*) scale.

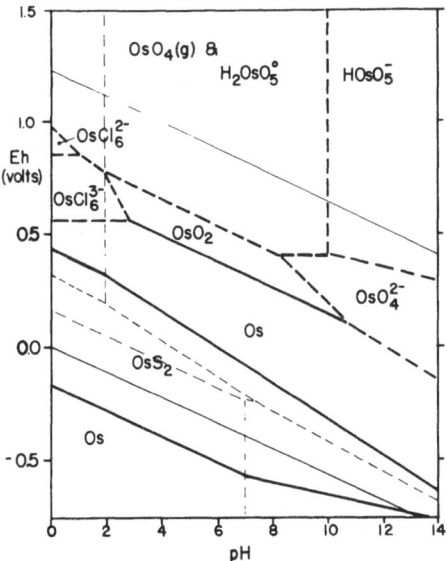

FIG. 5. *Eh*–pH diagram for Os at 25°C and $\Sigma \text{Os} = 10$ ppb, $\Sigma \text{S} = 0\cdot1\,m$, $\Sigma \text{Cl}^- = 1\cdot0\,m$. All lines have the same significance as those in Fig. 2. Data are from Llopis & Colom (1976c) and Westland (1981). Note the extremely small fields of predominance of the OsCl_6^{3-} and OsCl_6^{2-} ions but the relatively large fields of predominance of OsO_4 (g) and the osmate species at high *Eh*.

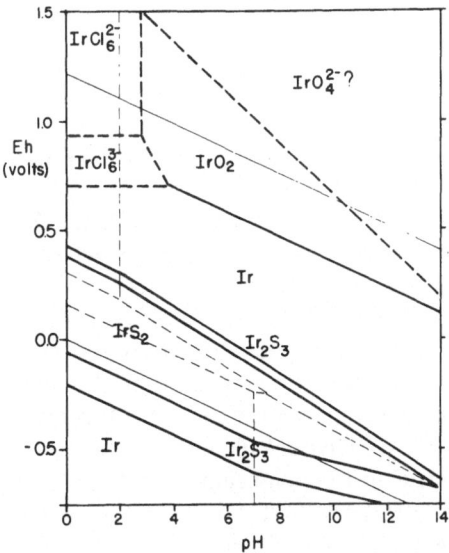

FIG. 6. *Eh*–pH diagram for Ir at 25°C and $\Sigma \text{Ir} = 10$ ppb, $\Sigma \text{S} = 0\cdot1\,m$, $\Sigma \text{Cl}^- = 1\cdot0\,m$. All lines have the same significance as in Fig. 2. Data are from Llopis & Colom (1976b) and Westland (1981). Iridium appears to be the least soluble of the PGE, and is only somewhat less soluble than gold.

order of thermodynamically predicted solubilities for the PGE as chloride complexes at 25°C is: Pd > Pt > Os > Ir.

The above discussion of PGE solubility has been limited to a temperature of 25°C. The geologist, however, is interested in PGE solubility in much higher temperature solutions. Unfortunately the lack of thermodynamic data even at 25°C precludes, at this time, a calculation of the equilibrium speciation and solubility of Ru, Rh, Ir and Os at elevated temperatures. However, sufficient data are available for Pt and Pd to allow such calculations for these two metals up to at least 300°C. The stability constants for the $PtCl^+$, $PtCl_2^0$, $PtCl_3^-$, and $PtCl_4^{2-}$ complexes are available at 25 (Elding, 1978) and 60°C (Mestre *et al.*, 1982). These data have been extrapolated to 300°C (Mountain & Wood, 1988) using the isocoulombic principle.

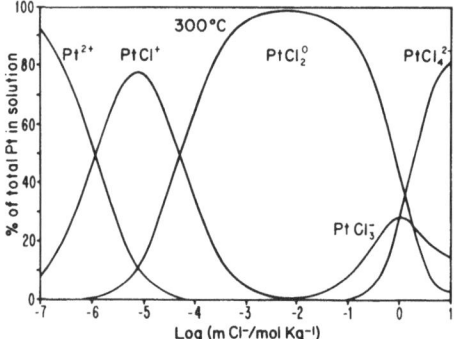

FIG. 7. Percentage distribution of Pt chloride complexes as a function of free chloride concentration at 300°C. Data were obtained by extrapolation of the data in Elding (1978) and Mestre *et al.* (1982); see Mountain & Wood (1988) for details. Note the expanded field of stability of the $PtCl_2^0$ species.

The distribution of Pt species calculated using the extrapolated stability constants is illustrated in Fig. 7. At 300°C the field of predominance of $PtCl_2^0$ is greatly expanded. This is consistent with accumulating evidence that neutral complexes become increasingly important at higher temperatures where the lowered dielectric constant of water makes the separation of charge more difficult (Crerar *et al.*, 1985 and references therein). At 1 molal chloride $PtCl_4^{2-}$ is no longer predominant, but $PtCl_4^{2-}$, $PtCl_3^-$ and $PtCl_2^0$ are approximately equal in concentration. Nevertheless, the solubility of Pt as $PtCl_4^{2-}$ still provides an order of magnitude estimate of the total Pt solubility as chloride complexes for $\Sigma Cl^- = 1 \cdot 0$ or greater. It is assumed that this is also the case for Pd. We have calculated solubilities of Pt and Pd minerals as MCl_4^{2-} complexes at 300°C in $1 \cdot 0$ molal chloride solutions using the data given in Barner & Scheuerman (1978) and Mills (1974) and the results are plotted on log fO_2–pH diagrams in Fig. 8. The *Eh*–pH diagrams at 25°C of Fig. 2 are reproduced in Fig. 8 as log fO_2–pH diagrams for ease of comparison with the 300°C diagrams. Solubilities of Pt and Pd greater than 10 ppb appear to be obtained as chloride complexes only under oxidizing, acidic conditions, even at 300°C. By neglecting MCl^+, MCl_2^0 and MCl_3^- complexes, the calculated solubilities most likely slightly

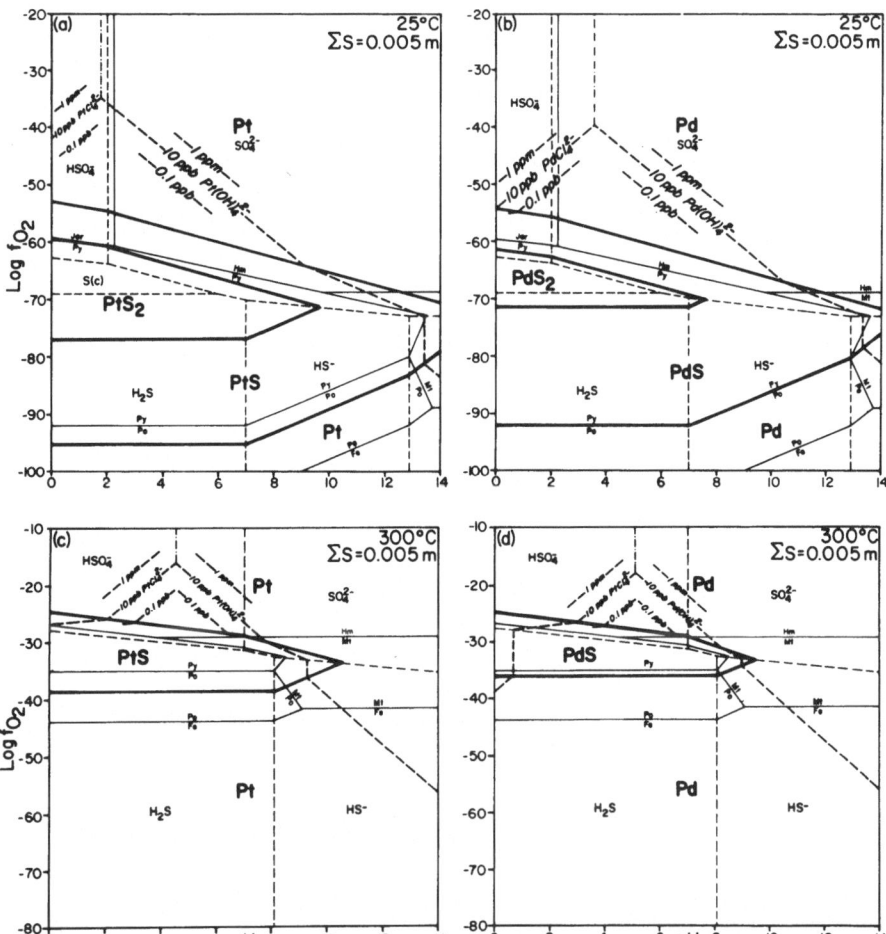

FIG. 8. Log fO_2 versus pH diagrams showing the stability fields of Pt and Pd native metal and sulphides and their solubilities as MCl_4^{2-} and $M(OH)_4^{2-}$ complexes. Light dashed lines represent boundaries between predominant dissolved sulphur species ($\Sigma S = 0.005\,m$). Light solid lines separate stability fields of Fe minerals, while heavy solid lines separate those of Pt or Pd minerals. Heavy dashed lines represent solubility contours of: (a) Pt, 25°C; (b) Pd, 25°C; (c) Pt, 300°C; and (d) Pd, 300°C as MCl_4^{2-} at $\Sigma Cl^- = 1$ molal and as $M(OH)_4^{2-}$. The 1 ppm and 0·1 ppb lines are only partially shown for clarity. At 25°C goethite may be the stable phase in place of hematite but the stability fields for both are nearly identical. The jarosite field for $\Sigma K = 10^{-3}\,m$ is shown (Brown, 1971). Note that hematite is the stable phase in place of jarosite and goethite at 300°C. Data sources include Barner & Scheuerman (1978), Mills (1974), Mountain & Wood (1988), Robie *et al.* (1978) and Westland (1981).

underestimate the true solubilities, but the above conclusion will not be significantly affected.

Figure 9 demonstrates that the solubility of Pt and Pd as tetrachloro complexes first decreases, passes through a minimum at about 200°C and then appears to increase with increasing temperature. It might be speculated that the solubility will increase dramatically at temperatures greater than 400°C because of the decrease in the dielectric constant of water and the subsequent increase in electrostatic attraction between metal and ligand. Also, an increase in solubility with respect to temperature above 300°C is expected because the MCl_2^0 complexes should predominate over an even larger range of chloride concentration. Thus the solubility as MCl_4^{2-} should be a small fraction of the total solubility.

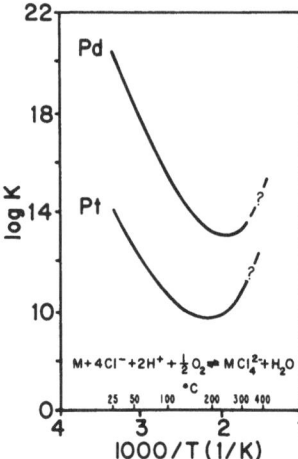

FIG. 9. Log K versus $1/T$ for the dissolution of Pt and Pd metal as MCl_4^{2-} complexes. Data from Barner & Scheuerman (1978).

Hydroxide and Oxyanionic Species

Figures 5 and 6 demonstrate the contribution of oxyanionic species to the solubility of Os and Ir at 25°C. These species appear to be significant only under highly oxidizing and alkaline conditions and therefore probably have little geologic significance at least at 25°C.

It might be expected that Pt and Pd may form hydroxy complexes, however, stability constants are available in the literature only for $Pd(OH)^+$ and $Pd(OH)_2^0$ at 25°C (Table 2). These are among the most stable hydroxy complexes known for any metal (Baes & Mesmer, 1976). Therefore Pt and Pd should exhibit some solubility in alkaline solutions. In order to estimate the stability constants for $Pd(OH)_3^-$ and $Pd(OH)_4^{2-}$ the method outlined by Belevantsev et al. (1982) was employed (Mountain & Wood, 1988). Because there are no direct measurements of stability constants for platinum hydroxide complexes available in the literature, the stability constant of $Pt(OH)_4^{2-}$ at 25°C had to be indirectly calculated from data given by Peshchevitskiy *et al.* (1962) for the hydrolysis of the MCl_4^{2-} complex. The remaining cumulative stability constants were then calculated using the method of Belevantsev *et al.* (1982) (Mountain & Wood, 1988).

The relative importance of hydroxide versus chloride complexing at 25°C is illustrated for both Pt and Pd in Fig. 10. Calculations based on the data of Peshchevitskiy *et al.* (1962) indicate that no predominance field exists for mixed hydroxychloro Pt complexes. It can readily be seen that the $M(OH)_4^{2-}$ complexes predominate over most of the geologically reasonable portion of the diagrams. At a chlorinity of 1 molal, the boundary between $Pt(OH)_4^{2-}$ and $PtCl_4^{2-}$ occurs at a pH of about 3.

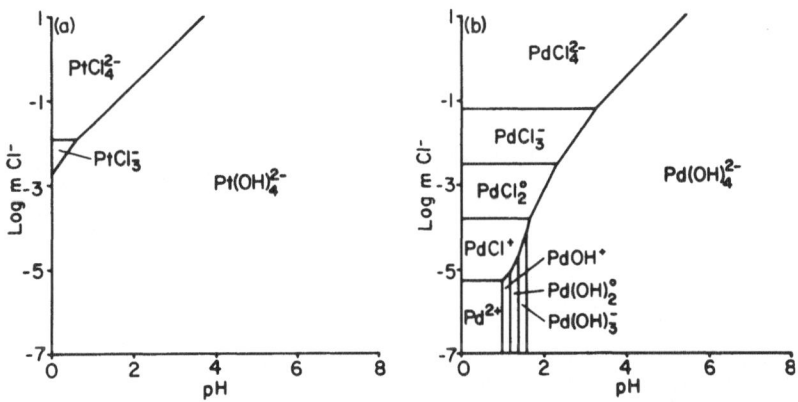

FIG. 10. Chloride versus hydroxide complexing at 25°C for (a) Pt, (b) Pd. Note the absence of a stability field for Pt^{2+} at non-negative pH and the small stability field for Pd^{2+}. Apparently, no field of predominance exists for mixed chloride–hydroxide complexes.

The effect of hydroxy complexing on the solubility of Pt and Pd at 25°C is illustrated in Figs 8(a) and 8(b). As a result of hydroxy complexing, Pt and Pd are appreciably soluble in near neutral to basic solutions under oxidizing to slightly reducing conditions. Furthermore, both Pt and Pd have nearly the same solubility as hydroxide complexes.

To obtain an estimate of the concentration of hydroxy complexes at 300°C, β_4 for both Pt and Pd hydroxy complexes was extrapolated from 25°C (Mountain & Wood, 1988). Calculated solubilities as $M(OH)_4^{2-}$ at 300°C are illustrated in Figs 8(c) and 8(d). The uncertainty of the extrapolated β_4's is probably of the order of ± 3 log units. However, this rather large uncertainty does not affect the conclusion that $M(OH)_4^{2-}$ complexes appear to predominate in near neutral to basic solutions at high temperature as well as low and could be responsible for Pt and Pd transport under the appropriate conditions. In any event, it is necessary to more accurately quantify the effects of hydroxide complexing for Pt and Pd transport through direct experimentation. Preliminary experiments by S. Wood and A. Mucci have found solubilities of Pt as high as 3 ppm in NaOH solutions at pH = 11–14.

Ammonia

The stability constants for Pt and Pd complexes with ammonia are relatively large (Table 2) and ammine complexing might be expected to be significant in the transport of these metals. Concentrations of ammonia in hydrothermal solutions may range from as high as $10^{-1.5}$ to less than 10^{-4} molal (Barnes, 1979). Figure 11 illustrates the solubility of Pt and Pd as ammine complexes at 25°C assuming the tetramine complexes to be the most stable. Under most conditions, the calculated solubility of Pt and Pd as ammine complexes is significantly less than the calculated solubility as hydroxyl complexes. Therefore, although ammonia can form quite soluble complexes with both Pt and Pd under near neutral to basic, oxidizing conditions, it is likely that hydroxide complexes of these metals predominate over the ammine complexes under such conditions.

Sulphur-containing Ligands

Thiosulphate, sulphite, polysulphide

All three of these ligands are thermodynamically unstable with respect to either bisulphide, sulphate or native sulphur under all possible conditions and should therefore not exist at high concentrations in natural systems if equilibrium is attained (Barnes, 1979). Nevertheless, there appear to be some conditions where thiosulphate and polysulphide may exist metastably for significant periods of time. Because of the high stability of PGE complexes with thiosulphate, and probably also polysulphide, these ligands may contribute under certain limited circumstances to PGE mobility in aqueous solutions.

According to Goldhaber (1983), the oxidation of a sulphide mineral, such as pyrite, may result in the formation, in order of increasing pH, of either sulphate, tetrathionate ($S_4O_6^{2-}$) or thiosulphate ($S_2O_3^{2-}$). The more basic the pH, the longer the thiosulphate persists before being further oxidized to sulphate. Apparently, sulphite (SO_3^{2-}) cannot exist in the presence of other sulphur species and is likely to be much less important than thiosulphate. The solubilities of Pt and Pd as $Pt(S_2O_3)^{6-}$ and $Pd(S_2O_3)^{6-}$ respectively are depicted in Fig. 12. Solubilities of at least 10 ppb are attained only in near neutral to basic solutions at 25°C. Under hydrothermal conditions, it is expected that the kinetic barriers which allow thiosulphate to persist metastably will be more readily overcome and so the importance of thiosulphate as a ligand is restricted to low temperatures. As discussed by Webster (1986) for the case of Au, thiosulphate complexing would be most important during low temperature oxidation of sulphide deposits in carbonate or carbonatized host rocks.

The PGE also tend to form complexes with polysulphides. The following have been reported (Pittwell, 1965): $Pd^{IV}S_3^{2-}$, PdS_{11}^{2-}, PtS_{15}^{2-} and IrS_{15}^{2-}. Polysulphide can exist metastably up to 240°C (Giggenbach, 1974) and can be formed by several geological processes including oxidation of sulphides or H_2S, dissolution of native sulphur, or acidification of thiosulphate or tetrathionate solutions (Murowchick & Barnes, 1986). Unfortunately there are no data on the stability of the above-mentioned polysulphide complexes, but such complexes should be quite strong and could play a role in transporting PGE in relatively low temperature environments.

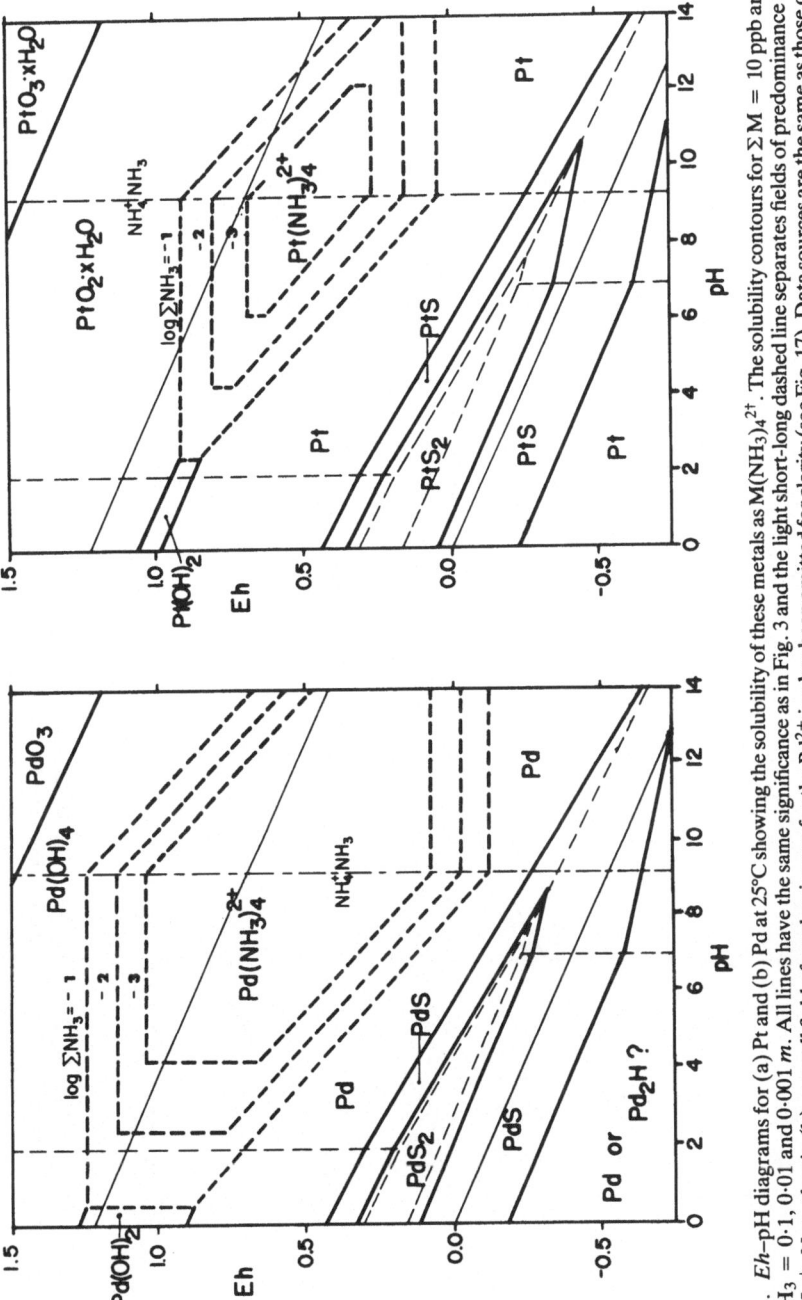

FIG. 11. *Eh*–pH diagrams for (a) Pt and (b) Pd at 25°C showing the solubility of these metals as $M(NH_3)_4^{2+}$. The solubility contours for $\Sigma M = 10$ ppb are given at $\Sigma NH_3 = 0 \cdot 1$, $0 \cdot 01$ and $0 \cdot 001$ m. All lines have the same significance as in Fig. 3 and the light short-long dashed line separates fields of predominance of NH_3 and NH_4^+. Note that in (b), a small field of predominance for the Pd^{2+} ion has been omitted for clarity (see Fig. 17). Data sources are the same as those of Fig. 2 plus the stability constants listed in Table 2.

Bruce W. Mountain, Scott A. Wood

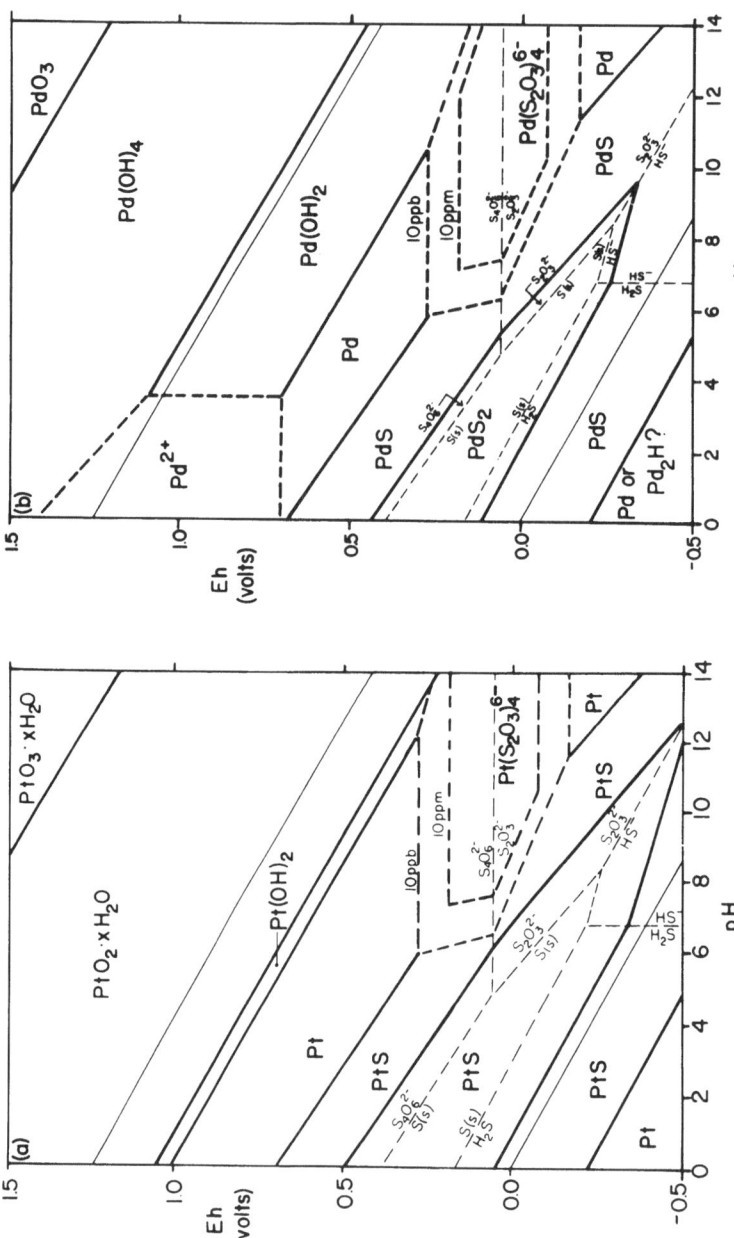

FIG. 12. *Eh*–pH diagrams for (a) Pt and (b) Pd at 25°C showing the solubility of these metals as thiosulphate complexes. Solubility contours are given at ΣM = 10 ppb and 10 ppm at ΣS = 0.1 m. All other lines have the same significance as in Fig. 3, however, the sulphur species have been drawn assuming that equilibria involving sulphate and other sulphur species are kinetically inhibited as discussed in the text.

Additional discussion of sulphite, thiosulphate and polysulphide as ligands for the PGE is given by Plimer & Williams (this volume).

Bisulphide

Because Pt^{2+} and Pd^{2+} are soft ions they should form strong complexes with the soft bisulphide anion as do Au^+ and Hg^{2+}. A comparison of Au with Pt and Pd (Fig. 13) suggests that bisulphide complexes of Pt and Pd should be nearly as strong as the cyanide complexes. Inasmuch as the stability constants for Pt and Pd bisulphide complexes have not yet been measured even at 25°C, it is necessary to estimate them. Hancock *et al.* (1977) have demonstrated that the stability constants of Pt, Au and Pd are remarkably well correlated. Figure 14(a) shows the relationship between $\log \beta_4(Pt^{2+})$ and $\log \beta_4(Pd^{2+})$. A similar correlation is demonstrated for $\log \beta_2(Au^+)$ versus $\log \beta_4(Pt^{2+})$ in Fig. 14(b). Using the stability constant for $Au(HS)_2^-$ at 25°C (Seward, 1984) and Fig. 14, the stability constants for $Pt(HS)_4^{2-}$ and $Pd(HS)_4^{2-}$ can be estimated with confidence (Table 2). It is assumed that the tetrabisulphide complex represents a significant portion of Pt and Pd dissolved as bisulphide complexes at geologically reasonable sulphur concentrations (10^{-3}–10^{-1} m). Neglect of complexes of lower ligation number makes any calculation of Pt or Pd solubility due to bisulphide complexing a minimum estimate.

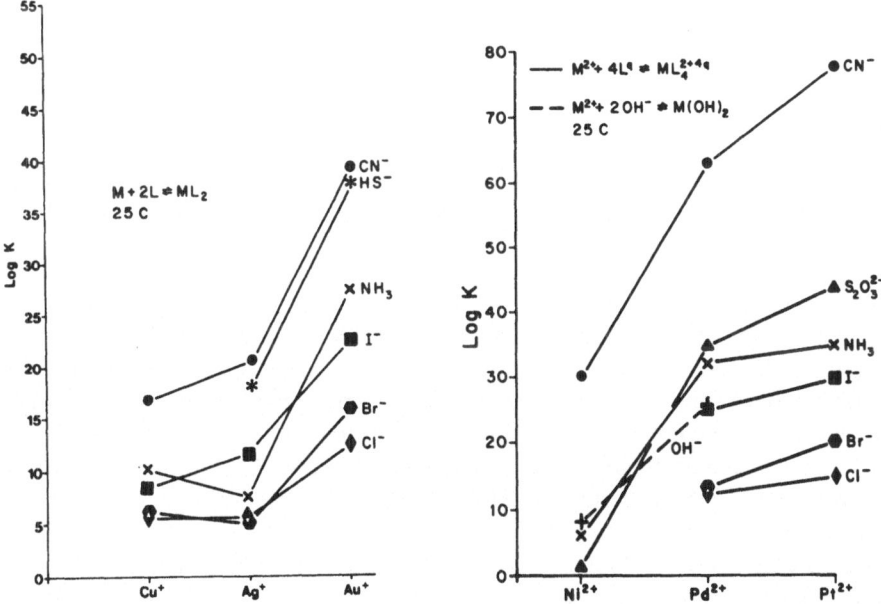

FIG. 13. Comparison of stability constants for Cu^+, Ag^+ and Au^+ with those of Ni^{2+}, Pd^{2+} and Pt^{2+}. Data are from Sillen & Martell (1971), Hancock & Evers (1976), Smith & Martell (1976), Högfeldt (1982), and Seward (1984). Note the similarity in the order of the stability constants of gold and platinum complexes with a variety of ligands.

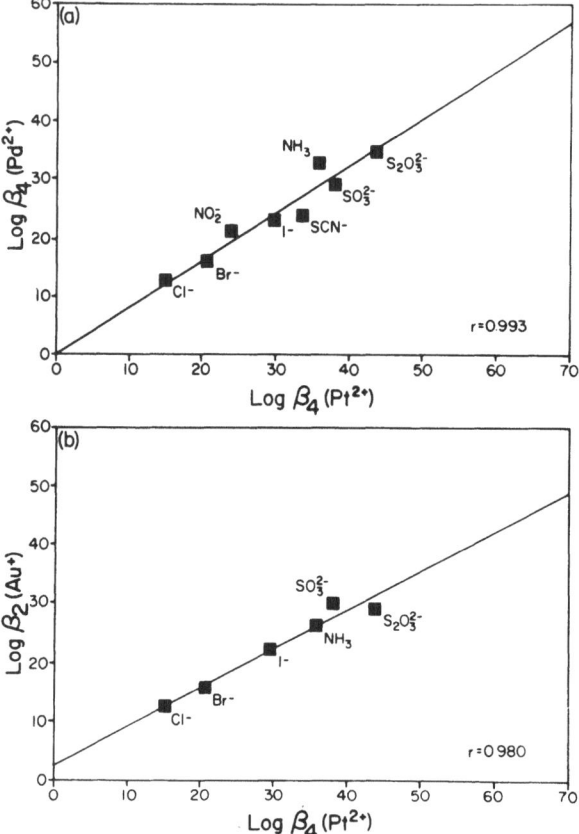

FIG. 14. Linear free energy relationships (Hancock et al., 1977) involving the cumulative stability constants of platinum, palladium and gold. (a) log $\beta_4(Pd^{2+})$ versus log $\beta_4(Pt^{2+})$: (b) log $\beta_2(Au^+)$ versus log $\beta_4(Pt^{2+})$.

Comparison of the stability constants in Table 2 shows that $Pt(HS)_4^{2-}$ and $Pd(HS)_4^{2-}$ appear to be among the more stable complexes listed. Thus it might be expected that Pt and Pd solubility as $M(HS)_4^{2-}$ complexes would be relatively high. However, the maximum calculated solubility attained is approximately $10^{-20} m$ at 25°C. This is due to the extremely low activity of free Pt^{2+} ion in the region of the diagram where HS^- is dominant. The solubility of Pd as $Pd(HS)_4^{2-}$ is nearly the same as for Pt.

The possibility of bisulphide or sulphide complexes at 25°C, however, is not to be discounted in spite of the above calculations. Westland (1981) states that Pt is soluble in alkaline sulphide solutions, possibly as $[Pt(HS)_x(OH)_{4-x}]^{2-}$ complexes. Recent experiments (Mucci & Wood, unpublished data) have yielded Pt concentrations of between 10 ppb and 1 ppm after several months for Pt metal in contact with $0.005-1.0 m$ Na_2S solution. These data indicate that perhaps other

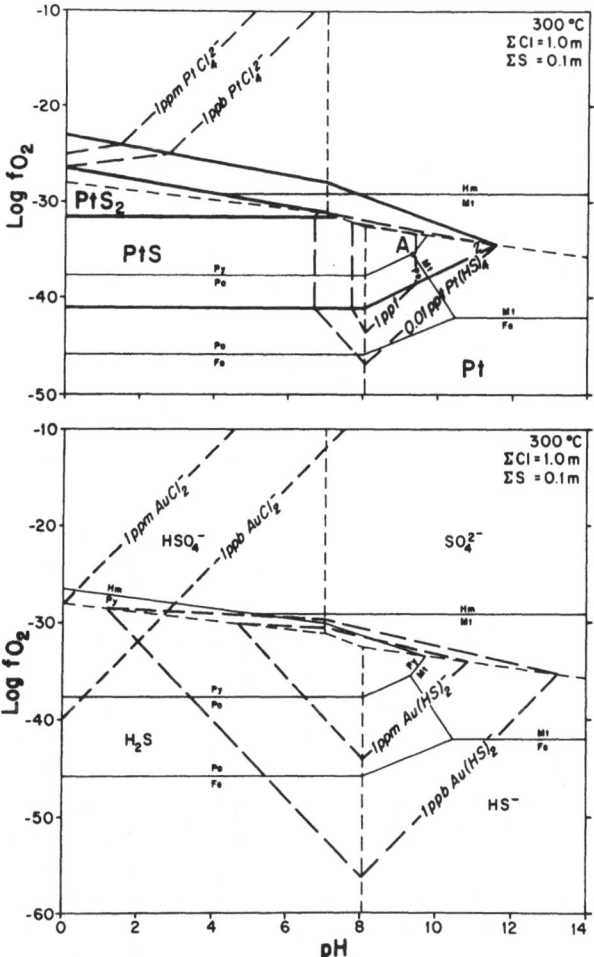

FIG. 15. Solubilities of gold and platinum as chloride and bisulphide complexes at 300°C, $\Sigma S = 0\cdot1\,m$ and $\Sigma Cl^- = 1\cdot0\,m$. Similar to Fig. 8 except contours for hydroxide complexes have been omitted. Note that Pt and Au solubilities as chloride complexes are quite similar but that the solubility of gold as a bisulphide complex greatly exceeds that of Pt. Data sources for Pt are the same as Fig. 8. Additional data sources for gold include Seward (1973, 1984) and Helgeson (1969).

species, which have not been taken into account, may be stable and contribute to Pt and Pd solubility.

The stability constants for the Pt and Pd tetrabisulphide complexes were extrapolated to 300°C in the same manner used for the hydroxy complexes (Mountain & Wood, 1988). Figure 15 shows that at $0\cdot1\,m$ total sulphur the solubility of Pt as $Pt(HS)_4{}^{2-}$ attains a value of $10^{-11}\,m$ (approx. 1 ppt) or greater over a limited region of $\log fO_2$–pH space near and to the left of the pyrite + pyrrhotite + magnetite triple point (point A, Fig. 15). At $\Sigma S = 0\cdot5\,m$, the solubility of Pt is approximately 1 ppb at the same fO_2 and pH as point A on Fig. 15. It is apparent that relatively high total sulphur concentrations are required to attain sufficient solubilities for ore transport. The calculated solubilities may represent only a portion of the total solubility as other species such as $Pt(HS)_2^0$, hydroxy complexes or mixed hydroxy–bisulphide complexes may be important in this region. The solubility of Pt as $PtCl_4^{2-}$ at $1 \cdot 0 \, m \, \Sigma Cl^-$ at point A is only $10^{-12 \cdot 8} \, m$ (3×10^{-18} ppb). Clearly chloride complexing cannot be an effective means of Pt transportation under these conditions, whereas some form of thiocomplexing could be effective. Palladium solubility as bisulphide at 300°C is approximately the same as that for Pt, as was the case at 25°C. As temperature increases above 300–350°C the importance of bisulphide complexing is expected to diminish because of the expansion of the H_2S field at the expense of the HS^- field (Crerar & Barnes, 1976).

Some bisulphide complexes of the other PGE are known or suspected to exist as well (Pittwell, 1965), e.g. $Os(HS)_6^{2-}$, $Ru(HS)_6^{3-}$, $Ir(HS)_6^{3-}$. However, there are evidently no thermodynamic data on these species. It is expected that these species would tend to hydrate and hydrolyse under normal geological conditions, giving rise to greater complexity in their chemistry.

COMPARISON OF GOLD AND PLATINUM SOLUBILITIES

It is of interest to compare the relative solubilities of Au and Pt under similar conditions. The solubilities of Au and Pt as chloride and bisulphide complexes are shown in Fig. 15 at a temperature of 300°C. The solubilities of Pt and Au as chloride complexes seem to be of the same order of magnitude. The 1 ppm contour of $PtCl_4^{2-}$ is only slightly higher than that for $AuCl_2^-$. However, the solubility of Pt drops off more rapidly as either pH increases or fO_2 decreases. Also, the solubility of Pt is decreased at very low pH by the formation of the sulphide, which does not occur for gold. On the other hand, the solubility of Au as a bisulphide complex appears to be many orders of magnitude greater than that of Pt. This is probably due to the more rapid decrease in Pt^{2+} activity as conditions become more basic and reducing. As mentioned previously, the solubility of Pt as bisulphide complexes may well be considerably higher than that calculated here. Nevertheless, it is unlikely that Pt would approach Au in solubility as a bisulphide.

The data thus suggest that separation of Au and Pt during deposition should be less efficient when the metals are transported as chloride rather than as bisulphide complexes. However, if gradients in pH, temperature, fO_2 or ΣS are quite steep, Au and Pt could possibly be deposited together from bisulphide solutions.

THE EFFECTS OF Se, Te AND As ON PGE SOLUBILITY

In addition to the complexes dealt with above, there remains the possibility that ligands containing selenium, tellurium, arsenic, antimony, bismuth or some combination of these with or without sulphur, may play a role in PGE transport. Many of the PGE minerals involve these elements (Cabri, 1981) and several complexes are known between Pt or Pd and ligands with either Se or Te as donor atoms (Livingstone, 1965). However, there are no stability constant data available for geologically pertinent complexes involving any of these elements and therefore it is not possible to evaluate them at this time. Furthermore, it seems unlikely that any of these elements could form complexes with the PGE which are sufficiently stronger than the bisulphide complexes to be able to overcome their naturally low abundances, relative to sulphur, in hydrothermal fluids. On the contrary, it seems that these metals have a very strong capacity to immobilize the PGE, because calculations suggest that the PGE arsenide, selenide and telluride minerals are much more stable than the corresponding sulphides.

VAPOUR TRANSPORT OF THE PGE

Another way of attacking the problem of PGE transport in aqueous fluids is to calculate the volatility of these elements under geological conditions (Wood, 1987). Such calculations provide lower bounds to the concentration of PGE that might be expected in an aqueous phase inasmuch as possible interactions between water molecules and PGE vapour species, such as solvation and ionization, are neglected. The ability of several gaseous PGE oxide, chloride and fluoride species to transport PGE were estimated using available thermodynamic data, at temperatures of 800 and 1200 K. The following conclusions can be drawn:

(1) The volatility of PGE, whether in the form of gaseous oxides, chlorides or fluorides, is strongly fO_2-dependent. The more oxidizing the conditions, the more volatile the PGE.

(2) Relatively high temperatures (equal to or greater than 1200 K) are necessary for significant vapour transport, i.e. greater than 10 parts per trillion (ppt) PGE in aqueous vapour.

(3) Only Os may be transported as an oxide in the vapour (OsO_4) at levels greater than 10 ppt at geologically reasonable fO_2 and then only at the fO_2 of the hematite–magnetite buffer or greater.

(4) Ir is not appreciably volatile as IrF_6 (g) in the presence of water under geological conditions. This is the only volatile fluoride for which thermodynamic data are available; however, it would appear that volatile fluorides are not effective in PGE transport due to the high stability of the HF molecule.

(5) Ruthenium and palladium metal are somewhat volatile as $RuCl_{3(g)}$ and $PdCl_2$ (g) respectively (Figs 16 and 17). Concentrations of these in the vapour may attain 0·1 ppt Ru and 10 ppt Pd at 1200 K, $\log f_{HCl} = 100$ bars and the fO_2 of quartz–fayalite–magnetite. Volatilities as chlorides depend directly on f_{HCl} and indirectly on f_{H2O} as indicated by the equations

$$Pd(s) + 2HCl(g) + \tfrac{1}{2}O_2(g) = PdCl_2(g) + H_2O(g)$$
$$Ru(s) + 3HCl(g) + \tfrac{3}{2}O_2(g) = RuCl_3(g) + \tfrac{3}{2}H_2O$$

Higher temperatures, higher f_{HCl} or lower f_{H2O} all favour greater volatility as chlorides.

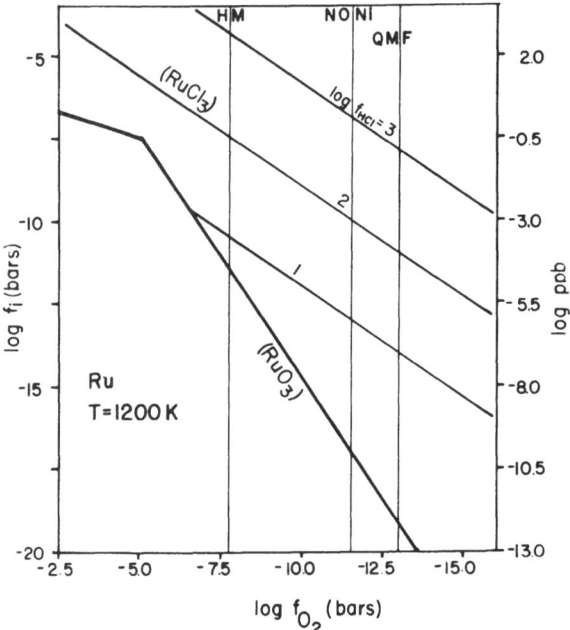

FIG. 16. Plot of log fugacity of Ru vapour species versus log fO_2 at several values of log f_{HCl}. Temperature = 1200 K, log f_{H2O} = 1000 bars. Heavy line labelled (RuO$_3$) represents the volatility of Ru(s) and RuO$_2$ (s) as the RuO$_3$ gaseous species. Medium lines represent volatility of Ru metal as RuCl$_3$ (g). Light lines show positions of the oxygen buffers, hematite–magnetite (HM), nickel–nickel oxide (NONi) and quartz–fayalite–magnetite (QFM). The data were taken from Huebner (1971) and Barin *et al.* (1975, 1977). See Wood (1987) for calculational details.

(6) High sulphur fugacities result in the formation of PGE sulphides from the native metal and the volatilities of the PGE are thereby reduced in the presence of significant S_2 gas.

(7) PGE-bearing vapours would deposit their metals upon a sufficient temperature, f_{O2} or f_{HCl} drop or upon an increase of f_{S2} or f_{H2O}.

CONCLUSIONS

The chemical nature and thermodynamics of Pt and Pd minerals and aqueous species suggest that Pt and Pd may be mobile as chloride, hydroxide or bisulphide complexes, depending on pH, f_{O2}, temperature and ligand concentration. Chloride complexing appears to be important only under highly oxidizing, acidic and saline conditions between 25 and 300°C but may become more important over a wider range of conditions at higher temperature (>400°C). The dominant chloride species between 25 and 300°C are PtCl$_4{}^{2-}$ and PdCl$_4{}^{2-}$ but the neutral species, PtCl$_2{}^0$ and PdCl$_2{}^0$, are expected to predominate at higher temperatures.

Hydroxide complexes appear to predominate over chloride complexes at near neutral pH even at high chlorinities (5 *m* Cl$^-$) and may be responsible for significant

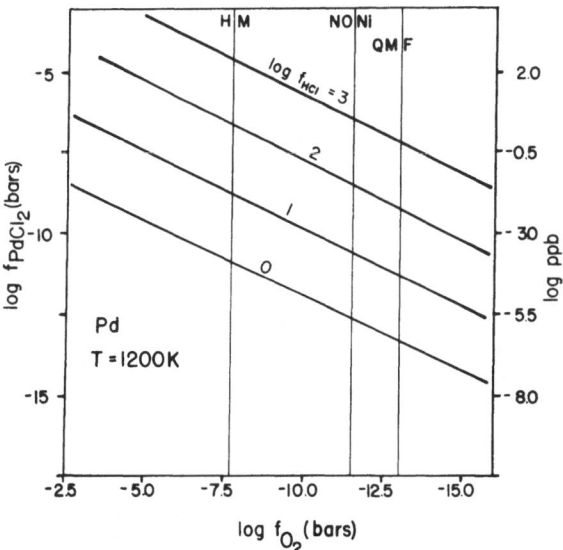

FIG. 17. Plot of log f_{PdCl_2} in equilibrium with Pd metal versus log fO_2 at several values of log f_{HCl}. Temperature = 1200 K, log f_{H_2O} = 1000 bars. Light lines show positions of oxygen buffers as in Fig. 16. Data from Barin *et al.* (1975, 1977), Huebner (1971) and Kubaschewski & Alcock (1979). See Wood (1987) for constructional details.

Pt and Pd solubilities at near neutral to basic pH. Bisulphide complexing may be important under similar pH conditions, high total sulphur and intermediate temperatures (250–400°C). Mixed bisulphide–hydroxide complexes could produce even higher solubilities than either bisulphide or hydroxide complexes alone.

Under the appropriate, low temperature conditions, the PGE may also be significantly mobile as thiosulphate, polysulphide or ammine complexes. On the other hand, the effect of As, Se and Te seems to be primarily to reduce the solubility of the PGE.

Finally, under magmatic conditions Os may be mobile as an oxide vapour species and Ru and Pd as a chloride vapour species. These volatilities are highly dependent on temperature, $f_{O_2}, f_{S_2}, f_{HCl}$, etc.

It is apparent from this study that more geological and geochemical research is needed to better define characteristics of supposed hydrothermal PGE occurrences. In addition, more experimental studies of the solubilities of PGE minerals and the thermodynamics of dissolved PGE species are clearly required.

ACKNOWLEDGEMENTS

This study was supported by a Natural Sciences and Engineering Research Council (NSERC) of Canada grant A0841 to S. Wood. B. Mountain gratefully acknowledges receipt of an NSERC Postgraduate Scholarship. We thank P. Williams and J. Bowles for constructive reviews of the manuscript.

REFERENCES

Baes, C. F. Jr & Mesmer. R. E. (1976). *The Hydrolysis of Cations.* John Wiley, New York, 489 pp.

Ballhaus, C. G. & Stumpfl, E. F. (1985a). Graphite, platinum and the C–O–H–S system. In *Fourth International Platinum Symposium*, Toronto, Abstracts; *Can. Min.*, **23**, 293–4.

Ballhaus, C. G. & Stumpfl, E. F. (1985b). Fluid inclusions in Merensky and Bastard Reefs, Western Bushveld Complex. In *Fourth International Platinum Symposium*, Toronto, Abstracts; *Can. Min.*, **23**, 294.

Barin, I., Knacke, O. & Kubaschewski, O. (1975). *Thermochemical Properties of Inorganic Substances.* Springer, New York, 921 pp.

Barin, I., Knacke, O. & Kubaschewski, O. (1977). *Thermochemical Properties of Inorganic Substances—Supplement.* Springer, New York, 861 pp.

Barner, H. E. & Scheuerman, R. V. (1978). *Handbook of Thermochemical Data for Compounds and Aqueous Species.* John Wiley, New York, 156 pp.

Barnes, H. L. (1979). Solubilities of ore minerals. In *Geochemistry of Hydrothermal Ore Deposits*, ed. H. L. Barnes. Wiley, New York, pp. 404–60.

Belevantsev, V. I., Kolonin, G. R. & Peshchevitskiy, B. I. (1982). Use of stepwise ligand replacement in estimating the stability constants of mixed complexes in geochemically important systems. *Geochem. Int.*, **19**, 169–79.

Boudreau, A. E. & McCallum, I. S. (1985). Evidence for mineral reactions and metasomatism by silica-undersaturated Cl-rich fluids in the main Pt-Pd zone, Stillwater Complex, Montana. In *Fourth International Platinum Symposium*, Toronto; Abstracts, *Can. Min.*, **23**, 293–4.

Boudreau, A. E. & McCallum, I. S. (1986). Investigations of the Stillwater Complex: III. The Picket Pin Pt/Pd deposit. *Econ. Geol.*, **81**, 1953–75.

Boudreau, A. E., Mathez, E. A. & McCallum, I. S. (1986a). The role of Cl-rich fluids in the Stillwater Complex. *GAC-MAC Prog.*, *with Abstr.*, **11**, 47.

Boudreau, A. E., Mathez, E. A. & McCallum, I. S. (1986b). Halogen geochemistry of the Stillwater and Bushveld Complexes: Evidence for transport of the platinum-group elements by Cl-rich fluids. *J. Petrol.*, **27**, 967–86.

Bowles, J. F. W. (1986). The development of platinum-group minerals in laterites. *Econ. Geol.*, **81**, 1278–85.

Brown, J. B. (1971). Jarosite-goethite stabilities at 25°C, 1 atm. *Mineral. Dep.*, **6**, 245–52.

Cabri, L. J. (1981). The platinum-group minerals. In *Platinum-Group Elements: Mineralogy, Geology, Recovery*, ed. L. J. Cabri. CIM Special Volume 23, pp. 83–150.

Chang, J. C. & Garner, C. S. (1965). Kinetics of aquation of aquopentachloroiridate (III) and chloride anation of diaquotetrachloroiridate (III) anions. *Inorg. Chem.*, **4**, 209–15.

Cousins, C. A. (1973). Notes on the geochemistry of the platinum group elements. *Geol. Soc. South Africa Trans.*, **76**, 77–81.

Cozzi, D. & Pantani, F. (1958). The polarographic behavior of rhodium (III) chlorocomplexes. *J. Inorg. Nucl. Chem.*, **8**, 385–98.

Crerar, D. A. & Barnes, H. L. (1976). Ore solution chemistry V. Solubilities of chalcopyrite and chalcocite assemblages in hydrothermal solution at 200° to 350°C. *Econ. Geol.*, **71**, 772–94.

Crerar, D., Wood, S., Brantley, S. & Bocarsly, A. (1985). Chemical controls on solubility of ore-forming minerals in hydrothermal solutions. *Can. Min.*, **23**, 333–52.

Dillon-Leitch, H. C. H., Watkinson, D. H. & Coats, C. J. A. (1986). Distribution of platinum-group elements in the Donaldson West Deposit, Cape Smith Belt, Quebec. *Econ. Geol.*, **81**, 1147–58.

Elding, L. I. (1972). Palladium(II) halide complexes, I. Stabilities and spectra of palladium (II) chloro and bromo aqua complexes. *Inorg. Chim. Acta*, **6**, 647–51.

Elding, L. I. (1978). Stabilities of platinum(II) chloro and bromo complexes and kinetics for anation of the tetraaquaplatinum(II) ion by halides and thiocyanate. *Inorg. Chim. Acta*, **28**, 255–62.

Fuchs, W. A. & Rose, A. W. (1974). The geochemical behavior of platinum and palladium in the weathering cycle in the Stillwater Complex, Montana. *Econ. Geol.*, **69**, 332–46.

Giggenbach, W. (1974). Equilibria involving polysulfide ions in aqueous solutions up to 240°C. *Inorg. Chem.*, **13**, 1724–30.

Goldhaber, M. B. (1983). Experimental study of metastable sulphur oxyanion formation during pyrite oxidation at pH 6–9 and 30°C. *Am. J. Sci.*, **283**, 193–217.

Hancock, R. D. & Evers, A. (1976). Formation constant of $Pd(CN)_4^{2-}$. *Inorg. Chem.*, **15**, 995–6.

Hancock, R. D., Finkelstein, N. P. & Evers, A. (1977). A linear free-energy relation involving the formation constants of palladium(II) and platinum(II). *J. Inorg. Nucl. Chem.*, **39**, 1031–4.

Hartley, F. R. (1973). *The Chemistry of Platinum and Palladium*. John Wiley, New York, 544 pp.

Helgeson, H. C. (1969). Thermodynamics of hydrothermal systems at elevated temperatures and pressures. *Am. J. Sci.*, **267**, 729–804.

Högfeldt, E. (1982). Stability constants of metal-ion complexes, Part A: Inorganic Ligands, IUPAC Chemical Data Series, No. 21: Pergamon Press, New York, 310 pp.

Huebner, J. S. (1971). Buffering techniques for hydrostatic systems at elevated pressures. In *Research Techniques for High Pressure and High Temperature*, ed. J. C. Ulmer. Springer, New York, pp. 123–78.

Kubaschewski, O. & Alcock, C. B. (1979). *Metallurgical Thermochemistry*. Pergamon Press, New York, 449 pp.

Livingstone, S. E. (1965). Metal complexes of ligands containing sulphur, selenium or tellurium as donor atoms. *Quart. Rev. (London)*, **19**, 386–425.

Livingstone, S. E. (1973). The second- and third-row elements of group VIIIA, B and C. In *Comprehensive Inorganic Chemistry*, *Vol. 3*, ed. J. C. Bailar Jr, H. J. Emeleus, R. Nyholm & A. F. Trotman-Dickenson. Pergamon Press, New York, pp. 1163–370.

Llopis, J. F. & Colom, F. (1976a). Platinum. In *Encyclopedia of Electrochemistry of the Elements*, *Vol. VI*, ed. A. J. Bard. Marcel Dekker, New York, pp. 169–219.

Llopis, J. F. & Colom, F. (1976b). Iridium. In *Encyclopedia of Electrochemistry of the Elements*, *Vol. VI*, ed. A. J. Bard. Marcel Dekker, New York, pp. 221–34.

Llopis, J. F. & Colom, F. (1976c). Osmium. In *Encyclopedia of Electrochemistry of the Elements*, *Vol. VI*, ed. A. J. Bard. Marcel Dekker, New York, pp. 235–51.

Llopis, J. F. & Colom, F. (1976d). Palladium. In *Encyclopedia of Electrochemistry of the Elements*, *Vol. VI*, ed. A. J. Bard. Marcel Dekker, New York, pp. 253–75.

Llopis, J. F. & Tordesillas, I. M. (1976). Ruthenium. In *Encyclopedia of Electrochemistry of the Elements*, *Vol. VI*, ed. A. J. Bard. Marcel Dekker, New York, pp. 277–98.

McCallum, M. E. Loucks, R. R. Carlson, R. R., Cooley, E. F. & Doerge, T. A. (1976). Platinum metals associated with hydrothermal copper ores of the New Rambler Mine, Medicine Bow Mountains, Wyoming. *Econ. Geol.*, **71**, 1429–50.

Mestre, J. M., Montagut, X., Ruiz, M. & Victori, L. (1982). Calculo de las constantes de estabilidad de los clorocomplejos de Platino(II) en medio acuoso. *Afinidad*, **39**, 484–9.

Miano, R. R. & Garner, C. S. (1965). Kinetics of aquation of hexachloroosmate(IV) and chloride anation of aquopentachloroosmate(IV) anions. *Inorg. Chem.*, **4**, 337–42.

Mihalik, P., Jacobsen, J. B. E. & Hiemstra, S. A. (1974). Platinum-group minerals from a hydrothermal environment. *Econ. Geol.*, **69**, 257–62.

Mills, K. C. (1974). *Thermodynamic Data for Inorganic Sulphides, Selenides and Tellurides*. Butterworths, London, 845 pp.

Mountain, B. W. & Wood, S. A. (1988). Chemical controls on the solubility, transport and deposition of platinum and palladium in hydrothermal solutions: A thermodynamic approach. *Econ. Geol.*, **83**.

Murowchick, J. B. & Barnes, H. L. (1986). Marcasite precipitation from hydrothermal solutions. *Geochim. Cosmochim. Acta*, **50**, 2615–29.

Murray, R. C., Jr & Cubicotti, D. (1983). Thermodynamics of aqueous sulfur species to 300°C and potential-pH diagrams. *J. Electrochem. Soc.*, **130**, 866–9.

Pearson, R. G. (1963). Hard and soft acids and bases. *J. Am. Chem. Soc.*, **85**, 3533–9.

Peshchevitskiy, B. I., Ptitsyn, B. V. & Leskova, N. M. (1962). Hydrolysis of the tetra-chloroplatinate ion. *Izvetsia Akademia Nauka, SSSR, Seria Khimia*, **11**, 143.

Pittwell, L. R. (1965). Thiometallates of the group-eight metals. *Nature*, **207**, 1181–2.

Poulsen, I. A. & Garner, C. S. (1962). A thermodynamic and kinetic study of hexachloro and aquopentachloro complexes of iridium(III) in aqueous solutions. *J. Am. Chem. Soc.*, **84**, 2032–7.

Razin, L. V. (1968). Problem of the origin of platinum metallization of forsterite dunite. *Int. Geol. Rev.*, **13**, 776–88.

Robie, R. A., Hemingway, B. S. & Fisher, J. R. (1978). Thermodynamic properties of minerals and related substances at 298.15 K and 1 bar (10^5 Pascals) pressure and at higher temperatures. *U.S. Geol. Surv. Bull.*, **1452**, 456 pp.

Rowell, W. F. & Edgar, A. D. (1986). Platinum-group element mineralization in a hydrothermal Cu–Ni sulfide occurrence, Rathbun Lake, Northeastern Ontario. *Econ. Geol.*, **81**, 1272–7.

Seward, T. M. (1973). Thiocomplexes of gold and the transport of gold in hydrothermal ore solutions. *Geochim. Cosmochim. Acta*, **37**, 379–99.

Seward, T. M. (1984). The transport and deposition of gold in hydrothermal systems. In *Gold '82: The Geology, Geochemistry, and Genesis of Gold Deposits*, ed. R. P. Foster. Geological Society of Zimbabwe Special Publication, Vol. 1, pp. 165–81.

Sillen, L. G. & Martell, A. E. (1971). Stability of metal-ion complexes. Chemical Society of London, Special Publication No. 25, 865 pp.

Smith, R. M. & Martell, A. E. (1976). *Critical Stability Constants: Vol. 4: Inorganic Complexes*. Plenum Press, New York, 257 pp.

Sohnge, P. G. (1945). The geology of the Messina copper mines and surrounding country, Pretoria. Geological Survey of South Africa, Mem. 40, 282 pp.

Stumpfl, E. F. (1974). The genesis of platinum deposits: further thoughts. *Miner. Sci. Engng*, **6**, 120–41.

Volborth, A. & Housley, R. M. (1984). A preliminary description of complex graphite, sulfide, arsenide, and platinum-group element mineralization in a pegmatoid pyroxenite of the Stillwater Complex, Montana, U.S.A. *Tschermaks Mineral. Petrog. Mitt.*, **33**, 213–30.

Wagner, P. A. (1929). *Platinum Deposits and Mines of South Africa*. C. Struik (Pty) Ltd, Capetown, South Africa, 338 pp.

Watkinson, D. H., Dahl, R. & McGoran, J. (1986). The Coldwell Complex platinum-group-element deposit: 2. Relationships of platinum-group-elements to pegmatitic biotite-bearing gabbro and the role of a fluid phase. *GAC-MAC Prog. with Abstr.*, **11**, 142–3.

Webster, J. G. (1986). The solubility of gold and silver in the system $Au–Ag–S–O_2–H_2O$ at 25°C and 1 atm. *Geochim. Cosmochim. Acta.*, **50**, 1837–46.

Westland, A. D. (1981). Inorganic chemistry of the platinum-group elements. In *Platinum Group Elements: Mineralogy, Geology, Recovery*, ed. L. Cabri. Canadian Institute of Mining and Metallurgy Special Volume 23, pp. 7–18.

Wolsey, W. C., Reynolds, C. A. & Kleinberg, J. (1965). Complexes in the rhodium (III)-chloride system in acid solution. *Inorg. Chem.*, **2**, 463–8.

Wood, S. A. (1987). Thermodynamic calculations of the volatility of the platinum-group elements (PGE) at magmatic temperatures. *Geochim. Cosmochim. Acta*, **51**, 3041–50.

9

New Mechanisms for the Mobilization of the Platinum-Group Elements in the Supergene Zone

Ian R. Plimer

Department of Geology, Newcastle University, Newcastle, New South Wales 2308, Australia

&

Peter A. Williams[*]

Department of Chemistry, University College, PO Box 78, Cardiff CF1 1XL, UK

ABSTRACT

The supergene chemistry of the PGE is dominated by the reaction chemistry of the elements themselves. Formation of complex co-ordination compounds in solution is necessary for mobilization and differentiation in oxide zones. Previous suggestions concerning this complex chemistry involving chloride ion are applicable only in very unusual circumstances. It is clear that the natural ligand set available for complex formation is severely restricted. The oxidation of the PGE in the presence of lower oxidation state oxyanions of sulphur represents a feasible mechanism for the dispersion of the elements in the supergene environment. Calculations indicate that this is easily accomplished for Pt and Pd. Applications to secondary enrichment, nugget formation and potential geochemical exploration methods are indicated.

INTRODUCTION

The chemistry of the PGE is voluminous as it occupies a central part in our understanding of inorganic and co-ordination chemistry. Extensive compilations of data and reactions are available (Wilkinson *et al.*, 1987) as are reviews concerning the geochemistry of the elements (Cabri, 1981; Mountain & Wood, this volume). Much *less* information is available concerning the supergene geochemistry and mineralogy of the PGE. This is in part due to their limited, if perhaps un-

---

[*] Correspondence may be addressed to either author.

recognized, economic importance in surficial environments, and in part to the general notion that the elements are so noble as to have a negligible aqueous geochemistry. However, this latter, commonly held belief is incorrect. Furthermore, if geochemical techniques are to be applied in the search for new PGE deposits, supergene chemistry needs to be properly elucidated. It is also clear that such chemistry may be important in the development of placer-type deposits and, potentially, other economically viable enrichments in the oxide zone.

MOBILITY AND ENRICHMENT OF PGE IN SURFACE ENVIRONMENTS

Although the literature is not extensive, enough has been reported to conclusively prove that the PGE are mobile under oxidizing conditions. Emmons (1917) pointed to the possibility some time ago, but the trace element analyses and mineralogical studies necessary to establish the fact are much more recent. A few particular examples are worth mentioning.

McGoldrick & Keays (1981) have reported the distribution of certain PGE in the No. 1 shoot of the Perseverance N1 deposit, W. Australia. Both Pt and Pd are enriched near the base of the oxide zone, but Ir and Au show little variation with depth. Differentiation as well as mobility in oxidizing groundwaters is clearly demonstrated. In the No. 2 Showing deposit of the Stillwater Complex, Pt is relatively immobile in upper soil horizons, whereas the reverse is true for Pd (Fuchs & Rose, 1974). Pd is enriched in B soil horizons and accumulates in *Pinus flexilis*.

Several reports indicate retention and concentration by weathering of all PGE in lateritic-type profiles. On the other hand, Keays & Davison (1976) have noted both Pd and Ir enrichment in lower sections of the oxide zone (and in the pyrite–violarite zone) at Kambalda, W. Australia. As far as conventional lateritic profiles are concerned, Travis *et al.* (1976) have provided data for Gilgarna Rocks, W. Australia. Pd is depleted at the surface and strongly enriched in the base of the laterite. Au and Ir also show distributions in accord with secondary mobilization. Bowles (1986, this volume, and references therein) has drawn attention to the possible mobilization and differentiation of the PGE in alluvial and eluvial deposits in Sierra Leone. Conspicuous depletion of Pd (and, possibly, enlarged nugget formation) is common to all such environments (Cousins & Kinloch, 1976) and secondary alterations in other analogous deposits has been elegantly reviewed by Stumpfl & Tarkian (1976).

Finally, mention should be made of the New Rambler Mine, Wyoming, USA (McCallum *et al.*, 1976). Here, in the oxide zone, no Pd minerals are found, whereas Rh is *strongly* enriched in the uppermost oxide zone ore (*vide infra*) and Pt is enriched to a lesser extent at a greater depth. Several other examples could have been cited, but these few serve to establish the fact that the PGE do migrate in the supergene zone, especially when associated with weathering sulphide ores; to quote Stumpfl & Tarkian (1976), 'the platinum-group elements . . . are neither inert nor insoluble'.

POTENTIAL REACTION CHEMISTRY IN THE
SUPERGENE ENVIRONMENT

The platinoids occur mainly as native elements, sulphides or sulphosalts (Cabri, 1981) and the formation of the former are of seminal importance during the oxidation of the latter (Westland, 1981). This is due to the fact that *reduction* of the PGE to *elements* is favoured overall during oxidation of sulphide to sulphate. For eqn (1), standard potentials

$$MS_2(s) + 8H_2O(l) \rightarrow M(s) + 2SO_4^{2-}(aq) + 16H^+(aq) + 12e \qquad (1)$$

at 298·2 K are +0·540 and 0·468 V for M = Ru and Os, respectively. A value of +0·480 V is known for eqn (2) under the same conditions.

$$Ir_2S_3(s) + 12H_2O(l) \rightarrow 2Ir(s) + 3SO_4^{2-}(aq) + 24H^+(aq) + 18e \qquad (2)$$

Similar thermodynamic data permit an evaluation of the PtS_2/PtS and PdS_2/PdS boundaries. The MS_2 species are desulphurized below potentials necessary for the formation of the simple sulphides analogous to cooperite. Then, for eqn (3), E^0 (298·2 K) values

$$MS(s) + 4H_2O(l) \rightarrow M(s) + SO_4^{2-}(aq) + 8H^+(aq) + 6e \qquad (3)$$

are +0·468 and +0·484 V for M = Pd and Pt, respectively. Other relevant thermo-dynamic data are tabulated by Westland (1981). It is significant that these potentials are *all* considerably lower than those for the oxidation of the PGE to produce chloride complexes in oxidation states II and IV (Milazzo *et al.*, 1978; Westland, 1981). Thus we might expect that a reasonably common natural situation during oxidation of such sulphide species and congeners would be the production of the elements themselves in a finely divided form and therefore more susceptible to chemical attack. In any case, as a first-order approach to subsequent considerations, the above fact leads to a significant simplification; the supergene chemistry of the PGE may be viewed from the standpoint of the chemistry of the elements themselves. Certain phases are, incidentally, most resistant to oxidation, and include sperrylite, $PtAs_2$, and the Pt–Fe alloys. With respect to the latter, this is almost certainly a kinetic phenomenon; such may also be the case with sperrylite, although this species has a marked thermodynamic stability in aqueous solution (Mountain & Wood, this volume).

Very little work has been reported on the supergene *mineralogy* of the PGE by way of evaluating the thermodynamic arguments in a practical sense. However it is significant to recall the observation of Wagner (1929) that cooperite is altered at and near the surface to metallic platinum. Extension of these kinds of considerations to sulphosalts is limited by the general paucity of experimental results concerning their oxidation. Some very recent results are worth mentioning. Buckley (1987) has reported that cobaltite, CoAsS, initially oxidizes via loss of As (as As_2O_3 then arsenate) to produce some CoO, but mainly CoS_2. If this pattern of reaction should prove to be general in the oxide zone, the course of oxidation of species such as irarsite, IrAsS, osarsite, OsAsS, ruarsite, RuAsS, as well as hollingworthite, platarsite, daomanite and the like will ultimately proceed along the same lines as those summarized above for simple sulphides.

In the natural environment, the potential ligand set is severely restricted for the PGE. Exotic ligands are ruled out and ligands other than H_2O itself *must* be sought because of the elements' noble natures (redox potential measurements establish this contention: Milazzo *et al.*, 1978). Furthermore, reference to this same compilation and to the general chemistry of the PGE indicates that the usual 'hard' ligands which abound in the natural environment cannot bring the PGE into solution. Carbonate, phosphate, fluoride and silicate species for example, do not lower the redox potentials of the PGE to values which can be realistically envisaged. Compounds of these species are, incidentally, most often stabilized only in the presence of other exotic, unnatural ligands such as phosphines.

Convoluted arguments might be advanced for 'soft' ligands such as CN^-, SCN^-, Br^-, I^- and organic donors, as has been the case for supergene Au chemistry (Kinkel & Lesure, 1968; Lakin *et al.*, 1974), but these suffer when placed in the perspectives of natural and likely availability. Great attention has been paid to chloride ion as a complexing agent. However, it is easily demonstrated that chloride is unlikely to be important, save in exceptional circumstances, and *certainly* it should be noted in passing, chloride ion complexation must be insignificant at depth because of the low redox potentials which obtain in such environments. Some illustrative calculations and data are given in the next section.

CHLORIDE COMPLEXES

For the sake of brevity, and because of the spread of data readily available for PGE, we focus here and subsequently on the chemistry of Pt and Pd except where special points are warranted. With respect to chloride complexes, the appropriate potentials may be evaluated for the oxidation of Pt and Pd at various a_{Cl^-} values to form MCl_4^{2-}, MCl_3^-, MCl_2^0 and $MCl^+(aq)$ species. As expected, the potentials are lowest in concentrated chloride. With respect to eqn (4), $E = E^0 + 0.0296$ $\log a_{MCl4^{2-}}$ with $a_{Cl^-} = 1$ and

$$MCl_4^{2-}(aq) + 2e \rightarrow M(s) + 4Cl^-(aq) \tag{4}$$

$E^0 = +0.495$ and $+0.755\,V$ for Pd and Pt, respectively. While it is true that other complexes are formed at lower a_{Cl^-}, it is easily calculated that the existence of complex chloride species has little effect on Pt or Pd concentrations in even oxidized groundwaters except under very unusual conditions of low pH and very high redox potential (Bowles, 1986; Mountain & Wood, this volume).

We may evaluate the importance of chloride complex formation using the data of Milazzo *et al.* (1978). First, as is clearly the case for the MCl_4^{2-} species above, it is true that at no $[Cl^-]$ or $[MCl_x^{2-x}]$ values will platinum be more easily oxidized than palladium. Secondly, as $[Cl^-]$ values fall, species other than MCl_4^{2-} predominate in aqueous solution, as expected from a consideration of equilibrium constants. Thus at $[Cl^-] = 10^{-3}$ M (corresponding to a weakly saline groundwater), the major Pd species in solution are $PdCl_2^0$ and $PdCl_3^-$, and $[PtCl_4^{2-}] \approx [PtCl_3^-] \approx [PtCl_2^0]$, although the potentials necessary to produce them are inevitably higher.

Irrespective of what individual chemical species provides the intimate mechanism of oxidation, it is next instructive to pick a representative value of [Pt, Pd] concentrations deemed appropriate for their mobility in the natural environment, and for this purpose we choose a value of 10^{-9} M (*ca.* 1 ppb) for an individual chloride species. In very saline groundwaters, such as *do* obtain in current lateritic environments (Mann, 1984), chloride ion concentrations reach about 10^{-1} M. From the above E^0 data, it can be calculated that Pd will oxidize to $PdCl_4^{2-}$ (predominantly) at $E = +0.347$ V. Pt will oxidize similarly to $PtCl_4^{2-}$ at $E = +0.607$ V. Naturally it is appropriate to ask whether these potentials are achieved in nature. Reference should be made to the work of Baas Becking *et al.* (1960), supplemented by more recent results tabulated by Thornber (1975). With few exceptions, measured pH/Eh values in the natural environment lie considerably below the potential corresponding to the oxidative decomposition of water and are probably related to the pH-dependent O_2/H_2O_2 couple important in the overall four-electron reduction of oxygen (Sato, 1960; Behar *et al.*, 1970). Under these conditions, Pd may be separated from Pt when the latter is restricted in reactivity to pH values less than 4. Such salinities are rarely encountered in nature; values of $[Cl^-]$ equal to 10^{-3} M are more representative in general. Here, $E(298.2$ K$)$ values are equal to $+0.545$ V for Pd (with $PdCl_2^0$(aq) dominant) and $+0.820$ V for Pt (with $PtCl_2^0$(aq) and $PtCl_3^-$(aq) the dominant oxidized chloro species). Under these circumstances the pH fields are restricted to less than 4 and 0 (!) for Pd and Pt, respectively.

Thus although these conditions might be encountered in nature rarely (apparently limited to lateritic terrains) and could therefore be responsible for some differentiation of Pd from Pt, they cannot be invoked as being general in describing the supergene complex chemistry of the two elements. We also note the existence of the mineral $Pd_4Bi_5Cl_3$ (Karpenkov *et al.*, 1981), but stress that the appropriate conditions for its information are of limited occurrence.

In passing, it also ought to be mentioned that all available thermodynamic data rule out the possibility of chloride complex formation and transport of the PGE in deep-seated igneous and hydrothermal environments, up to a few hundred degrees centigrade (Mountain & Wood, this volume). This is simply a consequence of the necessary high potentials not being able to be achieved in such environments. It is true that reactivity patterns may be altered at higher temperatures, but their possible importance is purely speculative. The association of chloride-rich fluid inclusions with PGE in the Merensky Reef (Ballhaus & Stumpfl, 1986) may be merely fortuitous.

Given that the natural ligand set is severely restricted (Williams, 1987) we have been inevitably drawn towards an examination of 'soft' donors comprising the intermediates formed via the oxidation of sulphide (or via the biologically mediated reductions of sulphate), in order to account for mobility in the supergene zone.

COMPLEX FORMATION WITH LOWER VALENCY SULPHUR OXYANIONS

Comparisons between the supergene chemistry of Au and the PGE are inevitable. Supergene mobilization of the former has long been recognized (Emmons, 1917) and its chemistry the subject of many studies. Chloride complexes are even less likely for Au, because of the even higher potentials required. Recently, it has become clear that thiosulphate and perhaps sulphite complexes are important in neutral to slightly basic solution and their formation and participation in the mobilization of Au in the oxide zone is now widely accepted (Listova *et al.*, 1966; Boyle, 1979; Mann, 1985; Webster, 1986).

As it turns out, the chemistry of the PGE in this respect closely mirrors that of Au. Thiosulphate, sulphite and perhaps polythionate/thionite complexes are all known (Wilkinson *et al.*, 1987) and a few which are of special relevance to geochemical considerations (not containing exotic ligands) are listed in Table 1.

TABLE 1
Simple inorganic complexes of Pt and Pd containing thiosulphate and sulphite donors

$[Pt(S_2O_3)Cl_2]^{2-}$	
$[Pt(S_2O_3)_2]^{2-}$ (*cis* and *trans*)	$[Pd(S_2O_3)_2]^{2-}$
$[Pt(S_2O_3)_3]^{4-}$	
$[Pt(S_2O_3)_4]^{6-}$	$[Pd(S_2O_3)_4]^{6-}$
$[Pt(SO_3)_2Cl_2]^{4-}$	
$[Pt(SO_3)_2]^{2-}$	$[Pd(SO_3)_2(H_2O)_2]^{2-}$
$[Pt(SO_3)_4]^{6-}$	$[Pd(SO_3)_4]^{6-}$
	$[Pd(SO_3)(H_2O)_3]^{0}$

These kinds of complexes are attractive, geochemically speaking, because they do *not* require extreme conditions for their formation.

Intermediate valency sulphur oxyanions are present in groundwaters associated with oxidizing sulphides (Goldhaber, 1983, and references therein) and have been shown to be persistent in a variety of natural waters (White *et al.*, 1963; Miyake, 1965; Wilson, 1966; Cline & Richards, 1969). It thus remains to establish whether or not such complexes do indeed have any geochemical significance.

Fortunately, a timely report of certain redox potential measurements has appeared (Hancock *et al.*, 1977) and these results are summarized in Table 2. They demonstrate that the standard potentials in the presence of SO_3^{2-} are close to zero volts, and that the oxidation of Pd and Pt in the presence of $S_2O_3^{2-}$ is thermodynamically favoured under standard conditions at 298·2 K. A sample calculation is instructive.

E^0 for the Fe(III)/Fe(II) couple is $+0·77$ V at 298·2 K, Fe(III) being a ubiquitous oxidant in the natural environment, especially in oxidizing sulphides. Thus, in the usual way E^0 and log K for eqn (5) equal $+0·94$ V and 31·8, respectively. K is equal to

$$Pt(s) + 4S_2O_3^{2-}(aq) + 2Fe^{3+}(aq) \rightarrow [Pt(S_2O_3)]_4^{6-}(aq) + 2Fe^{2+}(aq) \qquad (5)$$

$[Pt(S_2O_3)_4{}^{6-}] [Fe^{2+}]^2 / [S_2O_3{}^{2-}]^4 [Fe^{3+}]^2$ (ignoring activities for the sake of convenience). In order to simplify the expression, we may take $[Fe^{3+}] = [Fe^{2+}]$ and evaluate the amount of Pt dissolved *at equilibrium* in the presence of known amounts of thiosulphate. Even for $[S_2O_3{}^{2-}] = 10^{-8}$ M, $[Pt(S_2O_3)_4{}^{6-}] = 0.63$ M! Comparable results are found for the palladium system and for sulphite. Allowing a considerable latitude as to the limits of the assumptions (activities, iron concentrations and attainment of equilibrium) it seems inescapable to conclude that reduced sulphur oxyanions must play a role in the mobilization of Pt and Pd in oxidizing PGE-rich sulphide ores. The possibility appears to have been entirely overlooked in the geochemical literature. It would however immediately explain some of the enigmatic enrichment patterns reported above and should be applicable over a pH range from about 5 to 9.

TABLE 2
Electrode potentials for some Pt (II) and Pd (II) complexes of sulphite and thiosulphate

Reaction		E^0 *(298·2 K)*/V^a
$[Pd(SO_3)_4]^{6-} + 2e$	$\rightleftharpoons$ Pd + $4SO_3{}^{2-}$	+0·058
$[Pd(S_2O_3)_4]^{6-} + 2e$	$\rightleftharpoons$ Pd + $4S_2O_3{}^{2-}$	−0·116
$[Pt(SO_3)_4]^{6-} + 2e$	$\rightleftharpoons$ Pt + $4SO_3{}^{2-}$	+0·001
$[Pt(S_2O_3)_4]^{6-} + 2e$	$\rightleftharpoons$ Pt + $4S_2O_3{}^{2-}$	−0·170

^{a}Error estimated at ±6 mV.

Of course rate or kinetic phenomena must play an important role in this chemistry and therein may be part of the reason for the frequently observed differentiation of Pt and Pd in oxide zones and placer deposits. With respect to the latter, thiosulphate and sulphite complexes may be important in a fashion akin to the geochemistry associated with roll-type uranium deposits (Granger & Warren, 1969; Goldhaber *et al.*, 1978; Goldhaber, 1983). It is certainly worth noting here as well that the metastable ligands will have extra stability conferred upon them by virtue of complexation as in the case with other related sulphur analogues (Schenk, 1987). In this way, problems concerning metastability could be easily overcome.

We also note the possibility of formation of polysulphide species as intermediates (sulphide, plus sulphur) in the transport of PGE, although this chemistry may be more pertinent to the primary environment (Gillard & Wimmer, 1978, and references therein; Schmidt & Hoffmann, 1979; Krause *et al.*, 1982, and references therein).

RHODIUM OXIDE SOLUBILITY

The above-mentioned chemistry may be applicable to all of the PGE. However a peculiarity of the behaviour of Rh(III) in aqueous solution is of special relevance to its geochemistry. This is that it forms a remarkably insoluble hydroxide, long the bane of synthetic co-ordination chemists. Using modern reliable potential measurements (Milazzo *et al.*, 1978), K_{sp} (298·2 K) for $Rh(OH)_3(s)$ is equal to $3·7 \times 10^{-39}$. Then for eqn (3), $\log K = 49·1$, at the same temperature.

$$RhCl_6^{3-} + 3H_2O \rightleftharpoons Rh(OH)_3(s) + 6Cl^- + 3H^+ \qquad (6)$$

It could be seen that even for very high $[Cl^-]$ and very low $[RhCl_6^{2-}]$ values, the latter is thermodynamically unstable in water with respect to $Rh(OH)_3(s)$. Naturally, rate affects are important here (especially so in the laboratory) with the formation, slowly, of increasingly hydrolysed species. In geochemical terms we must conclude that although Rh may be relatively easily oxidized in aqueous solution containing chloride ion, it will be quickly and profoundly immobilized by the precipitation or coprecipitation of the hydroxide. This fact may serve to explain the distribution of Rh in the New Rambler mine (*vide supra*) but more analytical data concerning Rh are most desirable.

FUTURE OUTLOOK AND PRELIMINARY STUDIES

A brief survey of the known chemistry of the PGE, with special reference to Pd and Pt, indicates that novel mechanisms involving sulphur donors can explain their supergene chemistry in several environments commonly encountered in nature. Exploration geochemical applications are immediately apparent and in this respect Rh and Pd may be the most suitable target elements. The former would appear to be very immobile in the surficial environment and the latter the most mobile of the PGE. Furthermore, the chemistry does explain several puzzling distributions of the PGE in oxidized profiles which have been reported in the past particularly those which are related to oxidizing sulphide—hosted ores or protores. The alteration of PGE minerals in placers and their differentiation, migration and accumulation of nuggets may also be other consequences of this chemistry, especially in view of the fact that associated groundwaters frequently possess the appropriate redox potential regime although biological mediation of the oxidation of sulphide might be important.

We have recently begun a detailed sampling and analytical programme to try to evaluate the relevance of this chemistry in PGE-bearing deposits. This is linked to water analyses and computer modelling of associated groundwater chemistry. In the laboratory, preliminary experiments have shown that Pt and Pd are mobilized during the oxidation of pyrite under appropriate co-ordinations, using a leaching cell similar to that described by Granger & Warren (1969). Detailed results of these laboratory and related field studies will be published shortly.

REFERENCES

Baas Becking, L. G. M., Kaplan, I. R. & Moore, D. (1960). Limits of the natural environment in terms of pH and oxidation–reduction potentials. *J. Geol.*, **68**, 243–84.

Ballhaus, C. G. & Stumpfl, E. F. (1986). Sulfide and platinum mineralization in the Merensky Reef; evidence from hydrous silicates and fluid inclusions. *Contrib. Mineral. Petrol.*, **94**, 193–204.

Behar, D., Czapski, G., Rabini, J., Dorfman, L. M. & Schwarz, H. A. (1970). Acid dissociation constant and decay kinetics of the perhydroxyl radical. *J. Phys. Chem.*, **74**, 3209–13.

Bowles, J. F. W. (1986). The development of platinum-group minerals in laterites. *Econ. Geol.*, **81**, 1278–85.

Bowles, J. F. W. Further studies of the development of platinum-group minerals in the laterites of the Freetown Layered Complex, Sierra Leone. This volume, pp. 273–302.

Boyle, R. W. (1979). The geochemistry of gold and its deposits. *Bull. Geol. Surv. Canada*, **280**, 584 pp.

Buckley, A. N. (1987). The surface oxidation of cobaltite. *Aust. J. Chem.*, **40**, 231–9.

Cabri, L. J. (ed.) (1981). *Platinum-Group Elements: Mineralogy, Geology, Recovery.* Canadian Institute of Mining and Metallurgy, Special Volume 23, Montreal, 267 pp.

Cline, J. D. & Richards, F. A. (1969). Oxygenation of hydrogen sulfide in seawater at constant salinity, temperature, and pH. *Environ. Sci. Technol.*, **3**, 838–43.

Cousins, C. A. & Kinloch, E. D. (1976). Some observations on textures and inclusions in alluvial platinoids. *Econ. Geol.*, **71**, 1377–98.

Emmons, W. H. (1917). The enrichment of ore deposits, *US Geol. Surv. Bull.*, **625**, 530 pp.

Fuchs, W. A. & Rose, A. W. (1974). The geochemical behaviour of platinum and palladium in the weathering cycle in the Stillwater Complex, Montana. *Econ. Geol.*, **69**, 332–46.

Gillard, R. D. & Wimmer, J. L. (1978). Inorganic optical activity. *J. Chem. Soc., Chem. Comm.*, 936–7.

Goldhaber, M. B. (1983). Experimental study of metastable sulfur oxyanion formation during pyrite oxidation at pH 6–9 and 30°C. *Am. J. Sci.*, **283**, 193–217.

Goldhaber, M. B., Reynolds, R. L. & Rye, R. O. (1978). Origin of a south Texas roll-type uranium deposit: II. Sulfide petrology and sulfur isotope studies. *Econ. Geol.*, **73**, 1690–705.

Granger, H. C. & Warren, C. G. (1969). Unstable sulfur compounds and the origin of roll-type uranium deposits. *Econ. Geol.*, **64**, 160–71.

Hancock, R. D., Finkelstein, N. P. & Evers, A. (1977). A linear free-energy relation involving the formation constants of palladium(II) and platinum(II). *J. Inorg. Nucl. Chem.*, **39**, 1031–4.

Karpenkov, A. M., Rudashevskii, N. S. & Shamskaya, N. I. (1981). A natural chloride of palladium and bismuth—the phase of composition $Pd_4Bi_5Cl_3$. *Zapiski Vses. Mineralog. Obsh.*, **110**, 86–91 (in Russian).

Keays, R. R. & Davison, R. M. (1976). Palladium, iridium and gold in the ores and host rocks of nickel sulfide deposits in Western Australia. *Econ. Geol.*, **71**, 1214–28.

Kinkel, A. R. Jr & Lesure, F. G. (1968). Residual enrichment and supergene migration of gold, southeastern United States. *US. Geol. Survey Prof. Paper*, **600D**, D174–D178.

Krause, R. A., Kozlowski, A. W. & Cronin, J. L. (1982). Polysulfide chelates. *Inorg. Synth.*, **21**, 12–16.

Lakin, H. W., Curtis, G. C., Hubert, A. E., Shacklette, H. T. & Doxtader, K. G. (1974). Geochemistry of gold in the weathering cycle. *US Geol. Surv. Bull.*, **1330**, 80 pp.

Listova, L. P., Vainshtein, A. Z. & Ryabinina, A. A. (1966). Dissolution of gold in media forming during oxidation of some sulfides. *Metallogeniya Osad. Osad.-Metamorf. Porod, Akad. Nauk SSSR, Lab. Osad. Polez. Iskop.*, 189–99 (in Russian).

Mann, A. W. (1984). Mobility of gold and silver in lateritic weathering profiles: some observations from Western Australia. *Econ. Geol.*, **79**, 38–49.

Mann, A. W. (1985). Redistribution of gold in the oxidized zone of some Western Australian deposits. In *Gold-mining, Metallurgy and Geology*, Australian Institution of Mining and Metallurgy Symposium No. 39, pp. 447–58.

McCallum, M. E., Loucks, R. R., Carlson, R. R., Cooley, E. F. & Doerge, T. A. (1976).

Platinum metals associated with hydrothermal copper ores of the New Rambler Mine, Medicine Bow Mountains, Wyoming. *Econ. Geol.*, **71**, 1429–50.

McGoldrick, P. J. & Keays, R. R. (1981). Precious and volatile metals in the Perseverance nickel deposit gossan: implications for exploration in weathered terrains. *Econ. Geol.*, **76**, 1752–63.

Milazzo, G., Caroli, S. & Sharma, V. K. (1978). *Tables of Standard Electrode Potentials.* John Wiley, Chichester, 421 pp.

Miyake, Y. (1965). *Elements of Geochemistry.* Maruzen, Tokyo, 475 pp.

Mountain, B. W. & Wood, S. A. Solubility and transport of platinum-group elements in hydrothermal solutions: Thermodynamic and physical chemical constraints, this volume, pp. 57–82.

Sato, M. (1960). Oxidation of sulphide ore bodies. I. Geochemical environments in terms of *Eh* and pH. *Econ. Geol.*, **55**, 928–61.

Schenk, W. A. (1987). Sulfur oxides as ligands in coordination compounds. *Angew. Chem. Int. Ed. Engl.*, **26**, 98–109.

Schmidt, M. & Hoffmann, G. G. (1979). Zur synthese phosphinkomplexierter d^8-metalltetrasulfide (Ni(II), Pd(II) und Pt(II)) mit natriumpolysulfiden. *Z. Naturforsch.*, **34b**, 451–5.

Stumpfl, E. F. & Tarkian, M. (1976). Platinum genesis: new mineralogical evidence. *Econ. Geol.*, **71**, 1451–60.

Thornber, M. R. (1975). Supergene alteration of sulphides. II. A chemical study of the Kambalda nickel deposits. *Chem. Geol.*, **15**, 117–44.

Travis, G. A., Kemp, R. R. & Davison, R. M. (1976). Platinum and iridium in the evaluation of nickel gossans in Western Australia. *Econ. Geol.*, **71**, 1229–43.

Wagner, P. A. (1929). *The Platinum Deposits and Mines of South Africa.* Oliver and Boyd, Edinburgh, 338 pp.

Webster, J. G. (1986). The solubility of gold and silver in the system $Au–Ag–S–O_2–H_2O$ at 25°C and 1 atm. *Geochim. Cosmochim. Acta*, **50**, 1837–45.

Westland, A. D. (1981). Inorganic chemistry of the platinum-group elements. In *Platinum-Group Elements: Mineralogy, Geology, Recovery*, ed. L. J. Cabri. Canadian Institute of Mining and Metallurgy, Special Volume 23, Montreal, pp. 5–18.

White, D. E., Hem, J. D. & Waring, G. A. (1963). Chemical composition of subsurface waters. US Geol. Surv. Prof. Paper, 440F, 67 pp.

Wilkinson, G., Gillard, R. D. & McCleverty, J. A. (eds) (1987). *Comprehensive Coordination Chemistry.* Pergamon Press, Oxford, 7 volumes.

Williams, P. A. (1987). Geochemical and prebiotic systems. In *Comprehensive Coordination Chemistry*, ed. G. Wilkinson, R. D. Gillard & J. A. McCleverty. Pergamon Press, Oxford, Chapter 64, 37 pp.

Wilson, S. H. (1966). Sulfur isotope ratios in relation to vulcanological and geothermal problems. *Bull. Vulcanol.*, **29**, 671–90.

10

Spinel Non-stoichiometry as the Explanation for Ni-, Cu- and PGE-enriched Sulphides in Chromitites

A. J. NALDRETT* & J. LEHMANN

GIS, BRGM-CNRS, 1A Rue de la Férollerie, 45071 Orléans Cedex 2, France

ABSTRACT

High concentrations of PGE are associated with many chromitite layers in layered intrusions. The R-factor argument of Campbell et al. (1983) can explain the difference between the PGE concentrations in chromitite-hosted and Merensky-type PGE ores and the concentrations in magmatic sulphides that are less enriched in PGE. The very high concentrations of Ni and Cu also characteristic of chromitite-hosted sulphides are not explicable in this way. This has led Gain (1985) and von Gruenewaldt et al. (1986) to suggest that the original mass of these sulphides has been greatly reduced by the removal of Fe and S, with the consequent enrichment of other metals.

We suggest that the mechanism of Fe-loss from sulphide is that of Fe transferring to fill vacancies that exist in chromite crystallizing from basaltic magma.

Thermodynamic data for the reaction

$$4/3 Fe_2O_3{}^{(spinel)} + 1/3 FeS = Fe_3O_4{}^{(spinel)} + 1/6 S_2 \qquad (1)$$

indicate that, on cooling from 1150 to 930°C, if the fS_2 is buffered to that of sulphide cooling as a closed system, Fe–Ni–S sulphide containing 20 mole % NiS in equilibrium with chromite of the composition of the UG-2 can lose Fe to the chromite, using a high proportion of the available vacancies in the chromite at 1150°C. Depending on the mass ratio of the sulphide to chromite, the NiS content of the sulphide can be more than doubled.

Mass balance calculations indicate that if the sulphides of the UG-2 originally had the composition of those of the Merensky Reef (i.e. concentrations of 10·9% Ni, 4·65% Cu), augmenting this to their present concentrations would require the loss of 1578 ppm Fe, which would represent a gain in the mole fraction of Fe in the chromite of $1·8 \times 10^{-3}$.

A simple model for spinel, including the various end-members with vacancies (i.e.

*Present address: Department of Geology, University of Toronto, Toronto, Canada M5S 1A1.

those involving Al_2O_3 and Fe_2O_3), has been used to calculate the activities of these sesquioxides necessary for the spinel to contain a certain number of vacancies. When this model is used to calculate the vacancies needed to take up the Fe required to explain the enrichment in Cu, Ni and PGE in the Mandaagshoek section of the UG-2 (Gain, 1985), the calculated activities of Al_2O_3 and Fe_2O_3 are of the same order as those measured experimentally for basaltic melts at 1180°C and fO_2 of less than the Ni–NiO buffer. The Mandaagshoek section is one of the most sulphide-rich encountered in the Bushveld chromitite layers, and, therefore, requires more Fe to be removed than most. Thus the model will also account for all of the chromitites containing less sulphide than this area of the UG-2.

INTRODUCTION

The tenors of sulphides from the UG-2 and other chromitite layers in layered complexes tend to be extremely high in PGE, Ni and Cu in comparison with the tenor of sulphides within the Merensky Reef (Fig. 1).

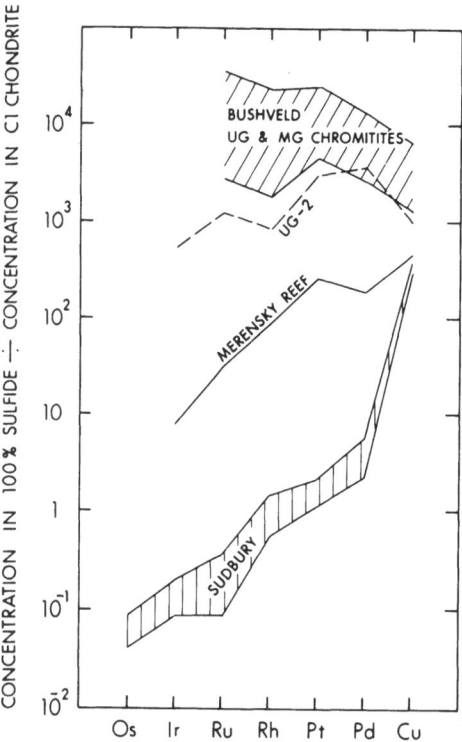

FIG. 1. Chondrite-normalized plot of PGE concentrations in sulphides of the UG-2 chromitite of the Bushveld Complex compared with those of the Merensky Reef. Data on UG-2 from Gain (1985), that on Merensky Reef from Naldrett & Cabri (1976). Normalization factors are those compiled from a variety of sources by Naldrett & Duke (1980).

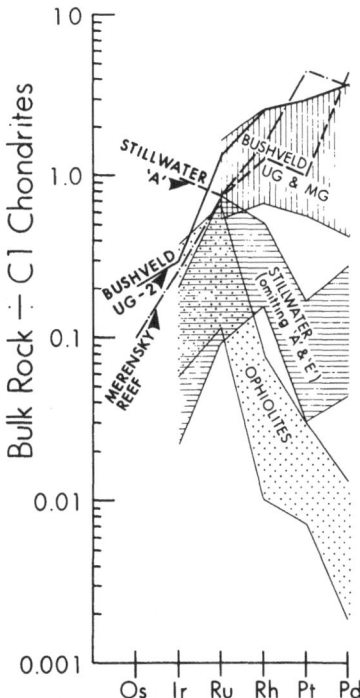

Fig. 2. Chondrite normalized plot of PGE concentrations in bulk rock samples of chromitite zones in the ophiolites of California and Oregon (Page *et al.*, 1986), Newfoundland (Page & Talkington, 1984), the Polar Urals (Page *et al.*, 1983), New Caledonia (Page *et al.*, 1982*a*), Oman (Page *et al.*, 1982*b*), the Stillwater Complex (Page *et al.*, 1985*a*), and the Bushveld Complex (von Gruenewaldt *et al.*, 1986; Gain, 1985).

Many chromitites, including those found in ophiolite complexes, are enriched in Ru, Ir and Os, with respect to ores that do not contain large amounts of chromite. This is seen in Fig. 2 which is a comparison between the bulk compositions (as contrasted with the sulphide tenors shown in Fig. 1) of the Merensky Reef and chromitite samples. It is thought that this enrichment is specifically related to the presence of chromite. This is not the problem considered in this paper.

Figure 2 illustrates the contrasting profiles of the chromitites from stratiform intrusions and those from ophiolite complexes; those from stratiform intrusions, while showing the same high Ru, Ir and Os that is shown by those from ophiolites, are also rich in Pt and Pd. Naldrett *et al.* (1987) pointed out that this can be explained as a consequence of the layered intrusion chromitites containing a small amount of PGE-rich sulphide in addition to the PGE concentrated directly by the chromite. They used the chromitites of the Stillwater Complex to illustrate their argument. The high Pt and Pd of the 'A' chromitite in comparison with the others is readily apparent in Fig. 3. Figure 4 illustrates the good match for the 'A' chromitite obtained by adding 0·06% of sulphide from the J-M Reef to the 'J' chromitite. This suggests that the difference between the 'A' and the others is very likely due to the

A. J. Naldrett, J. Lehmann

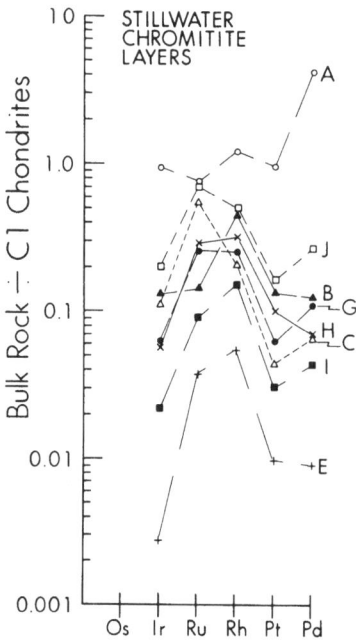

FIG. 3. Chondrite normalized plot of PGE concentrations in bulk rock samples of chromite layers from the Stillwater Complex (data from Page *et al.*, 1985*a*). Diagram after Naldrett *et al.* (1987).

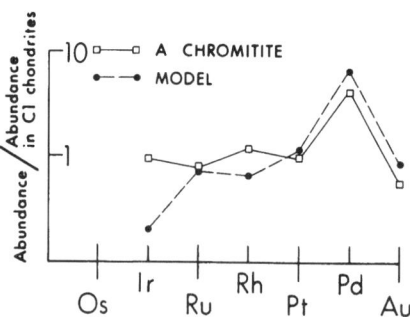

FIG. 4. Chondrite normalized plot of PGE content of the Stillwater 'A' chromitite and a model composition obtained by adding 0·06% of J-M Reef sulphide to the Stillwater 'J' chromitite. Diagram after Naldrett *et al.* (1987).

former containing more sulphide than the others. This interpretation is supported by the studies of McLaren & de Villiers (1982), Gain (1985), Hiemstra (1985) and von Gruenewaldt *et al.* (1986), who all note the association between sulphides and high PGE in the UG-2 chromitite of the Bushveld Complex. Thus, there seems to be little doubt that the high Pt and Pd in the chromitites of layered complexes is related to their sulphide content.

However, as can be seen from Fig. 1, the sulphides found in the chromitites of layered complexes are not simply enriched in PGE, but are significantly enriched in these metals with respect to a chromite-poor stratiform PGE-rich concentration from the same complex, the Merensky Reef. This is part of the specific problem that forms the subject of this paper.

In a number of recent papers, proponents of a magmatic origin for PGE-rich layers in stratiform intrusions have used the 'R-factor' argument to explain the difference between the PGE tenors of the Merensky-type ores and those of normal magmatic Ni–Cu sulphides (Campbell *et al.*, 1983; Naldrett *et al.*, 1986). The argument is based on an assumed very high sulphide melt-silicate magma partition coefficient of the order of 10^5 for PGE (Campbell & Barnes, 1984) and measured coefficients of the order of 10^2 for Ni, Cu and Co (Rajamani & Naldrett, 1978). Because of the high coefficients, the PGE concentration in the sulphides becomes as much a question of the ratio of the mass of magma with which the sulphide has equilibrated to the mass of sulphide as the value of the partition coefficient itself. Thus changes in mass ratio have a major effect on PGE concentrations, but very little effect on the concentrations of the base metals with their much lower coefficients. We believe that this mechanism is also important in the case of chromitite-hosted ores. Indeed, as seen in Fig. 1, the sulphides of the UG-2 are enriched in PGE by factors of 10–15 over those of the Merensky Reef. We would attribute much of this enrichment to the development of even higher 'R-factors' than was the case for the Merensky Reef. However, as von Gruenewaldt *et al.* (1986) have pointed out, chromite-hosted ores also have their base metal tenors enhanced by factors of 2–13 over the levels normally considered to be the maximum possible for sulphides in equilibrium with basaltic magma. Hence there is a problem in seeking to explain all aspects of their high metal tenors with respect to those of Merensky-type ores with very high 'R-factors'.

When pointing this out, von Gruenewaldt *et al.* (1986) suggested that part of the enhancement in metal tenors was due to sulphides with normal base metal tenors losing Fe to chromite, which then lost Mg to silicates and/or basaltic magma. The purpose of the present paper is to propose and illustrate an alternative reaction, in which vacancies present in high-temperature chromite take up the Fe, with the loss of S to the surroundings.

In order to illustrate that our suggested solution is viable, it is necessary to do the following:

(i) Demonstrate that the reaction is viable thermodynamically.
(ii) Estimate the amount of Fe that must be lost by the sulphides within any given chromitite.

(iii) Demonstrate that sufficient vacancies may exist, within the chromitite in question, to take up the required amount of Fe.

THERMODYNAMICS OF THE REACTION

The reaction under consideration has the form shown below:

$$\tfrac{4}{3}Fe_2O_3^{(spinel)} + \tfrac{1}{3}FeS = Fe_3O_4^{(spinel)} + \tfrac{1}{6}S_2 \qquad (1)$$

and

$$\frac{[Fe_2O_3]^{4/3}}{[Fe_3O_4]} = \frac{(fS_2)^{1/6}}{e^{-\Delta G/RT}[FeS]^{1/3}} \qquad (i)$$

At first glance it would seem that, knowing fS_2 and $[FeS]$ of the sulphide and $[Fe_3O_4]$ of the spinel, $[Fe_2O_3]$ could be calculated at two temperatures and the difference related to the change in vacancies of the spinel. However, a calculation of this type at a given temperature requires an accurate knowledge of the relation between the composition and thermodynamic parameters of the sulphides as well as of the spinel, together with the ΔG for the reaction. Inadequacies in these data led us to approach the problem in another way. From eqn (i), assuming that ΔG can be approximated by $(\Delta H - T\Delta S)$ over a range of temperatures, the activities of Fe_2O_3 ($[Fe_2O_3]_2$) and Fe_3O_4 ($[Fe_3O_4]_2$) after the sulphide–chromite assemblage has cooled through a temperature interval (from T_1 to T_2) can be related to the activities of FeS ($[FeS]_{1\ and\ 2}$) and the fugacities of S_2 ($(fS_2)_{1\ and\ 2}$) at the start and finish of the cooling respectively, and the activities of Fe_2O_3 and Fe_3O_4 at the start ($[Fe_2O_3]_1$ and $[Fe_3O_4]_1$), by the expression:

$$\left(\frac{[Fe_2O_3]_2}{[Fe_2O_3]_1}\right)^{4/3} \times \frac{[Fe_3O_4]_1}{[Fe_3O_4]_2}$$

$$= \exp\left(\frac{\Delta H}{R}\left(\frac{1}{T_2} - \frac{1}{T_1}\right)\right) \times \left(\frac{(fS_2)_2}{(fS_2)_1}\right)^{1/6} \times \left(\frac{[FeS]_1}{[FeS]_2}\right)^{1/3} \qquad (ii)$$

This equation has then been used as described below.

Parameters at Start and Finish of Reaction (1)

Reaction (1) results in a change in the concentrations of Fe^{2+}, Fe^{3+} and vacancies in the spinel with falling temperature. The physical picture to which we have applied it is shown in Fig. 5. Sulphides occur in chromitite, and remain in equilibrium with the chromitite as cooling proceeds. The sulphur fugacity within the chromitite is buffered through the magma. The magma is assumed to start with the same fugacity as that of the sulphides in the chromitite. However, its sulphur fugacity is assumed to change with falling temperature along the same path as would be followed by

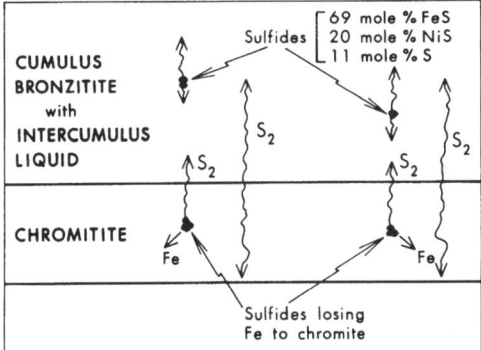

FIG. 5. Physical model of the cooling of a sulphide-bearing chromitite layer in a layered intrusion. Sulphides within the chromitite lose Fe to vacancies in the chromite. The resulting rise in fS_2 (due to the increasing S/metal ratio of the sulphides) is compensated for by sulphur diffusing into the overlying bronzite cumulates (which also contain interstitial liquid) where it reacts with Fe to form FeS. The fS_2 in the overlying rocks is assumed to be that of small amounts of sulphide within these rocks, which originally had the same composition as that in the chromite, but which have not lost Fe.

sulphides of the same initial composition as those in the chromitite but which were cooling as a closed system; that is along the same path as sulphides which did not have the opportunity to react with chromite. This is an attempt to approximate how fS_2 will change with temperature in a cooling basaltic magma. The philosophy behind the assumption is that it is likely that the fS_2 in basaltic magma changes with temperature in much the same way as it does in sulphide that has crystallized in equilibrium with the same basaltic magma, since signs of reaction between the two are not normally observed in natural samples.

We have chosen as a starting composition an Fe–Ni–S sulphide containing 20 mole %NiS, which, if one ignores Cu for the moment, corresponds closely to the composition of sulphides found in the Merensky Reef, and represents about the maximum Ni content to be expected in magmatic sulphides in equilibrium with basaltic (as opposed to komatiitic) magma.

Because we have much better data for solid sulphides than for Fe–Ni–S liquids, we have chosen 1150°C as the upper temperature limit of the calculation, which means that we have solid sulphide just below its liquidus temperature. The calculation has been terminated at 930°C, at which temperature Scott *et al.* (1974 and unpublished data) have determined the variation in the activity of FeS ([FeS]) and fugacity of S_2 across the Mss.

Scott *et al.* (1974) report that the monosulphide solid solution in the Fe–Ni–S system is a regular solid solution in which fS_2 varies strongly with changing metal/sulphur ratio, but much less with the changing Ni/Fe ratio (particularly where the mole % NiS < 20) at constant metal/sulphur atomic ratio. [FeS] also varies strongly with the changing metal/sulphur ratio, but varies proportionately with N_{FeS} at constant metal/sulphur atomic ratio. Thus [FeS] and fS_2 for Mss containing 20 mole % NiS and $N_{(Fe+Ni)S} = 0\cdot94$ at 1150°C can be estimated from values in the pure Fe–S system (for $(Fe + Ni)S$ in the $(Fe + Ni)S–S_2$ system (see Toulmin & Barton, 1964)).

Also in accordance with our suggested model, Mss containing 40 mole % NiS (which is close to the composition of sulphides in the UG-2, if one again ignores Cu) was taken as the final composition at 930°C.

If Fe is simply removed from the sulphide within the chromitite, its sulphur content will rise, for example along a path through the Mss similar to that marked 'A' in Fig. 6, and the fS_2 will also rise. However, as noted above, we have assumed that instead of the fS_2 rising unduly, sulphur has migrated into the adjacent interstitial basaltic magma, being buffered, as the temperature falls and the reaction proceeds, by this magma along the $fS_2–T$ path described above. The sulphide composition will therefore follow the path marked 'B' in Fig. 6 and the

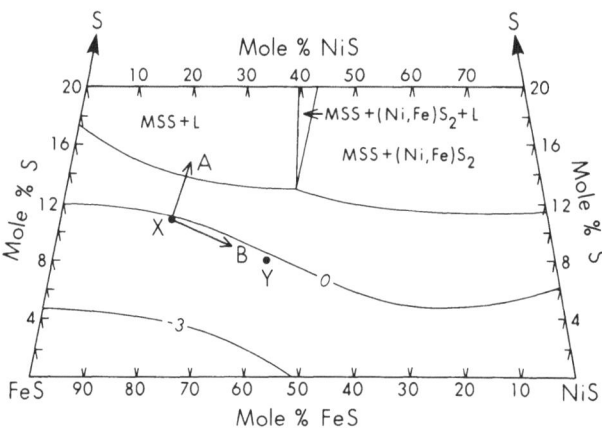

FIG. 6. The 930°C isotherm through the Mss field of the Fe–Ni–S system showing the fS_2 isobars at 10^{-0} and 10^{-3} bars. Data from Scott & Naldrett (unpublished data).

final composition will be given by the intersection of this path with the 40 mole % NiS line ('Y' in Fig. 6). (Note that Fig. 6 is the 930°C isotherm. In reality, any compositional change will be tempered by the fact that the reaction is occurring over a range of temperature, although the actual path followed in the Fe–Ni–S system will be that marked 'A' (assuming no buffering of fS_2) and close to that marked 'B' (assuming buffering). The [FeS] for this composition has then been obtained from Scott et al.'s data. These input data indicate that [FeS] decreases from 0·56 to 0·43 and fS_2 from $10^{1.02}$ to $10^{-0.4}$ between 1150 and 930°C.

Use of a Spinel Model to Relate Activities to Spinel Composition

The approach taken here towards using eqn (ii) has been to assume a certain mole fraction of vacancies in the spinel at 1150°C, use Lehmann & Roux's (1986) model to calculate the activities of components (in particular $[Fe_2O_3]_1$ and $[Fe_3O_4]_1$) for a spinel of the composition of that in the UG-2 (Table 1) with an assumed concentration of vacancies, solve the equation to obtain $[Fe_2O_3]_2^{4/3}/[Fe_3O_4]_2$, and use the model again to calculate the number of vacancies at the finish.

TABLE 1

Data on UG-2 chromite

	1	*2*	*3*	*4* ( )	[ ]
Mg	9·68	0·468	−1 796	0·376	0·044
Fe^{2+}	21·12	0·532	−2 021	0·421	0·053
Fe^{3+}	4·86	0·105	0	0·093	0·006
Al	18·04	0·722	−11 000	0·105	0·317
Cr	46·14	1·173	−24 000	0·002	0·580
Ti	0·95				
#			−3 000	0·002	$3·897 \times 10^{-4}$

= vacancies, ( ) = data for tetrahedral and [ ] = data for octahedral sites.
Column 1. Composition in wt % oxides. From Z. Johan, unpublished data.
Column 2. Idealized composition of stoichiometric spinel calculated on the basis of 4 oxygens and ignoring TiO_2.
Column 3. Octahedral site preference energies used in calories per mole.
Column 4. Site occupancy fractions at 1150°C assuming a vacancy content of 3×10^{-3} per mole.
As an example of the use of Lehmann & Roux's (1986) Table 7, one possible set of equations from which the activities of $Fe_{8/3}O_4$ $(=Fe_2O_3)^{4/3}$ and Fe_3O_4 can be calculated are shown below, where ( ) and [ ] indicate tetrahedral and octahedral sites respectively, and the subscript o indicates the pure end member and the values given are for the case of an assumed vacancy content of 3×10^{-3} per mole.

$$\text{Activity of } Fe_3O_4 = \frac{(Fe^{2+})[Fe^{3+}]^2}{(Fe^{2+})_o[Fe^{3+}]_o^2} = 1·54 + 10^{-4}$$

$$\text{Activity of } Fe_{8/3}O_4 = \frac{(\#)^{1/3}(Fe^{3+})^{2/3}[Fe^{3+}]^2}{(\#)_o^{1/3}(Fe^{3+})_o^{2/3}[Fe^{3+}]_o^2} = 3·21 \times 10^{-6}$$

Note: Standard states for these calculations were the pure end-member spinels, in contrast to the standard states of the liquid sesquioxides used for Figs 8 and 9. Note also that the activity of Fe_2O_3 corresponding to an activity of $Fe_{8/3}O_4$ of $3·21 \times 10^{-6}$ is $7·6 \times 10^{-5}$.

   The Lehmann–Roux model (1986) is a simple solid solution model which incorporates defect end members. Following classical procedure, the Gibbs free energy of solution is written by them as the sum of an ideal contribution and an excess part. The ideal contribution is the sum of three terms: (i) the configurational entropy of mixing, corresponding to the random distribution in each of the two sites (octahedral and tetrahedral) of the constituents (i.e. Fe^{2+}, Mg, Al, Cr, Fe^{3+} and vacancies in each site); (ii) a linear combination of the appropriate site occupancy numbers multiplied by the relevant octahedral site preference energies and (iii) a linear combination of the product of the end-member mole numbers and their chemical potentials in a state corresponding to a direct configuration, which relates to the mechanical mixing of direct components. The sum of terms (ii) and (iii) represents the non-configurational Gibbs free energy change occurring when mixing 'direct' end members (i.e. end members in which all octahedral sites are occupied by trivalent cations) to make a solid solution with a given internal distribution of each of the cations between the two sites (i.e. with given site occupancy numbers).

They obtain the partitioning of the different constituent ions between the two sites by minimizing the Gibbs free energy of the solution with constant numbers of moles, according to mass action equations of the form expressed below:

$$(Fe^{2+}) + [Mg] = [Fe^{2+}] + (Mg)$$

where square brackets indicate octahedral and parentheses tetrahedral sites.

TABLE 2
Sources of free energy data for reactions involving sulphides

$4Fe_3O_4^{(spinel)} + O_2 = 6Fe_2O_3^{(sesq)}$	Robie et al. (1978)
$Fe_2O_3^{(sesq)} = Fe_2O_3^{(spinel)}$	Lehmann & Roux (1986)
$Fe + 1/2S_2 = FeS^{(troilite)}$	Toulmin & Barton (1964)
$3Fe + 2O_2 = Fe_3O_4^{(spinel)}$	Robie et al. (1978).

The excess Gibbs free energy function was restricted by Lehmann & Roux to the contribution due to the reciprocal nature of the solution. They have shown that, in the case of a spinel solid solution with defect end members, this function cannot be, as it normally is, a polynomial, but must be a rational fraction. The coefficients of this function can be calculated from the ΔGs of reactions between the various end members of the solution, for example:

$$FeFe_2O_4^{(spinel)} + Al_2O_3^{(spinel)} = FeAl_2O_4^{(spinel)} + Fe_2O_3^{(spinel)}$$

We have not taken into account the corresponding reaction regarding Cr_2O_3 because of the lack of the necessary thermodynamic data. However, the results of Ulmer (1969) on the very limited solid solution of excess Cr_2O_3 in spinel imply that this element is not likely to have a major effect on the vacancy content.

For the temperatures and spinel compositions involved in this study, we have found that in so far as Al and Fe^{3+} are concerned the contribution of the excess Gibbs free energy is rather small. Since we are considering ratios of activities, and it is only the small changes in activity coefficient with temperature which would have a bearing on our calculation, we have assumed that the ratio of activities is equal to the ratio of the ideal part, neglecting the excess part.

As an example of the use of the spinel model in the calculation of activities of Fe_2O_3 and Fe_3O_4, we show in Table 1 the site occupancy fraction of the different cations plus vacancies (Ti has been ignored) for one assumed vacancy concentration at 1150°C as derived from the octahedral site preference energies given in the same table. The relevant activities have then been obtained from the site occupancy fractions as shown in Table 7 of Lehmann & Roux (1986). These are also listed for the UG-2 spinel composition in Table 2. Of course, these only relate to the one particular assumed concentration of vacancies in the spinel that is illustrated in this table. For each concentration of vacancies, there will be a different set of activities.

Solution of Equation (ii)

Returning to eqn (ii), the values of $[Fe_2O_3]_1$ and $[Fe_3O_4]_1$ needed for its solution have been obtained as described above. Values for $[FeS]_{1 \text{ and } 2}$ and $(fS_2)_{1 \text{ and } 2}$ obtained as outlined above have also been used in the equation. Equation (ii) has then been solved to give $[Fe_2O_3]^{4/3}/[Fe_3O_4]_2$. Since this ratio is an implicit function of the progress of the reaction (i.e. of the transfer of Fe from sulphide to chromite), it was necessary to solve it by an iterative method, at each step of which the site occupancy fraction was calculated.

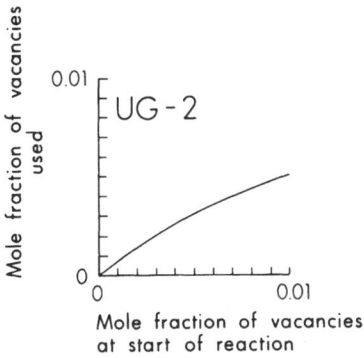

FIG. 7. Mole fraction of vacancies consumed in the reaction plotted against the assumed mole fraction of vacancies present initially (at 1150°C).

Because, as noted above, imperfections in the thermodynamic data meant that it was not possible to calculate precisely the number of vacancies at the start, this procedure has been repeated for a range of assumed initial concentrations of vacancies. In Fig. 7 the calculated mole fraction of vacancies consumed in the reaction (i.e. the assumed initial concentration of vacancies minus the concentration at the lower temperature calculated using eqn (ii) and the spinel model) is plotted against the mole fraction of initial vacancies for a reaction involving chromite of the composition of that of the UG-2, and the values of FeS activity and fS_2 at start and finish outlined above. The fact that the number of vacancies consumed is positive indicates that, within the limits of the thermodynamic data, the reaction will proceed.

All of the data for the calculations on spinels discussed here are given in Lehmann & Roux (1986) and have largely been extracted from the experimental data of Viertel & Seifert (1980), Dieckmann (1982) and Ishii *et al.* (1982). Data for reactions involving spinel and sulphide are from the sources shown in Table 2.

It should be stressed at this stage that while, because of the availability of thermodynamic data, we have investigated the reaction between solid sulphide and chromite, in nature the reaction is likely to start at higher temperature when the sulphides are still liquid, as soon as they come into contact with chromite. This implies that the reaction will proceed more easily than our calculations indicate. Natural sulphides contain Cu, an element that has been ignored in our treatment due again to the lack of thermodynamic data. The presence of Cu will affect the values of the activities that we have derived for pure Fe–Ni–S sulphides, and as such will modify the precise quantitative application of our calculations. However, as we stress elsewhere, the quality of the thermodynamic data used in the calculations do not justify their precise quantitative application.

ESTIMATE OF THE AMOUNT OF Fe EXCHANGE

In order to estimate the amount of Fe that must have been transferred from the sulphide to the chromite, we have used the data of von Gruenewaldt et $al.$ (1986) and Gain (1985) for the chromitites of the middle and upper groups of the Bushveld Complex.

We have taken their sulphur values, and assumed that the sulphur is associated with metals in a weight ratio of $1:1.54$ sulphur/metal, which is the S:Fe ratio of a sulphide with $N_{FeS} = 0.94$ in the FeS–S_2 system (a normal magmatic sulphide composition). This assumption gives rise to the sulphide contents for the rock

TABLE 3

Sulphides in the middle and upper group chromitites of the Bushveld Complex (data from von Gruenewaldt et $al.$, 1986 and Gain, 1985)

	I ppm S	II ppm sulph.	III wt% Cu	IV ppm sulph.	V factor	VI ppm Fe
UG-3A	72	183	19·7	775	4·2	359
UG-3	46	117	25·6	664	5·5	320
UG-2	926	2 356	9·8	4 965	2·1	1 578
UG-1	99	251	12·8	688	2·7	265
MG-4A	17	43	60·5	559	13·0	313
MG-3	34	86	24·4	451	5·2	222
MG-2B	62	157	15·3	934	3·3	472
MG2A/B	112	284	10·6	647	2·3	220
MG-2A	63	160	22·5	774	4·8	373
MG-1B	13	33	54·5	387	11·7	357
MG-1A	46	117	20·5	516	4·4	242

I = present S; II = present amount of sulphide; III = Cu in present sulphides; IV = original amount of sulphide if Cu was 4·65%; V = factor of upgrading; VI = Fe lost from original sulphide expressed as ppm of the bulk rock.

shown in column II of Table 3. Assuming that all of the Cu reported in von Gruenewaldt *et al.*'s analyses is in the sulphide, the wt% Cu in the sulphide is as shown in column III. This is extremely high, in most cases impossibly high for magmatic sulphide in equilibrium with basaltic magma.

Sulphides of the Merensky ore contain 4·65 wt% Cu. If one assumes that this is normal, and that sulphides of this composition were originally deposited with the chromitites, but that the Cu has been enriched as a result of these sulphides losing Fe and S in the manner proposed in our model, the original amount of sulphide present in each of them calculates to be that shown in column IV. The factor by which Cu and other metals apart from Fe have been upgraded in the sulphides is shown in column V, and the Fe lost to the chromite as a result of the reaction in column VI. Although crude, these calculations serve to show that in most cases the amount of Fe that is lost from sulphide to chromite, and that must therefore be accounted for by our model, is in the range of 200–500 ppm.

The UG-2 at Mandaagshoek has been sampled more thoroughly (Gain, 1985) and contains much more sulphide, so that the errors are proportionately lower. The figures for this can therefore give a more precise idea of the Fe transfer required in this particular case. According to Gain (1985), the average Ni and Cu contents for the sulphides are 22·5 and 9·8 wt% respectively, which implies an Fe content of 28·4 wt% for an Fe–Ni–Cu sulphide in which the $N_{(Fe+Ni+Cu)S} = 0.94$.

If one assumes, as before, that Merensky-type sulphides, containing 10·9 wt% Ni, 4·65 % Cu and 45·05 % Fe, were the prototype for this ore, and that no Ni or Cu have been lost, both Ni and Cu turn out to have been increased by the same factor of 2·1. The amount of Fe that must have been lost to achieve this is 1578 ppm. When this quantity is calculated as the mole fraction of a rock composed of 100 wt% chromite this amounts to 1.8×10^{-3}. Our calculations, expressed in Fig. 7, indicate that not all of the available vacancies in the chromite are utilized in the reaction. As seen from the figure, slightly less than 3×10^{-3} vacancies per mole are required in the chromite to make 1.8×10^{-3} per mole available for the reaction in order that it can explain the composition of the UG-2.

CALCULATION OF NUMBER OF VACANCIES IN CHROMITE

The third section of this paper involves the investigation of whether the UG-2 chromite is likely to contain enough vacancies to account for Fe losses of the order estimated above.

As discussed above, the critical components in the chromite are Al_2O_3 and Fe_2O_3, since these are known to substitute in spinel considerably in excess of stoichiometric proportions. Obviously, the higher the activities of Al_2O_3 and Fe_2O_3 in the magma from which the spinel has crystallized, the higher their activities and the greater their substitution above stoichiometry in the spinel. We have chosen to examine this by using the spinel model described above to calculate the activities of Al_2O_3 and Fe_2O_3 necessary to provide approximately the required number of vacancies.

The ΔGs for the reactions

$$Al_2O_3^{(spinel)} = Al_2O_3^{(corundum)}$$

and

$$Fe_2O_3^{(spinel)} = Fe_2O_3^{(hematite)}$$

have been calculated using the data of Viertel & Seifert (1980) and Dieckmann (1982). The calculated activities, using the liquid sesquioxides as the standard states, in a spinel of UG-2 composition containing 3×10^{-3} vacancies per mole are respectively 2×10^{-3} and 3×10^{-5} for Al_2O_3 and Fe_2O_3.

These values compare well with activities determined experimentally for basaltic liquids of a wide range of naturally occurring compositions at fO_2 of less than the Ni–NiO buffer. The calculated necessary activities are marked by the large crosses in Figs 8 and 9 which show experimentally determined activities of Al_2O_3 and Fe_2O_3 respectively. In general, the crosses coincide with, or fall below the activities determined for the natural basalts. Activities in natural basaltic magmas higher than those calculated as necessary will give rise to more vacancies in chromite, thus increasing the effectiveness of our model. However, we should caution the reader here that, in making these calculations, we have not been able to use ratios of activities, as we did in solving eqn (ii) above. Because they required an actual estimate of the activity and not merely the activity variation between a starting and a finishing state, the numerical results here are much less precise, and therefore the conclusions drawn from them less certain than were those drawn from the solution of eqn (ii).

CONCLUSION

Three main points have emerged as a result of our studies. Firstly, we have shown that the available thermodynamic data are consistent with the hypothesis that when chromite containing vacancies are in equilibrium with Fe–Ni sulphide at 1150°C, and the sulphur fugacity is buffered to that of a basaltic magma in equilibrium with sulphide, cooling will cause Fe to be drawn from the sulphide into the chromite to fill these vacancies, resulting in the diminution in the mass of sulphide and the consequent increase in the tenor of most other metallic constituents of the sulphide.

Secondly, mass balance calculations indicate that the amount of Fe transfer required to explain the increase in the Cu content of many chromite-hosted ores over and above that of those found in the Merensky Reef is generally between 200 and 400 ppm. The Mandaagshoek section is unusual, even for the UG-2, in having a very large amount of sulphide and requiring the transfer of over 1500 ppm Fe.

Thirdly, the available thermodynamic data coupled with Lehmann & Roux's (1986) adaptation of a classical spinel model indicate that the activities of Al_2O_3 and Fe_2O_3 in basaltic magma needed to produce the required number of vacancies may be comparable with those found experimentally to exist in such magma at the appropriate temperatures and realistic oxygen fugacities.

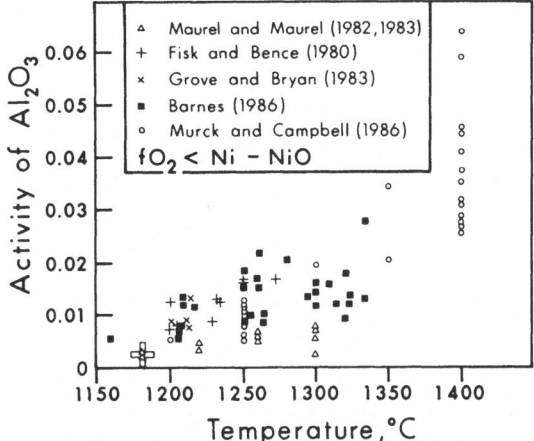

FIG. 8. Variation in the activity of Al_2O_3 (determined from the composition of the co-existing spinel) with temperature for basaltic liquids at fO_2 of less than the Ni–NiO buffer. The cross indicates the minimum activity of Al_2O_3 necessary for the model proposed in this paper to proceed. Data on Al_2O_3 activities from Lehmann *et al.* (in preparation).

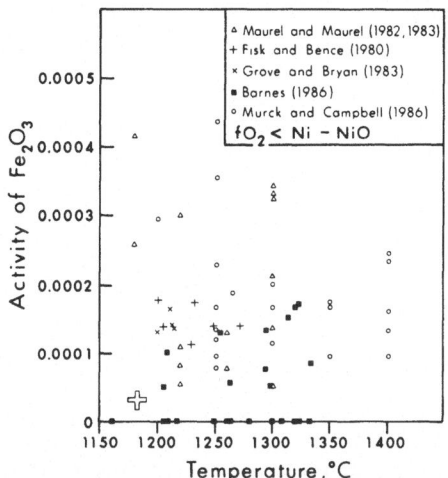

FIG. 9. Variation in the activity of Fe_2O_3 (determined from the composition of the co-existing spinel) with temperature for basaltic liquids at fO_2 of less than the Ni–NiO buffer. The cross indicates the minimum Fe_2O_3 activity necessary for the model proposed in this paper to proceed. Data on Fe_2O_3 activities from Lehmann *et al.* (in preparation).

ACKNOWLEDGEMENTS

This paper was written while one of us (A.J.N.) held a 1-year post as 'Chercheur Associé' with the CNRS, Orléans, France. He is most grateful to Dr Zdenek Johan for making this possible. We first appreciated the importance of sub-liquidus reactions between chromite and sulphides as a result of discussions with Professor von Gruenewaldt of the University of Pretoria who made available to us his, at the time, unpublished manuscript on the compositions of sulphides in chromitite layers of the Bushveld Complex. We thank Professor S. D. Scott and Miss E. C. Gasparrini of the University of Toronto for allowing us to quote some of their unpublished data, acquired in a joint study with A.J.N. Our colleague, Dr J. Roux, has been extremely kind in checking our calculations and helping us to be sure that the claims that we make in this article are realistic, and are not extrapolated beyond the bounds set by the input data.

REFERENCES

Campbell, I. H. & Barnes, S.-J. (1984). A model for the geochemistry of platinum-group elements in magmatic sulfide deposits. *Canad. Mineral.*, **22**, 151–60.

Campbell, I. H., Naldrett, A. J. & Barnes, S.-J. (1983). A model for the origin of platinum-rich horizons in the Bushveld and Stillwater Complexes. *J. Petrol.*, **24**, 133–65.

Dieckmann, R. (1982). Defects and cation diffusion in magnetite (IV): Non-stoichiometry and point defect structure of magnetite: $Fe_{(3-d)}O_4$. *Ber. Bunsenges. Phys. Chem.*, **86**, 112–18.

Gain, S. B. (1985). The geologic setting of the platiniferous UG-2 chromitite layer on the farm Mandaagshoek, Eastern Bushveld Complex. *Econ. Geol.*, **80**, 925–43.

Hiemstra, S. A. (1985). The distribution of some platinum-group elements in the UG-2 chromitite layer of the Bushveld Complex. *Econ. Geol.*, **80**, 944–57.

Ishii, M., Hiraishi, J. & Yamanaka, T. (1982). Structure and lattice vibration of Mg–Al spinel solid solution. *Phys. Chem. Minerals*, **8**, 64–8.

Lehmann, J. & Roux, J. (1986). Experimental and theoretical study of $(Fe^{2+}, Mg) (Al, Fe^{3+})_2O_4$ spinels: Activity–composition relationships, miscibility gaps, vacancy contents. *Geochim. Cosmochim. Acta*, **50**, 1765–83.

McLaren, C. H. & de Villiers, J. P. R. (1982). The platinum-group chemistry and mineralogy of the UG-2 chromitite layers of the Bushveld Complex. *Econ. Geol.*, **77**, 1348–86.

Naldrett, A. J. & Cabri, L. J. (1976). Ultramafic and related mafic rocks: Their classification and genesis with special reference to the concentration of nickel sulfides and platinum-group elements. *Econ. Geol.*, **71**, 1131–58.

Naldrett, A. J. & Duke, J. M. (1980). Tectonic settings of some Ni-Cu sulfide ores: Their importance in genesis and exploration. Geol. Assoc. Can. Spec. Paper 20, pp. 633–57.

Naldrett, A. J., Gasparrini, E. C., Barnes, S. J., von Gruenewaldt, G. & Sharpe, M. R. (1986). The upper critical zone of the Bushveld Complex and the origin of Merensky-type ores. *Econ. Geol.*, **81**, 1105–17.

Naldrett, A. J., Cameron, G. M., von Gruenewaldt, G. & Sharpe, M. R. (1987). The formation of stratiform PGE deposits in Layered Intrusions. In *Origins of Igneous Layering*, ed. Ian Parsons. NATO Advanced Research Workshop, D. Reidel, Dordrecht, 313–97.

Page, N. J & Talkington, R. W. (1984). Palladium, platinum, rhodium, ruthenium and iridium in peridotites and chromitites from ophiolite complexes from Newfoundland. *Canad. Mineral.*, **22**, 137–49.

Page, N. J, Cassard, D. & Haffty, J. (1982a). Palladium, platinum, rhodium, ruthenium and iridium in chromitites from the Massif du Sud and Tiebaghi Massif, New Caledonia. *Econ. Geol.*, **77**, 1571–7.

Page, N. J, Pallister, J. S., Brown, M. A., Smewing, J. D. & Haffty, J. (1982b). Palladium, platinum, rhodium, iridium and ruthenium in chromite-rich rocks from the Samail ophiolite, Oman. *Canad. Mineral.*, **20**, 537–48.

Page, N. J, Aruscavage, P. J. & Haffty, J. (1983). Platinum-group elements in rocks from the Voikar-Syninsky ophiolite complex, Polar Urals, USSR. *Mineral. Deposita*, **18**, 443–55.

Page, N. J, Zientek, M. L., Czamanske, G. K. & Foose, M. P. (1985a). Sulfide mineralization in the Stillwater Complex and underlying rocks. In *Stillwater Complex*, eds. G. K. Czamanske & M. L. Zientek. Montana Bureau of Mines and Geology, Special Publication No. 92, pp. 93–6.

Page, N. J, Zientek, M. L., Lipin, B. R., Raedeke, L. D., Wooden, J. L., Turner, A. R., Loferski, P. J., Foose, M. P., Moring, B. C. & Ryan, M. P. (1985b). Geology of the Stillwater Complex exposed in the Mountain View area and on the west side of the Stillwater Complex. In *Stillwater Complex*, eds G. K. Czamanske & M. L. Zientek. Montana Bureau of Mines and Geology, Special Publication No. 92, pp. 147–210.

Page, N. J, Singer, D. A., Moring, B. C., Carlson, C. A., McDade, J. M. & Wilson, S. A. (1986). Platinum-group element resources in podiform chromitites from California and Oregon. *Econ. Geol.*, **81**, 1261–71.

Rajamani, V. & Naldrett, A. J. (1978). Partitioning of Fe, Co, Ni and Cu between sulfide liquid and basaltic melts and the composition of Ni-Cu deposits. *Econ. Geol.*, **73**, 82–93.

Robie, R. A., Hemingway, B. S. & Fisher, J. R. (1978). Thermodynamic properties of minerals and related substances at 298.15 K and 1 bar (10^5 pascals) and at higher temperatures. US Geol. Surv. Bull. 1452.

Scott, S. D., Naldrett, A. J. & Gasparrini, E. (1974). Regular solution model for the Fe_{1-x} S-Ni_{1-x} S (mss) solid solution. Int. Min. Assoc. Ninth General Meeting, West Berlin and Regensberg. Collected Abstracts, pp. 172.

Toulmin, P. III & Barton, P. B. Jr (1964). A thermodynamic study of pyrite and pyrrhotite. *Geochim. Cosmochim. Acta*, **28**, 641–71.

Ulmer, G. C. (1969). Experimental investigation of chromite spinels. *Econ. Geol. Mon.*, **4**, 114–31.

Viertel, H. U. & Seifert, F. (1980). Thermal stability of defect spinels in the system $MgAl_2O_4$-Al_2O_3. *N. Jahrb. Mineral Abh.*, **140**, 89–101.

Von Gruenewaldt, G., Hatton, C. J., Merkle, R. K. W. & Gain, S. B. (1986). Platinum-group element-chromitite associations in the Bushveld Complex. *Econ Geol.*, **81**, 1067–79.

11

The Potential for Hydrothermal Platinum Deposits

JOHN W. LYDON

Geological Survey of Canada, 601 Booth Street, Ottawa, Ontario, Canada K1A 0E8

The cosmic and mantle abundance ratio of platinum:gold is about 10:1. Although, insofar as can be determined, the geochemical behaviour of platinum is similar to that of gold, the ratio of platinum:gold recovered by mining operations is about 1:20. The shortfall is in platinum recovered from mineral deposits in which the ore components have been transported in aqueous solutions at temperatures below 400°C.

Gold can be transported in significant concentrations as aqueous complexes, of which the most important inorganic ligands in natural waters are Cl^-, HS^- and OH^-. Platinum forms aqueous complexes with Cl^-, but under the low pO_2 conditions of most ore-forming solutions, platinum is orders of magnitude less soluble than gold. Platinum sulphides are known to be soluble in alkaline bisulphide solutions, suggesting complexing as thioanions (Westland, 1981). The slopes of solubility contours of platinum minerals in bisulphide solutions can be calculated as a function of pO_2 and pH for given stoichiometries of platinum thioanions, without knowing values for their associational constants. By assuming a range of possible stoichiometries, the maximum range of highest platinum solubilities can be identified and compared to the solubility patterns of gold for the same physicochemical conditions.

Comparisons of calculated maximum solubility fields for gold and platinum indicate that during ore precipitation involving oxidation of neutral to alkaline sulphurous solutions, platinum will be precipitated at lower pO_2 than gold, thereby potentially effecting a spatial separation of the two metals within a single orebody. This effect may explain the wide range of Pt:Au ratios in known deposits of hydrothermal affinity. Redox fronts are among the most favourable environments for both platinum and gold enrichment, where precipitation of the two metals may be effected either by reduction of oxic solutions, in which the metals are carried as chloride complexes, or by oxidation of anoxic solutions, in which the metals are transported as thioanions. The known platinum and gold associations with such deposit types as sediment-hosted uranium, 'red-bed' copper and Kupferschiefer-type, seem to illustrate this geochemical characteristic. Pore fluids during the initial stages of serpentinization of fresh ultramafic rocks have a high capacity to dissolve gold and platinum, but this capacity decreases as serpentinization progresses.

John W. Lydon

Because platinum readily forms arsenides but gold does not, the amount of platinum that can be transported in arsenic-bearing hydrothermal solutions is very low compared to gold.

REFERENCE

Westland, A. D. (1981). Inorganic chemistry of the platinum-group elements. In *Platinum-Group Elements: Mineralogy, Geology, Recovery*, ed. L. J. Cabri, The Canadian Institute of Mining and Metallurgy Special Volume 23, pp. 5–18.

12

The Use of Mantle Normalization and Metal Ratios in Discriminating between the Effects of Partial Melting, Crystal Fractionation and Sulphide Segregation on Platinum-Group Elements, Gold, Nickel and Copper: Examples from Norway

Sarah-Jane Barnes

Sciences de la Terre, Université du Quebéc, 555 Boulevard de l'Université, Chicoutimi, PQ, Canada G7H 2B1

R. Boyd, A. Korneliussen, L-P. Nilsson, M. Often

Norges Geologiske Undersøkelse, Postboks 3006, N-7002 Trondheim, Norway

R. B. Pedersen & B. Robins

Geologisk Institutt, Universitet i Bergen, N-5000 Bergen, Norway

ABSTRACT

The distribution of noble metals, Ni and Cu in mafic and ultramafic rocks is thought to be controlled by sulphides, chromite, olivine and platinum-group minerals (PGM). One method for presenting noble metal, Ni and Cu data focuses on the sulphide control by recalculating the data to 100% sulphides and presenting the data chondrite normalized. The relative importance of the influence of sulphides, chromite, olivine and PGM on the noble metals, Ni and Cu is examined here using two alternative methods.

Firstly, the metals can be plotted in the order: Ni, Os, Ir, Ru, Rh, Pt, Pd, Au and Cu and mantle normalized. The noble metals have a much higher partition coefficient into sulphides than Ni or Cu. Therefore, if sulphides have segregated from the magma or are retained in the mantle during partial melting, the magma and all rocks that subsequently form from it will be depleted in noble metals relative to Ni and Cu and the metal patterns will have an overall trough shape. Conversely any rocks containing these sulphides will be enriched in noble metals relative to Ni and Cu and the metal patterns will have an arch shape (characteristic of Pt reefs). Chromite-rich rocks tend to be enriched in the elements Os, Ir and Ru relative to the magma from which they form, therefore if chromite crystallizes from a magma or is retained in the mantle during partial melting, the magma and any rocks that subsequently form from it will be depleted in Os, Ir and Ru and the metal pattern will have a positive slope from Os to Pd, flat from Pd to Cu and have a positive Ni anomaly. The rocks containing cumulate chromite will be enriched in Os, Ir and Ru relative to the other elements and the metal patterns have an overall negative slope with a down turn at Ni (as seen in podiform chromitites). Olivine concentrates Ni and under some conditions Ir. Therefore if olivine crystallizes from a magma or is retained in the mantle during partial melting, Ni and Ir will be depleted in the magma and in any rocks that subsequently form from it. The metal patterns have an overall positive slope from Os to Pd, flat from Pd to Cu and no Ni anomaly. The cumulate will be enriched in Ni and Ir and tend to have a flat metal pattern (e.g. in dunites from komatiites and ophiolites).

A second approach to presenting noble metal, Ni and Cu data is suggested because while the effects of sulphide, chromite and olivine control can be seen on the metal patterns it is easier to distinguish these effects visually using metal ratio diagrams, Pd/Ir versus Ni/Cu and Ni/Pd versus Cu/Ir.

These two approaches are useful petrogenetic and exploration tools. Rocks which have trough-shaped metal patterns formed from magmas which were depleted in noble metals, possibly by sulphide segregation. These rocks do not make good exploration targets although there is the potential of a noble-metal deposit stratigraphically below them. Rocks which are not enriched or depleted in Ni and Cu relative to the noble metals formed from a magma that has not segregated sulphides and therefore a noble metal deposit could lie stratigraphically above these rocks.

INTRODUCTION

Naldrett *et al.* (1979) showed that if noble metal (Os, Ir, Ru, Rh, Pt, Pd, Au) analyses of sulphides are chondrite normalized and then plotted in order of decreasing melting point, smooth curves are obtained. The use of such chondrite normalized curves permits the distinction between sulphides that segregated from primitive magmas such as komatiites, and sulphides which segregated from more fractionated magmas, such as tholeiites. In their review of the noble metal abundance data then available, Barnes *et al.* (1985) were able to extend this approach to silicate rocks. They showed that it is possible to distinguish between noble-metal patterns from silicate rocks thought to represent the upper mantle (flat noble-metal patterns), komatiites (slightly Pd-enriched noble-metal patterns), and podiform chromitites (Os-, Ir- and Ru-enriched noble-metal patterns). However, the noble-metal patterns from; layered intrusions, ocean-floor basalts, continental-flood basalts, alkaline rocks and the non-tectonized portions of ophiolites are not unique and cannot be easily distinguished from each other. This is because the shape of the noble-metal pattern of a rock is influenced by the following factors:

(a) Whether any sulphides are present in the rock: noble metals tend to con-
    centrate in sulphides, therefore the noble-metal pattern of any rock con-
    taining sulphides will tend to be dominated by that of the sulphides, which in

turn will be similar to that of the liquid at the time of sulphide saturation (Keays, 1982). Thus, despite the fact that the rock may be a cumulate it could have the same shaped noble-metal pattern as the silicate liquid at the time of sulphide saturation.

(b) If no sulphides are present it is important to consider whether the rock represents a cumulate, or liquid, composition. Olivine and chromite cumulates tend to be enriched in Os, Ir and Ru (Agiorgitis & Wolf, 1977, 1978; Keays, 1982; Oshin & Crocket, 1982; Page *et al.*, 1982*a,b*; Davies & Tredoux, 1985; Crocket & MacRae, 1986). The mechanism by which this enrichment takes place is a matter of debate. Some authors (e.g. Keays & Campbell, 1981; Barnes *et al.*, 1985; Davies & Tredoux, 1985) suggest that early in the crystallization history of a magma it becomes saturated in Os-, Ru-, Ir-bearing PGM (i.e. at the ppb level) and that these PGM crystallize before the olivine and chromite and are then included within the olivine and chromite. Support is lent to this suggestion by: the observation of Os-, Ir- and Ru-bearing PGM in chromite-rich rocks (Prichard *et al.*, 1981; Stockman & Hlava, 1984; Talkington *et al.*, 1984); the covariance of chromite and laurite compositions in the Bird River Sill (Ohnenstetter *et al.*, 1986) and by the work of Amosse *et al.* (1987) who showed that Ir has a far lower solubility (<50 ppb) than Pt (<600 ppb) in basaltic melts. In contrast some workers (Agiorgitis & Wolf, 1978; Brugmann *et al.*, 1985) suggest that Os, Ir and Ru enter the chromite and olivine by solid substitution. In any event, a rock containing cumulate olivine or chromite may have a less fractionated noble-metal pattern than the liquid from which it formed.

(c) The relative timing of crystal fractionation and sulphide saturation (Keays & Campbell, 1981; Barnes *et al.*, 1985; Lee & Tredoux, 1986; Barnes & Naldrett, 1987). As mentioned above Os, Ir and Ru tend to be enriched in the olivine and chromite cumulates. Therefore, if no sulphide saturation occurs the noble-metal patterns will become progressively fractionated. Once sulphide saturation occurs most of the noble metals will tend to enter the sulphide and any rock containing this sulphide will tend to inherit the noble-metal pattern at the time of sulphide saturation. If sulphide saturation occurs early in the magma's history and the sulphides do not settle out, then the magma is likely to retain a relatively primitive noble-metal pattern despite the fact that the silicate minerals have fractionated.

The simple noble-metal patterns are inadequate for discerning which processes have occurred because they do not show the ratio of Ni and Cu relative to the noble metals. Therefore, two new methods for considering noble metal data have been developed here:

(1) By normalizing the noble metal contents to average mantle values and adding Ni and Cu to either end of the noble-metal pattern the ratio of Ni and Cu to the noble metals may be illustrated. These patterns will be called

mantle normalized metal patterns. Because Ni, Cu and the noble metals have different partition coefficients into sulphides, olivine, chromite and PGM, the relative timing of sulphide saturation, sulphide removal and crystal fractionation may be deduced from the metal patterns;

(2) Although the effects of partial melting, crystal fractionation and sulphide saturation on the metal patterns can be observed, it was found to be easier to separate these effects visually using the metal ratio plots of Pd/Ir versus Ni/Cu and Ni/Pd versus Cu/Ir.

In each case the method will be outlined by using a literature data base and then demonstrated using new data from Norway.

IS NORMALIZATION TO 100% SULPHIDES ALWAYS JUSTIFIED?

Normalization to 100% sulphides assumes that most of the noble metals are present in the sulphide portion of the rock, and such a recalculation allows a more meaningful comparison between rocks with widely differing sulphide contents, by removing the dilution effect of the silicate phases. However, a number of assumptions are inherent in this procedure which are not necessarily valid, particularly in sulphide-poor rocks. Firstly, normalizing to 100% sulphides assumes that durng crystallization of the rocks *all* the noble metals entered a sulphide liquid containing 36–38% sulphur. In rocks containing chromite or olivine some Os, Ir and Ru may be present in the chromite or olivine, or as PGM. Secondly, Ballhaus and Stumpfl (1986) suggest that under certain circumstances the noble metals partition into a fluid phase rather than into sulphides. Thirdly, it assumes that the present sulphur content of the rock represents the original igneous value. Sulphur is an extremely mobile element during hydrous alteration of rocks. Gain (1985) has suggested that sulphur was removed from the UG-2 reef of the Bushveld during cooling of the intrusion, so even rocks that do not contain hydrous minerals may have suffered sulphur loss. Therefore, in order to avoid the potential errors that can arise from recalculation to 100% sulphides the data in this work has not been recalculated.

THE ORDER OF THE ELEMENTS ON A METAL DIAGRAM

Naldrett and Barnes (1986) added Cu and Ni next to Au on the chondrite normalized noble-metal pattern (e.g. Fig. 1(a)). It is suggested here that Ni values be placed to the left of Os, rather than to the right of Cu (Fig. 1(b)). Ni is compatible with most early crystallizing phases (e.g. olivine, pyroxene, spinel), Os, Ir and Ru tend to be enriched in olivine- and chromite-rich rocks, therefore, it is logical that Ni should be placed on the left-hand side of the diagram. Cu, on the other hand, is a relatively incompatible element in mafic rocks, and hence the position of Cu to the right of Au.

THE ADVANTAGE OF MANTLE NORMALIZATION
OVER CHONDRITE NORMALIZATION

Simply moving Ni to the left of Os, however, does not produce a smooth chondrite normalized curve for mantle lherzolites (Fig. 1(b)) because the ratio of Ni and Cu in the mantle to C1 chondrites is approximately 0·13 and 0·17 respectively, and the ratio of Ir and Pd in the mantle to Ir and Pd in C1 chondrites is about 0·008 15 (4·4/540). Therefore, if the metal values for a mantle derived rock are chondrite normalized the Ni chondrite values will tend to be 16 times (0·13/0·008 15) the Ir values and the Cu values will tend to be 21 times the Pd values, this gives rise to the trough-shaped metal pattern in Fig. 1(b). As most terrestrial rocks are derived from the mantle, it would be more realistic to normalize to mantle values than C1 chondrites. The argument against mantle normalization has been that there is no

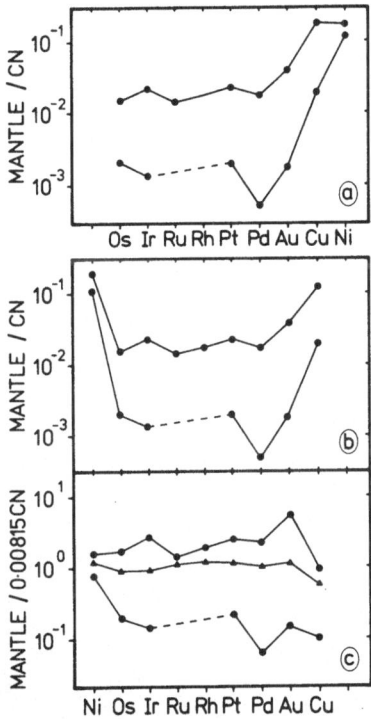

FIG. 1. (a) Conventional chondrite normalized metal patterns showing the range of rocks representing mantle (sources for mantle indicated in Appendix). Note that from Au to Ni the pattern is not flat. (b) Metal pattern of mantle fragments with Ni to the left of Os and Cu to the right of Au; chondrite normalized. Note the peaks at Ni and Cu which reflect the fact that there is 10–20 times more Ni and Cu chondrite normalized in the mantle than there are noble metals. If metal diagrams from ultramafic and mafic rocks are chondrite normalized they too will have peaks at Ni and Cu because they are mantle derived. (c) Revised metal pattern for the range of rocks representing mantle normalized to 2000 ppm for Ni 28 ppm for Cu and 0·008 15 × chondrites for the noble metals. Note that the metal patterns are now flat (triangles indicate the average value from the data base).

Sarah-Jane Barnes et al.

TABLE 1
Metal values

Location	Rock type	x̄[a]	n	Ni%	Cu	S	Os (ppb)	Ir	Ru	Rh	Pt	Pd	Au
World	Mantle fragments	b 114		0·2	0·002 8	0·020	4·2	4·4	6	2	9·2	4·4	1·4
	Estimate of mantle 0·00815 × C1[c]						4·2	4·4	5·6	1·6	8·3	4·4	1·2
Komatiites													
Karasjok	Volcaniclastics	a	24	0·12	0·012 4	0·081	2·1	1·5	<10	<2	11	11	6
		g	24	0·11	0·005 0	0·056	1·8	1·3	<10	<2	10	10	1·6
Rombak	Chlorite–actinolite	a	7	0·10	0·001 9	0·017	<5	1·2	<10	1·6	<13	15	0·8
	Fels	g	7	0·085	0·001 3	0·015	<5	0·8	<10	1·3	<13	14	0·7
Caledonide ophiolites													
Norway	Normal podiform	a	64	0·13	0·001 9	0·122	n.d.	n.d.	n.d.	n.d.	11	1·6	4
	chromitites	g	64	0·07	0·000 6	0·080	n.d.	n.d.	n.d.	n.d.	8	0·8	2·5
Norway	Enriched podiform	a	4	0·13	0·005 6	0·130	n.d.	n.d.	n.d.	n.d.	416	413	248
	chromitites	g	4	0·12	0·004 4	0·111	n.d.	n.d.	n.d.	n.d.	362	82	43
Leka	Stratiform chromitite		1	0·19	0·003 0	0·559	1070	1760	770	40	<10	<10	8
Leka	Chromite-rich dunite	a	15	0·20	0·002 7	0·303	7·2	7·8	24	2·4	11	17	2·9
		g	15	0·19	0·001 8	0·241	4·5	3·3	17	2	10	15	1·5
Fæøy	Ni-rich sulphides	a	6	2·27	1·55	44·92	221	205	203	187	1 503	4 066	34
		g	6	2·27	1·41	44·92	203	167	186	240	1 450	3 986	23
	Cu-rich sulphides	a	4	1·34	8·28	44·07	252	211	247	235	2 230	5 800	201
		g	4	0·99	7·09	43·95	170	103	177	149	1 630	5 432	163
	Weighted mean		10	2·10	2·69	44·77	225	206	210	195	1 626	4 360	62
Small possibly drift-related intrusions													
Reinfjord	Tholeiitic dykes	a	10	0·10	0·016 4	0·338	<2	0·5	<10	<2	5·3	6·4	1·2
		g	10	0·07	0·013 9	0·263	<2	0·3	<10	<2	3	6	1
Melvann	Alkaline dykes	a	12	0·08	0·024 2	0·615	<2	0·4	<10	<2	6·4	5	1·3

Locality	Rock type	mean	n	1	2	3	4	5	6	7	8	9	10
Lille Kufjord	Ultramafics	g	12	0·08	0·014 4	0·241	<2	0·3	<10	<2	6	4	1·2
		a	6	0·08	0·020 6	0·270	<2	0·6	<10	<1	<10	7	3
Hosanger		g	6	0·07	0·020 0	0·270	<2	0·5	<10	<1	<10	6	1·4
	Ni-rich sulphides	a	7	3·15	0·350	34·2	27·8	20·8	39	8	71	105	8·7
		g	7	2·96	0·280	33·9	25	19	37	7	63	92	6
	Cu-rich sulphides	a	3	0·62	7·72	30·96	28	23	62	<20	120[d]	[d]	341
		g	3	0·59	6·79	29·63	25	21	47	<20	60[d]	[d]	116
	Weighted mean		10	2·72	0·93	33·47	25	19·2	38	<8·3	62·7		17
Flåt	Ni-rich sulphides	a	6	2·90	0·27	35·47	26	29	38	11	78	87	19·6
		g	6	2·06	0·19	35·2	22	27	35	7	50	70	18
	Cu-rich sulphides	a	5	1·50	11·71	34·9	20	22	29	16	159[d]	151[d]	578
		g	5	1·07	8·40	34·8	20	22	22	6	40[d]	100[d]	441
	Weighted mean		11	1·94	1·22	35·15	22	26	33	7	49	74	71
Ertelien	Ni-rich sulphides	a	8	2·12	0·35	42·5	9·4	6·5	14	10	9	198	45·9
		g	8	1·76	0·23	42·3	2	1·7	8·8	6	7	179	17·5
	Cu-rich sulphides	a	7	0·91	10·24	35·57	<5	0·65	<10	7·7	603[d]	135[d]	1 319
		g	7	0·68	5·40	34·8	<5	0·6	<10	2	65[d]	70[d]	591
	Weighted mean		15	1·59	1·00	42·5	<2·5	1·5	<9	5·6	16	163	104
Large proterozoic layered intrusion													
Jotunheim	Pyroxenites	a	14	0·01	0·025	0·164	<2	0·3	<10	1	31	154	18·7
		g	14	0·01	0·023	0·136	<2	0·25	<10	0·9	30	150	16

[a] Because noble metals tend to have a log-normal distribution both arithmetic (a) and geometric (g) means are listed.
[b] Ni and Cu values from Sun (1982). S and noble metals from the sources listed in the appendix.
[c] C1 values listed in Naldrett (1981).
[d] Because of analytical difficulties the Pt and Pd numbers for these samples are not reliable.
n.d. = not determined.

reliable estimate of the mantle abundances for noble metals, and the mantle is heterogeneous so that any value chosen to represent it is somewhat arbitrary. Despite these drawbacks, the advantages from mantle normalization (i.e. normalizing to the probable source material) are significant. There are now over 100 published analyses, from 22 localities around the world, for the Ir and Pd contents of samples which are thought to represent mantle material. The average noble metal content of all the mantle material that has been analysed (spinel lherzolites, garnet lherzolites and harzburgites) is close to 0·008 15 × C1 chondrite (col. 1, Table 1). This value is similar to, but slightly higher than, the values obtained by Morgan (1986) for his estimate of the Os, Ir, Pd and Au contents of the upper mantle based on spinel lherzolites. Furthermore, 0·008 15 × C1 chondrite is also surprisingly close to Chou *et al.*'s (1983) estimate that the mantle has experienced an addition of 0·74% chondrite material by bombardment from meteorites after formation of the core. Therefore the mantle normalizing factors for noble metals used in this work are 0·008 15 times chondrite (row 2, Table 1). The mantle abundances for Ni and Cu quoted by Sun (1982) were used for these elements.

The result of this mantle normalized approach can be seen by comparing Figs 1(a), (b) and (c). Figure 1(a) shows the range of metal patterns for the mantle presented in the conventional fashion, note the sharp change in the slope of the curve at Cu. Figure 1(b) shows the metal pattern, still chondrite normalized but with Ni next to Os. Note the enrichment of Ni and Cu relative to the noble metals. Figure 1(c) shows the same metal patterns but normalized to mantle, note that the patterns are now smooth and almost flat.

In Fig. 2 the range of mantle normalized curves for the various rock types is shown. The range was defined by drawing an envelope around the values from the literature (see Appendix for the sources used).

The metal patterns for komatiites (Fig. 2(a)) are only slightly fractionated (Pd/Ir = 5–10) with the Ni and Cu mantle normalized values (mn) close to the mantle normalized values for noble metals (Ni/Ir_{mn} = 1·5–2; Cu/Pd_{mn} = 0·75–0·9). This produces a fairly smooth metal pattern. The very wide range (0·5–500 times mantle) of metal patterns for komatiites arises because the lower limit of the komatiite field is defined by olivine-spinifex textured komatiites (Munro Township and western Australia; Crocket & MacRae, 1986; Keays, 1982) which lack sulphides, whereas the upper limit is defined by massive sulphides in equilibrium with komatiites (Abitibi Greenstone Belt, Green & Naldrett, 1981).

High-MgO basalts (Fig. 2(b)) in this work are regarded as rocks derived from magmas containing 12–18% MgO. This division includes komatiitic basalt, high-MgO basalts and tholeiites but not boninites which have been included with low-TiO_2 basalts Fig. 2(e). The lower limit of the field is defined by the chill zone of Fred's Flow (a komatiitic basalt, Crocket & MacRae, 1986), and the upper limit is defined by the sulphides from Katiniq, Cape Smith Fold Belt (Barnes *et al.*, 1982); the Katiniq sulphides are thought to have formed in equilibrium with a komatiitic basalt. The noble metal portion of the patterns is slightly more fractionated (Pd/Ir = 20–30) than the komatiite curves (compare Figs 2(a) and (b)); the Ni/Ir_{mn} and Cu/Pd_{mn} ratios are close to 1 and produce an overall smooth metal curve.

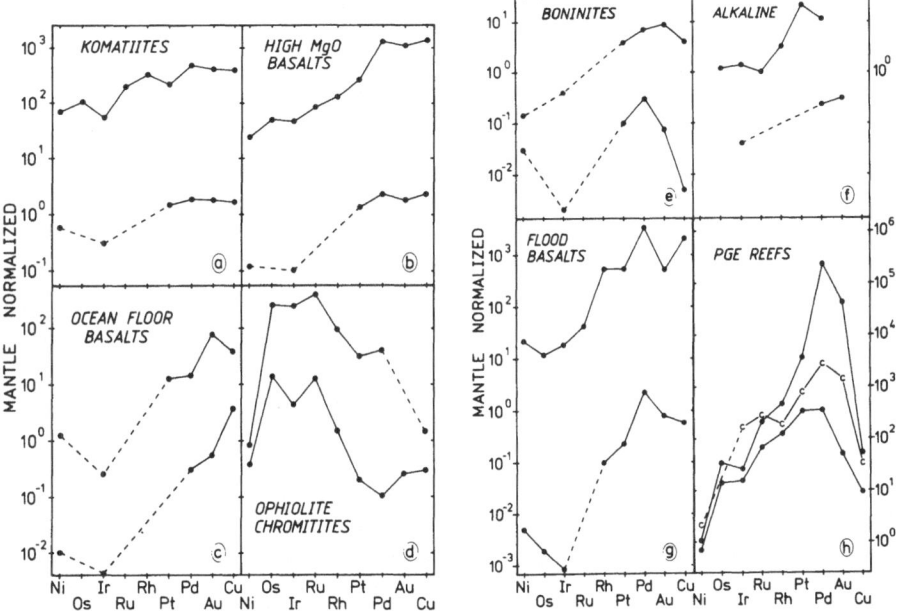

FIG. 2. Range of mantle normalized metal patterns for a variety of rock types using data from the literature: sources are indicated in the Appendix. (a) Komatiites and the sulphides associated with them. (b) High-MgO basalts and sulphides associated with them. (c) Ocean floor basalts. (d) Podiform chromitites from ophiolites. (e) Boninites and low-TiO$_2$ basalts. (f) Alkaline rocks. (g) Flood basalts and sulphides associated with them. (h) PGE reefs (with the Pt-enriched chromites from the Cliff locality of the Unst ophiolite shown as curve c).

Ocean floor basalts in this work include basalts from ridges, triple junctions and 'type I' lavas from the Thetford ophiolite (all the samples have flat or LREE depleted REE patterns). Metal patterns from ocean floor basalts (Fig. 2(c)) cover the range 0·003–3 times mantle. This range may be too narrow, because none of the samples from the localities considered here contain appreciable amounts of sulphides and none are from primitive ocean floor basalts (highest MgO content 9·2%). The noble metal portion of the patterns shows variable degrees of fractionation with Pd/Ir ratios from 20 to 200; Ni is enriched relative to Ir (Ni/Ir$_{mn}$ = 2–7) and Cu is enriched relative to Pd (Cu/Pd$_{mn}$ = 10–50). Consequently, unlike the metal patterns for more primitive rocks, ocean floor basalts do not have smooth metal patterns, but are trough shaped. This suggests that either some sulphides have been retained in the mantle during partial melting and these have preferentially retained the noble metals over Ni and Cu (Hamlyn *et al.*, 1985), or some sulphides segregated from the magma en route to the surface and these preferentially removed the noble metals (Hertogen *et al.*, 1980). If Hamlyn *et al.* (1985) are correct, then all ocean floor material should be depleted in noble metals. If Hertogen *et al.* (1980) are correct ocean floor basalts that have not segregated sulphides, possibly primitive ocean floor basalts, should not be noble metal depleted.

The wide range in Pd/Ir ratios from ocean floor basalts may be caused by the removal of Os, Ir and Ru either as PGM or by chromite crystallization, as has been suggested by Hertogen *et al.* (1980) and Oshin and Crocket (1986). The most remarkable feature of metal patterns from ophiolite chromites (Fig. 2(d)) is that they are strongly enriched in Os, Ir, Ru and have Pd/Ir ratios of $0·01–0·1$ and cover the range $0·1–200$ times mantle. Thus, the complementary metal patterns of ophiolite chromitites and ocean floor basalts can be understood if the chromitites represent the cumulate portion and the ocean floor basalt the fractionated silicate melt portions of a primitive mantle melt.

Some ocean floor material (e.g. the 'type I' basalts at the Thetford ophiolite) that has experienced chromite and olivine fractionation has relatively unfractionated metal patterns (Pd/Ir = 10–15). If sulphide saturation occurs early in the magma's history (i.e. before appreciable chromite or PGM crystallization), then the magma will inherit a primitive metal pattern. However, if chromite (or PGM) crystallization occurs before sulphide saturation, then the chromite or PGM will preferentially remove Os, Ir and Ru from the magma to produce a fractionated liquid and an Os–Ir–Ru-enriched cumulate. Using this reasoning the contradictions between the fractionated silicate geochemistry and the unfractionated noble metal patterns of Thetford 'type I' basalts, may be explained by sulphide saturation before chromite crystallization.

Boninites and low-TiO_2 lavas (Fig. 2(e)) exhibit a variable degree of noble metal fractionation (Pd/Ir ratio = 20–200) and cover the range $0·002–10$ times mantle. Ni tends to be enriched relative to Ir, Ni/Ir_{mn} = 2–10 and Cu is depleted relative to Pd, Cu/Pd_{mn} = $0·01–0·7$. Irivine and Sharpe (1982) suggest that one of the initial magmas to the Bushveld Complex was boninite-like. Hamlyn *et al.* (1985) have drawn attention to the similarity in the depletion of Cu relative to Pd in boninites and the Merensky reef. They suggest that the Cu depletion results from a previous melting event and that this is important in generating the type of magma from which Pt-reefs such as the Merensky reef of the Bushveld complex form.

Alkaline rocks (Fig. 2(f)) have relatively unfractionated metal patterns (Pd/Ir = 10–20), but the full shape of the metal pattern curve is not known because Ni and Cu values are not available for the samples in which the noble metal levels have been determined. Furthermore, the range of values indicated here may be too small as the data base consists of only 3 localities. The upper limit is based on kimberlites (Kaminskiy *et al.*, 1975) and the lower limit on basanites (Mitchell & Keays, 1981).

Flood basalts (Fig. 2(g)) tend to have fairly fractionated metal patterns (Pd/Ir = 100–200). The curves show a tendency towards enrichment of Ni and depletion in Cu relative to the noble metals. The lower limit is defined by the international standard BCR-1 (Govindaraju, 1984), and the upper limit by sulphides from Nor'ilsk (Naldrett, 1981). However, not all flood basalt related material is as fractionated, the Insizwa Intrusion is associated with the Karroo flood basalts, but the metal pattern is less fractionated than most other flood basalt related intrusions (Pd/Ir = 20, Lightfoot *et al.*, 1984). This anomaly may be explained by suggesting that the Insizwa magma became saturated in sulphides early in its history and that consequently the sulphides froze in the unfractionated metal pattern of the Insizwa magma.

The metal patterns for Pt-reefs are shown in Fig. 2(h). The UG-2 and Merensky reefs of the Bushveld Complex, the JM reef of the Stillwater Complex and the Roby zone of the Lac des Iles Intrusion are all considered in this group. Pt-reefs have also been reported from the Penikat intrusion in Finland (Alapieti & Lahtinen, 1986) and from the Great Dyke of Zimbabwe (Wilson & Prendergast, 1987), but insufficient noble metal data were available to include these localities here. The metal patterns show a variable degree of fractionation in the noble metals; from the UG-2 reef which is the least fractionated (Pd/Ir approximately 20) to the JM reef (Pd/Ir approximately 8000). The most distinctive feature of these metal patterns is the overall arched shape, which is a result of the sharp down turn at each end of the pattern at Ni and Cu respectively. One of the most distinctive features of Pt-reefs is the low ratio of Ni and Cu to noble metals (Ni/Ir$_{mn}$ <0·05, Cu/Pd$_{mn}$ <0·02). Interestingly one other rock type has metal patterns similar to Pt-reefs and this is the Pt-enriched chromitites from the Cliff locality of the Unst ophiolite (c on Fig. 2(h)).

METAL RATIO DIAGRAMS

An alternative approach to presenting noble metal data is to use metal ratio plots. If two metal ratios are plotted against each other, e.g. Pd/Ir versus Ni/Cu, then it is not necessary to recalculate the data to 100% sulphides nor to normalize the data to mantle or chondrite values.

Pd/Ir versus Ni/Cu

Figure 3(a) outlines the fields covered by the major magma suites on a plot of Ni/Cu versus Pd/Ir. Because the number of samples for which the noble metals and Ni and Cu have been determined is small the shape of these fields may change as more data are obtained. Nonetheless this diagram was found to be useful in outlining the processes that may effect the distribution of the noble metals, Ni and Cu. These two particular ratios were chosen because the metal patterns indicated that the Pd/Ir ratio increases as the magma suite becomes more evolved, while the Ni/Cu decreases (Figs 2 (a–g)). Therefore, a plot of these two ratios against each other produces a separation of the various magma suites with the most primitive (mantle material), at one end and the most evolved (continental flood basalts) at the other. The effect of variations in the degree of partial melting and the variations in composition of the source material on the Pd/Ir and Ni/Cu ratios may be estimated empirically by drawing a line through mantle, komatiites, high-MgO basalts, ocean-floor basalts, boninites and continental flood basalts. More detailed possible numerical models are presented elsewhere (Naldrett & Barnes, 1986; Barnes, 1987).

The fields for layered intrusions, and for the intrusive portions of ophiolites are also shown on Fig. 3(a). The effect of crystal fractionation on the Pd/Ir and Ni/Cu ratios can be examined by considering the komatiite and ophiolite data. Some olivine cumulates from komatiite flows (B on Fig. 3(a)) have lower Pd/Ir and higher Ni/Cu ratios than the lavas from which they crystallized. Olivine concentrates Ni in

Sarah-Jane Barnes et al.

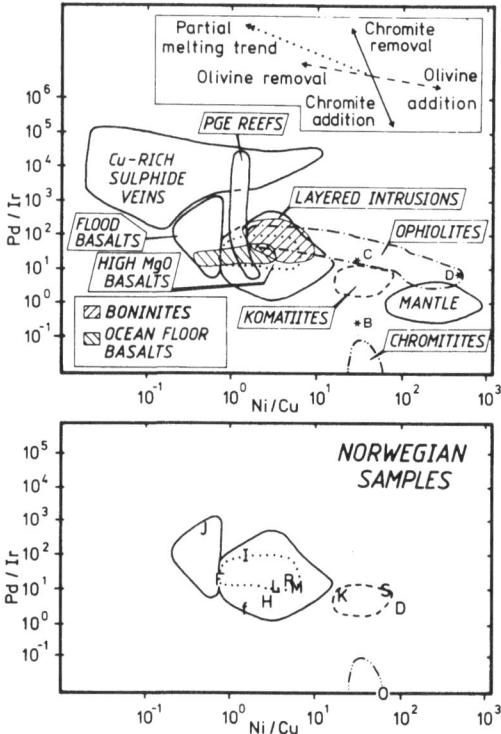

FIG. 3. (a) Metal ratio diagram of Pd/Ir versus Ni/Cu based on literature data for discriminating between the compositional fields of: mantle, komatiites and sulphides associated with them, high MgO-basalts and associated sulphides, ocean-floor basalts, boninites and low-TiO$_2$ basalts, flood basalts and sulphides associated with them, ophiolites, podiform chromitites from ophiolites, Pt-reefs, layered intrusions of unknown affinity, and Cu-rich sulphides. D = dunites from Thetford ophiolite, B = B-zones of komatiites and C = Cliff locality from Unst. The inset shows the displacement vectors on the diagram for the effects of partial melting of the mantle (dotted), olivine removal or addition (dashed) and chromite removal or addition (solid). (b) Diagram of the Pd/Ir and Ni/Cu ratios of Norwegian mafic and ultramafic rocks compared with selected compositional fields (others left out for the sake of clarity) from Fig. 3(a). Symbols: D = dunites from Leka; K = Karasjok komatiites; S = Rombak komatiites; O = Leka ophiolite chromitite; F = Fæøy ophiolite sulphides; R = Reinfjord Tholeiitic Intrusion; M = Melkvann Alkaline Intrusion; L = Lille Kufjord Tholeiitic Intrusion; J = Jotun pyroxenites; H = Hosanger sulphides; f = Flåt sulphides; I = Ertelien sulphides.

preference to Cu, therefore the higher Ni/Cu ratio in olivine-enriched portions of the flows is reasonable. Some of the olivine cumulates are enriched in Ir, thus the net effect of olivine crystallization is to displace the cumulate samples towards lower Pd/Ir and higher Ni/Cu ratios and to displace the fractionated liquid towards higher Pd/Ir ratios and lower Ni/Cu ratios (Fig. 3(a)).

Chromite crystallization has an effect similar to, but more intense than, olivine crystallization. The difference can be seen by examining the relative positions of chromitites and dunites within ophiolites. The dunites (D) show a large change in

Ni/Cu ratio from the initial liquid (presumed to lie in ocean floor basalt field) but a relatively small change in Pd/Ir ratio, the chromitites show a large change in Pd/Ir ratio and a small change in Ni/Cu ratio. Most of the variation in Pd/Ir and Ni/Cu ratios for komatiites, ophiolites, boninites, flood basalts, high-MgO lavas and layered intrusions may be accounted for by variation in conditions of partial melting and olivine or chromite crystallization.

Cu-rich sulphide veins commonly occur as footwall veins associated with Ni-Cu sulphide deposits, as at Sudbury (Hoffman *et al.*, 1979) and Kambalda (Lesher & Keays, 1984), but they may also occur as veins within the main ore zone as at Katiniq (Dillon-Leitch *et al.*, 1986) or Rathburn Lake (Rowell & Edgar, 1986). Cu-rich sulphide veins tend to be enriched in Au, Pd and at some localities Pt, but depleted in Os, Ir and Ru relative to the bulk composition of the Ni–Cu sulphide deposits with which they are associated. Consequently, the Cu-rich sulphide veins have low Ni/Cu ratios, high Pd/Ir ratios and form a distinct field on Fig. 3(a). Most authors consider these veins to have formed by hydrothermal remobilization of Cu, Au and Pd from the associated Ni–Cu sulphide deposit. However, the situation is complicated by the fact that many Ni–Cu sulphide deposits exhibit an internal compositional zonation with a Cu, Au, Pd, Pt enriched portion and a separate Os, Ir, Ru enriched portion (Kambalda, Keays *et al.*, 1981; Strachona & Leveck West, Naldrett *et al.*, 1982; Insizwa, Lightfoot *et al.*, 1984; Alexo, Barnes & Naldrett, 1986; Lillefjellklumpen, Grønlie, in press). The Cu-rich ore is displaced relative to the bulk ore in the direction of the Cu-rich sulphide veins but does not usually plot in Cu-rich sulphide vein field. Naldrett *et al.* (1982) suggested that at Sudbury the Cu-rich ore is partly fractionated sulphide liquid produced after the crystallization of monosulphide solid solution from an Fe–Ni–Cu sulphide liquid, and further-more, that the Cu-rich sulphide veins are the final fractionated product. In contrast Keays *et al.* (1981), suggest that the compositional zonation observed in the Ni–Cu sulphide ores at Kambalda is the result of hydrothermal remobilization. Regardless of which hypothesis is correct, it should be remembered when using these metal ratio plots that the composition of Cu-rich ores tends to be displaced towards lower Ni/Cu ratios and higher Pd/Ir ratios on Fig. 3(a) relative to the bulk ore.

The Pt-reefs show a narrow range of Ni/Cu ratios but have a wide range in Pd/Ir ratios. Although the Ni/Cu and Pd/Ir ratios of Pt-reefs from the Bushveld Complex lie in the field of layered intrusions, the JM-reef of the Stillwater Complex and the Roby zone of Lac des Iles Intrusion are extremely Pd-enriched and overlap with a field of Cu-rich sulphide veins. If Cu-rich sulphide veins form by hydrothermal remobilization, then the overlap of the JM reef and Roby zone with the field of Cu-rich sulphide veins could be interpreted as evidence of hydrothermal action in the formation of these rocks. However, as will be discussed below the Ni/Pd and Cu/Ir ratios of the JM reef and Roby Zone do not support such an interpretation. The Pt-enriched chromitites of the Unst ophiolite do not plot in the field of Pt reefs on this diagram (C on Fig. 3(a)), but in the field of ophiolites, which is consistent with their geological setting. The main difference between the Unst rocks and Pt reefs is the higher Ni/Cu ratio of the Unst rocks.

Ni/Pd versus Cu/Ir

Ni and Cu both have similar partition coefficients into sulphide liquid (Rajamani & Naldrett, 1978), therefore the segregation of a sulphide liquid from a silicate magma does not effect the Ni/Cu ratio of either the remaining silicate magma or of a cumulate which contains some sulphides. Partition coefficients for Pd and Ir into a sulphide liquid are unknown; estimates range from 1000 to 100 000 (Campbell & Barnes, 1984), but the partition coefficients are usually assumed to be similar. Therefore, as in the case of the Ni/Cu ratio, the Pd/Ir ratio of the silicate magma, or cumulate is not changed by the segregation of a sulphide liquid. For this reason the effect of sulphide removal from the silicate magma, or sulphide addition to the cumulate during the crystallization of a magma, is not visible on Fig. 3(a). In order to study the effect of sulphide saturation the Ni and Cu to noble metal ratio must be considered. Figure 4(a) shows the fields for some rock types on a plot of Cu/Ir versus Ni/Pd.

Because the partition coefficients for Cu and Ni into sulphides are much less than for noble metals, the segregation of sulphides from a silicate magma causes the silicate liquid to become depleted in noble metals relative to Ni and Cu; thus any cumulate containing these sulphides is enriched in noble metals relative to Ni and Cu. The Ni/Pd and Cu/Ir ratios of the remaining silicate magma increase, and the composition of any rocks that subsequently form from it are increased (Fig. 4(a)). The composition of the complementary cumulates containing the sulphides tend to be displaced to lower Ni/Pd and Cu/Ir ratios. However, the decrease in Ni/Pd and Cu/Ir ratios in the sulphides is also dependent on the amount of sulphides that segregate (i.e. the R-factor) as discussed by Keays and Campbell (1981) for Cu and Pd and Campbell and Barnes (1984) for Ni and Pt. If a small amount of sulphides segregates, then the Ni/Pd and Cu/Ir ratios in the sulphides are lower than those of the melt, but if a large amount of sulphides segregates the Ni/Pd and Cu/Ir ratios of the sulphides approach but never exceed those of the melt from which they formed. Thus any samples that plot above the fields outlined by the extrusive rocks on Fig. 4(a) probably formed from magmas that have segregated sulphides. These rocks will not make good exploration targets for noble metals.

The effect of variations in the conditions of partial melting on the Ni/Pd and Cu/Ir ratios, as illustrated by the trend of komatiites, high-MgO basalts, ocean-floor basalts and flood basalts is to decrease the Ni/Pd ratio and to increase the Cu/Ir ratio with decreasing degrees of partial melting. More detailed numerical modelling for partial melting may be found in Naldrett and Barnes (1986) and Barnes (1987).

Olivine crystallization increases the Ni/Pd ratio and decreases the Cu/Ir ratio of the cumulates (Fig. 4(a)). The fractionated liquid shows a complementary trend of an increase in Cu/Ir and decrease in Ni/Pd ratios. Chromite crystallization does not appreciably change the Ni/Pd ratio of the magma but it does increase the Cu/Ir ratio of the magma from which it fractionated because chromite or the PGM associated with it preferentially removes Ir from the liquid. Therefore chromite crystallization displaces the composition of the cumulate to lower Cu/Ir ratios but similar Ni/Pd ratios and the liquids to higher Cu/Ir ratios (Fig. 4(a)).

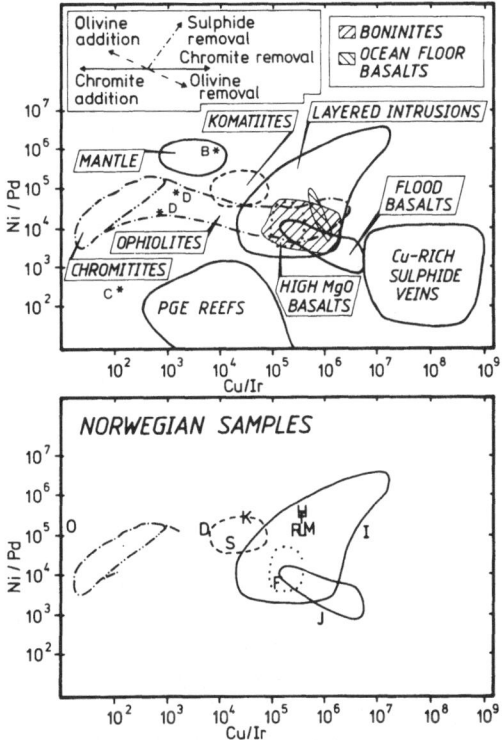

FIG. 4. (a) Metal ratio diagram of Ni/Pd versus Cu/Ir based on the literature data base. Compositional fields are, as on Fig. 3(a), distinguished, but notice the clear separation of PGE reefs and Cu-rich sulphides on this diagram. Inset shows the displacement vectors for olivine removal or addition (dashed), chromite removal or addition (solid) and sulphide removal (dot-dash), the partial melting trend has been left off for clarity. Symbols as on Fig. 3(a). (b) Diagram of the Ni/Pd and Cu/Ir ratios of Norwegian mafic and ultramafic rocks compared with selected compositional fields (others left off for clarity) based on the literature data shown on Fig. 4(a). Symbols as for Fig. 3(a).

The Cu-rich sulphide veins are enriched in Cu and Pd but depleted in both Ir and Ni, and plot in a distinct field on Fig. 4(a). However, on this diagram there is no overlap between the Cu-rich veins and the Pt-reefs as there was on the Ni/Cu versus Pd/Ir plot. The most distinctive feature of the Pt-reefs, their low base to noble metal ratio, is clearly illustrated on Fig. 4(a). Interestingly, the Pt-rich material from Unst ophiolite (C) plots close to the Pt-reef field.

THE NORWEGIAN EXAMPLES

Analytical Methods

For noble metal analysis the samples were divided into two categories based on the amount of chromite they contain. Samples containing <10% chromite were

TABLE 2

Comparison of metal values obtained for standards in this study with previously determined values

	Tv-84				*Ax-26*		*SARM7*	
Source	E	A	B	C	A	B	D	B
n	2	1	4	10	1	1	—	2
Os ppb	<2	n.d.	<2	<5	n.d.	<2	63	62
Ir	0·1	n.d.	0·23	2	n.d.	2·2	74	73
Ru	<10	n.d.	<10	<18	n.d.	<10	430	470
Rh	0·6	n.d.	<1	8·5	n.d.	9	240	204
Pt	2	<3	<10	51	43	35	3 740	3 475
Pd	<20	<2	<10	73	110	87	1 530	1 565
Au	5·9	10	1·2	12	28	8·8	310	275

A: Caleb (this work).
B: Becquerel (this work).
C: Barnes and Naldrett (1987).
D: Steele *et al.* (1975).
E: Barnes (1987).

analysed for noble metals by neutron activation analysis (NAA) after preconcentration into a Ni-sulphide bead by Becquerel of Toronto. Samples containing >10% chromite are not suitable for analysis by NAA because not all of the chromite melts during a normal fire assay and because incompletely dissolved Cr results in interference during the NAA. Those samples containing >10% chromite were analysed by atomic absorption using electrothermal atomization after preconcentration in a silver–lead bead, by Caleb Brett Ltd. This method is only suitable for the elements Pt, Pd and Au. Three samples were used to assess the performance of the laboratories. Tv-84 contains extremely low levels of all the noble metals: Becquerel and Caleb found this sample to contain noble metal levels at, or close to, their detection limit (Table 2) indicating that no significant contamination occurred. Ax-26 contains intermediate noble metal contents and the agreement between Becquerel and Caleb values and previously determined values is reasonable (Table 2). SARM7 contains high levels of the noble metals and Becquerel's values agree with the accepted values to within 10% for Os, Ir, Ru, Pt, Pd and to within 20% for Au and Rh (Table 2).

In NAA Cu can interfere with Pd and Rh determinations and the Crompton edge from Au with Pt determinations. Therefore, in Cu-rich samples, which are also Au-rich, it was not always possible to determine Rh, Pt and Pd.

Ni and Cu were determined by X-ray fluorescence for the komatiites, the Jotun pyroxenites and the Lille Kufjord rocks. For all other rocks Ni and Cu were determined by atomic absorption. S was determined by X-ray fluorescence.

Presentation of the Data
Table 1 presents the arithmetic mean for noble metals, Ni, Cu and S at each of the Norwegian localities considered here. However, noble metals tend to have a log-normal rather than a Gaussian distribution and therefore the geometric means

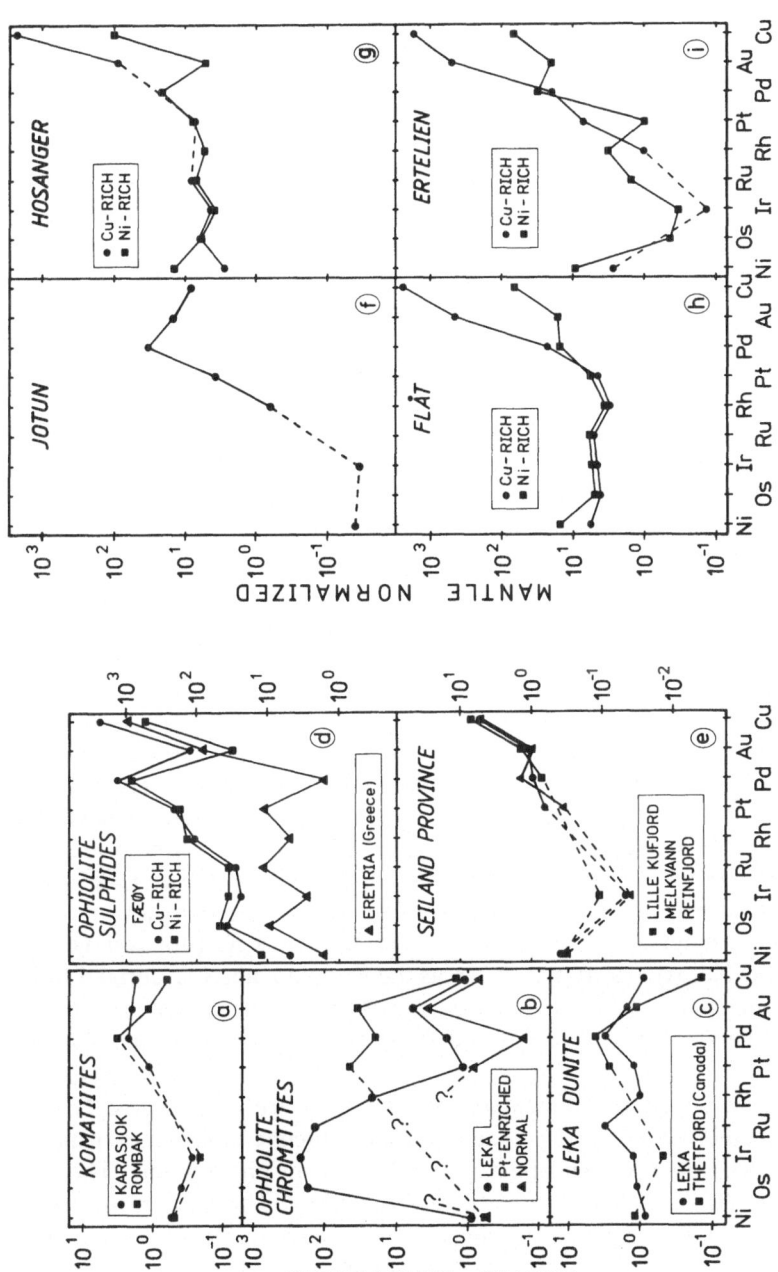

FIG. 5. Mantle normalized metal patterns for the Norwegian rocks sampled: (a) komatiites; (b) chromitites from ophiolites; (c) dunites from the Leka ophiolite compared with the dunite from Thetford mine ophiolite (Oshin & Crockett, 1982); (d) sulphides from ophiolites Fæøy compared with Eretria (Economou & Naldrett, 1984); (3) Seiland Province Intrusions; (f) Jotun; (g) Hosanger; (h) Flåt and (i) Ertelien.

are also presented in Table 1. On Figs 3(b), 4(b) and 5 the geometric means have been plotted. Ni, Cu and S were present in all rocks at greater than the detection limits, but this was not the case for the noble metals in some samples. At localities where some samples contained noble metals at less than the detection limit, the geometric mean was estimated using log-probability paper, and the arithmetic mean calculated using a standard statistical formulae (Bury, 1975 p. 279 eqn (8.8)).

At all of the Ni–Cu sulphide deposits two types of ore were found to be present, i.e. Cu-rich and Ni-rich. The mean for each of these ore types at each deposit is presented in Table 1 and plotted in Fig. 5. A simple average of the samples is probably not representative of the sulphide deposits as a whole. The weighted mean for each locality was calculated by weighting the two types of ores such that the Ni/Cu ratio of the weighted mean is the same as that reported for each sulphide deposit when it was mined (Boyd & Nixon, 1985). At most deposits the weighting was 90% Ni-rich ore 10% Cu-rich ore.

Komatiites

Pyroclastic komatiites of Proterozoic age (2085 ±85 Ma Sm/Nd; Krill *et al.*, 1985; Often, 1985) from the Karosjok Greenstone Belt of northern Norway (Fig. 6) have been analysed for noble metals (Table 1). The noble metal portion of the metal pattern is relatively flat (Pd/Ir = 8) and Ni and Cu are not fractionated relative to the noble metals, thus the overall shape of the metal pattern is smooth (Fig. 5(a)). Since these rocks are pyroclastic, the noble metal content of the rocks should be similar to that of the magma. It is interesting to note that the Karasjok komatiite metal patterns closely resemble the spinifex-texture komatiite metal patterns both in shape and level (compare lower curve on Figs 2(a) and 5(a)). Similarly the Ni/Cu, Pd/Ir, Ni/Pd and Cu/Ir ratios are similar to komatiites from the literature and on the metal ratio plots the Karasjok komatiites plot in the field of komatiites (K on Figs 3(b) and 4(b)). Noble-metal patterns from Proterozoic komatiites have been reported previously (Katiniq, Barnes *et al.*, 1982). These patterns are more fractionated than the Karasjok komatiites (compare upper curve on Figs 2(c) and 5(a)). The difference in metal patterns could be because the Karasjok komatiites formed from a magma containing approximately 25% MgO (the average MgO content of the Karasjok samples), whereas the Katiniq komatiites formed from a liquid containing 18% MgO. Tredoux *et al.* (1986) have also noted an antipathic relationship between MgO content and Pd/Ir ratio in lavas from the Kaapvaal craton.

In the eastern volcanosedimentary belts of the early Proterozoic Rombak window in northern Norway (Fig. 6) there are some discontinuous lenses of massive ultramafic rocks of unknown origin (Korneliussen, in preparation). The rocks presently consist of serpentine, chlorite and actinolite. The mean metal pattern (Fig. 5(a)) for the Rombak rocks is similar in shape and level to that of komatiites (Fig. 2(a)), except for a Au and Cu depletion in the Rombak samples. The overall similarity of the Rombak metal pattern with komatiite metal patterns suggests that these lenses may represent deformed and metamorphosed komatiites. Furthermore, the Rombak ultramafic rocks plot in the komatiite field on metal ratio plots (S on Figs 3(b) and 4(b)).

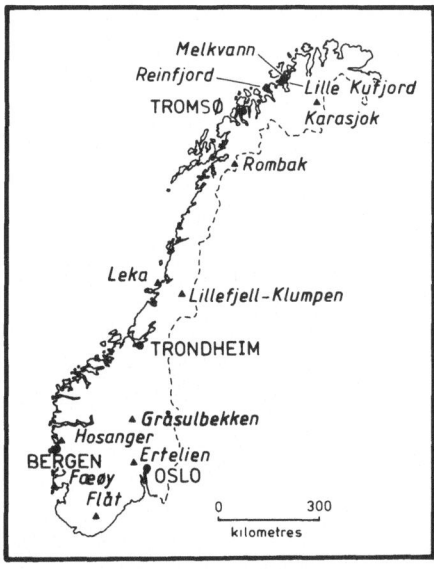

Fɪɢ. 6. Map showing the location of the Norwegian samples.

Caledonide ophiolites

Chromitites. High PGE values from podiform chromitites in ophiolites have been reported from Shetland (Gunn *et al.*, 1985; Prichard *et al.*, 1984, 1986), Ray-Iz and Kempirsay, USSR (Khvostova *et al.*, 1976). Ophiolite fragments are present throughout the Upper and Uppermost Allochthon of the Norwegian Caledonides (Gee *et al.*, 1985). Podiform chromitites from 19 of these ophiolite fragments (details of each locality are described in Nilsson, 1980) were analysed for Pt, Pd and Au. Only partial metal patterns can be drawn for the chromitites (Fig. 5(b)).

Chromitites from 15 of the fragments appear to contain noble metal levels similar to chromitites from around the world; the portion of the metal pattern that can be drawn has a similar shape to, and falls within, the range of podiform chromites from the literature (compare normal chromites on Figs 5(b) and 2(d)). These chromitites probably also exhibit the classic chromitite pattern with enrichment in Os, Ir and Ru.

However, chromitites from 4 localities contain Pt and Pd at >100 ppb which are much higher than that normally observed in podiform chromites from around the world (Fig. 2(d), and Pt-enriched on Fig. 5(b)). The shape of the available portion of the metal pattern resembles both the Pt-enriched metal pattern found in the podiform chromitite from the Cliff locality in the Shetland ophiolite and the Pt-reefs (Fig. 2(h)). The distinctive feature of these patterns is their arch shape which is a result of the high noble metal/base metal ratio. It is tempting to suggest that the overall metal pattern from these 4 Norwegian chromitites may have a similar shape and thus that they belong to the class of Pt-enriched podiform chromites.

Sixteen samples of chromite-rich dunites from the intrusive portion of the Leka ophiolite were analysed. Fifteen of the samples contain noble metal levels at 1–3 times mantle level and have essentially flat metal patterns (Fig. 5(c)). The overall shape and level of the metal pattern is similar to that observed for the dunite from the Thetford ophiolite of Quebec (Fig. 5(c)). These rocks are cumulates and therefore the flat metal pattern is not simply the result of sampling mantle; nor do the patterns represent trapped intercumulus liquid because that would be expected to resemble ocean-floor basalt which has Pd-enriched patterns (Fig. 2(c)). The metal patterns from the Leka chromite-rich dunites could represent the sum of the chromite pattern which is Os, Ir and Ru enriched and the ocean-floor basalt pattern (Pd-enriched). The presence of the cumulus olivine lowers the overall level of the pattern for all the elements except Ni. This interpretation is supported by the Pd/Ir versus Ni/Cu metal ratio diagram. The dunites plot on the Ni enriched side of the tie line between chromite and ocean floor basalt (D on Fig. 3(b)). The presence of cumulate olivine could have displaced the samples towards the Ni enriched side of the tie-line.

One sample from Leka which contains approximately 30% chromite has high values of Os, Ir and Ru; this sample exhibits a classic chromitite pattern with enrichment of Os, Ir and Ru (Fig. 5(b)). It is assumed that the metal pattern of this sample is dominated by the chromite. On the metal ratio diagrams this sample plots close to or within the field of chromites from ophiolites (O, Figs 3(b) and 4(b)).

Sulphides. There are two localities in Norway where massive sulphides associated with ophiolites contain ore grade levels of Pt. Both of these localities are too small to represent mineable deposits, but processes leading to their formation are of interest. Lillfjellklumpen is a massive sulphide lens lying concordantly between primitive MORB-type metabasalts and metagabbro in the allochthonous Gjersvik island arc complex (Grønlie, in press). Fæøy is a massive sulphide lens in the sheeted dyke complex of the Karmøy ophiolite (Boyd & Nixon, 1985). Only the Fæøy data will be discussed here as the Lillefjellklumpen data are presented elsewhere and are similar to the Fæøy data.

The massive ore at Fæøy contains two types of ore: Ni-rich and Cu-rich (Table 1). The metal patterns from both types are similar (Fig. 5(d)). Both increase steadily from Ni at 10 times mantle to Cu at 1000 times mantle and both have large negative Au anomalies. The Cu-rich ore is enriched in Cu, Au, Pd and Pt relative to the bulk ore. The metal patterns do not resemble any other metal patterns so far reported from ophiolites; in particular when compared with other ophiolites enriched in Pt (e.g. Cliff from the Shetland ophiolite, C on Fig. 2(h)) the Norwegian sulphides lack the characteristic arch shape. Neither do the Fæøy metal patterns resemble those of sulphides from the Erteria ophiolite, which have been attributed to hydrothermal action (Fig. 5(d); Economou & Naldrett, 1984). The Fæøy metal patterns closely resemble metal patterns from the sulphides which formed in association with komatiitic basalts (Katiniq sulphides from the Cape Smith Fold Belt of Canada, upper curve on Fig. 2(b)). Similarly, on the metal ratio diagrams the Fæøy sulphides plot in the field of high-MgO basalts (F on Figs 3(b) and 4(b)).

Massive sulphides within the intrusive portions of ophiolites are uncommon, but two possible mechanisms for their formation are: (a) they formed by hydrothermal fluids; (b) the sulphides segregated from the magma during cooling. The suggestion that the Fæøy sulphides formed from hydrothermal fluids may be discarded on the grounds that the metal patterns do not resemble metal patterns from hydrothermal sulphides (Eretria, Fig. 5(d); see also discussion by Grønlie, in press). At Fæøy there are two types of dykes present (Pedersen, in preparation), ocean ridge basalts and boninites. If hypothesis (b) is true it would be anticipated that the sulphides which formed in equilibrium with such magmas should have a metal pattern similar to those found in ocean floor basalts or boninites. The Fæøy metal patterns are not trough-shaped like ocean-floor basalt metal patterns (compare Figs 2(c) and 5(d)) or Cu-depleted like the boninites. This might be taken as evidence that they did not segregate from an ocean floor basalt or boninite. This contradiction may be resolved by suggesting that the material that erupts on the ocean floor may be depleted in noble metals by the prior (i.e. pre-eruption) removal of these elements in sulphides. The metal patterns from primary ocean floor magmas may in fact be less fractionated than those shown in Fig. 2(c) and primary ocean floor magmas may well resemble the metal patterns of high-MgO basalts.

Small possibly rift-related intrusions

Seiland Province rocks. The early Caledonide (500–550 Ma, Sturt & Roberts, 1978) Seiland Province intrusions range in composition from tholeiitic to alkaline in composition. These intrusions are thought to be related to the rifting phase of the Caledonide event (Pedersen *et al.* (submitted)).

Samples from three of these intrusions were analysed: Reinfjord, a tholeiitic intrusion; Melkvann, an alkaline intrusion and Lille Kufjord, an intrusion thought to represent alkaline magma that reequilibrated at high levels in the mantle to generate a late tholeiitic intrusion in the area where alkaline rocks predominate; Bennett *et al.* (1986) describe the Reinfjord and Melkvann intrusions in more detail. These intrusions present an opportunity to examine noble-metal patterns in rocks derived from partial melting of the mantle under various conditions.

The Reinfjord samples are wehrlites, pyroxenites and troctolites principally from dykes and the marginal zones of the intrusion. The Lille Kufjord samples are peridotites and pyroxenites. The samples from Melkvann are peridotites and pyroxenites principally from dykes. Approximately 0·5% S is present in all these rocks and the noble metal, Ni and Cu content of the rocks is probably controlled by these sulphides and the Ni content probably also depends on the cumulate olivine. Because the Reinfjord and Melkvann samples are mainly from dykes and the marginal zones of intrusions it is assumed that their metal patterns are similar to that of the magma which intruded. The metal patterns from all three intrusions are similar (Fig. 5(e)). The noble metals are relatively unfractionated with Pd/Ir ratios of 12–20. Ni is slightly enriched relative to Ir ($Ni/Ir_{mn} = 3–6$) and Cu is enriched relative to Pd ($Cu/Pd_{mn} = 3–6$). The level and shape of the noble metal portion of the metal patterns resembles that of high-MgO basalt (compare Figs 5(e) and 2(b)), but with Ni and Cu enriched. Similarly, on a plot of Pd/Ir versus Ni/Cu the Seiland intrusions plot in the field of high-MgO basalts (R, M and L on Fig. 3(b)), but on the Cu/Ir versus Ni/Pd plot the Seiland intrusion plots above the field of high-MgO basalts. The combination of high Ni/Pd and high Cu/Ir ratios suggests that the noble metals have been scavenged from the magma prior to the development of the present rocks, possibly by sulphides segregated from the magma earlier. The combination of low Pd/Ir ratios and high Ni/Cu ratios and high Ni/Pd and Cu/Ir ratios in the Seiland intrusions suggests that they formed from fairly primitive magmas that had experienced sulphide segregation prior to the development of these rocks.

The metal patterns for samples from the Reinfjord and Lille Kufjord Intrusions suggest that these rocks formed from magmas similar to high-MgO basalts and this is geologically reasonable since these intrusions are tholeiitic. However, the Melkvann intrusion is thought to have formed from an alkaline magma, and therefore, should be compared with a similar type of intrusion. There are, as yet, no noble-metal patterns available in the literature with which to compare it. Rocks of the alkaline family that have been analysed are: kimberlites from the USSR, which show similar PGE fractionation to the Melkvann samples, but at levels an order of magnitude higher (upper curve on Fig. 2(f)); and basanites (lower curve on Fig. 2(f)) from western Australia which also show a relatively unfractionated noble-metal pattern, but at levels slightly lower than the metal patterns from the Melkvann intrusion.

Despite the difference in partial melting conditions postulated for these three intrusions by Robins and Gardner (1975) their metal patterns are remarkably similar. This relationship confirms the conclusion, based on the rather limited data from the literature, that there is no obvious difference between the metal patterns from primitive alkaline rocks and those from high-MgO magmas (compare Figs 2(b) and 2(f)). Hence, primitive mantle melts whether tholeiitic or alkaline in composition have similar metal patterns.

Hosanger. The Hosanger intrusion is a sill of norite which may be part of the Anorthosite Complex within the Bergen Arc System (Boyd & Nixon, 1985). The sulphides occur either as matrix sulphides towards the base of the intrusion (Lien and Litvann ore) or as sulphide veins cutting the norite and gneissic country rock (Nonås ore). The ore from all three localities is similar, except for the presence of some Cu-rich ore from Nonas. In common with most Cu-rich ores the Nonas ore is enriched in Cu and Au relative to the bulk ore. Because of the analytical problems associated with Cu- and Au-rich samples it is not possible to say whether the Cu-rich ore is also enriched in Pt and Pd.

The noble metal portion of the metal patterns is relatively unfractionated Pd/Ir ratios 5–10. However, Cu and Ni are enriched relative to the noble metals ($Ni/Ir_{mn} = 4$, $Cu/Pd_{mn} = 5$) which gives the patterns their trough shape. The metal patterns resemble those of the Seiland intrusions and so are interpreted to have formed in the same fashion, that is by segregation from a high-MgO basalt that had not experienced olivine or chromite fractionation, but that had previously segregated some sulphides.

Ertelien and Flåt. The Ertelien deposit occurs at the margins of a small (600 m × 450 m) norite intrusion, located 40 km NW of Oslo (Fig. 6). The Flåt deposit occurs within a diorite dyke (4×2 km), 170 km SW of Oslo (Fig. 6). Neither intrusion has been dated, but on the basis of stratigraphic relationships they are thought to be part of a series of mafic intrusions emplaced into the gneissic craton between 1200 and 1370 Ma, possibly as the early rifting phase of Sveconorwegian Orogeny (900–1100 Ma) (Oftedahl, 1980). More details concerning the deposits are given in Boyd & Nixon (1985).

Both Ni-rich and Cu-rich ore types are present at Flåt. The Cu-rich ore is part of the massive sulphides. In common with most Cu-rich ores the Flåt Cu-rich ores are enriched in Cu, Au and Pd relative to bulk ore. The Pd value may not be reliable, because it was possible to determine Pd on only 2 of the 5 Cu-rich samples. The Flåt PGE are remarkably unfractionated (Pd/Ir = 2–5). However, as with the Seiland Province rocks, the Flåt patterns are enriched in Ni and Cu relative to the noble metals (Ni/Ir_{mn} = 1–2, Cu/Pd_{mn} = 4–132) which gives the metal patterns an overall trough shape. On the Ni/Pd versus Cu/Ir plot (Fig. 4(b)) the Flåt ore (f) plots above the fields of the extrusive rocks which indicates some sulphides were removed prior to the formation of the Flåt sulphides. The Flåt metal pattern does not resemble the metal patterns from any common magma type (compare Figs 2 (a–h) and 5(h)). The distinctive feature of the Flåt metal pattern is its low Pd/Ir ratio, this ratio is even lower than that found in komatiite magmas. On the Pd/Ir versus Ni/Cu diagrams (Fig. 3(b)) the position of the Flåt ore (f) suggests that the Flåt samples are enriched in chromite and/or olivine. But, the Flåt samples are massive sulphides which do not contain any olivine or chromite. The reason for the low Pd/Ir ratio is not clearly understood. Overall the Flåt metal pattern suggests that the Flåt sulphides formed in equilibrium with a magma that had not experienced olivine or chromite fractionation but that had experienced sulphide segregation.

At Ertelien both Ni-rich and Cu-rich ores are present: the Cu-rich ore is part of the massive sulphides and is enriched in Cu, Au and Pt and depleted in Ir relative to the bulk ore. The noble metal portion of the metal patterns is fractionated (Pd/Ir = 100–120) (Fig. 5(i)). Both ore types are enriched in Ni and Cu relative to noble metals (Ni/Ir_{mn} = 20–25, Cu/Pd_{mn} = 35–120) such that the metal patterns have a trough shape, similar to that observed in Seiland. The Ni/Cu ratio of the weighted mean of the Ertelien sulphides is higher than that of flood basalt related material but, on the other hand, the Pd/Ir ratio is higher than the ocean floor basalts, thus the Ertelien ore plots in neither field (I on Fig. 3(b)). The position of the Ertelien sulphides suggests that they formed from a primitive magma that had crystallized some olivine and chromite prior to the development of these sulphides. The trough shape of the metal pattern suggests that sulphides have been removed from this magma. This suggestion is reinforced by the position of the Ertelien ore (I) on the Ni/Pd versus Cu/Ir plot (Fig. 4(b)) where the Ertelien ore plots above the fields of extrusive rocks, suggesting that sulphides have been removed from the magma.

A remarkable feature of all these intrusions is the depletion of noble metals relative to Ni and Cu. This is interpreted as suggesting that sulphides have been removed from the magma, either by retention in the mantle during partial melting, or by segregation from the magma en route to surface. The authors favour the latter model and suggest that during early rifting, magmas may not have easy access to the surface and may pause often during emplacement into the crust; at each pause it is possible that sulphides could segregate from the magma. Another feature that these intrusions have in common is their small size. It is possible that small volumes of magma have difficulty intruding into the crust and therefore pause often allowing, at each pause, segregation of sulphides.

The Jotun Complex—a Large Mid-Proterozoic Mafic Intrusion
The Jotun Complex in southern Norway is a large (100×200 km) mafic intrusion of mid-Proterozoic age. It was metamorphosed to granulite facies between 900 and 1100 Ma and implaced in its present position during the Caledonide Orogeny (400 Ma) (Oftedahl, 1980). The size, age and mafic nature of this intrusion make it a target for a Pt-reef-type deposit. The target zone for finding a Pt-reef would usually be considered to lie near the boundary between the ultramafic and mafic zones of the intrusion. However, the polydeformed and polymetamorphosed nature of the Jotun Complex make it difficult to define the stratigraphy of the body, and hence, difficult to define any target zones. Magnetite-rich pyroxenites occur as small lenses 10–100 m across in the metagabbros of the north east portion of the intrusion. Fourteen samples of magnetite-rich pyroxenite from the area of Gråsulbekken have been analysed for noble metals (Table 1).

The noble metals from the pyroxenites are extremely fractionated (Pd/Ir ratio 770), but the pattern is enriched in PGE relative to Cu. The overall shape of the pattern most resembles that of flood basalts (compare Figs 2(f) and 5(f)). On the metal ratio diagrams the Jotun pyroxenites (J on Figs 3(b) and 4(b)) plot close to, or within, the field of flood basalts. The high Pd/Ir ratio of the Jotun pyroxenite rocks suggests that the liquid from which they formed was depleted in Ir, possibly by earlier chromite removal. Although the level (200 ppb) of PGE present in these rocks is an order of magnitude less than ore grade, the low Ni/Pd and Cu/Ir ratios suggest that no sulphides have segregated from the magma, hence the noble metals have not been scavenged from the magma and any sulphides that later segregated could be rich in noble metals, indicating that the Jotun Complex warrants further exploration.

CONCLUSIONS

By adding Ni to the Os end of a noble metal pattern and Cu to the Au end and then mantle normalizing the data rather than chondrite normalizing, the resulting metal patterns become powerful tools for investigating the following petrological problems in ultramafic and mafic rocks.

(1) If the rock represents mantle, a chromitite from an ophiolite, a Pt-reef, or if the rock formed from an unfractionated mantle derived melt, the metal patterns are distinctive and the origin of the rock can be established.

(2) Because the partition coefficient of Ni and Cu is lower than the noble metals into sulphides the separation of sulphides from a magma will produce a cumulate enriched in noble metals relative to Ni and Cu, leaving the magma

depleted in noble elements relative to Ni and Cu. Consequently, the cumulate will have arch-shaped metal patterns and the fractionated magma trough-shaped metal patterns. Any rock that subsequently forms from this fractionated magma will also have trough-shaped metal patterns.

(3) Chromite and, to a lesser extent, olivine both tend to concentrate Os, Ir and Ru, possibly as PGM inclusions, therefore the crystallization of chromite and olivine from a magma produces a cumulate enriched in Os, Ir and Ru and a magma depleted in Os, Ir and Ru. Any rock that subsequently forms from this fractionated magma will have metal patterns with steeper positive slopes than the original liquid (i.e. higher Pd/Ir ratios).

Because some workers may feel that the mantle abundance of the noble metals are poorly constrained they may object to the mantle normalization procedure to produce metal patterns. Furthermore, the presence of sulphides in a rock presents the dilemma of whether to recalculate the data to 100% sulphides or not. Both of these difficulties can be overcome by using the metal ratio diagrams Pd/Ir versus Ni/Cu and Ni/Pd versus Cu/Ir. These diagrams successfully separate rocks representing mantle, komatiites, high-MgO basalts, flood basalts, Pt-reefs, ophiolite chromitites and Cu-rich sulphide veins. The diagrams show the effects of Os, Ir and Ru removal by crystal fractionation. The Ni/Pd versus Cu/Ir diagram also shows the effects of sulphide removal as a trend markedly oblique to that due to olivine and chromite trends.

The combined use of mantle normalized metal patterns and metal ratio diagrams allows the effects of partial melting, sulphide segregation, chromite and olivine crystallization to be clearly recognized.

ACKNOWLEDGEMENTS

The analytical work for this project was financed by the EEC Raw Materials Program and Norges Geologiske Undersøkelse. Professor Frank Vokes kindly allowed access to the petrographic collection at the Technical University of Norway (NTH), Trondheim for samples from Ertelien, Faeøy, Flåt and Hosanger. Dr E. W. Sawyer is thanked for numerous corrections and suggestions to the various drafts of the text.

REFERENCES

Agiorgitis, G. & Wolf, R. (1977). The distribution of platinum, palladium and gold in Greek chromites. *Chem. Erde.*, **36**, 349–51.

Agiorgitis, G. & Wolf, R. (1978). Aspects of osmium, ruthenium and iridium contents in some Greek chromites. *Chem. Geol.*, **23**, 267–72.

Alapieti, T. & Lahtinen, J. (1986). Stratigraphy, petrology and platinum-group element mineralization of the early Proterozoic Penikat layered intrusion. *Econ. Geol.*, **81**, 1126–37.

Amosse, J., Allibert, M., Fischer, W. & Piboule, M. (1987). Etude de l'influence es fugacities d'oxygene et de soufre sur la differenciation des platinoides dans les magmas ultramafiques. Resultats preliminaires. *C.R. Acad. Sci. Paris*, t304 Seri II, **19**, 1183–5.

Ballhaus, G. G. & Stumpfl, E. F. (1986). Sulfide and platinum mineralization in the Merensky Reef: evidence from hydrous silicates and fluid inclusions. *Contrib. Mineral. Petrol.*, **94**, 193–204.

Barnes, Sarah-Jane (1987). Unusual nickel and copper to noble-metal ratios from the Rána Layered Intrusion, northern Norway. *Norsk geol. Tidsskrift*, **67**, 215–32.

Barnes, Sarah-Jane & Naldrett, A. J. (1987). Fractionation of the platinum-group elements concentrations in the Alexo Mine komatiite, Abitibi Greenstone Belt, northern Ontario. *Geol. Mag.*, **123**, 515–24.

Barnes, Sarah-Jane & Naldrett, A. J. (1987). Fractionation of the platinum-group elements and gold in some komatiites of the Abitibi greenstone belt, northern Ontario. *Econ. Geol.*, **82**, 165–83.

Barnes, Sarah-Jane, Naldrett, A. J. & Gorton, M. P. (1985). The origin of the fractionation of platinum-group elements in terrestrial magmas. *Chem. Geol.*, **53**, 303–23.

Barnes, Stephen J. & Naldrett, A. J. (1985). Geochemistry of the J-M reef of the Stillwater Complex, Minneapolis Adit Area. I Sulfide chemistry and sulfide-olivine equilibrium. *Econ. Geol.*, **80**, 627–45.

Barnes, Stephen J., Coats, C. A. J. & Naldrett, A. J. (1982). Petrogenesis of Proterozoic nickel sulfide–komatiite association: the Katiniq Sill, Ungava, Quebec. *Econ. Geol.*, **77**, 413–29.

Becker, R. & Agiorgitis, D. (1978). Iridium, osmium and palladium distribution in rocks of the Troodos Complex, Cyprus. *Chem. Erde.*, **37**, 302–6.

Bennett, M. C., Emblin, S. R., Robins, B. & Yeo, W. J. A. (1986). High-temperature ultramafic complexes in the north Norwegian Caledonides: I—Regional setting and field relationships. *Norges. Geol. Unders. Bull.*, **405**, 1–40.

Boyd, R. & Nixon, F. (1985). Norwegian nickel deposits: A review. *Geol. Surv. Finland, Bull.*, **333**, 364–94.

Brugmann, G. E., Arndt, N. T., Hofmann, A. W. & Tobschall, H.-J. (1985). Precious-metal abundances in komatiites and komatiitic basalts: implications for the genesis of PGE-bearing magmatic sulfide deposits. *Can. Mineral.*, **23**, 293–321.

Bury, K. V. (1975). *Statistical Models in Applied Sciences*. John Wiley, New York, 625 pp.

Campbell, I. H. & Barnes, Stephen J. (1984). A model for the geochemistry of the platinum-group elements in magmatic sulfide deposits. *Can. Mineral.*, **22**, 151–60.

Campbell, I. H. & Naldrett, A. J. (1979). The influence of silicate : sulfide ratio on the geochemistry of the magmatic sulfides. *Econ. Geol.*, **74**, 1503–5.

Chang, P.-K., Yu, C. M. & Chiang, C. Y. (1973). Mineralogy and occurrence of platinum-group elements in chromium deposits of northwest China. *Geochimica*, **2**, 76–85 (in Chinese).

Chou, C. L., Shaw, D. M. & Crocket, J. H. (1983). Siderophile trace elements in the earth's oceanic crust and upper mantle. *J. Geophys. Res.*, **88**, A507-518.

Clark, T. (1985). Precious metals in New Quebec. Ministere de l'Energie et des Res Quebec. Document de promotion.

Cowden, A., Donaldson, M. J., Naldrett, A. J. & Campbell, I. H. (1985). Platinum-group elements in komatiite-hosted Fe–Ni–Cu sulfide deposits at Kambalda, western Australia. *Econ. Geol.*, **81**, 1226–34.

Crocket, J. H. (1981). Geochemistry of the platinum-group elements. *Can. Inst. Min. Metall.*, *Spec. Iss.*, **23**, 47–64.

Crocket, J. H. & Chyi, L. L. (1972). Abundances of Pd, Ir, Os and Au in an alpine ultramafic pluton. *24th Inter. Geol. Cong.*, Section **10**, 202–9.

Crocket, J. H. & MacRae, W. E. (1986). Platinum-group element distribution in komatiitic and tholeiitic volcanic rocks from Munro Township, Ontario. *Econ. Geol.*, **81**, 1242–52.

Crocket, J. H. & Teruta, Y. (1977). Palladium, iridium and gold contents of mafic and ultramafic rocks drilled from the Mid-Atlantic Ridge, leg 37, Deep Sea Drilling Project. *Can. J. Earth Sci.*, **14**, 777–84.

Czamanske, G. K., Haffty, J. & Nabbs, S. W. (1981). Pt, Pd and Rh analyses and beneficiation of mineralized mafic rocks from the La Perouse Layered Gabbro, Alaska. *Econ. Geol.*, **76**, 2001–11.

Davies, G. & Tredoux, M. (1985). The platinum-group element and gold contents of the marginal rocks and sills of the Bushveld Complex. *Econ. Geol.*, **80**, 838–48.

Dillon-Leitch, H. C. H., Watkinson, D. H. & Coats, C. (1986). Distribution of platinum-group elements in the Donaldson West Deposit, Cape Smith Fold Belt, Quebec. *Econ. Geol.*, **81**, 1147–59.

Economou, M. I. & Naldrett, A. J. (1984). Sulfides associated with podiform bodies of chromite at Tsangli, Eretria, Greece. *Mineral. Deposita*, **19**, 289–97.

Fominykh, V. G. & Khvostova, V. P. (1970). Platinum content of Ural dunite. *Doklady Akad. Nauk. SSSR*, **191**, 443–5.

Gain, S. B. (1985). The geological setting of the platiniferous UG-2 chromitite layer on the farm Maandagshhoek, Eastern Bushveld. *Econ. Geol.*, **80**, 925–43.

Gain, S. B. & Mostert, A. B. (1982). The geological setting of the platinoid and base metal sulfide mineralization in the Platreef of the Bushveld Complex north of Potgietersrus. *Econ. Geol.*, **17**, 1395–1404.

Gee, D. G., Guezou, J.-C., Roberts, D. & Wolff, F. C. (1985). The central-southern part of the Scandinavian Caledonides. In *The Caledonide Orogen—Scandinavia and Related Areas*, ed. D. Gee & B. A. Sturt. John Wiley, New York, pp. 109–33.

Govindaraju, K. (1984). Compilation of working values and sample descriptions of 170 international reference samples of mainly silicate rocks and minerals. *Geostandards Newsletter*, **VIII**, 3–39.

Green, A. H. & Naldrett, A. J. (1981). The Langmuir volcanic peridotite-associated nickel deposits: Canadian equivalent of the Western Australian occurrences. *Econ. Geol.*, **76**, 1503–23.

Grønlie, A. (in press). Platinum-group minerals of the Lillefjellklumpen, nickel-copper deposit, Norway. *Norsk. Geol. Tidsskr.*

Gunn, A. G., Leake, R. C. & Styles, M. T. (1985). Platinum-group element mineralization in the Unst ophiolite, Shetland. Mineral Reconnaissance Programme Rep. British Geol. Surv., Vol. 73, 117 pp.

Hakli, T. A., Hanninen, E., Vuorelainen, Y. & Papunen, H. (1976). Platinum-group minerals in the Hitura nickel deposit, Finland. *Econ. Geol.*, **71**, 1206–13.

Hamlyn, P. R., Keays, R. R., Cameron, W. E., Warrington, E., Crawford, A. J. & Waldon, H. M. (1985). Precious metals in magnesian low-Ti lavas: Implications for metallogenesis and sulfur saturation in primary magmas. *Geochim. Cosmochim. Acta*, **49**, 1797–1811.

Hertogen, J., Janssen, M. J. & Palme, H. (1980). Trace elements in ocean ridge basalt glasses: implications for fractionation during mantle evolution and petrogenesis. *Geochim. Cosmochim. Acta*, **44**, 2125–43.

Hoffman, E. L., Naldrett, A. J., Van Loon, J. C. & Hancock, R. G. V. (1979). The noble-metal content of ore in the Levack West and Little Stobie mines, Ontario. *Can. Mineral.*, **17**, 437–51.

Hulbert, L. J. & von Gruenewaldt, G. (1982). Nickel, copper and platinum mineralization in the lower zone of the Bushveld Complex, south of Potgietererus. *Econ. Geol.*, **77**, 1296–1306.

Irivine, T. N. & Sharpe, M. (1982). Source rock compositions and depths of origin of Bushveld and Stillwater magmas. *Carnegie Inst. Washington Yrbk*, **81**, 294–303.

Jagoutz, E., Palme, H., Baddenhausen, H., Blum, K., Dreibus, G., Spettel, B., Lorenz, V. & Wanke, H. (1979). The abundance of major, minor and trace elements in the earth's mantle as derived from primitive ultramafic nodules. *Proc. 10th, Lunar Planet. Sci. Conf.*, NASA, Houston, pp. 2031–50.

Kaminskiy, F. V., Frantsesson, Ye. V. & Khvostova, P. (1975). First information on platinum metals (Pt, Pd, Rh, Ir, Ru, Os) in kimberlitic rocks. *Doklady Akad. Nauk. SSSR*, **219**, 190–3.

Keays, R. R. (1982). Palladium and iridium in komatiites and associated rocks: application to petrogenetic problems. In *Komatiites*, ed. N. T. Arndt & E. G. Nisbet. Allen, London, pp. 435–57.

Keays, R. R. & Campbell, I. H. (1981). Precious metals in the Jimberlana Intrusion, Western Australia: Implications for genesis of platiniferous ores in layered intrusions. *Econ. Geol.*, **76**, 1118–41.

Keays, R. R., Ross, J. R. & Woolrich, P. (1981). Precious metals in volcanic peridotite-associated nickel sulphide deposits in western Australia II: Distribution within the ores and host rocks at Kambalda. *Econ. Geol.*, **76**, 1645–74.

Keays, R. R., Nickel, E. H., Groves, D. I. & McGoldrick, P. J. (1982). Iridium and palladium as discriminants of volcanic-exhalative hydrothermal and magmatic nickel sulfide mineralization. *Econ. Geol.*, **77**, 1535–47.

Khvostova, V. P., Golvnya, S. V., Chernysheva, N. V. & Bukhanova, A. I. (1976). Distribution of the platinum-group metals in chromite ores and ultramafic rocks of the Ray-Iz massif (Polar Urals). *Geochem. Int.*, **13**, 35–9.

Krill, A. G., Bergh, S., Mearns, E. W., Often, M., Olerud, S., Olesen, O., Sanstad, J. S., Siedlecka, A. & Solli, A. (1985). New isotopic dates from the Precambrian crystalline rocks of Finnmark. *Norges Geol. Unders. Bull.*, **403**, 37–54.

Lahtinen, J. (1985). PGE-bearing copper-nickel occurrences in the marginal series of the early Proterozoic Koillismaa Layered Intrusion, northern Finland. *Geol. Surv. Finland, Bull.*, **333**, 165–79.

Latysh, I. K. & Buturlinov, V. N. (1970). Platinum-group metals in igneous rocks of the border zone between Donbas and Azov Region. *Geochem. Int.*, **7**, 620–3.

Leblanc, M. & Johan, Z. (1986). Un nouveau type de mineralisation platinifere: exemple des filons a arseniures de nickel et chromite du massif lherzolitique des Beni-Bousera (Maroc). *C.R. Acad. Sci. Paris*, t303, Serie II, **2**, 163–6.

Lee, C. & Tredoux, M. (1986). Platinum-group element abundances in the lower and middle critical zones of the Eastern Bushveld Complex. *Econ. Geol.*, **81**, 1087–96.

Lesher, C. M. & Keays, R. R. (1984). Metamorphically and hydrothermally mobilized Fe–Ni–Cu sulphides at Kambalda, Western Australia. In *Sulphide Deposits in Mafic and Ultramafic rocks*, ed. D. L. Buchanan & M. J. Jones. Institute of Mining and Metallurgy, London, pp. 62–9.

Lightfoot, P. C., Naldrett, A. J. & Hawkesworth, C. J. (1984). The geology and geochemistry of the Waterfall Gorge section of the Insizwa Complex with particular reference to the origin of the nickel-sulfide deposits. *Econ. Geol.*, **79**, 1857–79.

McCallum, M. E., Lovacks, R. R., Carlson, R. R., Cooley, E. F. & Docrye, T. A. (1976). Platinum metals associated with hydrothermal copper ores of the New Rambler mine, Medicine Bow Mountains, Wyoming. *Econ. Geol.*, **71**, 1429–50.

McLaren, C. H. & De Villiers, J. P. R. (1982). The platinum-group chemistry and mineralogy of the UG-2 chromite layer of the Bushveld Complex. *Econ. Geol.*, **77**, 1348–66.

Massey, N. W. D., Crocket, J. H. & Kabir, A. (1983). Ir in Keweenawan basalts of the Mamainse Point Formation, Ontario. *Can. Mineral.*, **21**, 655–60.

Mitchell, R. H. & Keays, R. R. (1981). Abundance and distribution of gold, palladium and iridium in some spinel and garnet lherzolites, Implications for the nature and origin of precious metal-rich intergranular components in the upper mantle. *Geochim. Cosmochim. Acta*, **45**, 2425–42.

Morgan, J. (1986). Ultramafic xenoliths: Clues to earth's late accretionary history. *J. Geophys. Res.*, **91** (B12), 12375–87.

Morgan, J. W., Wanderless, G. A., Petrie, R. K. & Irving, A. J. (1981). Composition of the earth's upper mantle—I. Siderophile trace elements in ultramafic nodules. *Tectonophysics*, **75**, 47–67.

Naldrett, A. J. (1981). Platinum-group element deposits. *Can. Inst. Min. Metall., Spec. Iss.*, **23**, 197–232.

Naldrett, A. J. & Barnes, S.-J. (1986). The fractionation of the platinum-group elements with special reference to the composition of sulphide ores. *Fortsch. Mineral. Petrol.*, **64**, 113–33.

Naldrett, A. J., Hoffman, E. L., Green, A. H., Chou, C-L. & Naldrett, S. R. (1979). The composition of Ni-sulfide ores, with particular reference to their content of PGE and Au. *Can. Mineral.*, **17**, 403–15.

Naldrett, A. J., Innes, D., Gorton, M. P. & Sowa, J. (1982). Compositional variations within and between five Sudbury ore deposits. *Econ. Geol.*, **77**, 1519–34.

Nilsson, L.-P. (1980). Undersøkelsel av ultramafiske bergarter og krommalm på strekeningen Råros-Feragen. Norges Geol. Unders. Rep. 1650/33A.

Oftedahl, C. (1980). The geology of Norway. *Norges Geol. Unders. Bull.*, **54**, 3–114.

Often, M. (1985). The early proterozoic Karasjok greenstone belt, Norway; a preliminary description of lithology, stratigraphy and mineralization. *Norges Geol. Unders. Bull.*, **403**, 75–88.

Ohnenstetter, D., Watkinson, D., Jones, P. C. & Talkington, R. (1986). Cryptic compositional variations in laurite and enclosing chromite from the Bird River Sill, Manitoba. *Econ. Geol.*, **81**, 1159–68.

Oshin, I. O. & Crocket, J. H. (1982). Noble metals in the Thetford mines ophiolite, Quebec, Canada. Part I Distribution of gold, iridium, platinum and palladium in the ultramafic and gabbroic rocks. *Econ. Geol.*, **77**, 1556–70.

Oshin, I. O. & Crocket, J. H. (1986). Noble metals in the Thetford mines ophiolite, Quebec, Canada. Part II Distribution of gold, silver, iridium, platinum and palladium in the Lac de l'Est volcano-sedimentary section. *Econ. Geol.*, **81**, 931–45.

Page, N. J & Talkington, R. W. (1984). Palladium, platinum, rhodium, ruthenium and iridium in peridotites and chromites from ophiolite complexes in Newfoundland. *Can. Mineral.*, **22**, 137–49.

Page, N. J, Rowe, J. J. & Haffty, J. (1972). Platinum metals lateral variations of platinum, palladium and rhodium in the Stillwater Complex, Montana. *Econ. Geol.*, **67**, 915–23.

Page, N. J, Myers, J. S. & Haffty, J. (1980). Platinum, palladium and rhodium in Fiskenaesset Complex, Southwest Greenland. *Econ. Geol.*, **75**, 907–15.

Page, N. J, Cassard, D. & Haffty, J. (1982a). Palladium, platinum, rhodium, ruthenium and iridium in chromitites from the Massif du Sud and Tiebaghi Massif, New Caledonia. *Econ. Geol.*, **77**, 1571–7.

Page, N. J, Pallister, J. S., Brown, M. A., Smewing, J. D. & Haffty, J. (1982b). Palladium, platinum, rhodium, iridium and ruthenium in chromite-rich rocks from the Samail ophiolite, Oman. *Can. Mineral.*, **20**, 537–48.

Page, N. J, Von Gruenewaldt, G., Haffty, J. & Aruscavage, P. J. (1982c). Comparison of platinum, palladium and rhodium distribution in some layered intrusions with special reference to the late differentiates (upper zone) of the Bushveld Complex, South Africa. *Econ Geol.*, **77**, 1405–18.

Page, N. J, Aruscavage, P. J. & Haffty, J. (1983). Platinum-group elements in rocks from the Voikar-Syninsky Ophiolite Complex, Polar Urals USSR. *Mineral. Deposita*, **18**, 444–55.

Page, N. J, Zientek, M. L., Czamanske, G. K. & Foose, M. P. (1985). Sulfide mineralization in the Stillwater complex and underlying rocks. Montana Bureau Mines and Geol. Spec. Pub. **92**, pp. 93–6.

Page, N. J, Singer, D., Moring, B., Carlson, A., McDade, J. & Wilson, S. A. (1986). Platinum-group element resources in podiform chromitites from California and Oregon. *Econ. Geol.*, **81**, 1261–71.

Papunen, H. (1986). Platinum-group elements in Svecokarelian nickel-copper deposits in Finland. *Econ. Geol.*, **81**, 1236–41.

Papunen, H. & Koskinen, J. (1985). Geology of the Kotalahti nickel-copper ore. *Geol. Surv. Finland Bull.*, **333**, 229–41.

Paul, D., Crocket, J. H. & Nixon, P. H. (1979). Abundances of palladium, iridium and gold in kimberlites and associated nodules. *Proc. 2nd Int. Kimberlite Conf.*, Vol. 1. American Geophysical Union. Washington, D.C., pp. 272–9.

Pedersen, R.-B. (in preparation) The Karmøy Ophiolite Complex.

Pedersen, R.-B., Furnes, H. & Dunning, G. (submitted). Norwegian ophiolite complexes reconsidered. *Geology.*

Prichard, H. M., Potts, P. J. & Neary, C. R. (1981). Platinum-group element minerals in the Unst chromite, Shetland Isles. *Inst. Mining Metall.*, **90**, 186–8.

Prichard, H. M., Potts, P. J. & Neary, C. R. (1984). Platinum and gold in the Shetland ophiolite. *Mining J.*, **303** (7772), 77.

Prichard, H. M., Neary, C. & Potts, P. J. (1986). Platinum-group minerals in the Shetland ophiolite. In *Metallogeny of Basic and Ultrabasic Rocks*, ed. M. J. Gallagher, R. A. Ixer, C. R. Neary & H. M. Prichard. Inst. Min. Metals, London, pp. 395–414.

Rajamani, V. & Naldrett, A. J. (1978). Partitioning of Fe, Co, Ni and Cu between sulfide liquid and basaltic melts and the composition of Ni-Cu sulfide deposits. *Econ. Geol.*, **73**, 82–93.

Razin, L. V. & Khomenko, G. A. (1969). Accumulation of osmium, ruthenium and other platinum group metals in chrome spinel in platinum bearing dunites. *Geochem. Int.*, **6**, 546–57.

Razin, L. V., Khvostov, V. J. & Noviko, V. A. (1965). Platinum metals in the essential and accessory minerals of ultramafic rocks. *Geochem. Int.*, **2**, 118–31.

Redman, B. A. & Keays, R. R. (1985). Archaean basic volcanism in the eastern goldfields province, Yilgarn Block, Western Australia. *Precambrian Res.*, **30**, 113–52.

Robins, B. & Gardner, P. M. (1975). The magmatic evolution of the Seiland province and Caledonian plate boundaries in northern Norway. *Earth Planet. Sci. Lett.*, **26**, 167–78.

Rowell, W. F. & Edgar, A. D. (1986). Platinum-group element mineralization in a hydro-thermal Cu-Ni sulfide occurrence, Rathbun Lake, Northeastern Ontario. *Econ. Geol.*, **81**, 1272–7.

Sharpe, M. (1982). Noble metals in the marginal rocks of the Bushveld Complex. *Econ. Geol.*, **77**, 1286–95.

Smirnov, M. E. (1966). The structure of Noril'sk nickel-bearing intrusions and the genetic types of their sulfide ores. All-Union Sci. Res. Inst. Mineral. Raw Materials (VIMS) Moscow (in Russian).

Steele, T. W., Levin, J. & Copelowitz, I. (1975). The preparation and certification of a reference sample of a precious metal ore. Nat. Inst. Met. Rep. 1696.

Stockman, H. W. (1982). Noble metals in the Ronda and Josephine peridotites. Unpublished, Ph.D., thesis, Massachusetts Institute of Technology, Cambridge, MA.

Stockman, H. W. & Hlava, P. F. (1984). Platinum-group minerals in Alpine chromites from southwestern Oregon. *Econ. Geol.*, **79**, 491–508.

Sturt, B. & Roberts, D. (1978). Caledonides of northern Norway. In Caledonian-Appalachian orogen of the North Atlantic Region, ed. P. Scheuk. Geol. Surv. Can. Paper 78-13, pp. 17–24.

Sun, Shen-Su (1982). Chemical composition and origin of the earth's primitive mantle. *Geochim. Cosmochim. Acta*, **46**, 179–92.

Talkington, R. W. & Lipin, B. (1986). Platinum-group minerals in chromite seams of the Stillwater Complex, Montana. *Econ. Geol.*, **81**, 1179–86.

Talkingon, R. W. & Watkinson, D. H. (1984). Trends in the distribution of the precious metals in the Lac des Iles Complex, northwestern Ontario. *Can. Mineral.*, **22**, 125–36.

Talkington, R. W., Watkinson, D. H., Whittaker, P. J. & Jones, P. C. (1984). Platinum-group minerals and other solid inclusions in chromite of ophiolite complexes: Occurrence and petrological significance. *Tschermaks. Mineral. Petrol. Mitt.*, **32**, 285–301.

Thompson, J. F. H., Nixon, F. & Siversten, R. (1980). The geology of Vakkerlien nickel prospect Kvikne, Norway. *Geol. Surv. Finland Bull.*, **52**, 3–22.

Tolmachev, I. I., Kalinin, S. K., Terekhovich, S. L., Syromyatnikon, N. G. & Zaravnyayeva, V. K. (1971). Distribution of Pd, Pt and Au in the Riphean and Cambrian igneous rocks of the Boshchekulsh region, northeastern Kazakhstan. *Geochem. Int.*, **8**, 194–9.

Tredoux, M., Davies, G., Lindsay, N. M. & Sellschop, J. P. F. (1986). The influence of temperature on the geochemistry of the platinum-group elements and gold. Geocongress 86 extended abstracts, pp. 625–8. Geological Society of South Africa, Johannesburg.

White, R. W., Motta, J. & De Araujo, V. A. (1971). Platiniferous chromitite in the Trocantins Complex, Niquelandia, Goias, Brazil. US Geol. Surv. Prof. Pap., 750-D, pp. 26–33.

Wilson, A. H. & Prendergast, M. D. (1987). The Great Dyke of Zimbabwe—an Overview. Guidebook for the 5th Magmatic Sulphides Field Trip pp. 23–8. Geological Society of Zimbabwe, Harare.

Zientek, M. L., Foose, M. P. & Mei, L. (1986). Palladium, platinum and rhodium contents of rocks near the lower margin of the Stillwater Complex, Montana. *Econ. Geol.*, **81**, 1169–78.

APPENDIX: SOURCES OF DATA FOR FIGS 2, 3 AND 4

Mantle: 1, 2, 3, 4, 6, 7, 8, 9, 10, 18, 20, 21.
Komatiites: 31, 33, 38, 52, 54, 64, 65, 76.
High-MgO basalts: 32, 33, 38, 55.
Ocean floor basalts: 1, 9, 16, 22, 66.
Continental flood basalts: 3, 6, 37, 38, 41, 58, 59, 60.
Boninites and low-TiO_2 basalts: 24, 66.
Alkaline rocks: 2, 3, 5, 25, 27.
Ophiolites: 6, 7, 8, 9, 10, 11, 12, 14, 15, 16, 17, 18, 19, 21, 63.
Layered intrusions: 34, 36, 38, 39, 41, 42, 44, 45, 46, 49, 50, 51, 53, 61, 62, 76, 77, 78.
Platinum reefs: 34, 38, 39, 40, 43, 47, 48.
Cu-rich sulphide veins: 68, 69, 71, 72, 75.

Refs: 1. Morgan *et al.* (1981); 2. Mitchell and Keays (1981); 3. Paul *et al.* (1979); 4. Jagoutz *et al.* (1979); 5. Kaminskiy *et al.* (1975); 6. Stockman (1982); 7. Crocket and Chyi (1972); 8. Oshin and Crocket (1982); 9. Becker and Agiorgitis (1978); 10. Page and Talkington (1984); 11. Page *et al.* (1982*b*); 12. Economou and Naldrett (1984); 13. Page *et al.* (1982*a*); 14. Agiorgitis and Wolf (1977, 1978); 15. Gunn *et al.* (1985); 16. Crocket and Teruta (1977); 17. White *et al.* (1971); 18. Khvostova *et al.* (1976); 19. Chang *et al.* (1973); 20. Fominykh and Khvostova (1970); 21. Page *et al.* (1983); 22. Hertogen *et al.* (1980); 23. Crocket (1981); 24. Hamlyn *et al.* (1985); 25. Tolmachev *et al.* (1971); 27. Latysh and Buturlinov (1970); 29. Razin and Khomenko (1969); 30. Razin *et al.* (1965); 31. Keays *et al.* (1982); 32. Redman and Keays (1985); 33. Crocket and MacRae (1986); 34. Keays and Campbell (1981); 35. Page *et al.* (1986); 36. Talkington and Watkinson (1984); 37. Massey *et al.* (1983); 38. Naldrett (1981); 39. Page *et al.* (1985); 40. Barnes and Naldrett (1985); 41. Sharpe (1982); 42. Hulbert and von Gruenewaldt (1982); 43. McLaren and de Villiers (1982); 44. Gain and Mostert (1982); 45. Page *et al.* (1982*c*); 46. Davies and Tredoux (1985); 47. Gain (1985); 48. Steele *et al.* (1975); 49. Page *et al.* (1980); 50. Czamanske *et al.* (1981); 51. Lahtinen (1985); 52. Hakli *et al.* (1976); 53. Papunen and Koskinen (1985); 54. Green and Naldrett (1981); 55. Barnes *et al.* (1982); 58. Smirnov (1966); 59. Lightfoot *et al.* (1984); 61. Thompson *et al.* (1980); 62. Clark (1985); 63. Leblanc and Johan (1986); 64. Barnes and Naldrett (1987); 65. Cowden *et al.* (1985); 66. Oshin and Crocket (1986); 68. McCallum *et al.* (1976); 69. Rowell and Edgar (1986); 71. Hoffman *et al.* (1979); 72. Lesher and Keays (1984); 75. Dillon-Leitch *et al.* (1986); 76. Papunen (1986); 77. Talkington and Lipin (1986); 78. Zientek *et al.* (1986); 79. Govindaraju (1984).

13

Noble Metal Geochemistry of some Ni-Cu Deposits in the Sveconorwegian and Caledonian Orogens in Norway

R. Boyd,[a] S.-J. Barnes[b] & A. Grønlie[a]

[a] *Geological Survey of Norway, PO Box 3006, N-7002 Trondheim, Norway*

[b] *Centre d'Etudes sur les Ressources Minérales, Université du Quebec, 555 Boulevard de l'Université, Chicoutimi, PQ, Canada G7H 2B1*

ABSTRACT

New noble metal analyses are presented for five Ni-Cu deposits in Norway and are considered along with published data from a further four deposits. By far the richest deposits are: Fæøy averaging 1790 ppb Pt and 4760 ppb Pd, and Lillefjellklumpen averaging 1800 ppb Pt and 3070 ppb Pd. Both are small massive sulphide bodies, the former associated with late stage dykes within the Karmøy ophiolite while the latter is associated with MORB-type tholeiites within the Gjersvik island arc complex. The Vakkerlien and Espedalen deposits contain respectively 1380 and 1000 ppb (PGE + Au) in 100% sulphides. Other deposits have low contents of PGE (<300 ppb in 100% sulphides) but several are locally enriched in Au, particularly Ertelien, in which Cu-rich samples average 1430 ppb Au. Data from the Norwegian deposits and from elsewhere indicate that Ni-Cu deposits in intrusions in orogenic belts tend to have low noble metal contents.

INTRODUCTION

Norway has a long tradition in mining and prospecting for Ni-Cu ores, aspects of which are reviewed in a recent paper (Boyd & Nixon, 1985). The mineral pentlandite was first described, though not named, from the Espedalen deposit (Scheerer, 1845) and for a brief period before the discovery of the lateritic ores in New Caledonia, Norway was one of the world's leading Ni producers. The last mines to be active, Flåt and Hosanger, closed in 1944 and 1945 respectively. Prospecting in the 1970s revealed one new deposit of significance, Vakkerlien (Thompson et al., 1980), and multiplied the reserves of another, Bruvann, tenfold (Boyd & Mathiesen, 1979). Pt and Pd analyses with detection levels of the order of 20–100 ppb were published for four Norwegian Ni-Cu deposits as long ago as 1932 Foslie & Johnson Høst, 1932) along with calculated grades for two deposits based on analyses of matte.

Knowledge of the platinum-group element (PGE) content of Norwegian orebodies did not advance, however, between 1932 and 1979 when the first modern PGE analyses, with detection levels of the order of 0·1–5 ppb, were published, for the Espedalen deposit, by Naldrett et al. (1979). Data have subsequently been published for the Vakkerlien deposit (Thompson et al., 1980), for the Lillefjell-klumpen deposit (Grønlie, 1985, 1986, 1987) and for the Bruvann deposit (Boyd et al., 1986, in press). This paper presents data from a further five deposits and will consider these with the results already published. These nine deposits include all the deposits in the country known to have (had) a tonnage of over 1000 metric tons Ni metal.

All the deposits except one, Fæøy, are Ni-dominated. Three, Hosanger, Flåt and Ertelien, while being completely dominated by Ni-Cu ore, also contain veins of Cu >> Ni mineralization which will be discussed separately.

GEOLOGICAL BACKGROUND

The location of the deposits is shown, on a simplified geological map of Norway (Fig. 1) and grades, tonnages, mineralogical and chemical data, are given in Table 1.

Two of the deposits considered in this paper are located in the Sveconorwegian (=Grenville) orogenic belt. The Flåt and Ertelien deposits occur in small plug-like intrusions (<5 km^2 in section) thought to immediately predate the Sveconorwegian orogeny. The other seven occur in the Caledonian orogenic belt, though Hosanger and Espedalen are pre-Sveconorwegian in age. Hosanger and Espedalen are associated with small intrusions, both of which probably originated within the Middle Proterozoic Jotun mafic intrusive complex, the largest preserved part of which is now exposed in the Jotun Nappe. The original dimensions of this intrusive complex must have been in excess of 300 km × 100 km implying a major magmatic event, probably associated with crustal extension. Of the deposits in the Caledonian belt, three, Bruvann, Skjækerdalen and Vakkerlien, can be described as synorogenic 'sensu lato' though at least for Bruvann and Vakkerlien, the term intra-orogenic, implying emplacement during a tensional episode, would be more precise while the remaining two, Lillefjellklumpen and Fæøy, are located in thrust slices of supra-subduction zone ophiolites (Dunning et al., 1986; Grønlie, in press).

Flåt. This deposit occurs within the Evje-Iveland amphibolite complex (Barth, 1947; Bjørlykke, 1947; Pedersen, 1975). It includes disseminated and massive mineralization and is Norway's second largest deposit.

Ertelien. This deposit is associated with a small noritic plug (600 m × 450 m). Mineralization includes massive, breccia and disseminated types near the margin of the plug (Johanssen, 1974).

Hosanger. This deposit is located within a noritic sill in the Anorthosite Complex, part of the Bergen Arc System, and possibly an outlier of the Jotun Nappe (Kvale, 1960). It experienced granulite facies metamorphism during the Sveconorwegian Orogeny (Sturt et al., 1975). The mineralization includes dis-

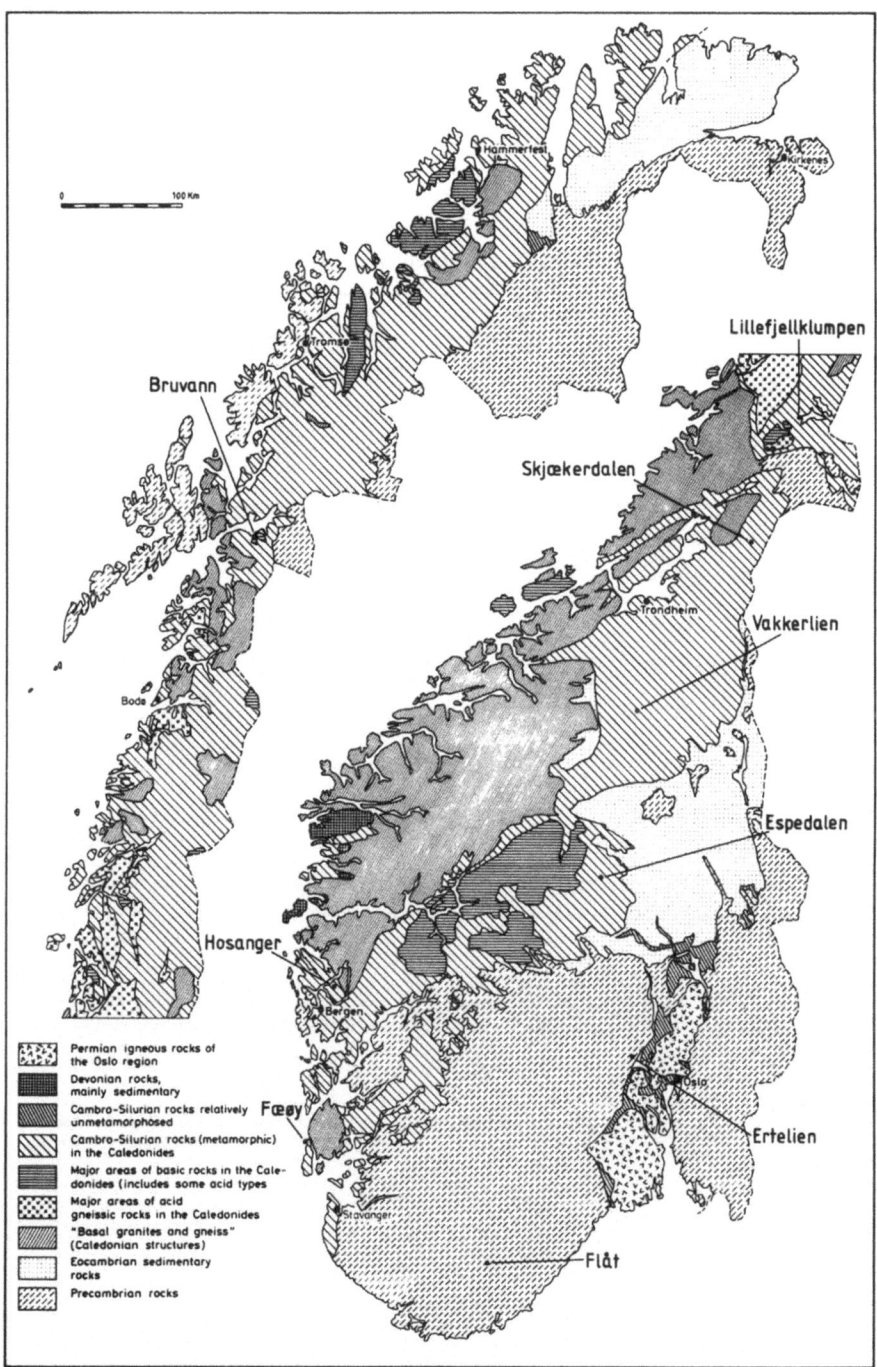

FIG. 1. Simplified geological map of Norway showing the location of the Ni-Cu deposits considered in this paper.

R. Boyd, S.-J. Barnes, A. Grønlie

TABLE 1

Tonnage, mineralogy and geochemical data for samples analysed for PGE

	Metric tons metal × 10³		Ave. grade %		Major sulphides	Trace minerals inc. known PGM		
	Ni	Cu	Ni	Cu			Ni	Cu
Bruvann (1)	142	34	0·32	0·08	po, pn, cp, py	Arsenopyrite gersdorffite niccolite molybdenite sphalerite No known PGM	9·0 (0·27–0·57)	2·1 (0·06–0·15)
Hosanger Ni-Cu	4·2	1·6	1·05	0·35	po, pn, cp, py	No data	3·2 (1·62–5·01)	0·4 (0·08–0·63)
Hosanger Cu-Ni	—	—	—	—		No data	0·6 (0·36–0·86)	7·7 (3·2–11·1)
Flåt Ni-Cu	19·5	12·2	0·75	0·47	py, po, pn, cp	Millerite violarite no known PGM	2·9 (0·27–6·06)	0·6 (0·06–2·71)
Flåt Cu-Ni	—	—	—	—		no data	0·8 (0·26–2·42)	7·2 (3·18–20·9)
Skjækerdalen	2·2	1·1	0·22	0·11	po, pn, cp	Bravoite linnaeite PGM not studied	3·1 (0·02–1·02)	2·7 (0·05–0·74)
Vakkerlien (2)	4	1·6	1·0	0·4	po, pn, cp, py	Gersdorffite violarite no known PGM	10·9 (n.a.)	2·2 (n.a.)
Espedalen (3)	1	0·4	1·0	0·4	po, pn, cp	Gersdorffite violarite no known PGM	6·6 (n.a.)	2·0 (n.a.)
Ertelien Ni-Cu	4·2	2·8	1·04	0·69	po, cp, py, pn	Gersdorffite violarite no known PGM	2·1 (0·5–3·47)	0·4 (0·09–1·44)
Ertelien Cu-Ni	—	—	—	—		Gersdorffite violarite no known PGM	0·7 (0·19–1·52)	11·8 (1·38–28·3)
Faeøy	0·8	1	2·1	2·63	po, cp, pn, py	Kotulskite temagamite sperrylite	1·9 (0·19–2·35)	4·2 (0·87–16·3)
Lillefjell-klumpen (4)	<0·1	<0·1	4·0	1·2	po, py, pn, cp	Merenskyite sperrylite moncheite temagamite electrum Ag-pentlandite	3·6 (2·53–4·74)	1·2 (0·30–2·85)

*Data unreliable because of analytical problems. n.a.: absolute values not available. Data taken from the following published sources: (1) Boyd et al. (1986); (2) Thompson et al. (1980); (3) Naldrett et al. (1979); (4) Grønlie (1985).

% Ni, Cu, ppb PGE + Au. Content in 100% sulphide
(range of absolute concentrations in parentheses)

Os	Ir	Ru	Rh	Pt	Pd	Au	Cu/Cu + Ni	Pd/Ir	n
—	6 (0·04–2·99)	—	—	52 (1·2–4·1)	70 (2–7)	1010 (1·0–55·5)	0·19	11·3	10
28 (10–37)	21 (6·6–28)	38 (<10–65)	7 (<2–18)	71 (29–120)	105 (36–185)	9 (3–14)	0·11	5·0	7
28 (17–47)	23 (<15–37)	59 (<20–110)	—ᵃ	—ᵃ	—ᵃ	342 (20–920)	0·93	—	3
23 (<2–38)	29 (0·6–54)	46 (<10–81)	11 (1–20)	68 (<10–150)	82 (21–174)	58 (7·3–290)	0·17	2·8	7
15 (<2–22)	19 (0·5–27)	53 (<10–68)	—ᵃ	—ᵃ	—ᵃ	650 (150–1 330)	0·90	—	4
—	—	—	—	100 (<3–7)	721 (<2–44)	536 (1–25)	0·47	—	7
39 (n.a.)	51 (n.a.)	62 (n.a.)	42 (n.a.)	788 (n.a.)	350 (n.a.)	49 (n.a.)	0·17	6·8	6
27 (n.a.)	26 (n.a.)	48 (n.a.)	36 (n.a.)	330 (n.a.)	250 (n.a.)	280 (n.a.)	0·23	9·6	6
7 (<3–32)	1·7 (0·3–4·1)	—	9 (<1–43)	273 (<10–2 400)	184 (75–400)	111 (2·2–430)	0·16	108	9
—	0·6 (<0·5–1·3)	—	10 (<2–33)	—ᵃ	—ᵃ	1 434 (78–4 000)	0·94	—	6
234 (31–490)	195 (6·4–410)	219 (72–480)	244 (23–440)	1 794 (640–5 400)	4 760 (2 600–8 000)	101 (6–310)	0·69	24	6
139 (23–250)	170 (14–290)	189 (<20–330)	214 (45–330)	1 799 (<100–5 900)	3 068 (410–4 700)	219 (8–1 910)	0·25	18	8

seminated to matrix sulphide near the base of the intrusion and veins of both Ni-
and Cu-rich sulphide (Bjørlykke, 1949). The deposit was mined intermittently from
1883 to 1945. Modern PGE data are presented here for the first time.

Espedalen. The Espedalen deposit is located in an ultramafic-mafic intrusion, one
of the second of three suites of mafic intrusives, in an outlier of the Jotun Nappe
(Heim, 1981). It includes disseminated and breccia mineralization. PGE data were
published by Naldrett *et al.* (1979).

Bruvann. This is Norway's largest Ni-Cu deposit and one of the largest Ni-Cu
sulphide deposits in Europe, the Soviet Union excluded. It is a predominantly
low-grade disseminated mineralization (Table 1), intercumulus to olivine and
orthopyroxene in ultramafic cumulates in the northwestern part of the Råna
intrusion (Fig. 1) (Boyd & Mathiesen, 1979).

Skjækerdalen. This deposit is associated with metagabbroic and ultramafic
fragments in magmatic breccias in a small but complex polymagmatic intrusion,
with maximum dimensions 4 km × 1 km in the Gula Group within the Caledonian
Trondheim Nappe (Fig. 1). PGE data are reported for this deposit for the first time.

Vakkerlien. The Vakkerlien intrusion and its associated Ni-Cu mineralization
have been described by Thompson *et al.* (1980) who also presented PGE data. The
mineralization includes vein-, stringer- and disseminated types, all dominated by
pyrrhotite, pentlandite and chalcopyrite, but with pyrite in most samples. The
mineralization occurs in the core of an ultramafic-mafic sill in the Gula Group
within the Caledonian Trondheim Nappe (Fig. 1).

Lillefjellklumpen. This small mineralization (<100 t Ni metal) forms a massive
sulphide lens lying concordantly between primitive MORB-type metabasalt and
metagabbro in the allochthonous Gjersvik island arc complex in the Caledonides
(Grønlie, 1985, 1986, in press). The mineralization shows variations in Cu/Cu + Ni
ratio but without development of the Cu-rich veins with Cu/Cu + Ni averaging
>0·9 found at Hosanger, Flåt and Ertelien. PGE data come from Grønlie (in press)
though it has long been known that the body is rich in PGE (Foslie & Johnson Høst,
1932).

Faeøy. This mineralization is also from an oceanic environment in that it forms a
massive mineralization within the dyke complex in the Karmøy ophiolite (Sturt *et
al.*, 1980). Recent work (Dunning *et al.*, 1986) suggests that the Karmøy ophiolite is
arc-related rather than 'normal' oceanic crust and gives a U/Pb age of 498 + 15/
−5 Ma for plagiogranite in the complex. Analytical data from this deposit indicate a
considerable range of Cu/Cu + Ni ratios, with values between 0·3 and 0·75 predominating. The data presented here are new though also this body was known to
have high PGE levels (Foslie & Johnson Høst, 1932; Boyd & Nixon, 1985).

ANALYTICAL METHODS

The samples from Skjækerdalen, Hosanger, Ertelien, Flåt and Faeøy were
analysed by Becquerel Laboratories of Toronto using INAA after preconcentration
in a nickel-sulphide bead. Values obtained for three blind standards are shown by

Barnes *et al.* (this volume) who compare the results obtained using this method with accepted values. Interference effects inherent in the method give unreliable results for Pd and Rh in samples with >5% Cu and unreliable Pt results in samples with high Au contents: these effects are generally coincident.

Ni and Cu were analysed by atomic absorption and S by X-ray fluorescence by Caleb Brett Ltd of Manchester.

RESULTS

It is a well-established convention that PGE data for Ni-Cu sulphide deposits are presented recalculated to 100% sulphide and then, in diagrammatic form, normalized to chondrite values with the PGE and Au plotted in order of decreasing melting point from Os to Au (Naldrett *et al.*, 1979; Naldrett, 1981). The case for recalculation to 100% sulphides is based on the assumptions that the PGE are collected by sulphide droplets, and that major disturbance of the concentrations by later alteration has not taken place, and on the advantage in comparing data from different deposits given by the removal of effects caused by variable content of gangue in the samples analysed. The assumptions may not hold for sulphide-poor rocks.

The samples analysed from the five Ni-Cu deposits for which new data are presented here were, with the exception of Skjækerdalen, all matrix or massive sulphides with >50% sulphide. The data in this paper are presented as ranges of absolute values and as content in 100% sulphides (Table 1) and are shown diagrammatically (Figs 2(A–D)) according to the convention established by Naldrett *et al.* (1979).

The samples analysed for PGE show variable degrees of agreement with available reserve figures in terms of their Cu/Cu + Ni ratio. In three cases, Hosanger, Flåt and Ertelien, this is partly due to the combination of Ni-Cu and Cu >> Ni mineralizations in the bulk figures for the deposits. The closest correspondence is for Bruvann as the ten samples on which the averages shown in Table 1 are based, are bulk samples of drillcore representing complete ore-zone intersections. In several cases the deposits are worked out and/or only partially accessible and it is not possible to ensure representative ore sampling.

The deposits have been divided into three groups based on their content of Pt and Pd in 100% sulphides.

Group 1. Bruvann, Hosanger and Flåt, three of the four largest Ni-Cu sulphide deposits in the country, have concentrations of PGE in 100% sulphides an order of magnitude below chondrite for all PGE except Pd, which is just over $0.1 \times$ chondrite (Fig. 2(A)). Concentrations of Pt and Pd are similar for all three and the concentrations of all PGE are similar for Hosanger and Flåt, while Bruvann has much lower concentrations ($<c. 0.01 \times$ chondrite) of Os, Ir, Ru and Rh. Few Ni-Cu deposits are recorded in the literature as having comparably low values of Pt and Pd: those known to the authors are Pipe (Naldrett *et al.*, 1979), Montcalm (Naldrett & Duke, 1980), Laukunkangas and Vammala (Papunen, 1986), St. Stephen, Lynn

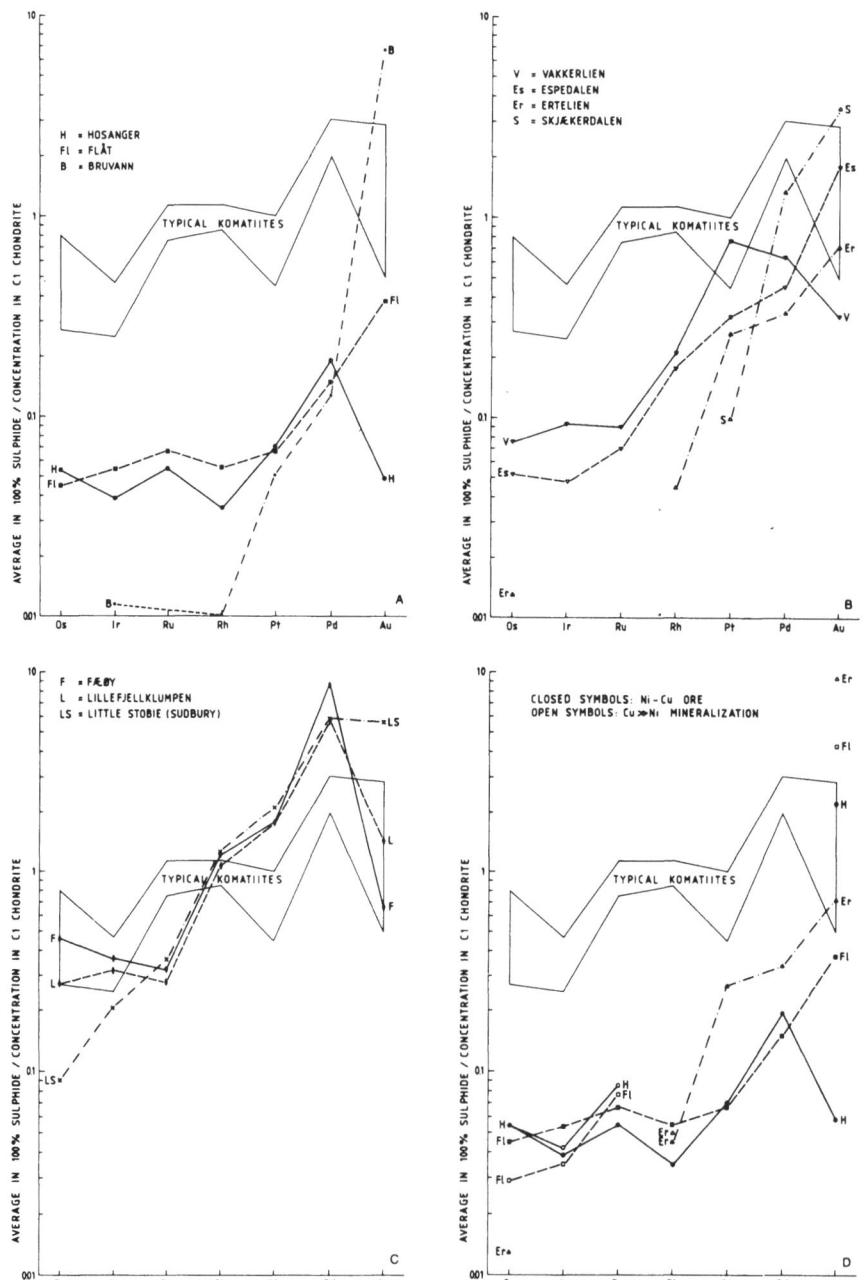

FIG. 2. Average chondrite-normalized noble metal contents in 100% sulphide in Norwegian Ni-Cu deposits. Published data are included from Naldrett *et al.* (1979) (Espedalen), Thompson *et al.* (1980) (Vakkerlien), Grønlie (1985) (Lillefjellklumpen), Boyd *et al.* (1986) (Bruvann). (A), (B), (C) show data from Ni-Cu mineralizations classified according to Pt- and Pd-levels with the field of komatiites and the average for Little Stobie (Naldrett, 1981) for reference. (D) shows data for Ni-Cu and Cu >> Ni mineralizations for Hosanger (solid line), Flåt (dashed line) and Ertelien (dashed-dot line).

Lake, Moak Lake and Renzy (Jonasson *et al.*, 1987). The Pd/Ir ratios are low for all three deposits and Hosanger and Flåt have, in general, patterns similar in form to the field shown by typical komatiites (Naldrett, 1981). The low PGE grades in these deposits are seemingly at variance with evidence of a relatively primitive magma from other indicators, for the case of Bruvann (Boyd *et al.*, 1987) and for mineralization in its host, the Råna intrusion in general (Barnes, 1987). Possible explanations are that the parental magmas were derived from an already depleted mantle or that PGE were removed in the crust but below presently accessible levels: both papers conclude that the evidence, particularly REE data (Barnes, 1986*a*), supports the second alternative.

The position of Au for Flåt and Hosanger refers to Au-poor Ni-Cu ore (Au is concentrated in Cu $\gg$ Ni mineralization in these deposits), while at Bruvann high Au values occur locally in disseminated Ni-Cu mineralization and no Cu-rich paragenesis is known to exist (based on a total of 28 500 m of drilling and several thousand assays).

Group 2 (Fig. 2(B)). Skjækerdalen, Vakkerlien, Espedalen and Ertelien contain Pt and Pd at higher levels than Group 1, but <chondrite, except for Pd at Skjækerdalen. The average Pd content at Skjækerdalen is however perhaps artificial, as one of the seven samples was unusually rich in Pd (4·7 ppm in 100% sulphide). Rh values from Espedalen and Vakkerlien are also markedly higher than those found in Group 1 deposits. For Espedalen, Ertelien and Skjækerdalen the contents of Os, Ir and Ru are similar to, or less than, those found in Group 1, with all deposits for which data was obtained showing flat patterns for these elements. Though the Pd/Ir ratios of Group 1 and Group 2 deposits overlap, they are generally higher in Group 2, especially for Ertelien, the Pd/Ir ratio from which is compatible with equilibration with a magma of composition between continental tholeiite and MORB (Naldrett & Barnes, 1986).

Group 3. Lillefjellklumpen and Fæøy (Fig. 2(C)) have Pt and Pd contents exceeding those of typical komatiites, while having concentrations similar to komatiites for Os, Ir, Rh and Au, and relatively low levels for Ru. The patterns for the two deposits are very similar. The concentrations of Pt and Pd are about half of those found in the ophiolite-hosted Illinois River massive sulphide deposit in Oregon (Foose, 1986) which includes both disseminated sulphides and pods of remobilized massive mineralization in deformed pyroxenitic and gabbroic cumulates. The levels for all PGE are about an order of magnitude lower than those found in the chromitite-associated Cliff mineralization in the Unst ophiolite, Shetland (Prichard *et al.*, 1986) which is located in a dunitic body in harzburgite tectonite. Chromitite-associated sulphide mineralization in the Eretria area of the Othris ophiolite has, however, much lower Pt and Pd levels (Economou & Naldrett, 1984) than those found at Lillefjellklumpen and Fæøy. The Ru, Rh, Pt and Pd concentrations at the two Norwegian deposits are close to those at Little Stobie, Sudbury (Naldrett, 1981).

Cu $\gg$ Ni mineralizations. Three of the deposits examined, Hosanger, Flåt and Ertelien, contain parageneses dominated by Cu (with Cu/(Cu + Ni) > 0·9) (Fig. 2(D)). No reliable data are available for Pt and Pd, and in two cases for Rh, in the

Cu-rich parageneses. The Ertelien Ni-Cu mineralization has exceedingly low contents of Os, Ir and Ru, with even lower contents of Os and Ir, and probably also of Ru, in the $Cu \gg Ni$ mineralization. At somewhat higher levels the same is true for Os and Ir at Flåt, while the Cu-Ni paragenesis at Hosanger has higher levels of Ir and Ru, and the same Os content as the Ni-Cu ore.

DISCUSSION

With the exception of the two deposits in Group 3, both from an 'ophiolitic' environment, there is no obvious relationship *within* the group of deposits discussed here between the apparent geotectonic environment or age of the deposits and their PGE or Ni-Cu grades. Group 1 contains three deposits, all with low $Cu/(Cu + Ni)$ and Pd/Ir ratios (<19 and $<11\cdot3$ respectively), but each from a different type of intrusion, only in the very broadest sense possibly, synorogenic. The Ni, Cu and PGE geochemistry, especially the ratios $Cu/(Cu + Ni)$ and Pd/Ir, of these deposits indicate derivation from relatively primitive magmas. The Pd/Ir ratios are of the same order as those found in Ni-Cu deposits related to komatiites (Barnes *et al.*, 1985). One of the more apparent relationships in Group 1 is the inverse 'correlation' between the content of $(Pt + Pd)$ and the size of the deposits. This might imply a low value of R, the ratio of silicate magma to the mass of sulphide melt equilibrating with it (Naldrett *et al.*, 1979), but the high content of Ni in 100% sulphides at Bruvann suggests that this is not the case for this deposit.

The deposits in Group 1 share their characteristics of low Pd/Ir ratio and very low PGE grade with several Ni-Cu deposits in other orogenic belts, e.g. in the Middle Proterozoic Svecokarelian belt in Finland (Papunen, 1986) and in the Appalachian Caledonides (Jonasson *et al.*, 1987). Even the more PGE-rich deposits located in orogenic belts have modest PGE grades, totalling $<c.$ 2000 ppb in 100% sulphides, except for Hitura at which secondary processes have enriched the noble metals (Häkli *et al.*, 1976). This suggests that there may be a common mechanism leading to depletion of PGE in mafic bodies emplaced in the upper crust in orogenic belts. The examples of Bruvann (Boyd *et al.*, 1987) and mineralizations at Råna in general (Barnes, 1986*a*) indicate that this is due to depletion of PGE in the crust rather than melting of a depleted mantle.

Vakkerlien and Espedalen, in Group 2, have $Cu/(Cu + Ni)$ and Pd/Ir ratios similar to the deposits in Group 1 while Ertelien, though it has a low $Cu/(Cu + Ni)$ ratio has a Pd/Ir ratio >100 and Skjækerdalen has both a higher $Cu/(Cu + Ni)$ ratio and a Pd/Ir ratio at least >100, implying a normal basaltic parent magma (Naldrett & Barnes, 1986).

The two deposits in Group 3 have higher $Cu/(Cu + Ni)$ ratios than the other Ni-Cu mineralizations except for Skjækerdalen, but have relatively low Pd/Ir ratios (18 and 24), well below the values shown for MORB by Barnes *et al.* (1985). Mineralizations of this type could represent the 'missing' Cu-Ni-bearing sulphide melt which Czamanske & Moore (1977), concluded had been depleted from MORB-type basalt.

Cu-rich veins and/or ore bodies are known from many Ni-Cu deposits, including several at Sudbury (Naldrett, 1981) as well as Katiniq, Alexo and Insizwa. Alternative explanations for this Cu enrichment have been:

—Hydrothermal mobilization of Cu, Pt, Pd and Au (not affecting the other PGE) (McCallum *et al.*, 1976; Keays *et al.*, 1981, 1982).
—Subsolidus diffusion (Naldrett & Kullerud, 1967; Hoffman *et al.*, 1979).
—Fractionation of PGE from the sulphide melt into monosulphide solid solution (MSS) (Distler *et al.*, 1977; Malevskiy *et al.*, 1977; Naldrett *et al.*, 1982).
—Preferential partition of Pt, Pd and Au into a high-temperature Cu-rich sulphide liquid. That such a liquid can exist was suggested by Hawley (1962) and demonstrated by Craig & Kullerud (1969).

The data for the three deposits shown on Fig. 2(D) are inconsistent with the first of these explanations being the sole reason for the Cu >> Ni parageneses in that the process resulting in this type of mineralization has not only a marked effect on the Au content but has also a systematic effect at the other end of the spectrum because the ratio (PGE)Cu >> Ni/(PGE)Ni-Cu shows an almost consistent increase from Os to Ir, Ir to Ru and Ru to Au for Flåt and Hosanger. The data available from Ertelien are also consistent with this pattern. The conclusion must be that the process causing formation of the Cu >> Ni mineralization has affected all the PGE + Au in a systematic manner and that any additional effect due to hydrothermal alteration has not disturbed the resulting pattern. A more detailed study of the mineralogy and field relationships of these ores than has been possible in this project would be required before drawing any further conclusions on which process caused the formation of the Cu >> Ni mineralization but we feel that the last explanation is the most probable.

ACKNOWLEDGEMENTS

This work has been supported by the European Economic Community Raw Materials Program (Contract no. MSN-2-0215-N(B)), by the Royal Norwegian Council for Scientific and Industrial Research (NTNF) and by the Geological Survey of Norway (NGU). We would like to thank Astri Hemming and the late Lars Holiløkk for help in preparing the figures and Gunn Sandvik Nielsen for typing the manuscript.

REFERENCES

Barnes, S.-J. (1986*a*). Investigation of the potential of the Tverrfjell portion of the Råna intrusion for platinum-group element mineralization. NGU-rapport nr. 86.021 (Geological Survey of Norway report no. 86.021), 169 pp.
Barnes, S.-J. (1986*b*). Platinum-group elements in the Tverrfjell portion of the Råna layered intrusion, Nordland, Norway. *Terra Cognita*, **6**, 559 (abs.).

Barnes, S.-J. (1987). Unusual base- to noble-metal ratios from the Råna layered intrusion, Nordland, Norway. *Norsk geol. tidsskr.*, **67**, 215–32.

Barnes, S.-J., Naldrett, A. J. & Gorton, M. P. (1985). The origin of the fractionation of platinum-group elements in terrestrial magmas. *Chem. Geol.*, **53**, 303–23.

Barnes, S.-J., Boyd, R., Korneliussen, A., Nilsson, L.-P., Often, M., Pedersen, R. B. & Robins, B. (1988). The use of mantle normalization and metal ratios in discriminating the effects of partial melting, crystal fractionation and sulphide segregation on platinum-group elements, gold, nickel and copper: examples from Norway. In *Geo-Platinum 87*, 113–43. Elsevier Applied Science Publishers, London.

Barth, T. F. P. (1947). The nickeliferous Iveland-Evje amphibolite and its relation. *Nor. geol. unders.*, **168a**, 71 pp.

Bjørlykke, H. (1947). Flåt nickel mine. *Nor. geol. unders. Bull.*, **168b**, 39 pp.

Bjørlykke, H. (1949). Hosanger nikkelgruve. *Nor. geol. unders. Bull.*, **172**, 38 pp.

Boyd, R. & Mathiesen, C. O. (1979). The nickel mineralization of the Råna mafic intrusion, Nordland, Norway. *Can. Mineral.*, **17**, 287–98.

Boyd, R. & Nixon, F. (1985). Norwegian nickel deposits: a review. In *Nickel-copper Deposits of the Baltic Shield and Scandinavian Caledonides*, ed. H. Papunen & G. I. Gorbunov, Geol. Surv. Finland. Bull., Vol. 333, pp. 363–94.

Boyd, R., McDade, J. M., Millard, H. T. & Page, N. J. (1988). Platinum and palladium geochemistry of the Bruvann nickel-copper deposit, Råna, North Norway. *Terra Cognita*, **6**, 559–60 (abs.).

Boyd, R., McDade, J. M., Millard, H. T. & Page, N. J. (1987). Platinum and metal geochemistry of the Bruvann nickel-copper deposit, Råna, North Norway. *Norsk geol. tidsskr.*, **67**, 205–14.

Craig, J. R. & Kullerud, G. (1969). Phase relations in the Cu–Fe–Ni–S system and their application to magmatic ore deposits. In *Magmatic Ore Deposits*, ed. H. D. B. Wilson. Economic Geology Monograph, Vol. 4, Economic Geology Publishing Company, pp. 344–58.

Crocket, J. H. (1974). Gold. In *Handbook of Geochemistry*, ed. K. H. Wedepohl, II-4. Springer-Verlag, Berlin.

Czamanske, G. K. & Moore, J. G. (1977). Composition and phase chemistry of sulphide globules in basalt from the Mid Atlantic ridge valley near 37°N latitude. *Geol. Soc. Amer. Bull.*, **88**, 587–99.

Distler, V. V., Malevskiy, A. Yu. & Laputina, I. P. (1977). Distribution of platinoids between pyrrhotite and pentlandite in crystallization of a sulphide melt. *Geochem. Int.*, **14**, 30–40.

Dunning, G. R., Krogh, T. E. & Pedersen, R. B. (1986). U/Pb zircon ages of Appalachian-Caledonian ophiolites. *Terra Cognita*, **6**, 155 (abs.).

Economou, M. I. & Naldrett, A. J. (1984). Sulfides associated with podiform bodies of chromite at Tsangli, Eretria, Greece. *Mineral. Deposita*, **19**, 289–97.

Foose, M. P. (1986). Setting of a magmatic sulfide occurrence in a dismembered ophiolite, southwestern Oregon. *US Geol. Survey. Bull.*, **1626-A**, 21 pp.

Foslie, S. & Johnson Høst, M. (1932). Platina i sulfidisk nikkelmalm. *Nor. geol. unders. Bull.*, **137**, 71 pp.

Grønlie, A. (1985). PGM-mineraliseringen ved Lillefjellklumpen nikkel-magnetkis forekomst, Nord-Trøndelag. In *Nye malmtyper i Norge*, ed. F. M. Vokes. BVLI, Trondheim, pp. 82–98.

Grønlie, A. (1986). Platinum-group minerals in the Lillefjellklumpen nickel-copper deposit, Norway. *Terra Cognita*, **5**, 560 (abs.).

Grønlie, A. (in press). Platinum-group minerals in the Lillefjellklumpen nickel-copper deposit, Norway. *Norsk geol. tidsskr.*

Gunn, A. G., Leake, R. C. & Styles, M. T. (1985). Platinum-group element mineralization in the Unst ophiolite, Shetland. Mineral Reconnaissance Programme Rep. Br. geol. Surv., No. 73, 117 pp.

Häkli, T. A., Hänninen, E., Vuorelainen, Y. & Papunen, H. (1976). Platinum-group minerals in the Hitura nickel desposit, Finland. *Econ. Geol.*, 71, 1206–13.

Hawley, J. E. (1962). The Sudbury ores: their mineralogy and origin. *Can. Mineral.*, 7, 1–207.

Heim, M. (1981). Basement/cover relations in the allochthon at Espedalen (Jotun-Valdres Nappe complex, S. Norway). *Terra Cognita*, 1, 51 (abs).

Hoffman, E. L., Naldrett, A. J., Alcock, R. A. & Hancock, R. G. V. (1979). The noble metal content of ore in the Levack West and Little Stobie mines, Ontario. *Can. Mineral.*, 17, 437–52.

Johanssen, G. A. (1974). Nikkel-, kobbermalmforekomster og bergarter i Tyristrand og Holleia. Cand. real. thesis. University of Oslo, 126 pp. (unpublished).

Jonasson, I. R., Eckstrand, O. R. & Watkinson, D. H. (1987). Preliminary investigations of the abundance of platinum, palladium and gold in some samples of Canadian nickel-copper ores: in Current Research, Part A, Geological Survey of Canada, Paper 87-1A, pp. 835-46.

Keays, R. R., Ross, J. R. & Woolrich, P. (1981). Precious metals in volcanic peridotite-associated nickel sulphide deposits in Western Australia, II. Distribution within the ores and host rocks at Kambalda. *Econ. Geol.*, 76, 1645–74.

Keays, R. R., Nickel, E. H., Groves, D. I. & McGoldrick, P. J. (1982). Iridium and palladium as discriminants of volcanic-exhalative and hydrothermal and magmatic nickel sulfide mineralization. *Econ. Geol.*, 77, 1535–47.

Kvale, A. (1960). The nappe area of the Caledonides in western Norway. *Nor. geol. unders. Bull.*, 212e, 43 pp.

Malevskiy, A. Yu., Laputina, I. P. & Distler, K. V. (1977). Behaviour of the platinum-group metals during crystallization of a sulfide melt. *Geochem. Inst.*, 14, 177–84.

McBryde, W. A. E. (1972). Platinum metals. In *Encyclopedia of Geochemical and Environmental Sciences*, ed. R. W. Fairbridge. Van Nostrand Reinhold, New York.

McCallum, M. E., Lovacks, R. R., Carlson, R. R., Cooley, E. F. & Docrye, T. A. (1976). Platinum metals associated with hydrothermal copper ores of the New Rambler mine, Medicine Bow Mountains, Wyoming. *Econ. Geol.*, 71, 1429–50.

Naldrett, A. J. (1981). Platinum-group element deposits. In *Platinum-group Elements: Mineralogy, Geology, Recovery*, ed. L. J. Cabri. Canadian Institute of Mining and Metallurgy Special Volume 23, pp. 197–232.

Naldrett, A. J. & Barnes, S.-J. (1986). The behaviour of platinum-group elements during fractional crystallization and partial melting with special reference to the composition of magmatic sulphide ores. *Fortschr. Mineral.*, 64, 113–33.

Naldrett, A. J. & Duke, J. M. (1980). Pt metals in magmatic sulfide ores. *Science*, 208, 1417–24.

Naldrett, A. J. & Kullerud, G. (1967). A study of the Strathcona Mine and its bearing on the origin of the nickel-copper ores of the Sudbury district, Ontario. *J. Petrol.*, 8, 453–531.

Naldrett, A. J., Hoffman, E. L., Green, A. H., Chen-Lin Chou & Naldrett, S. R. (1979). The composition of Ni-sulfide ores with particular reference to their content of PGE and Au. *Can. Mineral.*, 17, 403–15.

Naldrett, A. J., Innes, D., Gorton, M. J. & Sowa, J. (1982). Compositional variations within and between five Sudbury ore deposits. *Econ. Geol.*, 77, 1519–34.

Oshin, I. O. & Crocket, J. H. (1982). Noble metals in the Thetford mine ophiolite, Quebec, Canada, Part I. Distribution of gold, iridium, platinum and palladium in the ultramafic and gabbroic rocks. *Econ. Geol.*, 77, 1556–70.

Papunen, H. (1986). Platinum-group elements in Svecokarelian nickel-copper deposits, Finland. *Econ. Geol.*, 81, 1236–41.

Pedersen, S. (1975). Intrusive rocks of the northern Iveland-Evje area, Aust-Agder. *Nor. geol. unders. Bull.*, 322, 1–11.

Prichard, H. M., Potts, P. J. & Neary, C. R. (1981). Platinum-group element minerals in the Unst Chromite, Shetland Isles. *I.M.M. Trans.*, 90, B186–B188.

Prichard, H. M., Neary, C. R. & Potts, P. J. (1986). Platinum-group minerals in the Shetland Ophiolite complex. In *Metallogenesis of Basic and Ultrabasic Rocks*, ed. M. J. Gallagher, R. A. Ixer, C. R. Neary & H. M. Prichard. Institution of Mining and Metallurgy Symposium Volume, pp. 395–414.

Scheerer, T. (1845). Om nikkelens forekomst i Norge. *Nyt Mag. For Naturvitenskaperne*, **4**, 91–6.

Sturt, B. A., Skarpenes, O., Ohanian, A. T. & Pringle, I. R. (1975). Reconnaissance Rb/Sr isochron study in the Bergen Arc System and regional implications. *Nature*, **253**, 595–9.

Sturt, B. A., Thon, A. & Furnes, H. (1980). The geology and preliminary geochemistry of the Karmøy ophiolite, S.W. Norway. *Proc. Int. Ophiolite Symposium Cyprus*, 1979, Ministry of Agriculture and Natural Resources, Geological Survey Department, Cyprus, pp. 538–44.

Thompson, J. F. H., Nixon, F. & Sivertsen, R. (1980). The geology of the Vakkerlien nickel prospect, Kvikne, Norway. *Bull. Geol. Soc. Finland*, **52**, 3–22.

14

The Significance of Cumulus Chlorapatite and High-temperature Dashkesanite to the Genesis of PGE Mineralization in the Koitelainen and Keivitsa-Satovaara Complexes, Northern Finland

Tapani Mutanen, Ragnar Törnroos & Bo Johanson

Geological Survey of Finland, 02150 Espoo, Finland

Several PGE mineralizations are known from the 2435 Ma-old Koitelainen layered intrusion and the adjacent Keivitsa-Satovaara (K-S) complex. In Koitelainen, transient sulphide separation gave rise to sulphidic PGE mineralizations (Pd–Pt–Au) in the lower zone and upper main zone. PGE, without visible sulphides, occur in Lower Chromitites (Pd–Pt) in the lower zone, in ultramafic pegmatoid pipes (Pt) in the main zone, and in the Upper Chromitite layer (Ru–Pt–Pd) of the lower upper zone. A peculiar peridotite-mixed rock in the main zone is underlain by a Pt-enriched layer, a setting reminiscent of the Stillwater PGE reef. PGE removed by these mineralizations had little effect on the overall buildup of PGE in the rest of the magma; by far the most of PGE precipitated with the last sulphide saturation at 95% level, in magnetite gabbro (Pt–Au–Pd). In the upper part of the K-S complex, assimilation of sedimentary sulphides triggered the deposition of massive sulphides. PGE occur only in overlying meagre disseminations (Pt–Pd–Au), and reach the highest concentrations in chromite-rich rocks (Os–Pt–Pd) at the roof. The distribution of PGE in these intrusions has an unusual pattern. Disseminated and massive sulphides in lower parts, with low PGE_{sulph} (down to $\leq 10\,ppb$), are overlain by sulphide-poor mineralizations with high PGE_{sulph} (from 20–13 000 to ∞ ppm). This feature persists, and even increases during crystallization, the principal PGE concentrations being located at or near the roof. Association of PGE with primary Cl-minerals suggests that the odd behaviour of PGE was due to an association between PGE and chlorine.

Field and microscopic evidence indicate that contamination influenced and at times dictated the course of crystallization. In general, crystallization took place in a mixing zone between the acid roof melt and the mafic magma. 'Granitic' melt inclusions, common in olivine, indicate that addition of alkalis and H_2O shifted the melt composition into the olivine field, giving rise to basal olivine cumulates, and also to the peridotite-mixed rock. Daughter crystals of chlorapatite in these inclusions indicate that chlorine was an important contaminant since the early stages of crystallization. Chlorapatite resides in intercumulus sites in overlying pyroxene cumulates; finally, in peridotite-mixed rock big cumulus (crystals $\phi \leq 6\,mm$) of

159

chlorapatite turn up. In the K-S complex, chlorapatite likewise occurs as a daughter mineral in granitic melt inclusions of olivine in sulphide-rich peridotites. At the first occurrence of PGE, chlorapatite promptly appears as a cumulus phase, together with primary intercumulus dashkesanite, a Cl-hornblende.

We suggest that Cl (and Br), of sedimentary origin, formed melt-soluble complexes with PGE, which were thus not able to enter the sulphide liquid. The PGE were retained in the mafic liquid until the complexes became unstable, e.g. due to crystallization of chlorapatite. Decomposition of PGE-complexes governs the seemingly arbitrary, even unlikely stratigraphic distribution of PGE, with little correlation to sulphides.

15

The Shetland Ophiolite: Evidence for a Supra-subduction Zone Origin and Implications for Platinum-Group Element Mineralization

H. M. Prichard & R. A. Lord

Department of Earth Sciences, The Open University, Walton Hall, Milton Keynes MK7 6AA, UK

Chromite-rich samples from the ultrabasic and basic igneous complex in the Shetland Islands contain exceptionally rich PGE concentrations previously unrecorded in other ophiolite complexes and geochemically similar to those in stratiform complexes. This paper re-assesses the Shetland complex, with particular emphasis on the chrome-spinels and dykes, confirming that it forms the lower part of a Penrose ophiolite assembly. The podiform chromite is shown to have ophiolitic compositions and new electron microprobe analyses from chromite-rich samples indicate that there is nothing unusual about the compositions of the chrome-spinels in PGM-rich samples, which have similar compositions to barren samples. The uppermost part of the ophiolite complex consists of a swarm of dykes in gabbro and their geochemistry has boninitic affinities possibly indicating an origin for the complex from a hydrous magma source in a supra-subduction zone (SSZ). The relative roles of primary and secondary fluids in the concentration of the PGE in Shetland are difficult to determine because the chromite-rich lithologies characteristically lie in altered, highly serpentinized zones. Dunites associated with some of the PGE-rich chromite in lenses in the mantle harzburgite have been found to be PGE-rich and significantly these are some of the least altered of the ultramafic rocks. This suggests that the PGE concentrations are not solely related to alteration zones or secondary processes.

REFERENCES

Prichard, H. M., Potts, P. J. & Neary, C. R. (1984). Platinum and gold in the Shetland ophiolite. In industry in action: Exploration (Anon.) *Mining Journal*, **303**, 77.

Prichard, H. M., Neary, C. R. & Potts, P. J. (1986). Platinum-group minerals in the Shetland ophiolite. In *Metallogeny of basic and ultrasonic rocks*, eds M. J. Gallagher, R. A. Ixer, C. R. Neary & H. M. Prichard, Institution of Mining and Metallurgy, London, pp. 395–414.

Prichard, H. M., Potts, P. J., Neary, C. R., Lord, R. A. & Ward, G. R. (1987). Development of techniques for the determination of the PGE in ultramafic rock complexes of potential economic significance: mineralogical studies. Unpubl. report for the EEC, 170 pp.

16

Nickel-Copper and Precious Metal Mineralisation in the Caledonian Mafic and Ultramafic Intrusions of North-East Scotland

T. A. FLETCHER

Geology Department, Marischal College, Broad Street, Aberdeen AB9 1AS, UK

The Caledonian mafic and ultramafic intrusions of North-East Scotland are a suite of synorogenic, tholeiitic plutons, which are commonly layered and include xenolithic, contaminated and granular gabbroic varieties. They were emplaced 489 ± 17 mya and postdate the regional foliation in the country rock metasediments, but are themselves locally deformed by a set of regional ductile shear zones.

In the Huntly-Knock-Portsoy area, fragmented and poorly exposed bodies of layered peridotitic to gabbroic cumulates, xenolithic rocks and non-cumulate granular gabbros/norites $\pm$ biotite and quartz occur in an envelope of semi-pelites and pelites with calcareous and quartzitic horizons. In the cumulates the order of appearance of phases is ol-cr, ol-plag-cr, ol-plag-cpx, plag-opx-ol-cpx. Layering is common and the development of cyclic units may occur on the metre scale. The cumulates appear to grade into granular gabbros/norites, but are apparently cut by similar rocks elsewhere, suggesting a slightly younger age for some of the granular rocks.

Weakly disseminated Fe-Cu-Ni sulphides are widespread through the intrusions, but locally greater concentrations occur as:

(1) disseminated and submassive to massive lenses in a structurally complex, contact zone of norites and gabbros (often xenolithic and graphitic), metasediments and minor picrites;

(2) disseminated thin horizons within the cumulates, especially associated with olivine-rich layers in olivine gabbro cumulates;

(3) sulphide-, and graphite-rich pyroxenitic pegmatites;

(4) sulphide-rich veins following fractures and zones of intense deformation.

The sulphide assemblage, pyrrhotite with minor pentlandite, chalcopyrite, and pyrite and some of their textures suggest a primary magmatic origin with evidence of postmagmatic remobilisation by shearing, faulting and hydrothermal activity. The last is multiphase and involves quartz, amphibole, biotite, garnet, carbonate, chlorite, serpentine and graphite overprints.

Limited precious metal data from the contact zone indicate anomalous levels of Au, Ag and PGE within the submassive to massive sulphides. The potential for further PGE enrichment as stratabound disseminated mineralisation in the cumulate succession, and as possible remobilised concentrations within shear zones, is currently under investigation.

17

Platinum-Group Element Analyses of Serpentinites in the Eastern Central Alps, from Davos (Switzerland) to the Valmalenco (Italy)

DOROTHEE J. M. BURKHARD,[a] NORMAN J PAGE[b] & G. CHRISTIAN AMSTUTZ[a]

[a] *Mineralogisch-Petrographisches Institut der Universität Heidelberg, Postfach 104040, D-6900 Heidelberg, Federal Republic of Germany*

[b] *US Geological Survey, 345 Middlefield Road, Menlo Park, California 94025, USA*

ABSTRACT

Serpentinite occurrences in the Pennine nappes of the eastern Central Alps (between Davos, Switzerland and the Valmalenco, Italy) display varying metamorphic grades from north to south. In the north they consist of Mesozoic ophiolitic lherzolite tectonite and in the south they form a (pre-)Mesozoic ultramafic complex. Samples from both types of occurrences were analysed for Pd, Pt, Rh, Ru and Ir, and in some cases also for Cr, Cu, Co and Ni. Mean values are 3 ppb for Pd and 15 ppb for Pt; Rh, Ru and Ir are below the detection limits of 1, 20 and 100 ppb, respectively. Low Rh, Ru and Ir contents may be related to a low degree of partial melting, which is also reflected by a low abundance of chrome-spinel; a low ratio of Pd to Pt is probably typical for 'alpine' peridotites of the Alps. A relatively homogeneous platinum-group element pattern for serpentinites of both types of occurrences and the lack of any correlation with the mineralogy, petrography, or geochemical features is most likely an inherited pattern from the mantle sources or percentage of partial melting but might be related to a redistribution in connection with serpentinization and regional metamorphism.

INTRODUCTION

Serpentinites of varying metamorphic grades in the Pennine nappes of the eastern Central Alps, Switzerland and Italy, differ significantly; they are part of a Mesozoic ophiolite nappe in the north, whereas the Valmalenco serpentinite in the south is probably of pre-Mesozoic age and of unclear structural affiliation. Different types of pyroxene and spinel, alteration and contrasting serpentinization processes in the two types of occurrences support the genetic differences (Burkhard, 1987). Minor sulphide occurrences have been reported for both (Dietrich, 1972; De Capitani *et al.*, 1981; Burkhard, 1987). Rock samples were analysed for platinum-group elements (PGE) in order (1) to perform a reconnaissance study of the concentrations of PGE in alpine-age ophiolites; (2) to compare the PGE contents of the two types; (3) to investigate possible relations of the PGE content, metamorphic grade, and sulphide mineral content. Previous PGE investigations in the Alps have been restricted to the mafic Ivrea-Verbano complex of pre-alpine age in the southern Alps (Ferrario *et al.*, 1982; Garuti & Rinaldi, 1985).

GEOLOGY, MINERALOGY, METAMORPHISM

Serpentinized lherzolite and, to a lesser degree dunite, of the eastern Central Alps are part of the highly imbricated and tectonized Pennine nappes which are overlain to the east by Austroalpine units (Fig. 1). Between Davos and Engadin (Arosa-Platta nappe) the regional metamorphic grade increases from prehnite-pumpellyite to greenschist facies (Dietrich, 1969; Dietrich & Peters, 1971; Trommsdorff & Dietrich, 1980). Serpentinites of the Arosa-Platta nappe belong to an ophiolite sequence of Upper Jurassic to Upper Cretaceous age (Peters, 1963; Dietrich, 1967, 1969, 1979; Trommsdorff & Dietrich, 1980). The degree of serpentinization correlates with the progressive metamorphism (Burkhard, 1987). The Valmalenco serpentinites form an ultramafic complex (about 180 km^2). They are equilibrated in the greenschist facies and characterized by more than one metamorphic mineral generation and a pre-alpine foliation (Bucher & Pfeifer, 1973). Serpentinization may have occurred under stress forming antigorite directly (Burkhard, 1987). In the west, the Valmalenco serpentinite has overprinted contact metamorphism caused by a late Tertiary granodiorite, the Bergeller intrusion (Trommsdorff & Evans, 1972, 1977; Gautschi & Montrasio, 1978).

Primary minerals in the northern ophiolitic serpentinites include augite and a lamellar intergrowth of clino- and orthopyroxene (enstatite relicts survived only in the Davos-Arosa region), and two accessory spinels, an Al- and a Cr-rich variety (Burkhard, 1987; Burkhard & Amstutz, 1988). Minerals in the Valmalenco serpentinite form at least two generations. Diopside locally has lamellae filled with chlorite. Chrome-spinel is completely recrystallized and only recognized by the remains of several percent Cr_2O_3 (Burkhard, 1987; Burkhard & Amstutz, 1988). The mineral associations and the grade of serpentinization from the different areas are presented in Fig. 1.

Isotopic values of serpentine ($\delta^{18}O$ and δD) in the ophiolites (Arosa-Platta nappe) are typical for serpentinization in the presence of predominantly meteoric water (Davos-Arosa region) and/or regional metamorphic water (Oberhalbstein and Engadin). However, the isotopic signature of antigorite from the eastern part of the Valmalenco antigorite strongly suggests serpentinization in the presence of seawater. In the western part of Valmalenco, fluids related to the Bergeller intrusion may have influenced the isotopic composition of the antigorite (Burkhard, 1987; Burkhard & O'Neil, 1988).

Sulphides occur in both areas as accessory constituents, and locally as small masses. These deposits are situated mostly at tectonic boundaries and in shear

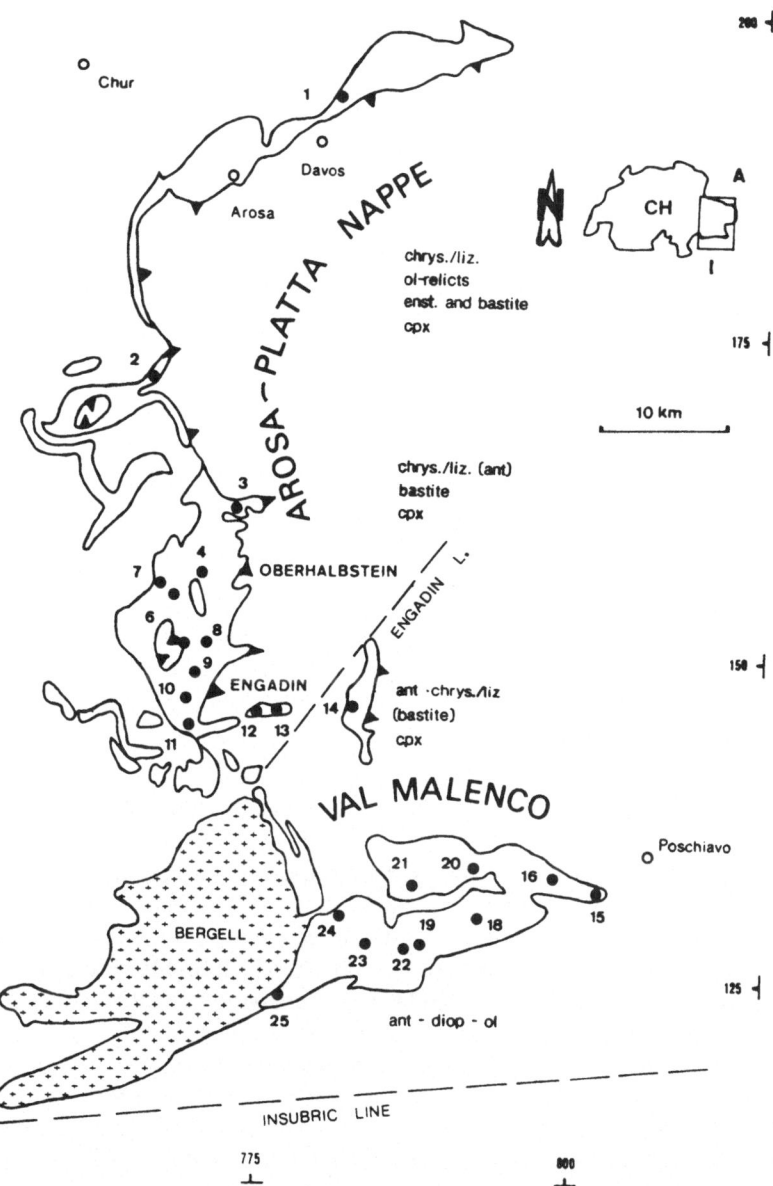

FIG. 1. Arosa-Platta nappe and the Val Malenco with the sampling localities in the serpentinites; dashed lines: faults; cross-pattern: granodiorite intrusion of the Bergell, Arosa-Platta nappe: 1, Davos (Da); 2, Tgant Ladrung (OH1); 3, Tinzener Ochsenalp (OH4); 4, Gruba (OH5); 6, Cotschens (OH10); 7, Muttans (OH11); 9, Schlucht Bivio (SchB); 10, Sur al Cant Fuorcla (OSCF); 11, Lunghin Area (Lungh); 12, Blaunca (Bl); 13, Grevasalvas (Grev); 14, Grialetsch (Grial). Val Malenco: 15, Selva; 16, Pass d'Ur (P.d'Ur); 18; Franscia (Fr.); 19, Chiesa; 20, Nero; 21, Alpe Senevedo (VM6); 22, above Chiesa (Prim); 23, Laghetti di Sassersa (Lagh); 24, Val Ventina (Vent); 25, Val Masino, Preda Rossa (V. Mas).

zones in the Arosa-Platta nappe (Dietrich, 1972), and are oriented parallel to the spinel and clinopyroxene layering in the Valmalenco (De Capitani *et al.*, 1981). Textural, mineralogical and geochemical constraints for a genetic interpretation are discussed in Burkhard (1987).

SAMPLING

Forty-eight serpentinite samples were selected for analyses from the eastern Central Alps, 24 from the ophiolitic serpentinites (Arosa-Platta nappe) and 24 from the Valmalenco mass. The samples from 21 localities can be subdivided into three groups, (a) without clinopyroxene, but generally containing some bastite, (b) with clinopyroxene, and (c) the sulphide occurrences (see Tables 1 and 2). Most samples represent original dunites or harzburgites; a few in group (b) may be classified as lherzolites (the ratio of the rock types presented here does not reflect the abundance in the field where lherzolites are by far predominant).

For comparison, 3 samples were analysed from the pre-Mesozoic Kraubath and Hochgrössen ultramafic and mafic complexes in the Eastern Alps (Austria) which were metamorphosed to the amphibolite facies (El Ageed *et al.*, 1980). Here, serpentinization is considered to be retrograde. All three samples were relatively chromite-rich, with 1–5 vol.%. Two more samples originate from the Montgenèvre ophiolite in the Western Alps (France), which is also part of the Pennine nappes. Both samples are lherzolites that have been metamorphosed to the pumpellyite-actinolite facies (Bertrand *et al.*, 1980); opaque minerals are scarce (also Nievegelt, personal communication).

ANALYTICAL METHODS AND RESULTS

The serpentinite samples were analysed by fire-assay—atomic absorption techniques for Pd, Pt and Rh described by Haffty *et al.* (1977), Simon *et al.* (1978), and Page *et al.* (1980) and fire-assay spectrographic techniques for Ir and Ru, according to Haffty *et al.* (1980). Detection limits for Pd, Pt, Rh, Ir and Ru are 1, 10, 1, 100 and 20 ppb respectively. Co, Cr, Ni and Cu were analysed by ICP-acid digestion.

Tables 1 and 2 list the analytical data by sample, locality, and mineral assemblage. All Rh, Ir and Ru values are below the detection limits and are not listed separately. Table 3 contains summary statistics for Pt and Pd that analysed above the detection limit. Co, Cr, Cu and Ni have contents typical for peridotites and serpentinites with the exception of the samples OH1/7, P.d'Ur7, P.d' Ur17 and Lagh5. For the Arosa-Platta nappe the mean Cr and Ni contents are 2507 and 2090 ppm which are not significantly different from 2262 and 2244 ppm for Cr and Ni from the Valmalenco area. Cr correlates directly with the presence of Cr-spinel; Co, Ni and Cu with the random occurrence of sulphides, such as pentlandite and chalcopyrite. These oxides and sulphides are distributed inhomogeneously and consequently so are Cr, Co, Cu and Ni.

The means of Pd and Pt for the different areas and groupings in Table 3 are not

TABLE 1

Palladium, platinum and cobalt, chromium, copper and nickel analyses of four serpentinite occurrences of the Alps; (a) Arosa-Platta nappe (Switzerland)

Sample field	Sample description	Pd	Pt	Co	Cr	Cu	Ni	Swiss coordinates
		(ppb)		(ppm)				
Da8	c/l, b, cpx	4	11	—	—	—	—	730.38/189.82
OH1/4	c/l, (sulph)	3	35	86	1 970	8	1 860	746.55/168.27
OH1/6	c/l, (cpx)	3	20	104	2 720	3	2 490	746.55/168.27
OH1/7	chl, mt, ilm	<1	16	114	346	9	97	746.55/168.27
OH1/9	c/l, b, later veins	5	27	120	3 140	11	2 880	746.55/168.27
OH4/1	c/l (sulph)	2	20	141	2 310	11	2 800	771.82/158.74
OH4/11	sulph, py	<1	<10	—	—	—	—	771.80/159.25
OH5/11	c/l, cpx	3	<10	—	—	—	—	769.20/153.79
OH5/12	c/l/a, sulph (po) ilv	2	<10	—	—	—	—	769.20/153.79
OH10/1	c/l	2	16	96	1 650	2 990	1 140	767.20/152.04
OH10/13	c/l, sulph	2	10	135	2 560	322	2 200	767.12/152.04
OH11/3	c/l	1	<10	—	—	—	—	765.93/152.76
OH11/8	c/l, cpx	4	12	—	—	—	—	765.93/152.76
SchB2	a (cpx)	4	24	105	2 780	15	2 170	bolder
OSCF3	a (sulph)	3	16	128	3 480	11	2 270	768.67/144.38
OSCF4	a (sulph)	2	28	113	2 900	10	2 670	768.67/144.38
Lungh11	a	4	<10	—	—	—	—	771.10/142.56
Lungh14	a, cpx (filty)	2	<10	—	—	—	—	771.00/142.42
Bl.1	a (c/l)	5	11	—	—	—	—	774.24/143.84
Bl.5	a (c/l), (cpx)	4	21	117	3 240	33	2 440	774.64/143.92
Bl.6	a/cpx	4	<10	—	—	—	—	774.40/143.98
Bl.11	a, sulph	4	10	—	—	—	—	774.50/143.65
Grev7	a (c/l)	4	<10	—	—	—	—	775.31/143.80
Grial3	a, trem	<1	<10	75	2 990	5	2 060	778.72/143.08

Analysts: Love, A. & Bradley, L. for Ir, Ru;
Gent, C. for Pd, Pt, Rh;
Briggs, S. & Moore, P. for Co, Cr, Ni, Cu.

a: antigorite; act: actinolite; b: bastite; c: chrysotile; carb: carbonate; chl: chlorite; cp: chalcopyrite; cpx: clinopyroxene; Cr-sp: Cr-spinel; diop (-r): diopside (relict); ilm: ilmenite; ilv: ilvaite; l: lizardite; lim: Fe-hydroxide; mt: magnetite; ol: olivine; po: pyrrhotite; py: pyrite; sulph: sulphide; trem: tremolite (in brackets: minor amounts, in the range of 1%). The sample/field numbers (column one) of Tables 1 and 2 refer to abbreviated location names given in the legend of Fig. 1.

significantly different at the 97·5% confidence level, using estimates based on the pooled standard deviation. The three samples with less than detection concentration of Pd from the Arosa-Platta nappe are not from ultramafic rocks (being either chlorite-rich and ilmenite-bearing (OH1/7), or actinolite-rich (Grial3) or consisting of almost only massive pyrite) and thus would not lower the mean values for the ultramafic samples from the nappe. The mean concentration of Pt in the Arosa-Platta nappe is 18·5 ppb and is significantly different from the mean of 14·1 for the Valmalenco serpentinite at the 95% level or less.

Dorothree J. M. Burkhard, Norman J Page, G. Christian Amstutz

TABLE 2

Palladium, platinum and cobalt, chromium, copper and nickel analyses of four serpentinite occurrences of the Alps; (b) Valmalenco (Italy); (c) Kraubath + Hochgrössen (Austria); (d) Montgenèvre (France)

Sample field	Sample description	Pd	Pt	Co	Cr	Cu	Ni	Swiss coordinates
		(ppb)		(ppm)				
Valmalenco (Italy and Switzerland)								
Selva1	a, schist (diop)	3	<10	—	—	—	—	800.50/130.70
Selva2	a, diop	4	21	102	2 430	4	2 090	800.50/130.70
P.d'Ur1	a, trem/act	3	<10	—	—	—	—	797.70/131.55
P.d'Ur4	a, diop	3	<10	—	—	—	—	797.70/131.55
P.d'Ur7	a, (lim)	4	17	13	1 890	1 270	897	797.70/131.55
P.d'Ur17	a, sulph (cp)	3	12	51	3 100	1 870	86	797.70/131.55
Fr.1	a, (carb)	5	14	—	—	—	—	790.20/129.10
Fr.4	a, at-layer	5	<10	—	—	—	—	790.20/129.10
NeroW2	a	3	13	121	2 520	23	2 360	784.85/134.00
VM6/4	a, ol, (diop)	3	10	100	2 180	17	1 830	782.65/133.40
Chiesa 1r	a, ol, schist	5	12	—	—	—	—	785.40/128.60
Chiesa 3a	a, diop	3	<10	—	—	—	—	785.77/127.90
Prim1	a, diop	3	13	107	268	20	2 190	784.31.126.00
Prim8	a, (carb)	15	22	—	—	—	—	784.31.126.00
Lagh 1	a, diop	3	10	105	3 020	35	2 230	781.80/126.90
Lagh 5	a, (sulph)	3	<10	428	852	18 000	393	781.90/126.95
Lagh 7	a, diop-r	3	13	137	3 440	76	2 930	781.95/126.95
Lagh 10	a, sulph	3	13	123	3 070	23	2 650	782.10/126.80
Lagh 2a	a, (diop)	11	16	301	2 110	2 230	7 030	781.95/127.00
Vent 1	ol, trem	6	13	—	—	—	—	779.50/129.75
Vent 3	ol, a, Cr-sp.	2	<10	—	—	—	—	779.70/124.55
V.Mas 5	trem, chl, tc	4	13	—	—	—	—	775.60/122.60
V.Mas 6	a, diop-r	2	<10	—	—	—	—	775.55/122.80
Kraubath + Hochgrössen (Austria)								
AKB 1	ol, c/l rgr.Cr	<1	14	—	—	—	—	45 15'N/14 57'E
AKB 2	ol, c/l rgr.Cr	4	13	—	—	—	—	45 15'N/14 57'E
AHB		<1	10	—	—	—	—	47 15'N/14 58'E
Montgenèvre (France)								
MG 2	c/l cpx	3	52	—	—	—	—	44 55'N/6 44'E
MG 7	c/l cpx	3	13	—	—	—	—	44 55'N/6 44'E

Analysts: as in Table 1. See also Table 1 for abbreviations.

INTERPRETATION AND COMPARISON

Samples with higher sulphide content of several percent (based on the copper content) and even a high sulphide percentage between 10 and 90% (OH4/11, OH5/12. Bl.11) do not have corresponding higher Pt and Pd values (Tables 1, 2 and 3). The variable Cu/Cu + Ni ratios (0·13–0·96) suggest that some of the sulphides may be of a primary magmatic origin (Burkhard, 1987). PGE contents also do not

TABLE 3

Summary statistics of platinum and palladium contents in serpentinites from the eastern Central Alps

Area	Palladium			Platinum		
	N	X	σ	N	X	σ
Arosa-Platta nappe	21	3·2	1·1	15	18·5	7·5
Valmalenco	23	4·3	3·0	15	14·1	3·5
Kraubath + Hochgrössen	1	4	—	3	12·3	2·1
Montgenèvre	2	3	—	2	32·5	—
Samples with Cu >300	6	4·2	3·4	5	14·2	3·0
Arosa-Platta nappe c/l assemblage	10	2·7	1·2	9	18·6	8·2
a assemblage	10	3·6	1·0	6	18·3	7·2
Valmalenco Area a assemblage without ol or diop	9	4·9	3·9	6	15·2	3·8
a with ol or diop	13	3·9	2·4	8	13·5	3·6

N = number of samples above detection limits, X = mean and σ = standard deviation in parts per billion

correlate with the variable presence of accessory sulphides or with the quantity of pentlandite present, which contains practically all the Ni. Pd is often concentrated in pentlandite (Crocket, 1981; Mitchell & Keays, 1981), but pentlandite from the Arosa-Platta nappe and the Valmalenco do not correlate with high Pd values, except for sample Lagh2a. Also, no positive correlation of sulphides with Pd can be recognized unlike that described by Dillon-Leitch *et al.* (1986) for massive and metamorphic sulphides relative to disseminated and net-textured sulphides from the Cape Smith Belt, Canada.

There is no correlation of the Pd and Pt contents with Cr; Cr-spinel occurs only as accessory and metamorphic relicts; but even samples with higher Cr contents (Vent 3, more than 5000 ppm, in Burkhard, 1987) do not have higher PGE values. The low Cr-spinel content may explain the low Ru and Ir (possibly in part the very low Rh values below detection limit) because laurite $[(Ru, Ir, Os) S_2]$ is commonly associated with chromite in ophiolites (for example Prichard *et al.*, 1986) but no platinum-group minerals could be observed in the samples discussed. The highest Pd and Pt contents (samples Prim8 and Lagh2a) cannot be correlated with the major mineralogy, the sulphide or chrome-spinel contents, or the rock chemistry (Table 3). The lack of correlation between the PGE and different rock types is shown by Table 3. Such lack of correlation was also reported in ophiolitic rocks from Newfoundland by Page & Talkington (1984).

The lack of correlation with sulphide and Cr-spinel contents or with any other mineralogical or geochemical feature with PGE contents and the relatively homogeneous distribution of the PGE in the Central Alps may suggest a certain degree of redistribution of the PGE related to the alteration processes, such as serpentinization and regional metamorphism.

Pd and Pt contents of the Arosa-Platta nappe and the Valmalenco area differ only by a small amount, but significantly; the Arosa-Platta nappe serpentinites have slightly lower Pd contents, while the more metamorphosed Valmalenco serpentinite has higher Pd contents. Compared to values of the East and West Alps (Kraubath + Hochgrössen and Montgenèvre, Table 3), the range for the Pd values is very similar whereas the Pt content is highest in the Montgenèvre serpentinite (prehnite-pumpellyite to greenschist facies) and lowest in the Kraubath + Hochgrössen occurrences (amphibolite facies). These data might indicate a possible correlation of slightly decreasing Pt values with increasing

TABLE 4

Palladium and platinum abundances from other areas (parts per billion; number of samples in parentheses). For comparison with data from the Alps in Table 3

Area	Palladium	Platinum	Reference
Ophiolitic ultramafics			
Medford region, Oregon			
Dunite	21·3 (99)	19·7 (99)	2
Peridotite	14·9 (321)	16·6 (321)	2
Serpentinite	10·4 (56)	14·4 (56)	2
Pyroxenite	32·0 (100)	20·4 (100)	2
Burro Mountain, California			
Harzburgite-Dunite			
(tectonite)	<4 (18)	<10 (18)	3
Voikar-Syinsky Complex			
UdSSR			
Dunite-harzburgite			
(tectonite)	3·9 (23)	7·3 (23)	4
Newfoundland			
Spinel lherzolite	4·6 (3)	7·4 (2)	5
Harzburgite	8·7 (6)	8·3 (3)	5
Dunite	4·5 (4)	3·1 (6)	5
Chomitite	5·8 (5)	35·0 (4)	5
Troodos (Cyprus)			
Harzburgite-dunite	8·1 (3)	—	6
Cumulate-dunite	0·9 (5)	—	7
Cumulate pyroxenite	13·0 (4)	—	7
Chromitites in ophiolites			
Semail (Oman)			
Chromitites as			
Pods or lenses in dunite	8 (19)	14 (20)	8
Thiebaghi + Massif du Sud			
(New Caledonia)	3·2 (9)	17 (43)	9
Guleman-Elazic Area			
(Turkey)	2·7 (7)	9·5 (19)	10

Table 4—*contd.*

Area	Palladium		Platinum		Reference
Sulphide-rich ultramafics					
Limassol Forest					
(Cyprus)	9·8	(2)	13·0	(2)	11
Mid-Atlantic ridge peridotites	41	(2)	35	(2)	12
'Non-Alpine' mafic-ultramafic complexes					
of the Southern Alps					
Ivrea-Verbano					
Basic Complex					
Sulphides					
associated with metasediments	40·9	(8)	93·8	(8)	13
Sulphides					
in pyroxenites	55·0	(4)	158·8	(4)	14
Pegmatoid					
clinopyroxenite layers	3·4	(7)	63·6	(7)	15

References: 1: This report; 2: Carlson *et al.* (1986); 3: Loney *et al.* (1971) 4: Page *et al.* (1983); 5: Page & Talkington (1984); 6: Agiorgitis & Becker (1979); 7: Becker & Agiorgitis (1978); 8: Page *et al.* (1982*a*); 9: Page *et al.* (1982*b*); 10, Page *et al.* (1984); 11, Foose *et al.* (1985); 12: Crocket & Teruta (1977); 13: Ferrario *et al.* (1982), Garuti & Rinaldi (1985).

metamorphic grade. One would suspect from the *Eh*–pH diagram for Pt and Pd (Westland, 1981) that both Pd and Pt would be depleted during progressive metamorphism. However, this would take a more extensive study than has been done so far and it would certainly raise the problem of the residence of these elements.

Abundances of PGE in tectonized and cumulate ultramafics, chromitites, and sulphide-rich occurrences from ophiolites of various ages and locations and from peridotites of the Mid-Atlantic Ridge are compared with those from the Arosa-Platta nappe and the Valmalenco (Tables and 4).

Pd and Pt contents of all Alpine samples are similar to other ophiolitic ultramafic rocks. Mid-Atlantic Ridge peridotites, however, have relatively high Pd and Pt contents as compared to the samples from the Alps and in comparison to many other ophiolitic ultramafics. Pd is very much at a lower level in the Alpine samples, whereas the Pt range is relatively high with values similar to those of some chromitites and sulphide-rich ultramafics.

The PGE distribution pattern for the Alpine samples may be the result of three independent processes; (a) mantle heterogeneity; (b) varying differentiation or partial melting and subsequent fractionation; (c) serpentinization and meta-morphism.

The serpentinized lherzolites of the (eastern Central) Alps represent the residue of a relatively undepleted upper mantle with a relatively low degree of partial melting (Ishiwatari, 1985; Burkhard, 1987). This relationship may explain the ratio of Pd and Pt to the (undetectable) Ir and Ru content, and possibly also in part, the very low Rh values.

It is significant that the two very different and contrasting serpentinite occurrences of the Arosa-Platta nappe and the Valmalenco have essentially similar PGE contents, characterized by the low Pd to Pt ratio and very low Rh values. The data from the East and West Alps, Kraubath + Hochgrössen and Montgenèvre respectively, also have a similar pattern, although Kraubath + Hochgrössen are much older. This PGE pattern of serpentinites of the Alps which differs slightly from that of other areas, may indicate a slight variation in the composition of the protomantle of the Alps.

SUMMARY

Samples of serpentine from Mesozoic ophiolites and a pre-Mesozoic ultramafic body in the Pennine nappes of the eastern Central Alps having different ore mineral content and metamorphic grade were analysed for their PGE content and in some cases also for Cr, Cu, Co and Ni. The characteristic features are:

—Cr, and Cu, Co, and Ni values are related only to the presence of variable amounts of chrome-spinel and sulphides, respectively.
—The PGE content is relatively low and Rh, Ru, and Ir are below the detection limit.
—Pd and Pt are not correlated with different rock types, such as lherzolite, harzburgite, or dunite, nor are they correlated with the chrome-spinel or sulphide content.

In comparison to sulphide-poor and/or sulphide-rich ophiolitic serpentinites of other areas the Pd to Pt ratio of these Alpine serpentinites is relatively low; Pd values are low, whereas Pt contents are similar even to those of some chromitites.

The low ratio of Pd to Pt and low Rh values are characteristic for the Alpine protomantle composition, and is probably independent of the geological setting and age.

The relatively homogeneous PGE distribution of all Alpine samples considered, and the lack of any correlation with the mineralogy, petrography and geochemical features is probably explained by inherited distribution patterns from mantle rock with the overprinting of a later small-scale redistribution in connection with serpentinization and regional metamorphism.

ACKNOWLEDGEMENT

The paper is part of a Ph.D. thesis of the first author at the University of Heidelberg. The help of the Studienstiftung des Deutschen Volkes which gave financial assistance and supported a visit of 1 year at the US Geological Survey at Menlo Park, California, is gratefully acknowledged.

REFERENCES

Agiorgitis, G. & Becker, R. (1979). The geochemical distribution of gold in some rocks and minerals of the Troodos Complex, Cyprus. *N. Jb. Mineral. Mh.*, 7, 316–20.

Becker, R. & Agiorgitis, G. (1978). Iridium, osmium and palladium distribution in rocks of the Troodos Complex, Cyprus. *Chemie d. Erde*, 37, 302–6.

Bertrand, J., Steen, D., Tinkler, C. & Vuagnat, M. (1980). The Mélangezone of the Col du Chenaillet (Montgenèvre Ophiolite, Hautes-Alpes, France). *Arch. Sci. Genèvre*, 33, 117–38.

Bucher, K. & Pfeifer, H. R. (1973). Metamorphose und Deformation der östlichen Malenco Ultramafite und deren Rahmengesteine, Prov. Sondrio N-Italien. *Schweiz. Mineral. Petrogr. Mitt.*, 53, 231–41.

Burkhard, D. J. M. (1987). Ore minerals and geochemistry of the serpentinites in the Eastern Central Alps (Davos to the Val Malenco) compared to occurrences in the Klamath Mountains (California and Oregon). Ph.D. thesis, University of Heidelberg, West Germany, *Geowiss. Abh.*, 12, 345 pp.

Burkhard, D. J. M. & Amstutz, G. C. (1988). Spinel-exsolution and the variety of Cr-spinel alteration in the Arosa-Platta nappe and the Val Malenco. *Mineral. Petrol.* (accepted).

Burkhard, D. J. M. & O'Neil, J. R. (1988). Mineralogy and contrasting serpentinization processes in the Eastern Central Alps (in prep., *Chem. Geol.*).

Carlson, C. A., Page, N. J & Carlson, R. R. (1986). Maps showing the distribution of platinum-group elements and gold in rocks from the western half of Medford and part of the adjacent Coos Bay 1°/342° Quadrangles, Southwestern Oregon. U.S. Geological Survey Miscellaneous Field Studies Map MF-1832, Scale 1.250 000, 2 sheets.

Crocket, J. H. (1981). Geochemistry of the platinum-group elements. In *Platinum-Group Elements: Mineralogy, Geology, Recovery*, ed. L. J. Cabri. Canadian Institute of Mining and Metallurgy, pp. 49–64.

Crocket, J. H. & Teruta, Y. (1977). Palladium, iridium and gold contents of mafic and ultramafic rocks drilled from the Mid-Atlantic Ridge, Leg 37. Deep Sea Drilling Project. *Can. J. Earth Sci.*, 14, 777–84.

De Capitani, L., Ferrario, A. & Montrasio, A. (1981). Metallogeny of the Val Malenco Meta-ophiolitic Complex, Central Alps. *Ofioliti*, 6(1), 87–100.

Dietrich, V. (1967). Vulkanismus in den oberen penninischen Decken Graubündens. *Geol. Rundschau*, 57, 246–64.

Dietrich, V. (1969). Die Ophiolithe des Oberhalbsteins (Graubünden) und das Ophiolithmaterial der Ostschweizerischen Molasseablagerungen. Ein petrographischer Vergleich. Europäische Hochschulschriften, XVII, 1, 180 pp.

Dietrich, V. (1972). Die sulfidischen Vererzungen in den Oberhalbsteiner Serpentiniten. *Geol. Schweiz. Geotechn.*, Serie 49, 128 pp.

Dietrich, V. (1979). Investigation of ophiolitic occurrences and ophiolitic detritus in the Eastern Alps. *Schweiz. Mineral. Petrogr. Mitt.*, 59, 179–80.

Dietrich, V. & Peters, T. (1971). Regionale Verteilung der Mg-Phyllosilikate in den Serpentiniten des Oberhalbsteins. *Schweiz. Mineral. Petrogr. Mitt.*, 51, 329–48.

Dillon-Leitch, H. C. H., Watkinson, D. H. & Coats, C. J. A. (1986). Distribution of platinum group elements in the Donaldson West Deposit, Cape Smith Belt, Quebec. *Econ. Geol.*, 81, 1147–58.

El Ageed, A. I., Saager, R. & Stumpfl, E. F. (1980). Pre-Alpine ultramafic rocks in the eastern Central Alps, Styria, Austria. Panayiotou (ed.). In *Ophiolites, Proc. Int. Ophiolite*

Symp., 1979, Cyprus. Geol. Surv. Dept. Ministry of Agriculture Natural Resources, Nicosia, Cyprus, pp. 601–6.

Ferrario, A., Garuti, G. & Sighinofi, P. (1982). Platinum and palladium in the Ivrea-Verbano Basic Complex, Western Alps, Italy. *Econ. Geol.*, **72**, 1548–55.

Foose, M. P., Economou, M. & Panayiotou, A. (1985). Compositional and mineralogic constraints on the genesis of ophiolite hosted nickel mineralization in the Pevkos Area, Limassol Forest, Cyprus. *Mineral. Deposita*, **20**, 234–40.

Garuti, G. & Rinaldi, R. (1985). Mineralogy of Melonite-Group PGM and other tellurides from the Ivrea-Verbano Basic Complex, Western Italian Alps. *Econ. Geol.*, **81**, 1213–17.

Gautschi, A. & Montrasio, A. (1978). Die andesitisch-basaltischen Gänge des Bergeller Ostrandes und ihre Beziehung zur Regional- und Kontaktmetamorphose. *Schweiz. Mineral. Petrogr. Mitt.*, **58**, 329–43.

Haffty, J., Riley, L. B. & Goss, W. D. (1977). Determination of iridium and ruthenium in geologic samples by fire assay and emission spectrography. *U.S. Geol. Survey Bull.*, **1445**, 45–8.

Haffty, J., Haubert, A. W. & Page, N. J. (1980). Determination of iridium and ruthenium in geologic samples by fire assay and emission spectrography. *U.S. Geol. Survey Prof. Pap.* 1129-G, pp. G1–G4.

Ishiwatari, A. (1985). Alpine ophiolites: product of low-degree mantle melting in a Mesozoic transcurrent rift zone. *Earth Planet. Sci. Lett.*, **76**, 93–108.

Loney, R. A., Himmelberg, G. R. & Coleman, R. G. (1971). Structure and petrology of the alpine-type peridotite at Burro Mountain, California, U.S.A. *J. Petrol.*, **12**, 245–309.

Mitchell, R. H. & Keays, R. R. (1981). Abundance and distribution of gold, palladium and iridium in some spinel and garnet lherzolites: implication for the nature and origin of precious metal-rich intergranular compounds in the upper mantle. *Geochim. Cosmochim. Acta*, **45**, 1245–62.

Page, N. J & Talkington, R. W. (1984). Palladium, platinum, rhodium, ruthenium and iridium in peridotites and chromites from ophiolite complexes in Newfoundland. *Can. Mineral.*, **22**, 137–50.

Page, N. J, Haffty, J. & Ahmad, Z. (1980). Palladium, platinum and rhodium concentrations in mafic and ultramafic rocks from the Zhab Valley and Dargai Complex, Pakistan, *U.S. Geol. Survey Prof. Pap.* 1124-F, 6 pp.

Page, N. J, Cassard, D. & Haffty, J. (1982a). Palladium, platinum, rhodium, ruthenium and iridium in the chromites from the Massif du Sud and Thiebaghi Massif, New Caledonia. *Econ. Geol.*, **77**, 1571–7.

Page, N. J, Pallister, J. S., Brown, M. A., Smewing, J. O. & Haffty, J. (1982b). Palladium, platinum, rhodium, iridium and ruthenium in chromite-rich rocks from Semail Ophiolite, Oman. *Can. Mineral.*, **20**, 537–48.

Page, N. J, Aruscavage, P. J. & Haffty, J. (1983). Platinum group elements in rocks from the Voikar-Syninsky Ophiolite Complex, Polar Urals. *Mineral Deposita*, **18**, 443–55.

Page, N. J, Engin, T., Singer, D. A. & Haffty, J. (1984). Distribution of platinum-group elements in the Bati Kef chromite deposit, Guleman-Elazig Area, Eastern Turkey. *Econ. Geol.*, **79**, 177–84.

Peters, T. (1963). Mineralogie und Petrographie des Totalp-serpentins bei Davos. *Schweiz. Mineral. Petrogr. Mitt.*, **43**, 531–685.

Prichard, H. M., Neary, C. R. & Potts, P. J. (1986). Platinum group minerals in the Shetland Ophiolite. In *Metallogeny of the Basic and Ultrabasic Rocks*, eds M. J. Gallagher, R. A. Ixer, C. R. Neary & H. M. Prichard, The Institution of Mining and Metallurgy, London, pp. 395–414.

Simon, F. O., Aruscavage, P. J. & Moore, R. (1978). Determination of platinum, palladium and rhodium in geologic material by fire-assay and atomic absorption spectroscopy using electrothermal atomization (abstr.). *Am. Chem. Soc. Natl. Mtg.*, 176th Miami Beach, Florida, Sept. 11–14.

Trommsdorff, V. & Dietrich, V. (1980). Alpine metamorphism in a cross-section between the Rhine and Valtellina valleys (Switzerland, Italy). In *Geology of Switzerland, Part B. geological excursions*, ed. Schweiz. Geol. Kommission, Wepf Basel, New York, pp. 317–34.

Trommsdorff, V. & Evans, B. W. (1972). Progressive metamorphism of antigorite schists in the Bergell tonalite aureole (Italy). *Am. J. Sci.*, **272**, 423–37.

Trommsdorff, V. & Evans, B. W. (1977). Antigorite-Ophicarbonate Contact Metamorphism in Valmalenco. *Contr. Mineral. Petrol.*, **62**, 301–12.

Westland, A. D. (1981). Inorganic chemistry of platinum-group elements. In *Platinum-Group Elements: Mineralogy, Geology, Recovery*, ed. L. J. Cabri. Canadian Institute of Mining and Metallurgy, pp. 7–18.

18

Further Data on Platinum–Palladium Minerals from the Ivrea-Verbano Sulphide Deposits

G. GARUTI

Istituto di Mineralogia e Petrologia, Università di Modena, Via S. Eufemia 19, I-41100 Modena, Italy

&

R. RINALDI

Dipartimento di Scienze della Terra, Università di Cagliari, Via Trentino 51, I-09100 Cagliari, Italy

The continuation of a systematic electron microprobe study of the platinum-group minerals occurring in ore samples from the Ivrea-Verbano basic complex (Western Italian Alps), has led to a better definition of the range of solid solution in the $NiTe_2$–$PdTe_2$–$PtTe_2$ ternary field.

The postulated existence of a complete solid solution between melonite and merenskyite is demonstrated beyond doubt by the present data; whereas a marked solubility gap seems to occur along the melonite-moncheite join. In fact compositions showing a wide range of Ni-Pt substitution are found only between palladium melonites and moncheites therefore suggesting the need for considerable Pd substitution to promote solid solution between these two phases.

The observed paragenetical associations suggest the existence of phase relations between relatively narrow compositional fields, corresponding to the coexistence of Ni-merenskyite, Pd-moncheite, and Pd-melonite. Pure melonite is never found associated with members pertaining to the merenskyite and moncheite fields.

The melonite compositions are generally more metal rich than merenskyite and moncheite. The excess metal has a positive correlation with the Ni content therefore confirming the postulated existence of a solid solution between melonite and NiTe.

The latest analytical data revealed that substitutions of Bi for Te and Pt for Ni in melonites reach up to 5 and 8 at.% respectively, thereby extending the previously observed ranges of substitutions.

The Pt-Pd-rich assemblage characterized by the presence of palladian melonite, merenskyite, and moncheite, was found to contain also sperrylite as a major Pt phase; the latter being the first documented occurrence at this locality.

19

Platinum-Group Elements and Au Distribution in Ni Arsenide-Chromite Veins from the Rifo-Betic Lherzolite Massifs (Morocco–Spain)

M. Leblanc

Centre Géologique et Géophysique, Université des Sciences et Techniques du Languedoc, 34060 Montpellier, France

&

F. Gervilla-Linares

Departamento de Mineralogia-Petrologia, Universidad de Granada, 18002 Granada, Spain

ABSTRACT

In the lherzolite massifs distributed along the Western Mediterranean Alpine belt from Southern Spain (Carratraca, Ojen, Ronda) to Northern Morocco (Beni-Bousera) there are about 50 veins of a very peculiar Ni-Cr mineralization including an association of niccolite and Zn-V-rich chromite with accessory arsenides, sulpho-arsenides, sulphides and graphite; the gangue minerals are orthopyroxene or cordierite. They are associated with pyroxenite dikes related to the late stages of upwelling and partial melting of the mantle peridotites. Mineral association and textural relations have led Oen (1973) to propose a high temperature magmatic segregation of an immiscible oxyarsenide liquid for the origin of these veins.

Ten samples of Cr-Ni ores from Morocco and Spain were analysed by fire-assay and neutron activation (XRAL, Don Mills, Ontario). They contain from 0·7 to 5·2 ppm platinum-group elements (PGE) with up to 1·5 ppm of Pt and of Pd. They have high gold contents (3·5–18 ppm) and there is a positive correlation between Au–Ni–As and PGE.

These Cr-Ni ores exhibit flat PGE patterns with an Os depletion. They show a large range of values from the Ni-poor ores, which have only 10 times the values obtained for mantle peridotites, to the Ni-rich ores which have chondritic values. However the PGE patterns are similar to those of their primitive mantle source rock, suggesting that they were unfractionated during the concentration processes.

From a single sample (Beni-Bousera) separated chromite displays a PGE pattern with a strong negative slope whereas the corresponding niccolite exhibits a strong positive slope. The chromite pattern closely resembles podiform chromitites, but Os-Ir values are 10 times higher; the niccolite pattern is similar to the trend of the

Merensky Reef. Thus there was a partitioning of PGE in the magma between chromite (Os-Ir) and niccolite (Pt-Pd). PGE minerals have not yet been observed.

A chromite pod and a spinel layer (Ronda) were analysed. The PGE trend of the chromite displays an Os depletion and a moderate negative slope, intermediate between the field of the podiform chromitites and the flat trend of the Ni-Cr ores. The spinel has a PGE trend characterized again by a negative slope from Os to Pd (330–30 ppb) but shows a strong Pt enrichment (230 ppm).

The authors feel that these rocks represent a new type of PGE and gold mineralization in mantle peridotites. The concentration of PGE, up to chondritic values, is ascribed to the segregation of an immiscible oxyarsenide liquid from a magma resulting from the partial melting of a mantle diapir.

INTRODUCTION

The concentration of PGE is generally ascribed to early magmatic processes in ultramafic-mafic complexes (Naldrett *et al.*, 1979; Cabri, 1981; Crocket, 1981; Cabri & Naldrett, 1984). In undepleted upper mantle material the PGE chondrite patterns are nearly flat and PGE absolute abundances are low (Jagoutz *et al.*, 1979; Mitchell & Keays, 1981). The main concentrations occur in large layered stratiform intrusions (Bushveld, Stillwater) and in Ni-Cu magmatic sulphide deposits (Norilsk, Sudbury). In both cases, the most abundant PGE are Pt, Pd and Rh, they are concentrated in sulphide and arsenide phases; the PGE patterns show a strong positive slope. Enrichment of Ru, Os and Ir relative to Pd, Pt and Rh have been reported in ophiolitic podiform chromitites (Page *et al.*, 1982; Page & Talkington, 1984) containing Ru-Os-rich laurite inclusions in chromite (Prichard *et al.*, 1981, 1986; Stockman, 1982; Stockman & Hlava, 1984), the PGE patterns display a strong negative slope.

A peculiar type of Ni-Cr mineralization in the mantle-derived alpine-type lherzolite massifs of Malaga (South Spain) and Beni-Bousera (North Morocco) shows a unique example of magmatic segregation of Ni-arsenide and chromite. Enriched PGE contents have been reported in the Beni-Bousera Ni-Cr ores (Leblanc & Johan, 1986). The purpose of this paper is (1) to evaluate the PGE contents of these Ni-Cr ores in South Spain and Morocco; (2) to define the PGE distribution between chromite and Ni-arsenide; (3) to discuss the PGE patterns with respect to undepleted mantle rocks, stratiform complexes and ophiolitic podiform chromitites.

GEOLOGIC SETTING

The lherzolite massifs of Malaga (Dickey, 1970; Loomis, 1972; Obata, 1980; Frey *et al.*, 1985) and Beni-Bousera (Kornprobst, 1969) belong to a discontinuous string of lherzolite massifs (Fig. 1), which extends along the internal zone of the Western Mediterranean Alpine belt from Southern Spain (Carractraca, Ojen, Ronda), through Northern Morocco (Ceuta, Beni-Bousera), to Algeria (Collo, Edough).

The Ronda (300 km) and Beni-Bousera (70 km) massifs are the most important ones.

Structurally, the Ronda massif is a 1·5 km thick slab of peridotites forming part of a thrust sheet (Tubia, 1985) that exhibits a zoned metamorphic aureole of kinzigite, migmatite, gneiss which grades into a sillimanite, staurolite, garnet and biotite bearing micaschist away from the massif (Loomis, 1972; Tubia, 1985). High temperature emplacement (900°C) of the peridotite sheet resulted in the mylonitization of the lowermost part of the peridotite slab and developed a syntectonic migmatization in the crustal rocks (Tubia & Cuevas, 1986).

The peridotites range from dunites to pyroxenites with lherzolites and harzburgites predominating; they exhibit tectonite textures which are typical of mantle

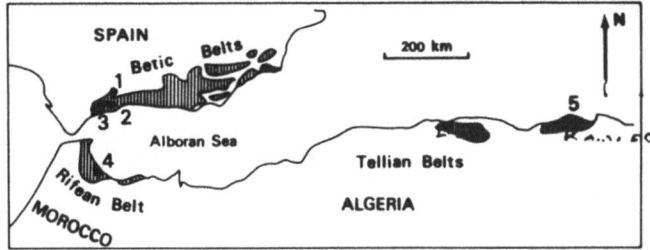

FIG. 1. Sketch map of the investigated lherzolite massifs (black) of the Western Mediterranean Alpine belt. 1, Carratraca; 2, Ojen; 3, Ronda; 4, Beni-Bousera; 5, Collo.

material (Darot, 1973). An incomplete internal zonation is observed from the NW border inwards (Obata, 1980): (i) strongly foliated and fine-grained garnet lherzolite; (ii) little deformed and coarse-grained spinel lherzolite; (iii) annealed plagioclase lherzolite. Two types of mafic layers occur as concordant bands of 1 cm to 3 m thickness, mainly within the garnet lherzolite (Dickey, 1970): (1) a tectonic-type, comprising websterite or enstatite, characterized by bright green chromian diopside; (2) a magmatic-type, varying in composition from garnet pyroxenite in garnet lherzolite to olivine gabbro in plagioclase lherzolite. On the basis of mineral compositions, Obata (1980) concluded that, after a prior differentiation by partial melting under mantle conditions, the peridotite was equilibrated at 1100–1200°C at 20–25 kb. The lherzolite zonation reflects variations in recrystallization pressure during dynamic cooling as the sub-solidus peridotite diapir ascended from 70 km depth to intrusion levels in the crust, where a contact metamorphism is developed. Frey *et al.* (1985) assume that this zonation resulted from variable degrees of melting and incomplete melt segregation during the diapir upwelling prior to sub-solidus recrystallization. The less fractionated melts are placed in the plagioclase lherzolite zone and the more fractionated ones occur in the garnet lherzolite zone (Suen & Frey, 1987). An extensive isotopic study revealing strong heterogeneities led Polvé & Allègre (1980) to suggest that the Western Mediterranean orogenic lherzolite massifs are complex tectonic sandwiches of mantle. The subsequent emplacement of the peridotite massifs by thrusting in the internal tectonic units of the Alpine Betic belt was associated with extensive serpentinization along the late tectonic contacts.

The Beni-Bousera peridotite outcrops in a NW-SE trending antiform and is surrounded by a metamorphic aureole including kinzigites and gneisses (Kornprobst, 1969). The peridotite displays tectonic textures (Reuber *et al.*, 1982) with a NNW-SSE stretching mineral lineation. The strong positive gravity anomaly associated with this massif suggests a connection at depth with the upper mantle (Bellot, 1985). The core of the massif consists mainly of spinel lherzolite displaying a thin pyroxenite banding (Kornprobst, 1969); the lherzolite grades outwards into harzburgite and dunite including abundant garnet-clinopyroxenite bands. The peridotite diapir originated from a relatively deep mantle zone (100 km) during a transcurrent process related to the Atlantic opening (Kornprobst & Vielzeuf, 1984). During the upwelling of the peridotite massif, partial melting along its borders resulted in a refractory residue of harzburgite and dunite; the melts crystallized at 1000°C and greater than 15 kb as garnet-clinopyroxenite bands (Kornprobst, 1969). The peridotite massif was emplaced into the continental crust along an Alpine shear zone, the corresponding shear-heating resulted in the development of a metamorphic aureole (Reuber *et al.*, 1982). The peridotites were strongly serpentinized along late tectonic contacts.

Ni-Cr MINERALIZATION

The occurrence of chromite–niccolite ores in the peridotite massifs of the province of Malaga, Spain, was first reported by Gillman (1986). About 40 Ni-Cr veins are recorded, mainly in the small massifs of Ojen and Carratraca. Approximately 10 Ni-Cr veins are also known in the Beni-Bousera massif, Morocco (Agard *et al.*, 1959; Leblanc, 1986).

The Ni-Cr ores occur in lenticular veins (Fig. 2), a few tens of metres in extent and less than 1 metre thick. They are generally too small for economic exploitation, and only a few thousand tons of nickel ore was produced. The Ni-Cr veins are concordant (Ronda) or discordant (Beni-Bousera) with respect to the peridotite foliation plane. They are often associated with orthopyroxenite dykes and lenses. Some samples exhibit parallel bands (Fig. 2) of Ni-Cr ore and of pyroxenite or cordierite-rich rock, whereas Ni-Cr nodules and veinlets are locally found in the pyroxenite dykes. Ni-Cr ores grade from almost pure chromitite to nickel-rich ore (10–15% Ni) even in the same Ni-Cr vein (Beni-Bousera). The chromite-rich ores display a banding and consist of euhedral chromite grains (0·15–0·5 mm) in an orthopyroxene gangue; rounded to euhedral niccolite inclusions and euhedral orthopyroxene inclusions are both found in the chromite grains. Oen (1973) described polyphase inclusions (niccolite–pyrrhotite–rammelsbergite–chalcopyrite) in the chromite of La Gallega (Ojen). In the Ni-rich ores, the chromite grains (20–30%) are rounded, embayed and cracked; they are surrounded by and their cavities and cracks are filled with polygonally textured niccolite; orthopyroxene is only found as small inclusions in the chromite together with niccolite inclusions.

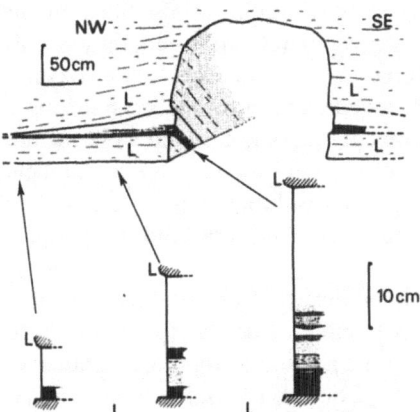

FIG. 2. Sketch of a concordant Ni-Cr vein from La Gallega (Ojen, Spain). Black: Chromite-(niccolite); white: cordierite gangue; L: surrounding lherzolite.

Oen *et al.* (1973, 1980) have carefully investigated the mineralogy of the Ni-Cr mineralization. The Zn-V-rich chromite and niccolite assemblage contains accessory arsenides (maucherite, orcelite, Ni-rich loellingite, Ni-rich safflorite, pararammelsbergite) (Fe, Co)-rich gersdorffite), sulphides (pyrrhotite, pentlandite, millerite, healzewoodite, polydimite, violarite, bornite, chalcopyrite), native elements (copper, gold), copper–gold alloys and graphite. Orthopyroxene and cordierite in some deposits are the main silicate gangue minerals; red-brown mica, cordierite, plagioclase are locally found in late pull-apart structures a few millimetres thick (Oen *et al.*, 1973). Cordierite occurs also as occluded patches or bands within the Ni-Cr ores, these cordierite-rich rocks grade laterally into the chromite-rich ore through a zone full of fine chromite inclusions and then a zone with symplectitic chromite–cordierite intergrowths.

Symplectitic chromite–cordierite intergrowths are regularly found attesting that cordierite, which is usually formed as a metamorphic mineral, crystallized here with chromite as a high temperature magmatic mineral. From a comparison of chromite composition with available experimental data, and considering the high temperature stability field of Mg-cordierite and phlogopite, together with paragenetical and textural evidence, Oen (1973) suggested that the chromite, cordierite and phlogopite association crystallized at magmatic temperatures between approximately 1300° and 1000°C at low oxygen fugacities and pressures between 10 and 5 kb. This is consistent with a magmatic crystallization of partial melting liquids during the late stages of the mantle diapir upwelling.

The Ni-arsenides included in the chromite grains are associated with Fe-Ni-Cu sulphides. The four phases observed (niccolite, pyrrhotite, rammelsbergite and chalcopyrite) are present in approximately constant proportion (Oen, 1973). The homogeneous composition of these inclusions may reflect the segregation of immiscible As-Ni-rich droplets during crystallization of the chromite grains and precludes a secondary hydrothermal alteration of magmatic sulphides to arsenides.

Oen (1973) stressed the role of liquid immiscibility in the formation of these Ni-arsenide ores, in a way similar to that generally ascribed to the genesis of magmatic Fe-Ni-Cu sulphides. Oen suggested that an oxyarsenide liquid coexisted with a separate orthopyroxene and chromite crystallizing magma. Small orthopyroxene grains and drops of Ni-As-rich liquid were included into chromite crystals. Later segregations of Ni-As-rich interstitial liquid included chromite grains, which were partly dissolved resulting in a rounded shape with embayments (Leblanc, 1986). Niccolite must have crystallized at liquidus temperature near 800°C (Oen, 1973).

Arsenic might possibly be derived from low-melting disseminated arsenides in the parental rocks as represented by accessory orcelite, niccolite and maucherite grains identified in the mantle magmatic mineral assemblages of the Beni-Bousera peridotite (Ramdohr, 1967; Lorand, 1983); the observation of Ni-arsenides associated with Fe-Ni-Cu sulphides in the high-pressure garnet pyroxenites from Beni-Bousera supports the hypothesis of a mantle origin for arsenic (Lorand, 1987). Late deuteric alterations (chlorite, serpentine, talc) postdate the Ni-Cr ore crystallization.

Small chromitite pods or schlierens, graphite veins and Fe-Ni-Cu sulphide mineralization also occur in the lherzolite massifs. Native gold and scarce grains of skutterudite and safflorite have been found during alluvial exploration in the Beni-Bousera massif (Aloui and Leblanc, 1971); alluvial PGE occurrences are known in the Malaga province (Orueta, 1919).

SAMPLES AND ANALYTICAL METHODS

Nine samples of Ni-Cr ores grading from chromite-rich (90% chromite) to Ni-rich ores (65% Ni As) have been collected in Beni-Bousera (3 samples), Ojen (4 samples) and Carratraca (2 samples). A sample of the separated chromite phase (sample 9b), including 1–3% arsenide inclusions, was obtained from the Ni-rich ore of Beni-Bousera (sample 9a). A massive spinel sample from a spinel-clinopyroxene-olivine banded peridotite and a chromitite from a foliated pod have been collected in El Arroyo de la Cala (Ronda massif). A sample of a chromitite pod from the lherzolite massif of Collo (Algeria) was also analysed.

These whole rock ore samples were analysed by X-Ray Assay Laboratories Ltd, Don Mills, Ontario. The procedure involves a fire-assay followed either by Neutron Activation or Plasma Mass Spectrometry. The detection limits are Pt, Pd, Ru and Re, 5 ppb; Os, 3 ppb; Rh and Au, 1 ppb; and Ir, 0·1 ppb. Talkington & Watkinson (1986) have obtained a standard deviation of approximately 10–15% for Pt, Rh, Os, Ir, Ru and 25% for Pd from a chromitite sample from Newfoundland analysed by X-Ray Assay Labs. Cr, Ni and As contents were analysed in nine samples by X-ray fluorescence, with a 0·012% detection limit. For the three other ore samples, chromite/niccolite ratios were inferred from microscopic observations. The Ni-arsenide contents of the investigated ores vary from 2 to 80% (Table 1). Complementary microprobe analyses of chromites and silicates of the Moroccan (Leblanc,

TABLE 1

Analytical data (PGE and Au: ppb; Cr, Ni and As: %)

No.	Rock type	Location	Re	Os	Ir	Ru	Rh	Pt	Pd	ΣPGE	Au	Cr	Ni	As	'NiAs'
1	Chromite pod	Ronda (SP)	5	58	180	320	40	190	130	920	1 600	30·9	0·61	0·01	0
2	Spinel band	Ronda (SP)	5	330	110	130	23	230	30	853	6	3·03	0·16	0·01	0
3	Cr (Ni) ore	Baeza. Ojen (SP)	5	12	50	95	12	75	60	304	48	27·20	2·00	0·05	1–2
4	Cr-Ni ore	La Gallega. Ojen (SP)	5	56	120	210	35	250	260	930	3 400	10·70	4·71	0·28	5–10
5	Cr-Ni ore	La Gallega. Ojen (SP)	50	110	170	270	49	310	330	1 240	3 700	13·70	7·76	10·10	20
6	Cr-Ni ore	La Gallega. Ojen (SP)	25	16	150	490	65	460	610	1 791	8 000	12·20	11·50	16·20	30
7	Cr-Ni ore	Los Jarales. Carratraca (SP)	100	68	210	360	91	410	540	1 679	15 000	12·90	11·10	12·80	25
8	Cr-Ni ore	Los Jarales. Carratraca (SP)	10	18	190	420	62	430	550	1 670	9 200	15·20	12·20	15·60	30
9a	Cr-Ni ore	Beni-Bousera (M)	360	400	620	1 000	240	1 500	1 400	5 160	18 000	8·32	28·20	37·10	65
9b	Separated Cr	Beni-Bousera (M)	—	1 200	780	160	12	90	120	2 350	5 800	—	—	—	2–3
	Calculated														
Ni	NiAs	Beni-Bousera (M)	—	140	180	940	230	1 980	1 850	5 320	12 300	—	—	—	—
10	Cr-Ni ore	Beni-Bousera (M)	—	350	300	700	—	1 600	1 500	4 530	11 000	—	—	—	80
11	Cr-Ni ore	Beni-Bousera (M)	260	57	77	140	27	190	220	712	11 000	—	—	—	10
12	Chromite pod	Collo (Algeria)	25	1 100	770	2 700	28	20	25	4 620	9	26·1	0·11	tr.	0

'NiAs': approximate volume percent of Ni-arsenides; — element not determined; SP: Spain; M: Morocco.

1986), Spanish (Gervilla *et al.*, 1987) and Algerian (Temagoult, current work) ores have been obtained at the University of Sciences of Montpellier (France).

RESULTS AND INTERPRETATION

Analytical results are listed in Table 1. The total PGE contents of the whole rock ore samples range from 300 to more than 5000 ppb. The Ni-Cr ores are the richest in PGE, with up to 1600 ppb Pt and 1500 ppb Pd, they constitute a newly recognized type of PGE mineralization (Leblanc & Johan, 1986). Their PGE contents are positively correlated with Ni (Fig. 3) and As contents.

One way to examine the PGE data is to use chondrite normalized plots as proposed by Naldrett *et al.* (1979). Chondrite concentrations (ppb) are those of chondrite C2 (McBryde, 1972) = Os, 700; Ir, 500; Ru, 1000; Rh, 200; Pt, 1200; Pd, 1500. Figure 4 shows the chondrite normalized patterns for Ni-Cr ores from Spain and Morocco. The PGE normalized values range from chondritic values to 0·1 times chondrite. This wide range is related to the NiAs contents of the samples with the highest PGE contents in niccolite-rich ores. The overall trends are flat, except for an Os depletion. Such a sharp change in Os to Ir is unusual except for some ophiolitic chromitites (Page *et al.*, 1982), it implies a different behaviour for Os and Ir. The richest Ni ore (sample 10) shows a slightly positive slope from Os to Pt, indicating a Pt enrichment. The Ni-poorest sample displays a stronger depletion of Rh, Pt, Pd than of Ir and Ru with respect to the average chondrite. These PGE

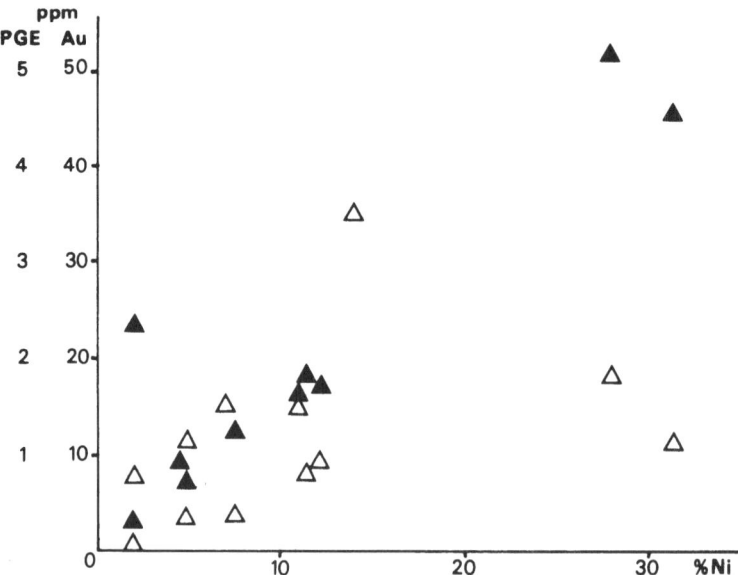

FIG. 3. Plot of PGE (total) versus Ni (closed symbols) and Au versus Ni (open symbols) from Table 1, including some previous analyses for gold and nickel.

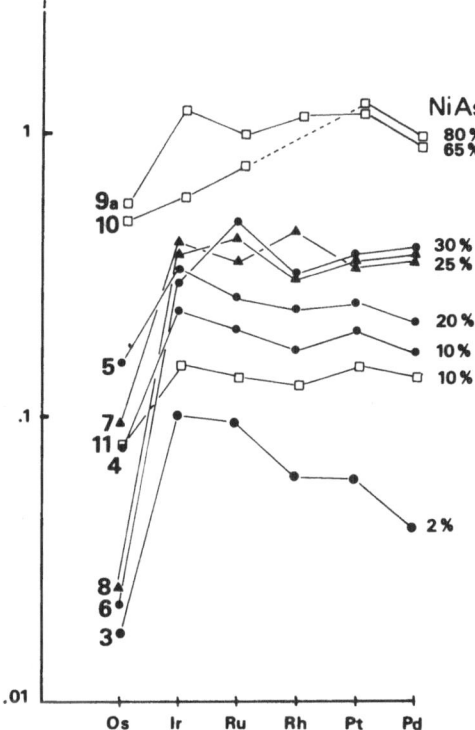

FIG. 4. Chondrite normalized PGE ratios for Ni-Cr ores with their approximate NiAs content (%). Open squares: Beni-Bousera; black dots: Ojen; black triangles: Carratraca.

contents are 10–100 times the values obtained for a lherzolite from Ronda (Stockman, 1982) which also displays a flat pattern typical of undepleted mantle material. The Ni-Cr ores have unfractionated patterns similar to those of their primitive mantle source-rock. PGE concentrations up to chondritic values, that is to say 100 times the peridotite contents, occur in the NiAs-rich ores.

Figure 5 concerns a Ni-rich ore (sample 9a) from Beni-Bousera with a flat and chondritic PGE pattern. The corresponding separated chromite (sample 9b) displays a strong depletion of Ru, Rh, Pt, Pd and a slight enrichment of Os and Ir with respect to the average chondrite. From the mineral composition of this ore sample (65–70% niccolite, 25% chromite, 5% silicate) and from the PGE contents of the whole rock and of the separated chromite, the PGE content of the corresponding niccolite has been calculated (Table 1). The calculated pattern for pure niccolite shows an opposite trend with a strong depletion of Os, Ir and a slight Pt, Pd enrichment. The mixing of the chromite-pattern with a 'negative' slope and the niccolite-pattern with a 'positive' slope, may result fairly well in the flat pattern of the whole rock. Pt and Pd are probably contained in the arsenide phase, as this is suggested by the PGE versus Ni correlation diagram (Fig. 3), and Os and Ir in the chromite phase. There was probably a PGE partition between the two main

M. Leblanc, F. Gervilla-Linares

crystallizing phases in the magma:chromite (Os and Ir rich) and niccolite (Pt and Pd rich). Pt-Pd concentration in niccolite could be facilitated by a solid solution between PdAs, PtAs and NiAs (Pearson, 1967), which all have the same NiAs structure (Hansen & Anderko, 1958). The calculated PGE pattern of niccolite is quite similar to that observed in the Merensky Reef (Hiemstra, 1979). The chromite pattern parallels the field of the ophiolitic podiform chromitites (Page *et al.*, 1982; Talkington, 1986), except for Pd and Pt which may indicate small niccolite inclusions in chromite grains. The overall PGE values of the chromite sample 9b are about 10 times higher than those of the ophiolitic chromitites: Os, Ir contents are over the chondrite average and are the highest reported for chromite-rich samples, this bears some resemblance to the chromite-rich rocks of Harolds Grave, Shetland (Prichard *et al.*, 1981, 1986). These high Os, Ir contents could be ascribed to laurite and osmium-rich alloys in chromite, but these PGE minerals have not yet been

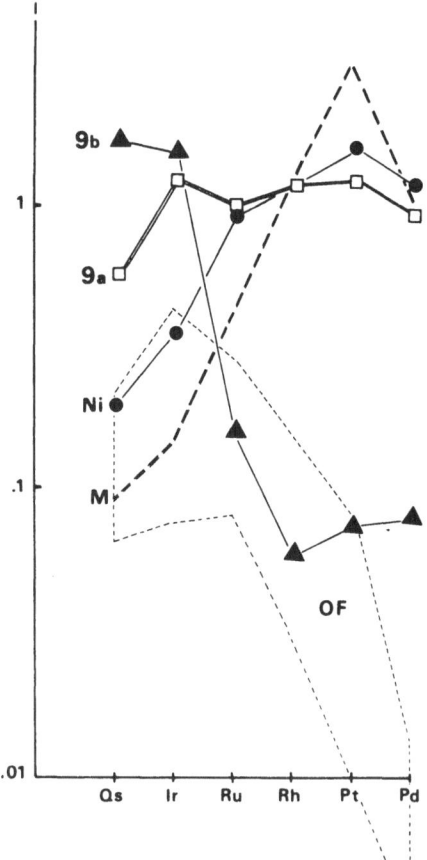

FIG. 5. Chondrite normalized PGE ratios for a Ni-Cr ore from Beni-Bousera (9a) and its separated chromite (9b) and the corresponding calculated niccolite (Ni), with reference (M) to the Merensky Reef composite pattern (Hiemstra, 1979) and (OF) the ophiolite field (Page *et al.*, 1982).

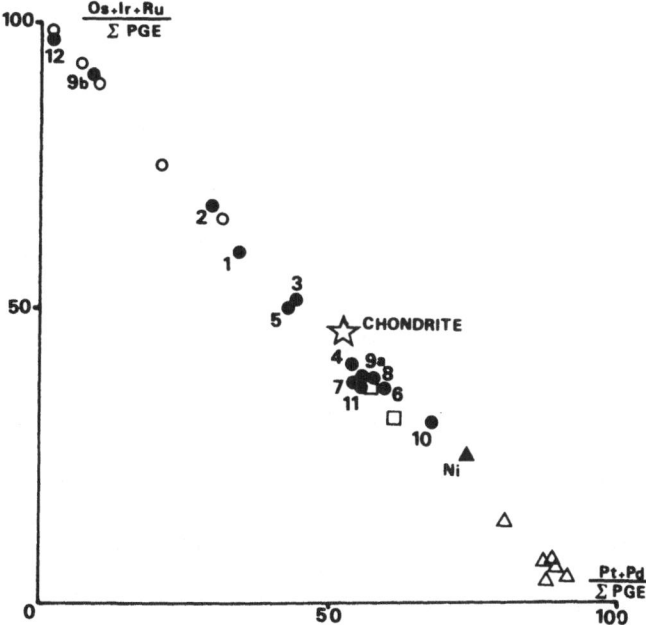

FIG. 6. PGE ratio in Ni-Cr ores separated chromite (black dots) and calculated niccolite (black triangles) with references to chondrite (empty star), komatiites (empty squares), magmatic stratiform (empty triangle), and ophiolitic podiform chromitites (empty circles). Chromite pod and spinel band are also plotted (black dots 1 and 2). Numbers refer to samples in Table 1. References are from Barnes *et al.*, (1985) and Legendre (1982).

observed. The inversely related patterns of PGE depletion and enrichment in ophiolitic podiform chromitite and stratiform chromitite are generally ascribed to major differences in the petrogenesis of the corresponding ultramafic-mafic complexes. Here, the opposite patterns are both found in the same rock sample. This suggests that PGE fractionation depends both on differences in solubility of PGE (Barnes *et al.*, 1985) and structure of the host minerals during the crystallization of Ni-Cr ores. The low solubility of Os, Ir and Ru leads to the early formation in the magma of Os–Ir alloys and RuS_2 which were included in the early crystallizing chromite (Legendre, 1982). The more soluble Pd, Pt and Rh were enriched in the magma and then concentrated in the immiscible oxyarsenide phase, in an analogous way as commonly observed in the case of a magmatic sulphide phase (Naldrett *et al.*, 1979, 1982). In theory, PdAs and NiAs show the same lattice spacing and may substitute for each other (Pearson, 1967); Pt and Pd ions may occupy the place of Ni ions in the NiAs structure (Hansen & Anderko, 1958). Figure 6 shows the PGE distribution between the Ir group (Os, Ir, Ru) and the Pd group (Pd, Pt, Rh) for the Ni-Cr ores. Most of the Ni-Cr ores show approximately chondritic and komatiitic Ir/Pd group ratios, but the Cr-rich ores and the separated chromite shows a trend towards the ophiolitic field (Ir-rich), whereas the Ni-rich ores and calculated niccolite show a trend towards the field of the stratiform magmatic complexes (Pd-rich).

Figure 7 is a plot of Cr_2O_3/R_2O_3 versus MgO/RO of chromites of the Ni-Cr ores from Spain and Morocco and includes earlier data from Oen *et al.* (1979). These chromites display a wide compositional range characterized by a decrease of Cr_2O_3/R_2O_3 with a decrease of MgO/RO. This trend parallels the magmatic trend of the chromites of the ultramafic-mafic stratiform complexes. The Ni-Cr ores plot in a peculiar field which overlaps partly, in its upper part, both the ophiolitic and stratiform fields and which extends widely downwards from the field of the stratiform complexes. In the upper part of the field of the Ni-Cr ores the chromites plot near the chromites of the UG2 and the Merensky platiniferous horizons of the Bushveld. The points representative of the chromites from the same deposit plot in the same restricted area within the Ni-Cr ore field. Cr-Mg-rich and Cr-Mg-poor chromite deposits may be found within the same massif although in different areas. This suggests that variations in chromite composition may be related to local differences in the composition of parental magmas and to the conditions of crystallization. Nevertheless in the same deposit (La Gallega, Ojen), the chromites of the Ni-Cr ores with cordierite show higher Al-Fe contents than those with orthopyroxene. With increasing amount of cordierite the chromites become less magnesian-rich and more ferrous-rich (Oen *et al.*, 1973), this suggests a magmatic evolutionary trend from the orthopyroxene to the cordierite crystallization. In each deposit, the late chromites appear to be enriched in Cr and Fe (Oen *et al.*, 1979) and could have re-equilibrated. The podiform chromitite of Ronda (sample 1) plots in the overlapping part of the three fields. The podiform chromitite of Collo (Algeria) is more clearly related to the ophiolitic field: it displays a large range of Cr_2O_3/R_2O_3 with an approximately constant MgO/RO ratio which corresponds to the typical compositional trend of chromites in podiform chromitites. The chromites of the Ni-Cr ores are obviously different from the accessory Cr-spinels in the surrounding peridotites; these latter Cr-spinels plot in the field of mantle lherzolites and display a partial melting trend characterized by a strong Cr_2O_3 enrichment. The aluminous spinels (up to 68% Al_2O_3) from a spinel band from Ronda (sample 2) also plot in the latter field and exhibit the same evolutionary trend as the accessory Cr-spinels in the lherzolite.

Figure 8 shows the PGE chondrite normalized values of investigated chromites and chromitites. There are differences in PGE abundances between samples, but all patterns display a negative slope for the chondrite normalized values. A similar PGE pattern with negative slope characterizes the ophiolitic chromitites. The chromitite pod from Ronda (sample 1) and the richest chromite Ni-Cr ore (sample 3) display a similar PGE trend, suggesting a common origin. They both show a PGE-pattern overlapping those of the ophiolitic podiform chromites (Page *et al.*, 1982; Talkington, 1986), but show a convex PGE curve due to lower Os and higher Rh, Pt, Pd values. The higher Rh, Pt, Pd values can be ascribed to Ni-arsenide micro-inclusions in chromite. The separated chromite (sample 9b) from the Beni-Bousera Ni-Cr ore shows a similar pattern, but with Os and Ir near chondrite values. These Os and Ir concentrations are high compared to those of ophiolitic

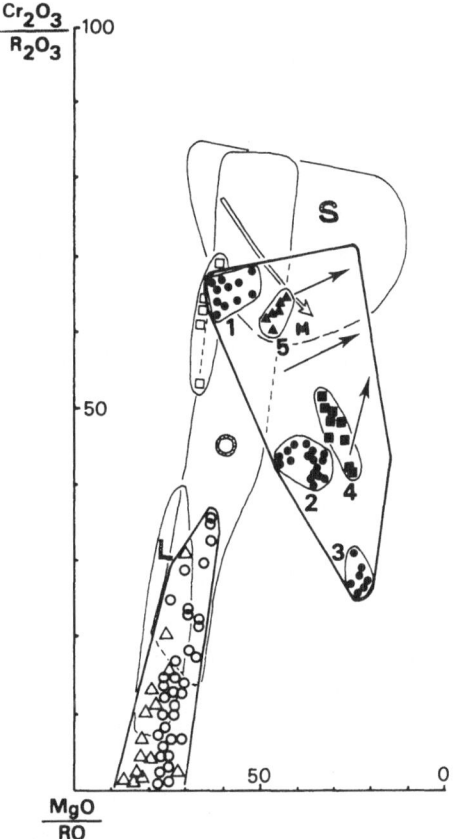

Fig. 7. Cr_2O_3/R_2O_3 versus MgO/RO for the chromites of the Ni-Cr ores. The chromites analysed in this work (black symbols) are from: (1) Mina Baeza (Ojen, Spain); (2) (3) La Gallega (Ojen, Spain), without and with cordierite, respectively; (4) Beni-Bousera (Morocco); (5) Arroyo de la Cala (Ronda, Spain). The Ni-Cr field includes previous data and arrows indicate compositional trend from early to late chromites in a same deposit (Oen *et al*, 1979). Open squares: chromite pod (Collo, Algeria); open circles: accessory spinels from the Beni-Bousera peridotite; open triangles: spinel bands from the Ronda peridotite. The reference fields are: L, lherzolite xenoliths (Frey & Green, 1974; Basu & McGregor, 1975); O, ophiolite (Leblanc *et al.*, 1980); S. stratiform magmatic complexes (Irvine, 1967) with the Bushveld magmatic trend (arrow) and the chromites (M) from the Merensky and UG2 platiniferous reefs (Waal de, 1975).

podiform chromites; the only ophiolite data in this range concerns the peculiar chromitites from Shetland (Prichard *et al.*, 1986). The chromitite pod of Collo, Algeria (sample 12), displays a nearly similar pattern and Os, Ir contents as the Beni-Bousera chromite (sample 9b), but the very high Ru content (2700 ppb) and low Rh, Pt, Pd values are striking features, resulting in a curve which crosses the ophiolitic field with a steep negative slope from Ru to Pd. The spinel band from Ronda (sample 2) has a PGE pattern falling partly within the ophiolitic field; once more a relatively high Os content is observed, while the positive Pt anomaly is also noteworthy.

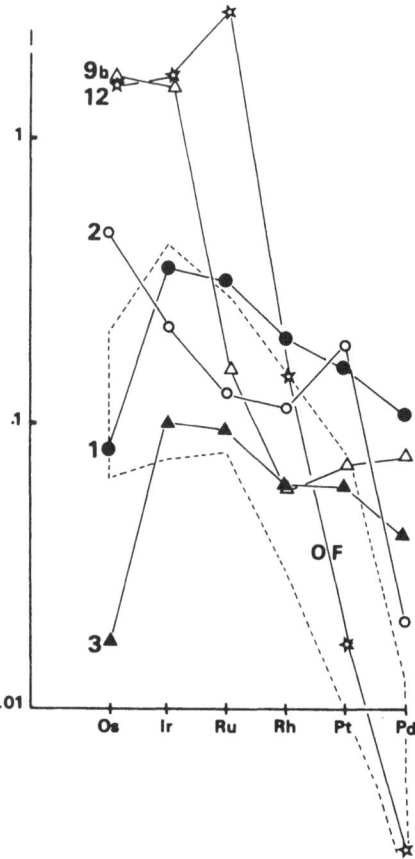

FIG. 8. Chondrite normalized PGE ratios for chromite and spinel samples with reference to the ophiolite field (OF). Black dots: chromite pod (Ronda); black triangles: chromite-rich Ni-Cr ore (Ojen); open triangles: separated chromite from Ni-Cr ore (Beni-Bousera); stars: chromite pod (Collo, Algeria); open circles: spinel bands from Ronda peridotite.

Gold contents range from 3·5 to 18 ppm in Ni-Cr ores analysed in this work (Table 1), but gold contents up to 35 ppm are indicated by previous analyses of Beni-Bousera ores (Leblanc, unpublished report, 1971). These high gold contents are orders of magnitude (1000–10 000) greater than normal lherzolite values (Dostal *et al.*, 1981; Mitchell & Keays, 1981). There is a positive correlation between Au and Ni (Fig. 3), and between Au and PGE contents. The link between gold and arsenide mineralization in many peridotite massifs has been emphasized by Leblanc (1986). In the Ni-Cr ores of Beni-Bousera, small inclusions (0·03–0·05 mm) of gold have been observed in niccolite (Leblanc, 1986) and Au–Cu alloys have been described in niccolite (Oen & Kieft, 1974; Vinogradova *et al.*, 1976). In chromite and spinel samples, gold contents are very low (<50 ppb), except for the chromite pod from Ronda (sample 1) which contains 1600 ppb Au and 0·61% Ni; this suggests a genetic link between the chromite pods and the Ni-Cr ores.

CONCLUSIONS

(1) Ni-Cr ores of the lherzolite massifs from the Alpine belt of South Spain and North Morocco contain 3–18 ppm gold and up to 1·5 ppm of platinum and of palladium. The highest contents are in NiAs-rich ores. This may represent a new type of gold and PGE mineralization in mantle peridotites.

(2) PGE of the Ni-Cr ores are unfractionated and display flat chondrite-normalized patterns similar to those of their primitive mantle rock-source; with respect to lherzolite PGE are enriched 100 times, up to chondrite values.

(3) In the Ni-Cr ores, there is a strong PGE partition between the two main mineral phases: chromite (Os, Ir) and niccolite (Pt, Pd). These two mineral phases display ophiolitic and stratiform PGE patterns, respectively, whereas the whole-rock exhibits a flat pattern. Oen (1973) proposed an early segregation of an immiscible oxyarsenide phase in a magma produced by partial melting of upwelling mantle lherzolites. In the partial melting liquid, the PGE were probably unfractionated. Os and Ir minerals crystallized early and were included within the chromite grains, whereas Pt and Pd were collected and concentrated in the immiscible Ni-arsenide phase.

(4) On a compositional diagram, chromites of the Ni-Cr ores plot in a peculiar field which extends widely outwards from the field of the stratiform ultramafic-mafic complexes and which is in its major part clearly separated from the ophiolite field. They closely resemble the negative slope PGE/chondrite pattern found in ophiolitic podiform chromites, but their Os, Ir or (and) Pt, Pd contents may be higher. Some chromite podiform bodies should be genetically related to the Ni-Cr ores.

ACKNOWLEDGEMENTS

The authors are indebted to Dr I. S. Oen and A. Temagoult for obtaining some complementary samples from Spain and Algeria. We thank Dr Oen for reviewing the manuscript and for improvement of the English text.

REFERENCES

Agard, J., Jouravsky, G. & Milliard, Y. (1959). Les gites minéraux (graphite, vermiculite, magnésite, nickel, cuivre, chrome) liés aux roches ultrabasiques et métamorphiques des Beni-Bousera (Rif Septentrional). *Mines et Géologie, Rabat*, **8**, 31–7.
Aloui, M. & Leblanc, M. (1971). Découverte de minéraux de cobalt (skuttérudite, safflorite) dans le massif ultrabasique des Beni-Bousera (Rif). *Notes Service Géologique Maroc*, **237**, 292.
Barnes, S.-J., Naldrett, A. J. & Gorton, M. P. (1985). The origin of the fractionation of platinum-group elements in terrestrial magmas. *Chem. Geol.*, **53**, 303–23.
Basu, A. R. & McGregor, I. D. (1975). Chromite spinels from ultramafic xenoliths. *Geochim. Cosmochim. Acta*, **39**, 937–45.
Bellot, A. (1985). Etude gravimétrique du Rif paleozoïque: la forme du massif des Beni-

Bousera. Thèse Doct-Ingenieur, Université Sciences Techniques Languedoc, Montpellier, France, 146 pp.

Cabri, L. J. (1981). The platinum-group minerals. In *Platinum-Group Elements: Mineralogy, Geology, Recovery*, ed. L. J. Cabri. Canadian Institute of Mining and Metallurgy Special Volume 23, Moscow State Univ. publ., pp. 83–150.

Cabri, L. J. & Naldrett, A. J. (1984). The nature of the distribution and concentration of platinum-group elements in various geological environments. *Proceedings of the 27th International Geological Congress*, vol. 10, pp. 17–46.

Crocket, J. H. (1981). Geochemistry of the platinum-group elements. In *Platinum-Group Elements, Mineralogy, Geology, Recovery*, ed. L. J. Cabri. Canadian Institute of Mining and Metallurgy Special Volume 23, pp. 47–64.

Darot, M. (1973). Cinématique de l'extrusion, à partir du manteau, des péridotites de la Sierra Bermeja (Serrania de Ronda, Espagne). *C.r. Acad. Sci. Paris*, **278**, 1673–6.

Dickey, J. S. Jr (1970). Partial fusion product, in Alpine-type peridotites: Serrania de Ronda and other examples. Mineralogy Society of America, Special Paper 3, pp. 33–49.

Dostal, J., Dupuy, C. & Leblanc, M. (1981). Distribution of gold and copper in ophiolites from New Caledonia. *Can. Mineral.*, **19**, 225–32.

Frey, F. A. & Green, D. H. (1974). The mineralogy, geochemistry and origin of lherzolite inclusion in Victorian basanits. *Geochim. Cosmochim. Acta*, **38**, 1023–99.

Frey, F. A., Suen, C. J. & Stockman, M. W. (1985). The Ronda high temperature peridotite: Geochemistry and Petrogenesis. *Geochim. Cosmochim. Acta*, **49**, 2469–91.

Gervilla, F., Torres-Ruiz, J. & Fenoll Hach-Ali, P. (1987). Las mineralizaciones de Cr-Ni de los macizos ultrabasicos de la provincia de Malaga (Sur de Espana). *Boltin Geol. Minero de Espana* (in press).

Gilman, F. (1986). Notes on the ore deposits of the Malaga serpentines. *Trans. Inst. Mining Metal., London*, **5**, 159–68.

Hansen, M. & Anderko, K. (1958). *Constitution of Binary Alloys*, Vol. 1. McGraw-Hill, New York, 305 pp.

Hiemstra, S. A. (1979). The role of collectors in the formation of the platinum deposits in the Bushveld complex. *Can. Mineral.*, **17**, 469–82.

Irvine, T. N. (1967). Chromian spinel as a petrogenetic indicator. Part 2, petrologic applications. *Can. J. Earth Sci.*, **4**, 71–103.

Jagoutz, E., Palme, H., Baddenhausen, M., Blum, K., Dreibus, G., Spettel, B., Lorenz, V. & Wanke, H. (1979). The abundance of major, minor and trace elements in the earth's mantle as derived from primitive ultramafic nodules. In *Early Solar System and Lunar Megolith*, ed. R. B. Merill, Proceedings 10th Lunar Planet. Sci. Conf., Vol. 2, pp. 2031–50.

Kornprobst, J. (1969). Le masif ultrabasique des Beni Bouchera (Rif interne, Maroc): étude des péridotites de haute température et de haute pression, et des pyroxénites, à grenat ou sans grenat qui leur sont associées. *Contribution Mineralogy Petrology*, **23**, 283–322.

Kornprobst, J. & Vielzeuf, D. (1984). Transcurrent crustal thinning: a mechanism of uplift of deep continental crust/upper mantle associations. In *Kimberlites II: the Mantle and Crust–Mantle Relationships*, Kornprobst edition, Elsevier, Amsterdam, pp. 347–59.

Leblanc, M. (1986). Co-Ni arsenide deposits with accessory gold in ultrabasic rocks from Morocco. *Can. J. Earth Sci.*, **23**(10), 1592–602.

Leblanc, M. & Johan, Z. (1986). Un nouveau type de minéralisation platinifère: exemple des filons à arseniures de nickel et chromite du massif lherzolitique des Beni-Bousera (Maroc). *C.r. Acad. Sci., Paris*, **303**, 163–6.

Leblanc, M., Dupuy, C., Cassard, D., Moutte, J., Nicolas, A., Prinzhoffer, A., Rabinovitch, M. & Routhier, P. (1980). Essai sur la genèse des corps podiformes de chromitite dans les peridotites ophiolitiques, etude des chromites de Nouvelle-Calédonie et comparaison avec celles de Mediterranée orientale. In *Ophiolites, Proceedings Int. Ophiolite Symp., Cyprus*, 1979, ed. A. Panayiotou, Geological Survey, Cyprus, pp. 691–701.

Legendre, O. (1982). Minéralogie et géochimie des platinoïdes dans les chromitites ophiolitiques. Comparaison avec d'autres types de concentrations en platinoïdes. Thèse Doct. 3° cycle, Universitaire Sciences, Paris VII, p. 157.

Loomis, T. P. (1972). Diapiric emplacement of the Ronda high temperature ultramafic intrusion, Southern Spain. *Geol. Soc. Amer. Bull.*, **83**, 2475–96.

Lorand, J. P. (1983). Les mineraux opaques des lherzolites à spinelle et des pyroxénites associées: une étude comparative dans les complexes orogéniques et dans les enclaves des basaltes alcalins. Thèse 3e cycle, Paris, p. 230.

Lorand, J. P. (1987). Sur l'origine mantellaire de l'arsenic dans les roches du manteau: exemple des pyroxénites à grenat du massif lherzolitique des Beni-Bousera (Rif. Maroc). *C.r. Acad. Sci., Paris*, **134**, 331–4.

McBryde, W. A. E. (1972). Platinum metals. In *The Encyclopedia of Geochemical and Environmental Sciences*, ed. R. W. Fairbridge. Van Nostrand Reinhold, New York, pp. 957–61.

Mitchell, R. H. & Keays, R. R. (1981). Abundance and distribution of gold, palladium and iridium in some spinel and garnet lherzolites—Implications for the nature and origin of precious metal-rich intergranular components in the upper mantle. *Geochim. Cosmochim. Acta*, **45**, 2425–42.

Naldrett, A. J., Hoffman, E. L., Green, A. H., Chou, C. L., Naldrett, S. R. & Alcock, R. A. (1979). The composition of nickel-sulfide ores, with particular reference to their content of platinum-group elements and gold. *Can. Mineral.*, **17**, 403–15.

Naldrett, A. J., Innes, D., Gorton, M. P. & Sowa, J. (1982). Compositional variations within and between five Sudbury ore deposits. *Econ. Geol.*, **77**, 1579–34.

Obata, M. (1980). The Ronda peridotite: Garnet-, Spinel-, and Plagioclase-lherzolite facies and the P-T trajectories of a high-temperature mantle intrusion. *J. Petrology*, **21**(3), 533–72.

Oen, I. S. (1973). A peculiar type of Cr-Ni-mineralizations: cordierite-chromite-niccolite ores of Malaga, Spain, and their possible origin by liquid unmixing. *Econ. Geol.*, **68**, 831–42.

Oen, I. S. & Kieft, C. (1974). Nickeline with pyrrhotite and cubanite exsolutions, Ni-Co-rich loellingite, and an Au-Cu alloy in Cr-Ni ores from Beni-Bousera, Morocco. *Neues Jahrbor Mineral*, 1974–1, 1–8.

Oen, I. S., Kieft, C. & Westerhof, A. B. (1973). Composition of chromites in cordierite and mica-bearing Cr-Ni ores from Malaga province, Spain. *Mineralogical Magazine*, **39**, 193–203.

Oen, I. S., Kieft, C. & Westerhof, A. B. (1979). Variation in composition of chromites from chromite-arsenide deposits in the peridotites of Malaga, Spain. *Econ. Geol.*, **74**, 1630–6.

Oen, I. S., Kieft, C., Burke, A. J. & Westerhof, A. B. (1980). Orcelite and associated minerals in the Ni–Fe–As–S system in chromitites and orthopyroxenites of Nebral, Spain. *Bulletin Minéralogie*, **103**, 198–208.

Orueta, D. (1919). Informe sobre el reconocimiento de la Serrania de Ronde. *Bol. Inst. Geol. Min. Esp.*, **40**, 200–33.

Page, N. J & Talkington, R. W. (1984). Palladium, platinum, rhodium, ruthenium and iridium in peridotites and chromitites from ophiolite complexes in Newfoundland. *Can. Mineral.*, **22**, 137–49.

Page, N. J, Cassard, D. & Haffty, J. (1982). Palladium, platinum, rhodium, ruthenium, and iridium in chromitites from the Massif du Sud and Tiebaghi Massif, New Caledonia. *Econ. Geol.*, **77**, 1571–7.

Pearson, W. B. (1967). *A Handbook of Lattice Spacings and Structures of Metals and Alloys*, Vol. 2. Pergamon Press, New York, p. 1446.

Polve, M. & Allegre, C. J. (1980). Orogenic lherzolite complexes studied by 87Rb-87Sr: a clue to understand the mantle convection processes. *Earth Planet. Sci. Letters*, **51**, 71–93.

Prichard, H. M., Potts, P. J. & Neary, C. R. (1981). Platinum-group element minerals in the Unst chromite, Shetland Isles. *Inst. Min. Metall.*, **90**, B186–B188.

Prichard, H. M., Neary, C. R. & Potts, P. J. (1986). Platinum-group minerals in the Shetland ophiolite. In *Metallogeny of Basic and Ultrabasic Rocks, Proceedings IMM Conference Edinburgh*, 9–12 April 1985, ed. Gallagher *et al.*, Institution of Mining and Metallurgy, London, pp. 395–414.

Ramdohr, P. (1967). A widespread mineral association, connected with serpentinisation. *N. Jahrb. Min.*, **107**, 241–65.

Reuber, I., Michard, A., Chalouan, A., Juteau, T. A. & Jermoumi, B. (1982). Structures and emplacement of the alpine-type peridotites from Beni-Bousera, Rif, Morocco: a polyphase tectonic interpretation. *Tectonophysics*, **82**, 231–51.

Stockman, H. W. (1982). Noble metals in the Ronda and Josephine peridotites. Ph.D. thesis, Massachusetts Institute of Technology, Cambridge, Mass., USA (unpublished), 180 pp.

Stockman, H. W. & Hlava, P. F. (1984). Platinum-group minerals in alpine chromitites from southwestern Oregon. *Econ. Geol.*, **79**, 491–508.

Suen, C. J. & Frey, F. A. (1987). Origin of the mafic and ultramafic rocks in the Ronda peridotite. *Earth Planet. Sci. Letters* (in press).

Talkington, R. W. & Watkinson, D. M. (1986). Whole rock platinum-group element trends in chromite-rich rocks in ophiolitic and stratiform igneous complexes. In *Metallogeny of Basic and Ultrabasic Rocks, Proceedings IMM Conference, Edinburgh*, 9–12 April 1985, ed. Gallagher *et al.*, Institution of Mining and Metallurgy, London, pp. 427–40.

Tubia, J. M. (1985). Significado de las deformaciones internas en las peridotitas de Sierra Alpujata (Malaga). *Estud. Geol.*, **41**, 369–80.

Tubia, J. M. & Cuevas, J. (1986). High-temperature emplacement of Los Reales peridotite nappe (Beltic Cordillera, Spain). *J. Struct. Geol.*, **8**, 473–82.

Vinogradova, R. A., Kroutov, G. A., Mikhailov, N. P., Roudachevsky, N. S. & Vialson, L. N. (1976). Sur les produits de transformation de la nickéline provenant des filons de chromite-nickéline du massif des Beni-Bouchera au Maroc du Nord. *Mines et Geol., Rabat*, **39**, 41–8.

Waal de, S. A. (1975). The mineralogy, chemistry and certain aspects of reactivity of chromitite from the Bushveld igneous complex. National Institute of Metallurgy, Johannesburg, Report 1709, 80 pp.

20

PGE Distribution in some Ultramafic Rocks and Minerals from the Bou-Azzer Ophiolite Complex (Morocco)

W. Fischer, J. Amossé

UA No. 69, Institut Dolomieu, 38031 Grenoble, France

&

M. Leblanc

Centre Géologique et Géophysique, Université des Sciences et Techniques du Languedoc, 34060 Montpellier, France

ABSTRACT

The Bou-Azzer ophiolite complex, of Upper Proterozoic age, is located in the central Anti-Atlas. The corresponding mantle peridotites, including chromitites pods, were fully serpentinized during a multistage history resulting in serpentinite minerals and accessory magnetites. Locally, chromite was altered to stichtite (Cr carbonate) during post-obduction weathering. The Bou-Azzer district is well known for its cobalt–arsenide mineralization in quartz—carbonate lenses located along the borders of serpentinite massifs and grading laterally, through a talc-carbonate zone, into serpentinites. These carbonate rocks (listwaenites) result from an hydrothermal metasomatic transformation of the ultramafic rocks. Listwaenites and cobalt arsenides may contain very high gold contents (>10 ppm).

The aim of this paper is to investigate the PGE distribution in some minerals derived from ultramafic rocks. The samples have been analysed by electro-thermal atomic absorption spectrophotometry after extraction and enrichment by coprecipitation with selenium and tellurium.

To allow comparison between samples, analytical data have been normalized against C1 chondritic concentrations.

PGE distributions can be related to the main formation steps of the Bou-Azzer ophiolite and its cobalt mineralization. It is proposed that the very different PGE patterns in chromitites and magnetites result from PGE distributions in the primary magmatic assemblages.

Chromitite patterns, with a negative slope, are characteristic of ophiolitic podiform deposits, and ascribed to PGE minerals like laurite or Ir-Ru alloys. On the other hand, magnetite patterns exhibit a positive slope and may represent the primary PGE distribution in magmatic Fe-Ni sulphide or arsenide in mantle peridotites.

During late hydrothermal processes, carbonatization and cobalt ore formation, it appears that PGE were less mobile than gold in the hydrothermal fluids. PGE contents in arsenides do not correlate with gold contents, and may have been deposited as discrete PGE minerals. PGE minerals have not yet been observed.

INTRODUCTION

Platinum-group element (PGE) concentrations are generally associated with chromitite and nickel sulphide occurrences in mafic and ultramafic rocks within large layered intrusions (Stillwater, Bushveld), or in greenstone belts of komatiitic affinity (Naldrett & Duke, 1980).

Within ophiolitic sequences, the PGE are poorly concentrated, except in podiform chromitites, where minor concentrations of Ru-Os-Ir have been reported (Legendre, 1982; Talkington & Watkinson, 1986). Stratiform and podiform chromitites are clearly distinguishable from each other. Stratiform chromitite deposits are Pt-Pd-Rh rich, and thus their chondrite normalized patterns display a strong positive slope, on the other hand podiform chromitite patterns display a strong negative slope (Talkington & Watkinson, 1986).

Flat patterns are characteristic of undepleted upper mantle materials (Jagoutz *et al.*, 1979; Mitchell & Keays, 1981; Leblanc & Gervilla-Linares, 1987).

Until recently, PGE deposits were considered as strictly magmatic processes but investigations on non-magmatic sulphides or arsenides and carbonatized ultramafic rocks, in veins, fractures and sheared zones suggest that PGE may be redistributed and concentrated by late-stage hydrothermal activity (Lesher & Keays, 1984; Dillon-Leitch & Watkinson, 1986; Rowell & Edgar, 1986). The close association of PGE, with remobilized portions of ore bodies, and the various PGE patterns of altered magmatic rocks, point towards the mobility of PGE and a rearrangement of the original magmatic distribution (Stumpfl, 1986; Barnes *et al.*, 1985).

The aim of this paper is to investigate the PGE distribution of minerals in serpentinites from the Bou-Azzer ophiolite complex (Morocco) and its associated gold bearing cobalt-arsenide deposits, according to the genetic model proposed by Leblanc & Billaud (1982).

GEOLOGICAL SETTING

The Bou-Azzer area is located in the central part of the Anti-Atlas (Fig. 1). The serpentinites and associated basic rocks have been interpreted as an Upper Proterozoic complex (788 ± 9 my) obducted onto the northern margin of the West African craton (2000 my) during the main Pan-African tectonic phase (Leblanc, 1981).

The Bou-Azzer ophiolite is 4–5 km thick and from bottom to top, consists of 2000 m of serpentinized peridotites, 500 m of layered gabbros, a large stock of quartz diorite, 500 m of basic lavas with pillow-lavas and poorly developed dyke swarms, and 1500 m of a volcano sedimentary series.

The serpentinites, which include chromitite pods, still show a tectonic fabric. Their average Ni content (2650 ppm) and the composition of the chromitites, locally altered to stichtite, are similar to those from mantle peridotites of modern ophiolite complexes (Leblanc *et al.*, 1984). Furthermore, their geochemistry, low Al_2O_3 and CaO contents, and some orthopyroxene relics suggest they were derived from dunites and harzburgites. These mantle peridotites were fully serpentinized during a multistage history resulting in a lizardite–chrysotile–magnetite mineral association. The progressive development of Ni-rich magnetite (1% NiO), with different shapes, octaedric and fibrous, left iron-poor serpentine minerals. The

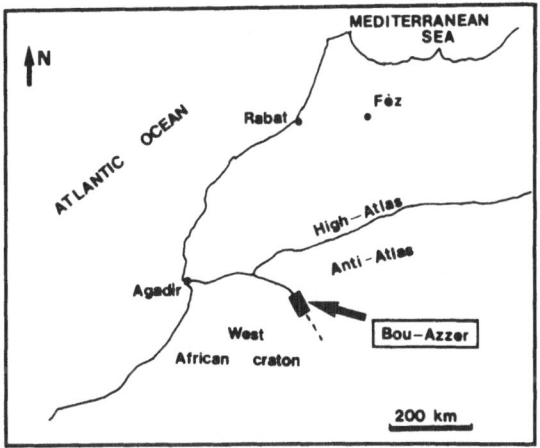

FIG. 1. Location sketch map of the Bou-Azzer ophiolite (Anti-Atlas, Morocco).

contact with the wall-rocks, which consist of gabbro and quartz-diorite, is marked by a black wall of chloritite and rodingite whereas the serpentinites at the contact contain talc, chlorite and carbonates.

The cobalt arsenide mineralization is located in a carbonate-quartz gangue along the borders of serpentinite massifs (Leblanc & Billaud, 1982). The cobalt ore bodies display different shapes depending on their location either along the sub-vertical tectonic contacts (lodes, veins, small stocks) or along the subhorizontal Infracambrian cover of the serpentinite massif (flat lenses, complex shells). Most of the ore bodies are located along Pan African and Hercynian tectonic structures and they display several phases of brecciation and crystallization. Quartz and grey calcite, coloured by micro-inclusions of hematite and magnetite, are the main gangue minerals; accessory chromite grains with a magnetite shell are randomly distributed and are similar to the accessory chromites in serpentinite in regard both to their shape and their geochemistry. Finally, talc, Mg-chlorite and serpentine minerals become abundant toward the serpentinite wall-rocks.

The ore minerals are disseminated to massive and suffered three brecciation and recrystallization phases. The Co-arsenides are distributed somewhat randomly in the inner part of the orebodies; the Fe arsenides are in the outer zones and the Ni arsenides are located close to the serpentinite wall-rocks or concentrated, with molybdenite and copper sulphides, along late shear zones.

SAMPLE PREPARATION

Chromitite samples were collected in two different pods of metre size. Two types of magnetites were collected in serpentinites, one occurs as centimetric octaedric crystals (MO), and the other as massive fibrous magnetite filling veinlets of 0·2–2 cm in thickness (MF). They were ultimately purified through handpicking under a binocular microscope. The samples were agate milled to less than 50 μm.

ANALYTICAL TECHNIQUES

PGE analyses were performed by the method of Amossé *et al.* (1986). This method involves extraction and enrichment by coprecipitation with selenium and tellurium. The detection is improved by electrothermal atomic absorption spectrophotometry. Extraction from rocks amounts to 90–100%. Detection limits are defined by sample sizes and these limits are less than 1, 0·5, 2, 0·5 and 0·5 ppb for Ru, Rh, Pt, Pd and Au respectively.

PGE DISTRIBUTION

Analytical results are listed in Tables 1 and 2, respectively for two chromitite samples (K43, K12) which are unaltered magmatic rocks and for minerals derived from ultramafic rocks. The total PGE, except gold contents, range from 20 ppb to 2500 ppb. The best way to compare the PGE distribution in the different samples is to use chondrite normalized plots as proposed by Naldrett *et al.* (1979). Chondrite concentrations are those of chondrite C1:Ru = 690 ppb, Rh = 200 ppb, Pt = 1020 ppb, Pd = 545 ppb and Au = 152 ppb (Naldrett & Duke, 1980). The corresponding chondrite normalized patterns are shown in Figs 2–5.

(a) PGE in Chromitites
The chromitite patterns (Fig. 2), with a negative slope from Ru to Pd-Au, are characteristic of podiform chromitites deposits (Talkington & Watkinson, 1986). These typical patterns are ascribed to the presence of platinum-group minerals (PGM), generally Ir-Os rich laurite (RuS_2) or Os-Ir alloys depending on the fS_2 during crystallization (Legendre, 1982).

In stichtite (ST), a Cr-carbonate mineral resulting from chromite alteration, PGE contents are very low. The pattern for stichtite (Fig. 4) is similar to the chromites.

TABLE 1
PGE concentrations in chromitites (ppb)

Sample number	Ru	Rh	Pt	Pd	Au
K 12	139	21·5	28·5	4·5	0·5
K 43	430	65	68	55	1

TABLE 2
PGE concentrations in ultramafic rock derived minerals (ppb)

Sample number	Rh	Pt	Pd	Au	Minerals
ST	3	10	4	2·5	Stichtite
MO	<0·5	7·5	6	10	Magnetite (octaedric)
MF	<0·5	21	76	3	Magnetite (fibrous)
T90	49	2 415	58	36 300	Skutterudite
T4	<0·5	20	7·5	45 000	Skutterudite
BA55	<0·5	19	5·5	13 700	Skutterudite
BA1	<0·5	20	68	50	Loellingite
BA47	2	20	17	18	Carbonate
BA49	<0·5	6·5	59	8	Hematite

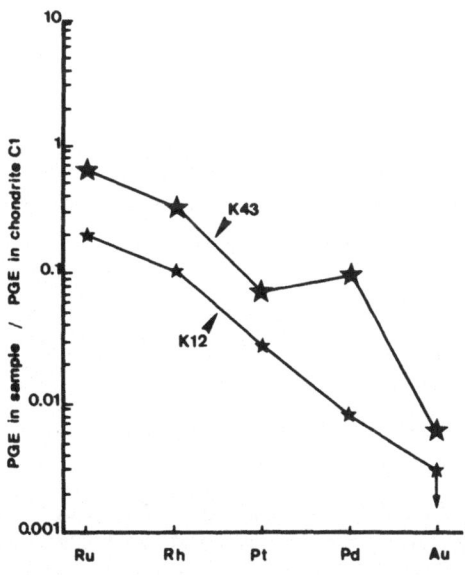

FIG. 2. Chondrite normalized patterns for two chromitite samples.

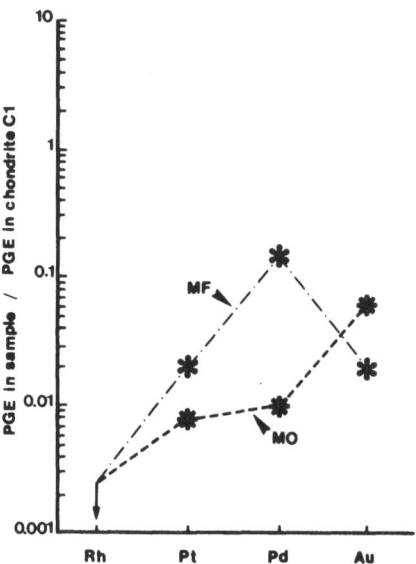

FIG. 3. Chondrite normalized patterns for two magnetite samples. MO, octaedric magnetite; MF, fibrous magnetite.

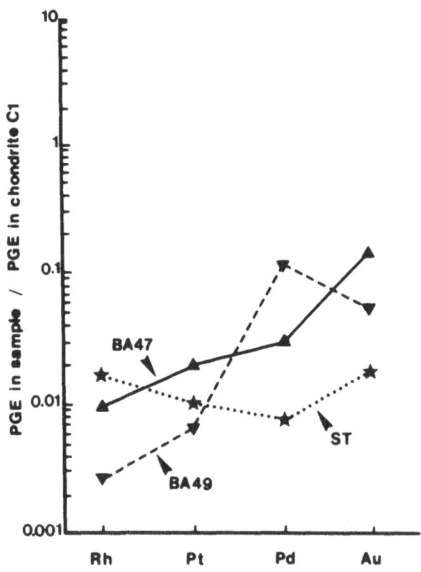

FIG. 4. Chondrite normalized patterns for hydrothermal carbonate (BA47) (listwaenites) and related hematite sample (BA49). In comparison, ST, the pattern for a stichite sample (Cr-carbonate).

(b) PGE in Magnetites
The patterns for magnetite (Fig. 3) are opposite to the chromites. There are no comparative data for magnetites in serpentinized ophiolitic rocks, but in ultramafic-mafic stratiform complexes, magmatic magnetite is an important gold and platinum carrier (Parry, 1984) and PGE occur in magnetic concentrates as discrete mineral grains, as inclusions in magnetite and as diffused atoms in the magnetite lattice (Rosenblum *et al.*, 1986).

(c) PGE in Carbonatized Ultramafic Rocks (Listwaenites)
Hydrothermal carbonate (BA47) and hematite (BA49) exhibit similar patterns (Fig. 4) and PGE contents (Table 2) than the magnetites.

(d) PGE in Arsenides
In mineralized carbonatized ultramafic rocks (1–10 ppm Au), pyrite and Co-arsenides (skutterudite) are the main gold bearing minerals (10–100 ppm Au; Buisson & Leblanc, 1986). In arsenides, PGE contents are low (Table 2), except for sample T90 (Pt = 2·4 ppm and Rh = 48 ppb) and do not correlate with gold contents. The Co-arsenides (T4, T90, BA55) all exhibit a similar pattern (Fig. 5)

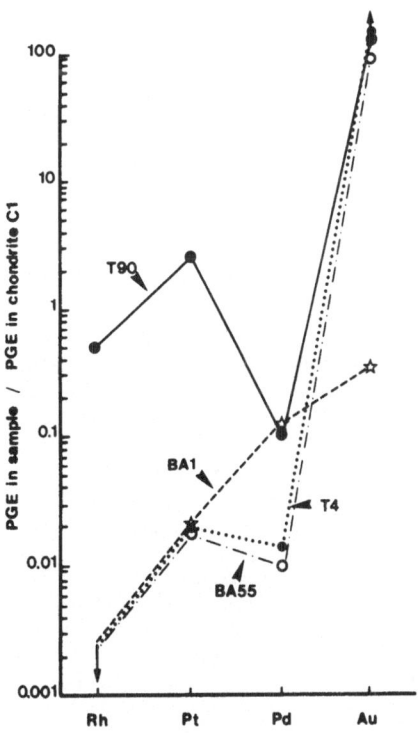

FIG. 5. Chondrite normalized patterns for arsenide samples. T90, T4, and BA55: Co-arsenides, BAl: Fe-arsenide.

characterized by a slight depletion in Pd and a strong enrichment in Au, in comparison with Fe-arsenide (loellingite, BA1). PGE distribution in Co-arsenides can be compared to the ophiolite hosted nickel mineralization in the Pevkos area (Cyprus), where PGE contents are low, normalized patterns are flat and show a strong enrichment in gold (Foose *et al.*, 1985).

INTERPRETATION AND DISCUSSION

Today, it is considered that chromitite pods can result from crystallization of a basaltic magma, in small chambers along conduits, feeding the main magmatic chamber (Lago *et al.*, 1982). Laurite and Os–Ir alloys are within chromite or silicate grains, their crystallization is earlier than that of chromite (Augé, 1986), and can be explained by a decrease in Ir solubility versus an increase in the intrinsic oxygen fugacity of the magma (Amossé *et al.*, 1987). So, the residual magma becomes enriched in Pt, Pd and Au.

During serpentinization, at low temperature (<250°) and involving large volumes of sea-derived fluids (Bonatti *et al.*, 1983), the primary mineral assemblages were destroyed resulting in lizardite (Table 3), chrysotile and magnetite. The most easily altered and the first minerals to be destroyed are the interstitial sulphides and arsenides. Probably Pt, Pd and Au have been concentrated in magmatic Ni-Fe sulphides or arsenides (Lorand & Pinet, 1984) during early magmatic processes, and are not related to PGE fractionation during chromite crystallization. Following the same processes indicated for gold (Buisson and Leblanc, 1987), PGE were probably leached from primary sulphides and arsenides by acidic and buffered solutions produced by sea-derived fluids (Bischoff

TABLE 3
Transformation of peridotite minerals during serpentinization and carbonatization

| | ←————Serpentinization————— ————Carbonatization————→ | | |
Mantle peridotites primary paragenesis		Serpentinites	Listwaenites
Silicates:	Olivine + Opx	Serpentine minerals + magnetite	Carbonate + hematite
	Interstitial sulphides + arsenides?	Secondary sulphides + magnetite	Hematite
Spinel:	Chromite	Chromite + stichtite	Chromite + fuchsite

& Seyfried, 1978), and transported as Cl, S or As complexes. Later serpentinization stages involving the destruction of olivine to form serpentinite minerals and magnetite, induced a decrease of oxygen fugacity and allowed PGE to be deposited and concentrated in magnetite.

Studies have shown that serpentinization hardly affects PGE distribution (Oshin & Crocket, 1982). However, gold can be depleted relative to the PGE (Keays & Davidson, 1976; Ross & Keays, 1979). Magnetite PGE patterns thus fairly well represent PGE distribution in their primary surrounding host-rocks.

During the latest stages of serpentinization, CO_2-rich acidic solutions would react with the serpentine and magnetite to produce talc and carbonate assemblages (Buisson & Leblanc, 1986). During this alteration, magnetite is hydrothermally destroyed, and iron enters into carbonate and the excess forms hematite (Table 3). The PGE patterns shown in Fig. 4 confirm the relationships of these rocks formed by metasomatic hydrothermal transformations with serpentine and magnetite. Hydrothermal carbonates (BA47) exhibit relatively flat patterns and the hematite pattern (BA49) fits very well with that of the fibrous magnetite. On PGE normalized patterns (Fig. 4) hydrothermal carbonates (BA47) are distinguishable from Cr-carbonates, stichtite (ST), resulting from chromite alteration.

According to the genetic model of gold bearing cobalt-arsenide deposits (Leblanc & Billaud, 1982; Buisson & Leblanc, 1986), cobalt and gold were concentrated into magnetite during serpentinization. Then, gold was released during hydrothermal destruction of magnetites, transported by Co-As-S-Cl-rich fluids and precipitated when entering into the reducing environment of the carbonate rocks. So, in view of PGE contents in magnetites (Table 2), which are greater than gold contents, interesting PGE contents may be expected in cobalt arsenides. But, analytical data show that PGE contents are low, except for sample T90 (Pt > 2 ppm) and do not correlate with gold. It appears that the PGE were less mobile than Au during this kind of hydrothermal alteration. These observations agree with the thermodynamic calculation of the solubility and transport of PGE in hydrothermal fluids (Mountain & Wood, 1987) which points out that the solubility of Pt and Pd as Cl⁻ complexes is low between 25 and 300°C, and that low concentration of dissolved As, Se and Te restrict the mobility of Pt and Pd in solution because of the large stability fields for Pt and Pd arsenides, selenides and tellurides. The fact that platinum readily forms arsenides and gold does not, explains that the amounts of platinum that can be transported in such arsenic-rich fluids is very low compared to gold (Lydon, 1987). Furthermore, the typical patterns for Co-arsenides (Fig. 5), which all exhibit a depletion in Pd, in comparison with Fe-arsenide, and their wide range of Pt tenors (20–2500 ppb) may be related to PGE deposition. So far, no PGM have been observed, but probably PGE occur as discrete PGM, such as sperrylite (PtAs₂) as described by Hudson and Donaldson (1984) at Kambalda. There, sperrylite is interpreted as a product of low temperature hydrothermal activity (300°C). These observations correspond to the moderate temperatures (150–300°C) assessed by Buisson and Leblanc (1986) for mineral assemblages and fluid inclusions.

CONCLUSIONS

(1) Chromites and magnetites, both in ultramafic rocks, display distinct PGE chondrite normalized patterns. chromites, which are primary magmatic minerals, display PGE patterns with negative slopes, characteristic of ophiolitic chromite deposits. Like gold, PGE were concentrated in magnetites during serpentinization. Magnetites PGE related patterns may represent the primary PGE distribution in magmatic Fe-Ni sulphides or arsenides.

(2) The PGE and Au concentrations observed in the Bou-Azzer cobalt arsenides and listwaenites (carbonatized serpentinites) are not related to magmatic evolution but to late hydrothermal processes (serpentinization followed by cobalt ore formation). Unlike gold, PGE were less mobile during carbonate alteration and cobalt arsenide ore formation.

(3) Ultramafic rocks are probably the source of Au and PGE concentrations, in the Bou-Azzer cobalt arsenide mineralization and listwaenites.

REFERENCES

Amossé, J., Fischer, W., Allibert, M. & Piboule, M. (1986). Méthode de dosage d'ultra-traces de platine, palladium, rhodium et or dans les roches silicatées par spectro-photométrie d'absorption atomique électrothermique. *Analysis*, **14**(1), 26–31.

Amossé, J., Allibert, M., Fischer, W. & Piboule, M. (1987). Etude de l'influence des fugacités d'oxygène et de sourfes sur la différenciation des platinoïdes dans les magmas ultramafiques. Résultats préliminaires. *C.r. Acad. Sci. Paris*, **304**, Série 2(19), 1883–5.

Augé, T. (1986). Platinum-group-mineral inclusions in chromitites from the Oman Ophiolites. *Bull. Minéral.*, **109**, 301–4.

Barnes, S.-J., Naldrett, A. J. & Gorton, M. P. (1985). The origin of the fractionation of platinum group elements in terrestrial magmas. *Chem. Geol.*, **53**, 303–23.

Bischoff, J. L. & Seyfried, W. E. (1978). Hydrothermal chemistry of seawater from 25° to 350°C. *Am. J. Sci.*, **278**, 838–60.

Bonatti, E., Simmons, E. C., Breger, D., Hamlyn, P. R. & Lawrence, J. (1983). Ultramafic rock/seawater interaction in the oceanic crust. *Earth Planet. Sci. Lett.*, **62**, 229–38.

Buisson, G. & Leblanc, M. (1986). Gold bearing listwaenites (carbonatized ultramafic rocks) in ophiolite complexes. In *Metallogeny of Basic and Ultramafic Rocks*, ed. M. J. Gallagher, R. A. Ixer, C. R. Neary & H. M. Prichard. The Institution of Mining and Metallurgy, London, pp. 121–32.

Buisson, G. & Leblanc, M. (1987). Gold in mantle peridotites from Upper Proterozoic ophiolites in Arabia, Mali and Morocco. *Econ. Geol.*, **82**, 2091–7.

Dillon-Leitch, H. C. H. & Watkinson, D. H. (1986). Distribution of platinum-group elements in the Donaldson West Deposit, Cape Smith Belt, Quebec. *Econ. Geol.*, **81**, 1147–58.

Foose, M. P., Economou, M. & Panayiotou, A. (1985). Compositional and mineralogic constraints on the genesis of ophiolite hosted nickel mineralization in the Pevkos Area, Limassol Forest, Cyprus. *Mineral. Deposita*, **20**, 234–40.

Hudson, D. R. & Donaldson, M. J. (1984). Mineralogy of platinum-group elements in the Kambalda nickel deposits, Western Australia. In *Sulphide Deposits in Mafic and Ultramafic Rocks*, eds D. L. Buchanan & M. J. Jones. The Institution of Mining and Metallurgy, London, pp. 55–61.

Jagoutz, E., Palme, H., Baddenhausen, M., Blum, K., Dreibus, G., Spettel, B., Lorenz, V. & Wanke, H. (1979). The abundance of major, minor and trace elements in the earth's mantle as derived from primitive ultramafic nodules. *Proceedings 10th Lunar Planet. Sci. Conf.*, Pergamon Press, New York, pp. 2031–50.

Keays, R. R. & Davidson, R. M. (1976). Palladium, iridium and gold in the ores and host rocks of nickel sulfide deposits in Western Australia. *Econ. Geol.*, **71**, 1214–28.

Lago, B. L., Rabinowicz, M. & Nicolas, A. (1982). Podiform chromite orebodies: a genetic model. *J. Petrol.*, **23**, 103–23.

Leblanc, M. (1981). The late Proterozoic ophiolites of Bou-Azzer (Morocco): evidence for Pan-African plate tectonics. In *Precambrian Plate Tectonics*, ed. A. Kroner. Elsevier, Amsterdam, pp. 435–51.

Leblanc, M. & Billaud, P. (1982). Cobalt arsenide orebodies related to an Upper Proterozoic ophiolite: Bou-Azzer (Morocco). *Econ. Geol.*, **77**, 162–75.

Leblanc, M. & Gervilla-Linares, F. (1987). PGE and Au distribution in Ni arsenide-chromite veins from the Rifo-Betic lherzolite massifs (Morocco–Spain). In *Geo-platinum 87*, eds H. M. Prichard, P. J. Potts, J. F. W. Bowles & S. J. Cribb. Elsevier Applied Science Publishers, London, pp. 181–98.

Leblanc, M., Dupuy, C. & Merlet, C. (1984). Nickel content of olivine as discriminatory factor between tectonic and cumulate peridotite in ophiolites. *Sciences Terra, Strasbourg*, **37**, 131–5.

Legendre, O. (1982). Minéralogie et géochimie des platinoïdes dans les chromitites ophiolitiques. Comparaison avec d'autres types de concentrations en platinoïdes. Thèse Doct. 3° cycle, Université Sciences, Paris VII, 157 pp.

Lesher, C. M. & Keays, R. R. (1984). Metamorphically and hydrothermally mobilized Fe-Ni-Cu sulphides at Kambalda, Western Australia. In *Sulphide Deposits in Mafic and Ultramafic Rocks*, eds D. L. Buchanan & M. J. Jones. The Institution of Mining and Metallurgy, London, pp. 62–9.

Lorand, J. P. & Pinet, M. (1984). L'orcelite des péridotites de Ronda (Espagne), Beni-Boussera (Maroc). Table Mountain et Blow-Me-Down (Terre Neuve) et du Pinde septentrional (Grèce). *Can. Min.*, **22**, 553–60.

Lydon, J. W. (1987). The potential for hydrothermal platinum deposits. In *Geo-platinum 87*, eds H. M. Prichard, P. J. Potts, J. F. W. Bowles & S. J. Cribb. Elsevier Applied Science Publishers, London, pp. 111–2.

Mitchell, R. H. & Keays, R. R. (1981). Abundance and distribution of gold, palladium and iridium in some spinel and garnet lherzolites. Implications for the nature and origin of precious metal-rich intergranular components in the upper mantle. *Geochim. Cosmochim. Acta*, **45**, 2425–42.

Mountain, B. W. & Wood, S. A. (1987). Solubility and transport of PGE in hydrothermal solutions: Thermodynamic and physical chemical constraints. In *Geo-platinum 87*, eds H. M. Prichard, P. J. Potts, J. F. W. Bowles & S. J. Cribb. Elsevier Applied Science Publishers, London, pp. 57–82.

Naldrett, A. J., Hoffman, E. L., Green, A. H., Chou, C. L., Naldrett, S. R. & Alcock, R. A. (1979). The composition of nickel-sulfide ores, with particular reference to their content of platinum-group elements and gold. *Can. Mineral.*, **17**, 403–15.

Naldrett, A. J. & Duke, J. M. (1980). Platinum metals in magmatic sulfide ores. *Science*, **208**(4451), 1417–24.

Oshin, I. O. & Crocket, J. H. (1982). Noble metals in Thetford Mines Ophiolites, Quebec, Canada. Part I: Distribution of gold, iridium, platinum and palladium in the ultramafic and gabbroic rocks. *Econ. Geol.*, **77**, 1556–70.

Parry, S. (1984). Abundance and distribution of palladium, platinum, iridium and gold in some oxide minerals. *Chem. Geol.*, **43**, 115–25.

Rosenblum, S., Carlson, R. R., Nishi, J. M. & Overstreet, W. C. (1986). Platinum-group elements in magnetic concentrates from the Goodnews Bay District, Alaska, US Geol. Survey Bulletin A 1660.

Ross, J. S. & Keays, R. R. (1979). Precious metals in volcanic-type nickel sulfide deposits in Western Australia. I: Relationship with the composition of the ores and their host rocks. *Can. Mineral.*, **17**, 417–35.

Rowell, W. F. & Edgar, A. D. (1986). Platinum-group element mineralization in a hydrothermal Cu-Ni sulfide occurrence, Rathbun Lake, Northeastern Ontario. *Econ. Geol.*, **81**, 1272–7.

Stumpfl, E. F. (1986). Distribution, transport and concentration of platinum group elements. In *Metallogeny of Basic and Ultramafic Rocks*, eds M. J. Gallagher, R. A. Ixer, C. R. Neary & H. M. Prichard. Institution of Mining and Metallurgy, London, p. 379–94.

Talkington, R. W. & Watkinson, D. H. (1986). Whole rock platinum-group element trends in chromite-rich rocks in ophiolitic and stratiform igneous complexes. In *Metallogeny of Basic and Ultramafic Rocks*, eds M. J. Gallagher, R. A. Ixer, C. R. Neary & H. M. Prichard. Institution of Mining and Metallurgy, London, pp. 427–40.

21

Exploration for Platinum-Group Elements in the Labrador Trough, Canada

FENTON SCOTT
*La Fosse Platinum Group, 17 Malabar Place, Don Mills, Ontario,
Canada M3B 1A4*

ABSTRACT

A reconnaissance survey of known regions of sulphide mineralisation in the Labrador Trough, Canada has revealed nine potentially economic platinum and palladium deposits. Five occur in ultramafic rocks whilst the other four are associated with mafic intrusions. Assay results as high as 68·5 g/t were discovered in association with chalcopyrite, pentlandite, pyrrhotite and pyrite in both mafic and ultramafic bodies. Ninety percent of the intrusives sampled did not contain detectable platinum values although copper- and nickel-bearing sulphides are common in all the areas examined.

It is noted that:

(1) The platinum/palladium ratio is usually higher in the deposits associated with ultramafic rocks compared with those in mafic rock units.

(2) The highest platinum values accompany rich chalcopyrite mineralisation.

INTRODUCTION

The Labrador Trough is an orthogeosyncline of early Proterozoic Aphebian age emplaced along the western margin of an Archean Craton (Fig. 1). The sedimentary succession has been repeatedly cut by intrusive and extrusive igneous events varying in composition from basaltic flows and tholeitic sills with silica-rich differentiates to ultramafic sills and coarsely crystalline pyroxenites. The Labrador Trough is dominated by a series of subparallel thrusts, dipping to the east, which separate several complexly folded thrust slices.

Massive and disseminated iron sulphide mineralisation is prevalent in sediments throughout the Labrador Trough accompanied by trace amounts of base metals. Base metal deposits are found in these rocks and occur as stratabound copper-zinc massive sulphides in the sediments and both massive and disseminated copper-nickel deposits associated with the mafic and ultramafic sills. Weathering of the sulphide horizons has led to the development of extensive gossans. This paper

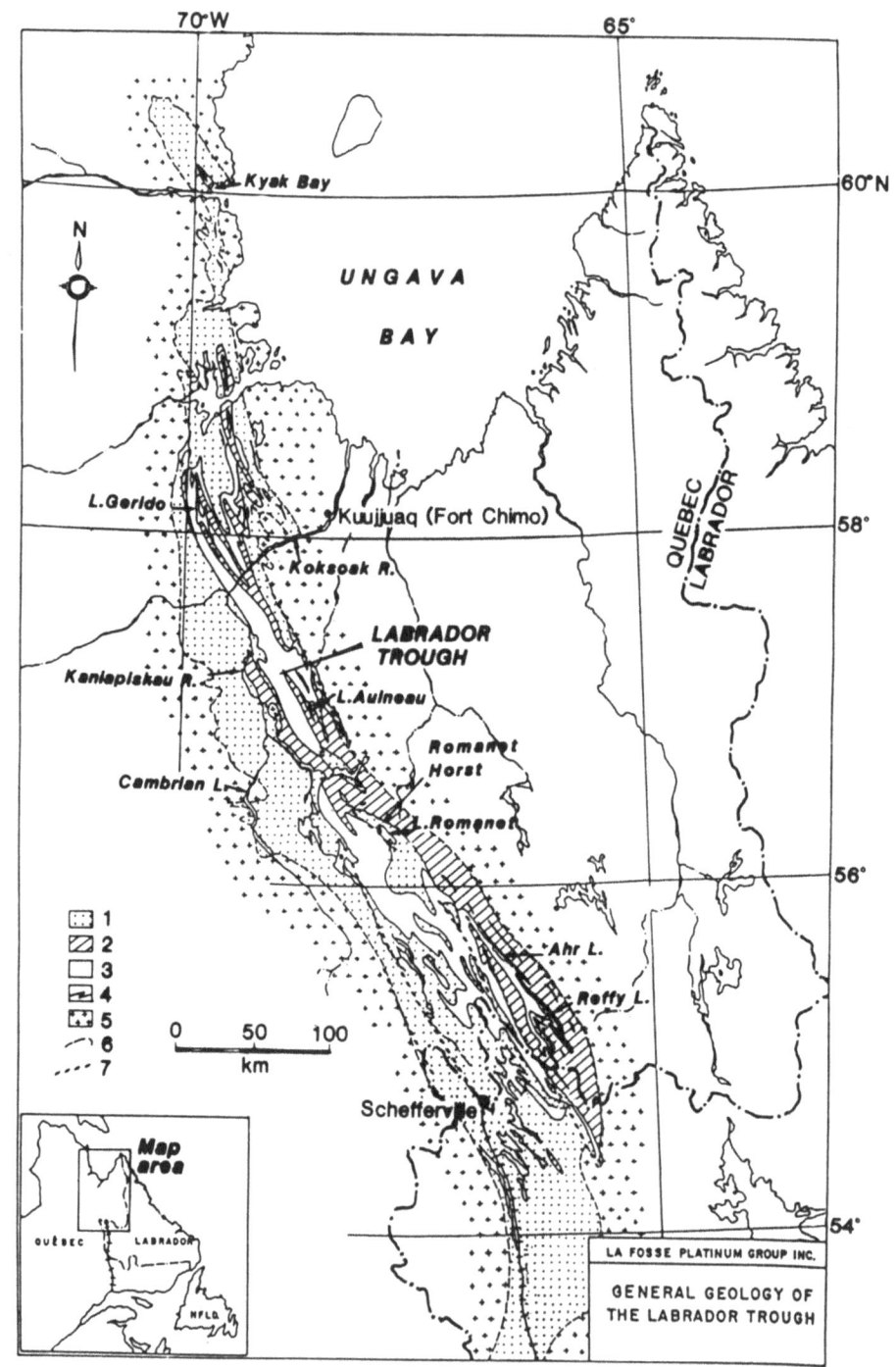

FIG. 1.

summarises the results of a reconnaissance survey for PGE targeted around known areas of sulphide mineralisation.

SULPHIDE MINERALISATION

The principal associations and deposits are:

(1) Stratabound copper-zinc-silver massive sulphide lenses which appear to be localised by fold structures and which may also contain lead mineralisation (Table 1).

TABLE 1

Deposit	Size (t)	Cu (%)	Zn (%)	Pb (%)	Au (oz/t)	Ag (oz/t)
Prudhomme No. 1	562 000	2·05	2·39	—	0·05	1·01
Prudhomme No. 2	4475 000	1·52	1·24	—	0·05	0·60
Frederickson	307 600	0·77	4·38	—	0·02	1·23
Jimmick	120 000	0·26	5·20	0·61	—	—

(2) Copper-nickel massive sulphides in ultramafic sills. These usually occur near the base of the sill and normally have a hanging wall halo of disseminated sulphides. Pyrrhotite is the dominant mineral (Table 2).

(3) Disseminated and massive sulphides in basic sills. The Cu/Ni ratio is usually higher than in the ultramafic rocks (5 instead of 2). This type of deposit has major tonnage possibilities although the total base metal content lies below 0·5%. Again pyrrhotite is the dominant suplhide (Table 3).

TABLE 2

Deposit	Size (t)	Cu (%)	Ni (%)
Chance Lake	550 000	0·66	0·89
Blue Lake	506 400	0·85	0·40
Pogo Lake	692 600	1·00	0·65
Centre	91 400	1·26	0·75

TABLE 3

Deposit	Size (t)	Cu (%)	Ni (%)
Leslie No. 2	765 000	1·56	0·33
Erickson No. 1	573 000	1·12	0·32
Erickson No. 2	75 000	1·00	0·25
Chrysler No. 2	580 000	1·79	0·48
Marymac No. 2	930 000	1·60	0·43
Lepage	790 000	2·76	0·66
Float	133 000	2·10	0·34
Redcliff	1068 000	2·09	0·51

(4) Disseminated copper and nickel sulphides in ultramafic sills. This mineralisation is widespread and occurs throughout the thickness of the unit. Grades are usually 0·30% or less.

(5) Copper-nickel veins in volcanics. These crosscut the sedimentary structures and usually occur near ultramafic units.

(6) Copper deposits in sediments. These deposits often contain bornite or chalcopyrite and are both stratigraphically and structurally controlled. They carry low silver and, occasionally, gold values.

PLATINUM-GROUP ELEMENT MINERALISATION

The reconnaissance survey had three aims:

(1) To locate platinum-rich, low sulphide layers in the layered mafic and ultra-mafic sills.
(2) To investigate the possible platinum content of the massive sulphide deposits associated with the gabbroic sills.
(3) To determine the platinum content of the massive sulphide deposits occurring near the base of the mafic and ultramafic sills.

In the low sulphide, layered sills, the highest value obtained was 470 ppb platinum plus palladium. Background values for these sills ranged up to 100 ppb.

So far, nine sulphide-rich PGE localities with potential economic importance have been found (Table 4). The copper content at these localities ranges from 1·2 to 12·8%. The occurrences are associated with both mafic and ultramafic sills. No specific platinum-group minerals have yet been identified.

TABLE 4

Prospect	Width (ft)	Pt (g/t)	Pd (g/t)	Cu (%)
Lac St. Pierre	5	4·5	2·0	1·6
Goose Pond	6	1·3	15·42	7·9
Chrysler	22	0·4	5·1	12·8
Lepage	10	1·3	4·6	2·4
Redcliff	6	0·4	1·8	6·6
Retty	10	0·4	1·5	—
Pogo	6	1·3	10·26	1·2
Centre	8	27·4	41·1	4·2
Blue	5	0·3	2·1	6·0

In four mafic associated locations the average platinum to platinum plus palladium ratio is 0·13, while in five ultramafic locations the average ratio is 0·30. The copper to copper plus nickel ratio of these deposits shows a similar variation (0·63 in the ultramafic deposits; 0·82 in the mafic deposits). As an exploration guide in this area, it seems that the highest platinum values occur near ultramafic sills especially those associated with higher copper mineralisation.

22

Platinum-Group Element Mineralisation and the Relative Importance of Magmatic and Deuteric Processes: Field Evidence from the Lac des Iles Deposit, Ontario, Canada

A. J. MACDONALD*

Ontario Geological Survey, 1028–77 Grenville Street, Toronto, Ontario, Canada M7A 1W4

ABSTRACT

The Lac des Iles (LDI) mafic and ultramafic intrusion is located approximately 80 km NNW of Thunder Bay, Ontario. Pd-Pt-Au-Ni-Cu mineralisation is hosted by mafic components of the intrusion (anorthositic gabbro, gabbro, noritic gabbro and norite). The principal zone of mineralisation is reported to be at the contact of two discrete gabbros (Dunning, 1979), which were intruded prior to the ultramafic portion of the LDI complex. Several other sulphide zones, with lower PGE content, are present within mafic rocks of the LDI intrusion.

Characteristics of sulphide-bearing, as opposed to barren, gabbro include: (1) variably altered mineralised gabbro: plagioclase to epidote, chlorite and sericite; clinopyroxene to fibrous amphibole, Pye (1968); (2) a marked variation in texture, from fine grained to pegmatoidal, with crystals up to 10 cm in length. In addition, quartz- and sulphide-bearing, cross-cutting, gabbroic pegmatite dykes often contain high PGE grades; (3) compositionally variable lithologies between pyroxenitic and anorthositic end members. Anorthosite is present as a constituent of (a) modal layering, (b) irregular segregations, locally cored by quartz, tourmaline and sulphides, and (c) cross-cutting dykes; (4) breccias with a generally gabbroic matrix are present in both PGE-rich and barren sulphide zones, consisting of gabbroic, pyroxenitic and anorthositic clasts, up to 3 m in diameter, locally encrusted by either pyroxenitic or pegmatoidal gabbro rinds; (5) the frequency of tonalite dykes within the gabbroic complex increases as one approaches sulphide-bearing zones. In places, tonalitic and gabbroic magmas have mixed, resulting in pegmantoidal gabbro, hosting sulphides; (6) sulphide zones containing richer grades of PGE are associated with a PGE-rich, mafic to ultramafic dyke that intruded, and locally mixed with the gabbro. This dyke cuts pegmatoidal layering in gabbro, but is cut by pegmatitic gabbro dykes. The tonalitic magma appears locally to have mixed with the ultramafic magma; (7) gabbroic rocks hosting PGE-bearing sulphides are characterised by

*Present address: Agip Resources Ltd, PO Box 7, Suite 2116, 130 Adelaide St. W., Toronto, Ontario, Canada M5H 395.

cuspate, pegmatoidal, comb layering, commonly with a pyroxenitic base, overlain by large plagioclase and pyroxene crystals perpendicular to the layer, indicating direction of crystallisation. The convexity of the pegmatoidal, cuspate layers and, hence, direction of crystallisation are consistent on an outcrop scale, defining solidification vectors within the gabbro at any one location. The margins of the mineralised zone, however, display opposing solidification vectors directed into the centre of the zone, suggesting that crystallisation progressed from the margins inwards.

Whereas the ultramafic intrusion into the gabbro host appears to be the fundamental control in concentrating economically significant PGE, the close association between mineralisation, alteration of silicate phases and the presence of pegmatoids and pegmatites, indicates the role played by volatiles during the latter stages of formation of the Lac des Iles deposit.

INTRODUCTION

The Lac des Iles (LDI) mafic/ultramafic complex is located approximately 80 km NNW of Thunder Bay, Northwestern Ontario (Fig. 1). The LDI complex is host to a Pd-Pt-Au-Ni-Cu deposit with an estimated reserve of 20·4 million t at 6·17 g/t total platinum-group elements (PGE), and a Pt:Pd ratio of 1:7 (*Northern Miner*, November 10, 1986). Madeleine Mines Ltd of Toronto announced that the deposit will be put into production employing open pit mining methods with the capability of milling approximately 2700 t/day (*Northern Miner*, July 27, 1987).

This contribution describes the results of mapping (1:1000 and 1:100) of recently exposed outcrops within and around economically significant PGE mineralisation at Lac des Iles. These observations supplement those obtained in previous studies (e.g. Pye, 1968; Macdonald, 1985) and are integrated into a geological framework that indicates magmatic and deuteric processes during mineralisation.

REGIONAL GEOLOGY

The Archean Lac des Iles complex lies within a granite/granite gneiss terrain of the Wabigoon Subprovince (Pye, 1968). To the east of the LDI complex, the Southern Wabigoon greenstone belt is subparallel to the boundary with the Quetico Subprovince to the south (Fig. 1). The locally intense deformation in this metasedimentary and metavolcanic rock and amphibolite grade metamorphism (Pye, 1968; Kaye, 1969) is not seen within the LDI intrusion. All the Archean rocks have been intruded by Proterozoic diabase dykes and sills.

The LDI composite intrusion is one of a number of similar complexes investigated by Sutcliffe (1986) which define a regional linear trend parallel to the Wabigoon–Quetico boundary. In the immediate LDI area, Sutcliffe and Sweeney (1985a) noted that several intrusive, tholeiitic, mafic and ultramafic complexes define a circular structure, approximately 30 km in diameter. Granite/gneissic

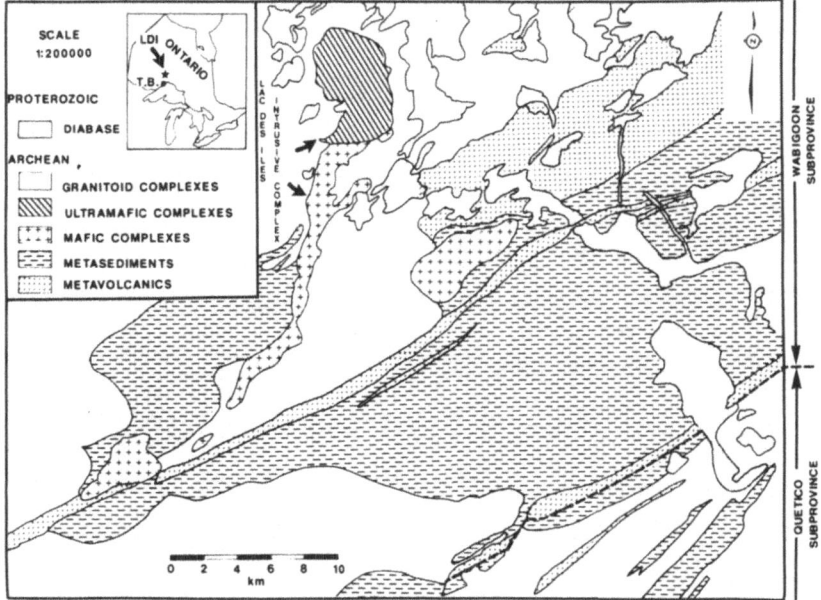

Fig. 1. Location of the Lac des Iles Complex, north of the Wabigoon-Quetico Subprovince boundary, 80 km NNW of Thunder Bay (T.B.), Northwestern Ontario. From Pye (1968) and Kaye (1969).

rocks are found around the margins of the circular structure, whereas a suite of granitic rocks, from hornblende tonalite to biotite granite is found in the centre. Field relationships suggest that the mafic and ultramafic rocks of the LDI intrusion are co-magmatic with the granitic rocks (Sutcliffe & Sweeney, 1985a; Sutcliffe, 1986).

PREVIOUS WORK

Several sulphide-bearing zones (A–H, Figs 2 and 5) were outlined during geological, geochemical and geophysical surveys by Gunnex Ltd (1963 and 1964). Subsequent diamond drilling in 1974 by Boston Bay Mines Ltd determined the general SSE strike of the main mineralised body, known as the Roby Zone (zones C, E and part of F).

Pye (1965, 1968) published the first geological maps and reports on the Lac des Iles area. Guarnera (1967) studied a small portion of the gabbro complex to the south of the Roby Zone. Dunning (1979) investigated Roby Zone mineralisation, examining diamond drill core and synthesising the results of geological mapping by Texasgulf Canada Limited. Watkinson & Dunning (1979), Cabri & Laflamme (1979), Dunning *et al.* (1981) and Cabri (1981) discussed the mineralogy of the PGE-bearing rocks, describing several Pt, Pd, Ni, Te, Bi, Sb, As and S bearing

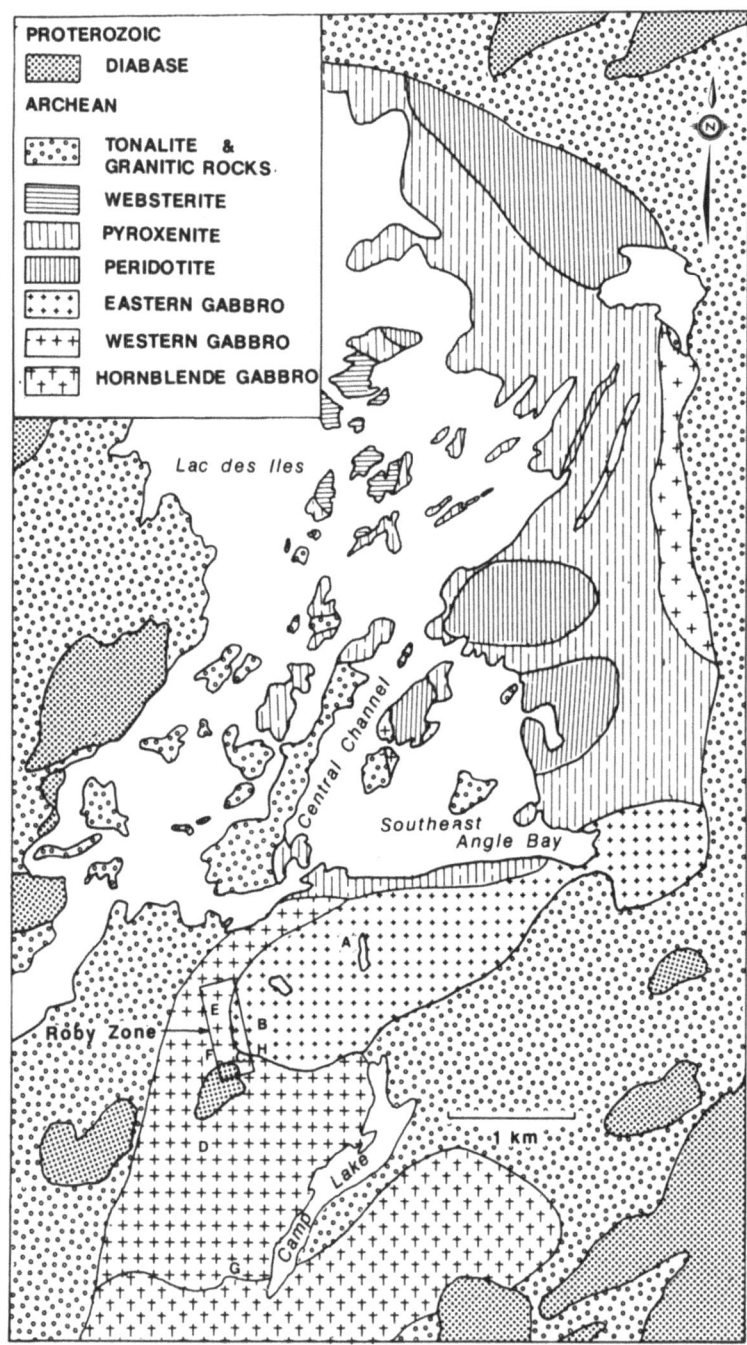

FIG. 2. The Lac des Iles Mafic/Ultramafic Complex. A–H are sulphide zones exposed at surface. Roby Zone (subsurface continuity, in part, of the E, C and F zones) contains the most significant PGE mineralisation. Modified from Pye (1968), Dunning (1969) and Sutcliffe and Sweeney (1986b).

phases, with vysotskite ((Pd, Ni)S) and kotulskite (Pd(Te, Bi)) as the most common platinum-group minerals (PGM). Naldrett (1981) reported an average composition of Roby Zone mineralisation: 900 ppb Pt, 1600 ppb Pd, 110 ppb Rh, 3 ppb Ru, 0·4 ppb, Ir, 0·2 ppb Os, 680 ppb Au, 0·172% Ni and 0·128% Cu. Talkington & Watkinson (1984) provided petrographic descriptions of mineralised rocks and investigated the variability of metal grades in different rock types.

Sutcliffe & Sweeney (1985*b*) re-mapped the LDI complex and Macdonald (1985) mapped the gabbroic portion of the LDI complex, in and around the Roby Zone. In 1986, Sweeney and Sutcliffe further described Roby Zone mineralisation and Linhardt & Sutcliffe (1986) investigated the northern, ultramafic portion of the complex. Fortescue & Webb (1986) conducted a geochemical survey of humus over the E Zone. Gupta *et al.* (1986) carried out a gravity survey in the area.

GEOLOGY AND STYLES OF MINERALISATION AT LDI

Ultramafic Complex

The LDI intrusive complex is divided into a northern ultramafic complex and a southern mafic complex (Figs 1 and 2). The ultramafic rocks consist of dish-shaped layers, from the base upwards, of dunite and wehrlite, wehrlite and clino-pyroxenite, websterite to gabbro norite and finally clinopyroxenite (Linhardt & Sutcliffe, 1986). During exploration by Texas Gulf Canada Ltd in 1975, small concentrations of chromite, up to 3·5% Cr_2O_3, were encountered in the ultramafic rocks and impersistent Cu-Ni sulphide occurrences were located in the Central Channel area and on the northeast shore of the Lake (Fig. 2), with maximum assays of 0·29% Cu, 0·14% Ni and 2600 ppb total PGE (Winfield & Workman, 1976). Subsequently, Sutcliffe & Sweeney (1985*a*) reported assays from the northeast shore of 2·3% Cu, 1·1% Ni, 2040 ppb total PGE and from South-east Angle Bay of 1·3% Cu, 1·6% Ni, 1585 ppb total PGE. Linhardt & Sutcliffe (1986) described several sulphide occurrences, with assays up to 2880 ppb total PGE. No significant tonnages have yet been defined for any of the occurrences within the ultramafic complex.

Mafic Complex

Dunning (1979) subdivided the mafic complex into eastern and western gabbro phases (Fig. 2). The eastern gabbro is dominated by anorthositic gabbro to gabbro, with only minor noritic gabbro, whereas the western gabbro contains a greater abundance of noritic, gabbro norite (20%) and 70% gabbro. The plagioclase composition is more Ca-rich in the western gabbro (An_{80-82}) than the eastern gabbro (An_{52-60}). Orthopyroxene is richer in Mg in the western gabbro (En_{72-79}), and richer in Fe in the eastern gabbro ($En_{62·5-64·5}$). Clinopyroxenes in both gabbros show similar Fe versus Mg relationships (Dunning, 1979).

The non-mineralised portions of the gabbroic complex at LDI, consist of monotonous sequences with most of the rocks containing cumulus plagioclase, and locally cumulus pyroxene (Dunning, 1979). Gabbro and anorthositic gabbro contain cumulus plagioclase and intercumulus pyroxene, now altered to amphibole. Norite and noritic gabbro are characterised by stubby, cumulus orthopyroxene. Contacts between the various gabbroic phases are gradational and irregular.

The gabbroic rocks commonly contain an igneous foliation, which dips steeply inwards and generally follows the margin of the intrusion. Modal layering is rare, although where present it also dips steeply and is subparallel to foliation. Both igneous foliation and modal layering within the gabbroic rocks describe an elongate, funnel shape (Macdonald, 1985). Deformation within the intrusion is minimal, restricted to 060°-striking faults with sinistral offsets of generally less than a metre.

The gabbroic rocks have undergone significant alteration, ascribed to deuteric processes by Pye (1968), resulting in partial to total saussuritisation of plagioclase, forming sericite, zoisite and chlorite, and uralitisation of clinopyroxene, forming plates and fibres of amphibole. Orthopyroxene tends to exhibit less alteration, although locally talc forms at the expense of orthopyroxene. Whereas Watkinson & Dunning (1979) suggested that secondary minerals in the LDI complex may indicate metamorphism to amphibolite facies, Talkington & Watkinson (1984) considered that deuteric alteration effects predominate.

Geology of the Mineralised Zones
Dunning (1979) ascribed a pyroxenitic unit to the western gabbro, suggesting that layers of pyroxenite occur at the contact with the eastern gabbro. This unit is especially significant because it is often host to economically significant PGE grades (up to 15 000 ppb total PGE). Detailed mapping (1:100) of newly exposed outcrops, however, reveals that the pyroxenite is a dyke, intrusive into anorthositic gabbro. The ultramafic dyke mixed chaotically with gabbro at the eastern margin of the E zone (Fig. 2); cobbles of gabbro are locally contained within pyroxenite (Fig. 3). Dunning (1979) and Talkington & Watkinson (1984) noted that the original pyroxenes of this unit are completely altered to fine grained amphibole, as are pyroxenes in mineralised gabbro described by Pye (1968). A second pyroxenite dyke was also found to intrude anorthositic gabbro in the vicinity of the C zone (Figs 4(a) and 4(b)). The constituent minerals in this dyke are less altered and consist of both cumulus orthopyroxene and clinopyroxene, with only minor (<10%) intercumulus plagioclase, i.e. the rock is a websterite. Gabbro adjacent to the websterite dyke is locally deformed in a ductile fashion, the degree of deformation decreasing with distance from the dyke. The gabbro host rock may have only been partially consolidated and then deformed plastically during websterite intrusion. Clots of pyroxene up to 5 cm in diameter, are present within gabbro adjacent to the websterite dyke, indicating limited geochemical interaction between the two rock types.

The websterite dyke in Fig. 4 is, in turn, cut by a tonalite dyke. Metasomatic feldspathisation of the websterite occurs up to 2 cm into the hanging wall of the tonalite dyke. This is a similar observation to those described by both Winfield &

Fɪɢ. 3. Pyroxenite dyke intruding anorthositic gabbro, east side of Roby Zone. Note inclusion of gabbro in pyroxenite. Pyroxenite is cut, in turn, by tonalite dyke. Hammer for scale.

Workman (1976) and Sutcliffe (1986) in mixing zones between both felsic and mafic/ultramafic magmas in the LDI area.

The pyroxenitic rocks were used during exploration by diamond drilling to outline the contact between the eastern and western gabbros. No unequivocal intrusive contacts however between the two gabbro phases have yet been observed in the field. A zone of considerable heterogeneity, several hundred metres wide in places and spatially associated with PGE mineralisation (Macdonald, 1985), may constitute a mixing zone between the eastern and western gabbros, as suggested by Sutcliffe & Sweeney (1985a). As these ultramafic rocks intrude the gabbro, rather than representing a marginal phase of a much larger gabbroic intrusion, the existence of two discrete, intrusive gabbroic bodies must be questioned.

Further work (Macdonald *et al.*, in preparation) to distinguish the petrogenetic affinity of the gabbro phase(s) at Lac des Iles is in progress to determine whether:

(a) two gabbro phases mixed (Dunning, 1979);
(b) in situ magmatic differentiation caused a more Mg-rich western gabbro and a more Fe-rich eastern gabbro;
(c) gabbroic and websteritic components mixed.

There is also disagreement as to the relative timing of intrusion for the eastern and western gabbros: Dunning proposed that the western gabbro intruded first,

(a)

(b)

FIG. 4. (a) Pyroxenite (websterite) dyke cutting modal layering in anorthositic gabbro, north of C zone. Hammer for scale. (b) Pyroxenite intrusion brecciating anorthositic gabbro, adjacent to (a). Hammer for scale.

while Sutcliffe & Sweeney (1985*a*) suggested the reverse. Both Dunning (1979) and Sutcliffe & Sweeney (1985*a*) agree that intrusion of the gabbroic complex is followed by intrusion of the ultramafic complex to the north. For the remainder of this paper, the gabbroic rocks are not subdivided into 'eastern' and 'western' bodies, as it is suggested that the geological basis for such a distinction is erroneous.

Lithological Complexity in the Mineralised Zones

The portion of the gabbroic complex that contains significant PGE mineralisation, is shown in Fig. 5. A magnetite-rich anorthositic unit is present in the eastern half of the area shown in Fig. 5. The unit describes a crude, partial annulus, which is subparallel to the strikes of foliation and modal layering. This constitutes a lithostratigraphic marker horizon within the gabbroic complex.

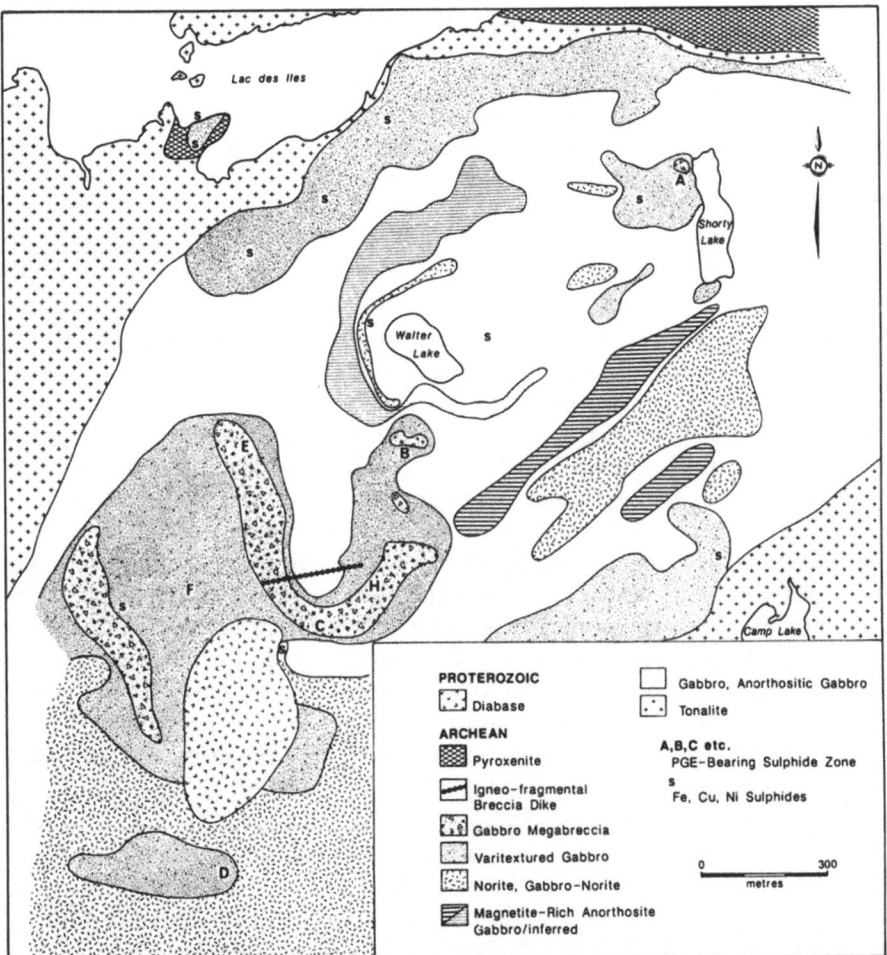

FIG. 5. Map of the Lac des Iles gabbro complex in the vicinity of the Roby Zone, modified from Macdonald (1985). The G zone is located 1 km due south of the D zone.

In contrast with monotonous, unmineralised gabbroic rocks, the degree of lithological complexity is considerable in the vicinity of sulphide-bearing zones, with an increase in rock types (a)–(f) described below.

The degree of lithological and textural variability is considerable, even on a hand specimen scale, and mapping of individual lithologies is precluded. The area containing these complexities was, therefore, mapped as one unit, termed varitextured gabbro (Fig. 5). The central portions of varitextured gabbro are dominated by zones of igneous breccia (gabbro megabreccia, Fig. 5).

(a) Tonalite. Tonalite dykes are present throughout the gabbroic complex, although their frequency increases in the vicinity of some sulphide zones. There are at least two generations of tonalite dykes, based upon cross-cutting relationships. The dykes locally contain a quartz (± tourmaline)-bearing core. In places, pegmatoidal gabbro, with a grain size up to 10 cm, develops in reaction zones where tonalite and gabbroic magmas have interacted, an observation which supports the hypothesis that both felsic and mafic magmas coexisted at Lac des Iles (Sutcliffe & Sweeney, 1985*a*).

(b) Anorthosite. In a zone of up to 100 m wide surrounding rocks that host PGE mineralisation, anorthosite develops as layers, subparallel to foliation and modal layering, as cross-cutting dykes, as cuspate lenses interleaved with gabbro and as subspherical pods.

(c) Pegmatoids and pegmatites. Sulphide-bearing zones are characterised by the development of pegmatoidal gabbro, and by cross-cutting pegmatitic gabbro dykes up to 1 metre wide which typically consist of plagioclase and pyroxene crystals, with grain size up to 10 cm. Pyroxenes are most commonly perpendicular to the pegmatite wall (Fig. 6) and lithologic contacts may be offset by pegmatites, suggesting that formation of pegmatites was accompanied by at least some open-space filling. Gabbro pegmatites commonly have pyroxene-rich margins and plagioclase-rich cores, locally quartz-bearing. They sometimes pass laterally into a dyke that contains little pyroxene, i.e. an anorthosite.

(d) Heterogeneous gabbro. Gabbro in sulphide-bearing zones, and especially in the Roby Zone, displays a complex compositional heterogeneity. Figure 7 is a map of a portion of the E zone, at the north end of the Roby Zone (Fig. 2). The map demonstrates some of the complexity encountered in mineralised zones, in contrast to the relatively monotonous, unmineralised gabbro. At this locality, layering dips steeply to the east. Pegmatoidal layering is developed locally and exhibits a cuspate morphology on horizontal surfaces, with the convex side generally to the west and southwest, often accompanied by lithologic changes from anorthositic gabbro, to pyroxenite, to pegmatoidal gabbro (Fig. 8) and the pegmatoidal layer in the lower part of Fig. 8 grades into a cross-cutting pegmatite dyke.

Individual pegmatoidal layers locally have a pyroxenitic base and large, up to 5 cm, crystals of interdigitating pyroxene and plagioclase oriented perpendicularly to the pyroxenite layer (Fig. 9). This type of layering has been termed informally 'Willow Lake-type layering' (Poldervaart & Tauberneck, 1959), or more generally 'Comb Layering' (Moore & Lockwood, 1973) which is particularly useful in determining relative direction of accumulation of crystals in a magma chamber, as is the

FIG. 6. Gabbro pegmatite dyke, E zone, northern Roby Zone. Note inward growth of pyroxene crystals from the dyke wall. Hammer handle for scale.

geometry of the curved comb layers. The pegmatoidal crystals in Fig. 9 grew in a magma chamber whose floor or wall consisted at that time of pyroxenite, with the convex side of the layer generally protruding away from the accumulating crystal pile and into the magma from which the crystals formed (cf. Moore & Lockwood, 1973, Fig. 6).

In exposures revealing clear intrusive relationships between pyroxenite and gabbro the ultramafic dyke cuts pegmatoidal layering within the gabbro. An example in Fig. 3 shows that the dyke, in turn, is cut by gabbro pegmatite dykes, establishing a relative chronology of the development of pegmatoidal layering, pyroxenite intrusion and development of the cross-cutting pegmatites.

(e) Igneo-fragmental breccia dyke. A poorly exposed, 5 m wide, easterly-striking dyke is present 100 m to the north of the C zone and in the vicinity of the F zone (Fig. 5). The dyke contains clasts and boulders, up to 50 cm in diameter, of gabbro, pegmatoidal gabbro, pyroxenite and tonalite, in a matrix of mafic material which consists of dolerite and comminuted gabbroic material. Locally the matrix of the igneo-fragmental breccia dyke is highly foliated.

(f) Gabbro megabreccia.   Within the core of the zone of varitextured gabbro, a complex, heterolithic breccia, with clasts locally in excess of 5 m in diameter, is associated with areas of richer mineralisation (e.g. Fig. 10). The gabbro megabreccia contains clasts of anorthosite, pegmatoidal gabbro, anorthositic gabbro, noritic gabbro, norite, pyroxenite and foliated dolerite; the last clast type is very similar in appearance to the matrix of the igneo-fragmental breccia dyke.

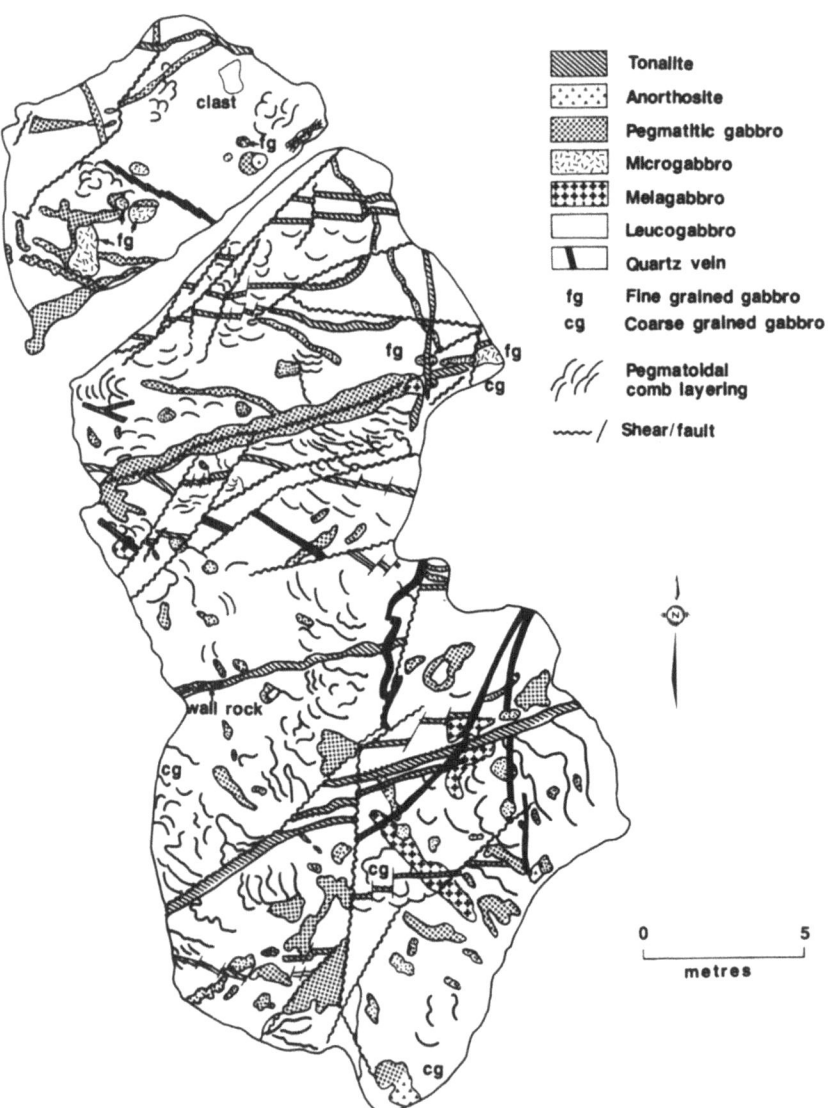

Fɪɢ. 7. Outcrop map of a portion of the E zone, northern Roby Zone.

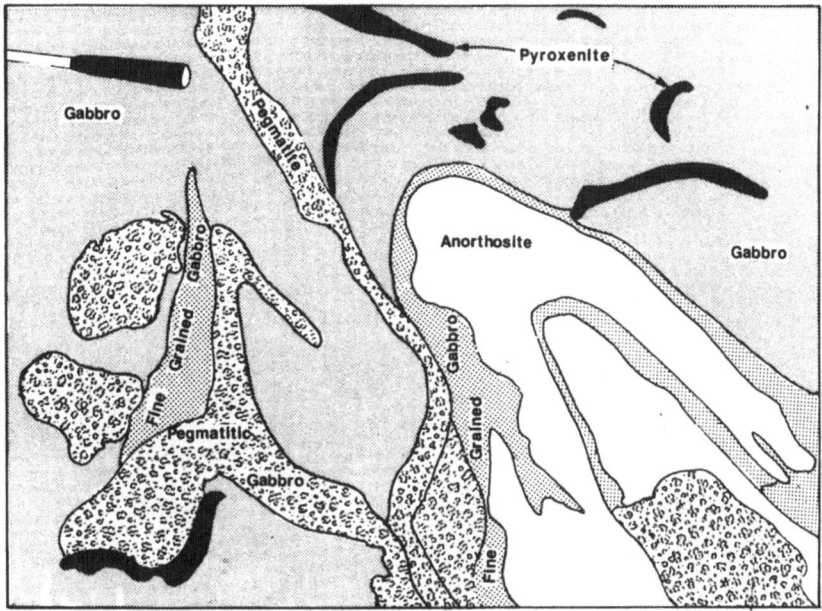

FIG. 8. Cuspate, pegmatoidal, comb layers, E zone, northern Roby Zone. Note that gabbro pegmatite dyke passes with continuity into pegmantoidal comb layer. Hammer handle for scale. Drawn from photograph.

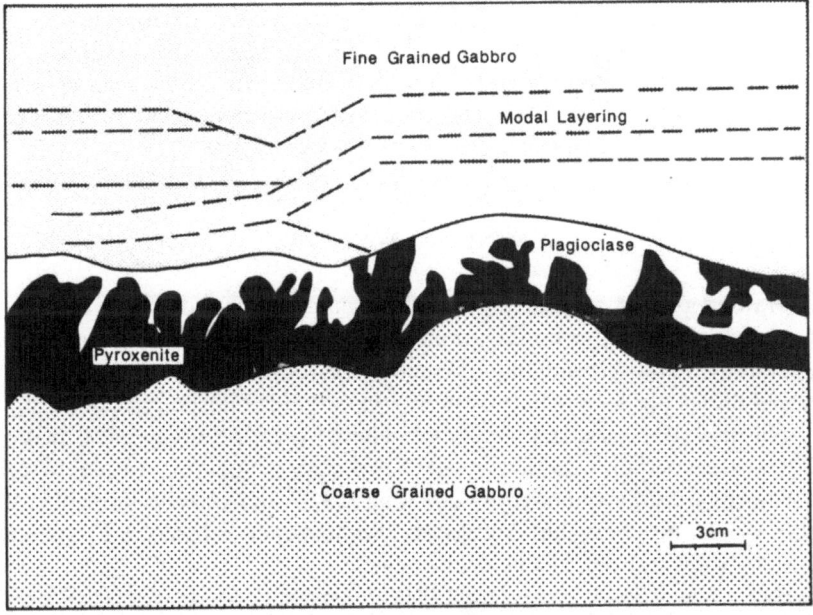

FIG. 9. Pegmatoidal comb layer, E zone, northern Roby Zone. Note pyroxenite base and interdigitated pyroxene and plagioclase crystals aligned perpendicularly to layer. Pen for scale. Drawn from photograph.

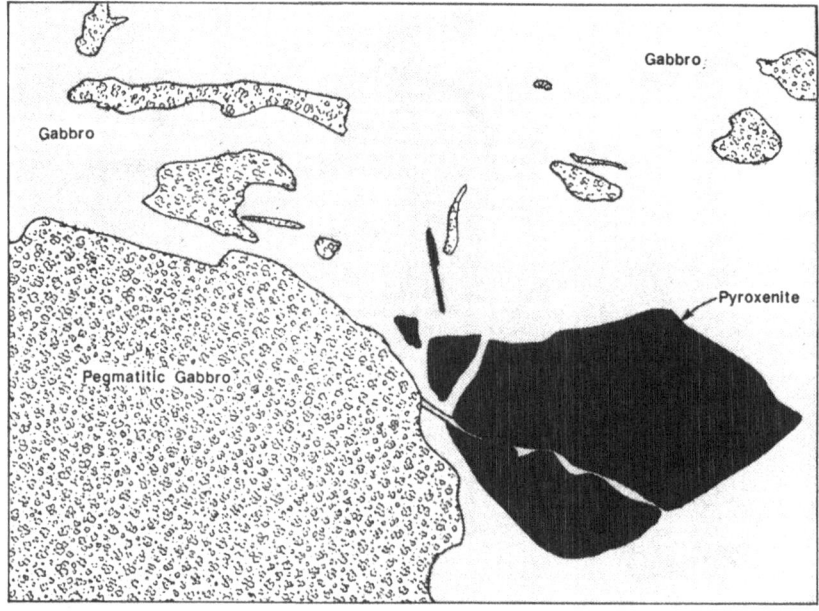

FIG. 10. Gabbro megabreccia, C zone. Pen for scale. Drawn from photograph.

Some clasts are composite, consisting of gabbro and anorthosite, while others contain varitextured gabbro of highly variable composition and grain size.

Some gabbro clasts within a gabbro matrix are coated by 'rinds' of either pyroxenite or pegmatoidal gabbro (Fig. 11), termed orbicular texture (Leveson, 1966; Moore & Lockwood, 1973), and are found in association with comb layering, the rinds being similar to the constituents of the comb layers.

MINERALISATION

In mineralised gabbro, sulphides (in decreasing order of abundance: pyrite, chalco-pyrite, pentlandite, pyrrhotite, millerite, violarite, sphalerite, galena and molybdenite) are most commonly disseminated throughout gabbroic or pyroxenitic hosts. Sulphides are locally net-textured, interstitial to silicate phases, or aligned along chlorite- or secondary amphibole-filled fractures. Frequently, however, sulphides are spatially associated with secondary amphibole within pyroxene pseudomorphs. PGM identified by Dunning (1979) and Cabri & Laflame (1979) include vysotskite, braggite, kotulskite, isomertieite, merenskyite, sperrylite, moncheite, stillwaterite and palladoarsenide.

Highest grades (up to 35 ppm total PGE + Au, Macdonald, 1985) are found in two environments: (a) associated with highly altered pyroxenite at the eastern margin of the Roby Zone, and (b) in cross-cutting gabbro pegmatites. Local sulphide accumulations which contain variable grades of PGE occur in pegmatitic

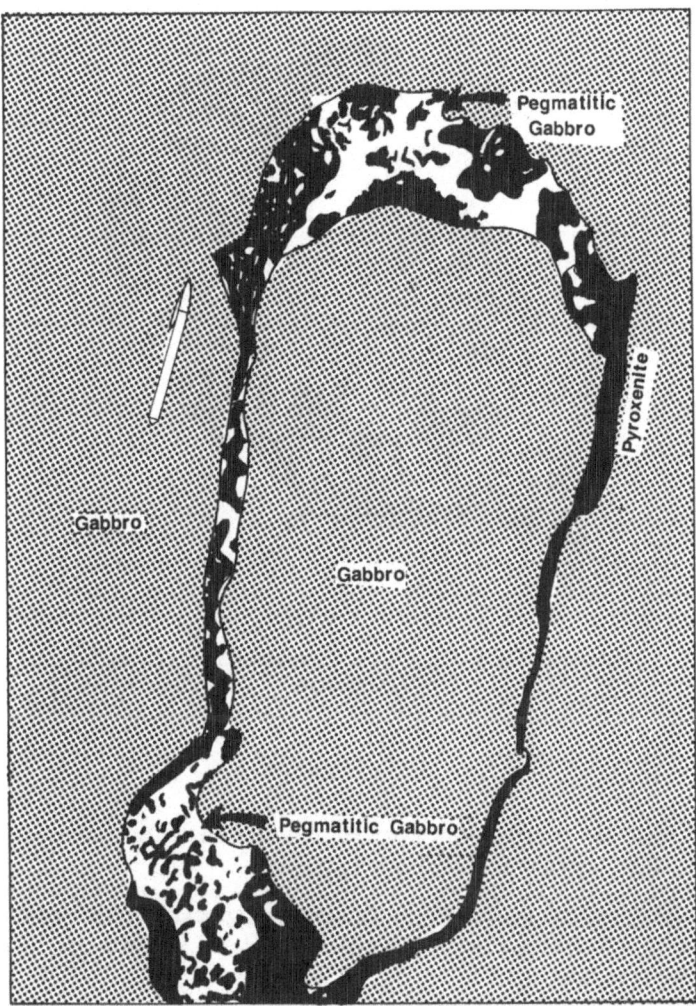

Fɪɢ. 11. Orbicular gabbro megabreccia clast, E zone, northern Roby Zone. Note rinds of both pyroxenite and pegmatoidal gabbro coating clast. Pen for scale. Drawn from photograph.

pods (Fig. 12), characterised by inward radiating pyroxene crystals, with plagioclase, quartz and sulphides in cores of the pods and appear similar to miarolitic cavities described by Watkinson (1987) in PGE-bearing gabbros of the Coldwell Complex, north of Marathon, Ontario.

The eight sulphide-bearing zones (A–H, Fig. 2) are not equally endowed with PGE. Only the C, E and F zones contain economically significant PGE grades (>1 ppm total PGE) although all sulphide zones contain approximately identical Cu and Ni grades (0·6–0·9% combined; Pye, 1968). This indicates that the grade of Pt + Pd + Au is independent of Ni and Cu contents, as concluded by Talkington & Watkinson (1984).

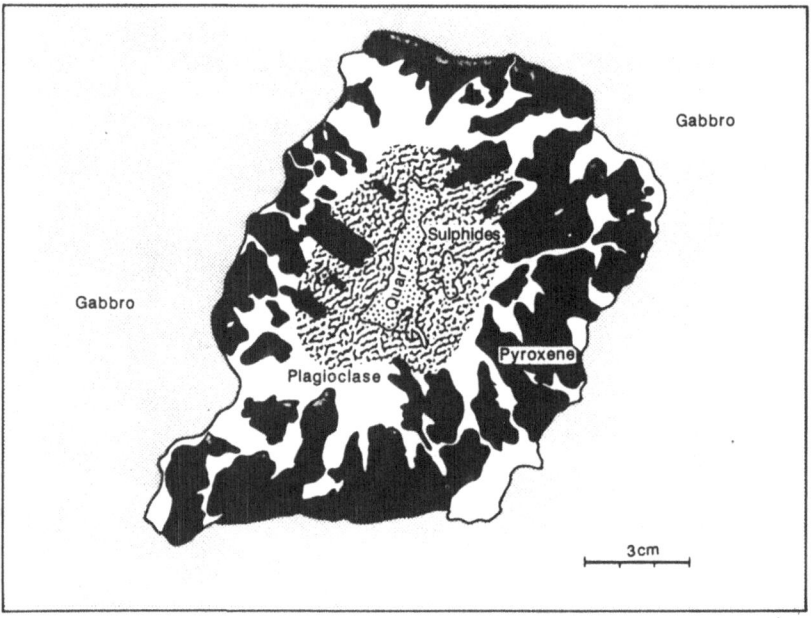

FIG. 12. Sulphide-rich, miarolitic pod in E zone, northern Roby Zone. Pen for scale. Drawn from photograph.

DISCUSSION

The eight sulphide-bearing zones at Lac des Iles share the following common characteristics not seen in unmineralised gabbro:

(a) a higher degree of deuteric alteration; plagioclase is altered to chlorite, sericite and zoisite, clinopyroxene to amphibole, and orthopyroxene, where present, to talc;

(b) a marked heterogeneity in composition, from anorthosite to pyroxenite;

(c) a wide variation in texture, from fine grained to pegmatoidal.

The close association between PGE-bearing sulphides and the development of pegmatoidal textures at Lac des Iles is by no means a unique occurrence. Similar observations in the Bushveld and Stillwater complexes have led to the suggestion that volatile transport of the PGE may be an integral component of the mineralising process in general (e.g. Stumpfl, 1974, 1986), and at Lac des Iles in particular, Talkington & Watkinson (1984), who proposed that 'filter pressing of inter-cumulus material permitted fluids to percolate upward through the mush of crystals, causing extensive deuteric alteration' and redistribution of metals. Whereas the development of pegmatoidal textures and deuteric alteration may be indicative of elevated volatile pressures during, and subsequent to, accumulation of the gabbroic rocks at Lac des Iles, these features are present in all sulphide-bearing zones, both PGE-rich and PGE-poor.

In addition, many of the sulphide zones display evidence of some interaction between gabbroic and tonalitic magmas. Irvine (1975) and Naldrett & Macdonald (1980) have suggested that addition of silica to a mafic magma may induce sulphide liquation, and subsequent scavenging of PGE from the silicate magma. Whereas tonalite contamination may account for some of the sulphide mineralisation, particularly those zones near the contact between gabbro and tonalite (e.g. A and G zones, Fig. 2), these zones do not contain the most significant PGE grades (>1000 ppb total PGE), and another factor must be sought to account for these elevated grades within the Roby Zone.

The three zones that carry highest PGE grades (E, C and F), in addition to the characteristics outlined above also display the following features:

(a) Cross-cutting pegmatite dykes.

The presence of pegmatite dykes, supports the hypothesis that volatiles played a greater role in zones containing higher PGE grades. Based upon cross-cutting relationships, pegmatite developed after formation of pegmatoidal comb layers, ultramafic intrusion and the formation of gabbro megabreccia. Pegmatite dykes are only found in the varitextured gabbro unit, and do not reflect a magmatic influx from an external source. It is more likely that the pegmatites reflect a late adjustment in a volatile-rich environment of plagioclase and pyroxene crystals that constituted the gabbroic host rock, an hypothesis supported by the relationships between pegmatoidal comb layers and cross-cutting pegmatite dyke in Fig. 8.

Subvertical, cuspate pegmatoidal comb layering exposed in the host rocks (Fig. 9), provides evidence of (1) crystallisation in the presence of a transient, elevated volatile content, (2) direction of solidification of the gabbro host rock, i.e. solidification vector. Azimuths of solidification vectors vary from zone to zone at Lac des Iles (Fig. 13). The vectors indicate that crystallisation of the gabbroic magma proceeded inwards from the margins of the varitextured gabbro unit (Fig. 5) in the vicinity of the C, F, E and H zones.

(b) Abundance of gabbro megabreccia.
(c) Proximity to an ultramafic intrusion

The websterite dyke that intrudes gabbro on the eastern margin of the Roby Zone, is commonly the locus of high PGE grades, approximately 15 000 ppb total PGE (Talkington & Watkinson, 1984). The ultramafic dyke may have intruded the LDI gabbroic complex before the gabbro was wholly consolidated, and locally intermingled with the gabbro on the eastern margin of the Roby Zone. Mixing of an ultramafic liquid and a magma of anorthositic affinity has been proposed as a process responsible for mineralisation in the J-M Reef of the Stillwater complex (Todd *et al.*, 1982; Campbell *et al.*, 1983; Irvine *et al.*, 1983) and in the Merensky Reef, Bushveld Complex (Irvine *et al.*, 1983; Irvine & Sharpe, 1986; Sharpe *et al.*, 1986). There is, however, some discussion as to whether ultramafic liquids pulsed into anorthositic magma, or vice versa (cf. Campbell *et al.*, 1983 and Sharpe *et al.*, 1986). At Lac des Iles, unequivocal field relationships are preserved demonstrating the intrusive relationship of an ultramafic dyke into a anorthositic gabbro host (Fig. 3).

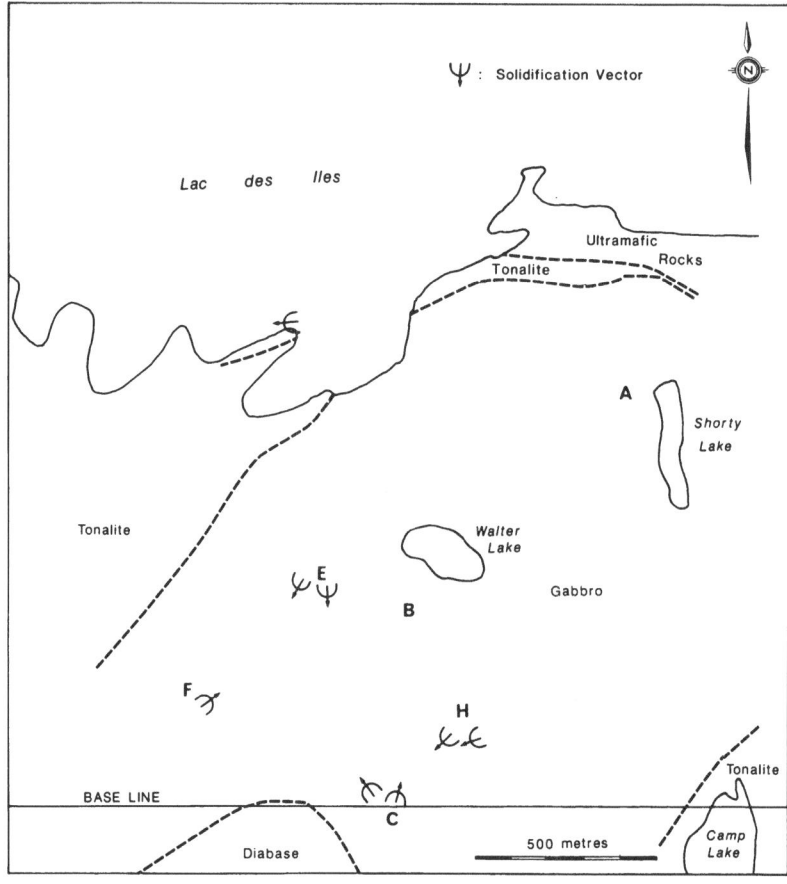

FIG. 13. Map of solidification vectors in varitextured gabbro, for part of the area in Fig. 5.

INTERPRETATION

Figure 14 summarises the main field relationships in the varitextured gabbro of the mineralised zone for the LDI.

Modal layering, comb layering and igneous foliation all dip steeply in the gabbroic rocks defining a crude, elongate, funnel-shaped intrusion. Crystal accumulation was not, therefore, dominated by simple gravity-driven crystal settling on the floor of a magma chamber. Rather, this type of layering forms by *in situ* crystallisation, as proposed by Moore & Lockwood (1973), Lofgren & Donaldson (1975) and discussed by Campbell (1978). Either comb layers were deposited from large volumes of volatile-rich fluid migrating along the interface between the crystal pile and magma (Moore & Lockwood, 1973; Lofgren & Donaldson, 1975; Petersen, 1985), or by nucleation in response to super-

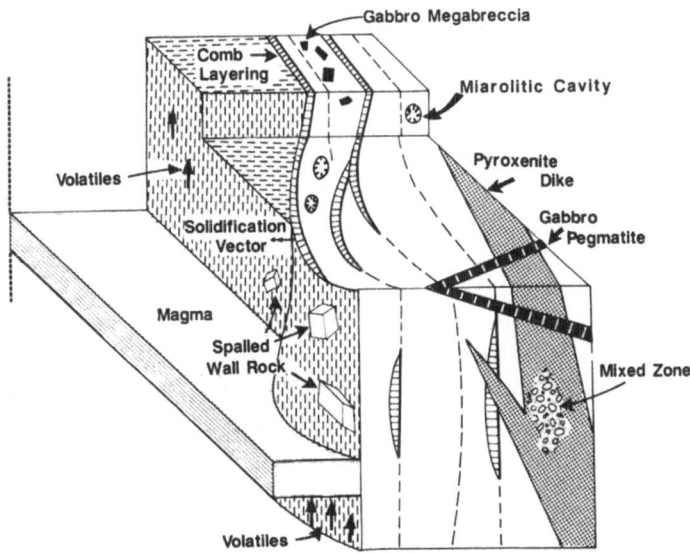

Fig. 14. Geometry of features characterising varitextured gabbro in mineralised zones at Lac des Iles, showing modal and comb layering, pyroxenite dyke, sulphide-bearing miarolitic pods, gabbro mega-breccia, cross-cutting gabbro pegmatites, and relationships with magma and volatiles. Comb layering concept modified from Moore and Lockwood (1973).

cooling of a solute-enriched boundary layer, although their arguments were addressed more towards branching, crescumulate crystal aggregates, the more simple 'teeth on a comb' style of layer described by Moore & Lockwood (1973), typical also of Lac des Iles layering. Whereas super-cooling of a solute-rich melt and crystallisation from a volatile-rich melt may be valid processes, the presence of pegmatites, spatially associated with, and apparently derived from volatile-rich, pegmatoidal comb layers, the marked deuteric alteration in zones containing comb layers, and the presence of miarolitic cavities (Fig. 12), supports an hypothesis which integrates both crystal accumulation and volatile activity.

Clasts within the gabbro megabreccia unit include all the lithologies observed within varitextured gabbro that surrounds the megabreccia (Fig. 7), suggesting that they are locally derived. Gabbroic clasts within the gabbro megabreccia (Fig. 11) display similar relationships to the orbicular rocks, also spatially associated with comb layers, described by Moore & Lockwood (1973), who consider that 'orbicules formed by precipitation of comb layers on bobbling inclusions suspended within the upward flowing fluid'.

Anorthosite layers, pods and dykes, are a prominent feature of varitextured gabbro. The origin of anorthosite layers remains enigmatic, but anorthositic dykes appear to reflect flow differentiation of gabbro pegmatites: pyroxene preferentially crystallises on the walls of pegmatite dykes leaving plagioclase-rich (anorthositic) material in the centre of the dyke. Consequently, as the pegmatitic fluid passes further from its point of origin, pyroxene is gradually removed leaving anorthosite.

Anorthositic pods similarly result from preferential crystallisation of pyroxene at the margin of the pod, leaving an anorthositic centre. Highest PGE grades (up to 35 000 ppb total PGE) are found in pegmatite dykes and pods, especially where quartz-bearing (Fig. 12).

Further evidence for high volatile levels during magmatism is provided by the igneo-fragmental breccia dyke (Fig. 5), with a mixed magmatic and comminuted rock matrix, that intruded prior to generation of the gabbro megabreccia, as demonstrated by clasts of the dyke's matrix found within the breccia.

In summary, the most PGE-rich rocks at Lac des Iles have formed within a gabbroic host that accumulated through in situ crystallisation in the presence of a transient, elevated volatile flux. Highest PGE grades are associated with (a) a websterite intrusion, and (b) cross-cutting gabbro pegmatite dykes, in the vicinity of the ultramafic intrusion. While a number of processes can be proposed to account for sulphide liquation and subsequent PGE transport, such as mixing of gabbroic and tonalitic magmas, and transport of PGE in deuteric fluids, intrusion of an ultramafic magma into a volatile-rich portion of anorthositic gabbro appears to be the fundamental process localising economically significant grades of mineralisation at Lac des Iles. Subsequent deuteric alteration and the development of cross-cutting pegmatite dykes, indicative of an elevated volatile component, are considered to be responsible for redistribution of the orthomagmatic Ni, Cu and precious metal mineralisation over widths locally in excess of 100 m.

ACKNOWLEDGEMENTS

I received considerable help in the field from M. Hine and G. Lawson, to whom I am most grateful. Discussions with a number of geologists greatly enhanced interpretations of outcrops. I would particularly like to acknowledge useful discussions with A. J. Naldrett, G. C. Wilson, J. F. H. Thompson and G. Brugmann. An earlier draft of the manuscript benefited greatly from the critical review of H. R. Williams, C. P. Barrie. G. Lawson drafted the diagrams and K. Gil prepared the photographs: their contribution is gratefully appreciated. An earlier version of this manuscript was much improved by the comments of Dr H. M. Prichard and two anonymous Geo-platinum 87 reviewers. This paper is published with permission of the Director, Ontario Geological Survey.

REFERENCES

Cabri, L. J. (1981). The platinum-group minerals. In Platinum-Group Elements: Mineralogy, Geology, Recovery, ed. L. J. Cabri. CIM Special Volume, pp. 83–150.
Cabri, L. J. & Laflamme, J. H. G. (1979). Mineralogy of samples from the Lac des Iles area, Ontario. CANMET, Energy, Mines and Resources Canada, Rep. 79–27.
Campbell, I. H. (1978). Some problems with the cumulus theory. Lithos, 11, 311–23.
Campbell, I. H., Naldrett, A. J. & Barnes, S.-J. (1983). A model for the origin of the platinum-rich sulfide horizons in the Bushveld and Stillwater Complexes. J. Petrol., 24(2), 133–65.

Dunning, G. R. (1979). The geology and platinum-group mineralization of the Roby Zone, Lac des Iles Complex, Northwestern Ontario. Unpublished M.Sc. thesis, Carleton University, Ottawa, 129 pp.

Dunning, G. R., Watkinson, D. H. & Mainwaring, P. R. (1981). Correlation of platinum-group elements, copper and nickel, with lithology in the Lac des Iles Complex, Canada. In *Proceedings of the International Symposium on Metallogeny of Layered Mafic/Ultramafic Intrusions*, Athens, Greece, International Geological Correlation Program, Project 169, pp. 83–102.

Fortescue, J. A. C. & Webb, J. R. (1986). An orientation geochemical study at the Lac des Iles Complex, District of Thunder Bay. In Ontario Geological Survey, Miscellaneous paper 132, pp. 217–219.

Guarnera, B. J. (1967). Geology of the McKernan Lake Phase of the Lac des Iles Intrusive, Thunder Bay District, Ontario. Unpublished M.Sc. thesis, Michigan Technical University, Houghton, 100 pp.

Gupta, V. K., Wadge, D. R., Nakashima, A. & Mark, P. (1986). Gravity studies of mafic and ultramafic intrusions in the Lac des Iles area, District of Thunder Bay. In Ontario Geological Survey, Miscellaneous Paper 132, pp. 220–221.

Irvine, T. N. (1975). Crystallisation sequences of the Muskox intrusion and other layered intrusions—II. Origin of chromitite layers and similar deposits of other magmatic ores. *Geochim. Cosmochim. Acta*, **39**, 991–1020.

Irvine, T. N. & Sharpe, M. R. (1986). Magma mixing and the origin of stratiform oxide ores in the Bushveld Complex. In *Geocongress 86*, Extended Abstracts, Johannesburg, Geological Society of South Africa, Marshalltown, RSA, pp. 599–601.

Irvine, T. N., Keith, D. W. & Todd, S. G. (1983). The J-M platinum-palladium Reef of the Stillwater Complex, Montana: II. Origin by double-diffusive convective magma mixing and implications for the Bushveld Complex. *Econ. Geol.*, **78**, 1287–34.

Kaye, L. (1969). Geology of the Eayrs Lake-Stearnes Lake Area, District of Thunder Bay. Ontario Department of Mines, Geological Report 77, 29 pp., accompanied by Map 2172.

Leveson, D. J. (1966). Orbicular rocks: a review. *Bull. Geol. Soc. Amer.*, **77**, 409–26.

Linhardt, E. & Sutcliffe, R. H. (1986). Petrological and economic geological studies on the Northern Lac des Iles Complex, District of Thunder Bay. Ontario Geological Survey, Miscellaneous Paper 132, pp. 80–81.

Lofgren, G. E. & Donaldson, C. H. (1975). Curved branching crystals and differentiation in comb-layered rocks. *Contributions to Mineralogy and Petrology*, **49**, 309–19.

Macdonald, A. J. (1985). The Lac des Iles platinum group metals deposit, Thunder Bay District, Ontario. Ontario Geological Survey, Miscellaneous Paper 126, pp. 235–241.

Macdonald, A. J., Brugmann, G. & Naldrett, A. J. (in preparation). Coeval felsic, mafic and ultramafic magmas: magma mixing during formation of PGE-rich Ni-Cu sulphides at Lac des Iles, Ontario, Canada. IGCP Project 161, 5th Magmatic Sulphides Conference, Harare, August 3–13, 1987.

Moore, J. G. & Lockwood, J. P. (1973). Origin of comb layering and orbicular structure, Sierra Nevada Batholith, California. *Bull. Geol. Soc. Amer.*, **84**, 1–20.

Naldrett, A. J. (1981). Platinum-group element deposits. In *Platinum-group Elements: Mineralogy, Geology and Recovery*, ed. L. J. Cabri. CIM Special volume 23, pp. 197–231.

Naldrett, A. J. & Macdonald, A. J. (1980). Tectonic settings of some Ni-Cu sulfide ores: their importance in genesis and exploration. In *The Continental Crust and its Mineral Deposits*, ed. D. W. Strangway. Geological Association of Canada. Special Paper No. 20, pp. 633–57.

Petersen, J. S. (1985). Columnar-dendritic feldspars in the Lardalite Intrusion, Oslo Region, Norway: 1. Implications for unilateral solidification of a stagnant boundary layer. *J. Petrol.*, **26**, 223–52.

Poldervaart, A. & Taubeneck, W. H. (1959). Layered intrusions of the Willow Lake type. *Bull. Geol. Soc. Amer.*, **70**, 1395–8.

Pye, E. G. (1965). Lac des Iles Area. Ontario Dept. of Mines, Map P. 275.

Pye, E. G. (1968). Geology of Lac des Iles area. Ontario Dept. of Mines, Report 64, 47 pp., accompanied by maps 2135, 2136.

Sharpe, M. R., Evensen, N. M. & Naldrett, A. J. (1986). Sm/Nd and Rb/Sr isotopic evidence for liquid mixing, magma generation and contamination in the Eastern Bushveld Complex. In *Geocongress 86*, Extended Abstracts, Johannesburg, Geological Society of South Africa, Marshalltown, P.SA, pp. 621–624.

Smith, A. R. & Sutcliffe, R. H. (1986). Geology of the Tib gabbro, Lac des Iles area, District of Thunder Bay. In Ontario Geological Survey, Miscellaneous Paper 132. pp. 76–79.

Stumpfl, E. F. (1974). The genesis of platinum deposits: Further thoughts. *Min. Sci. Engng.*, **6**, 120–41.

Stumpfl, E. F. (1986). Distribution, transport and concentration of platinum-group elements. In *Mineralization in Mafic and Ultramafic Rocks*, ed. M. J. Gallagher, R. A. Ixer, C. R. Neary & H. M. Prichard. Institution of Mining and Metallurgy, pp. 379–94.

Sutcliffe, R. H. (1986). Regional geology of the Lac des Iles area, District of Thunder Bay. In Ontario Geological Survey, Miscellaneous paper 132, pp. 70–75.

Sutcliffe, R. H. (1987). PGE geology of the Lac des Iles complex, Ontario. In *Seminar on Platinum-Group Elements*, ed. R. Goldie. Geoscience Canada, in preparation.

Sutcliffe, R. H. & Sweeney, J. M. (1985a). Geology of the Lac des Iles Complex, District of Thunder Bay. In Ontario Geological Survey, Miscellaneous Paper 126, pp. 47–53.

Sutcliffe, R. H. & Sweeney, J. M. (1985b). Lac des Iles Complex. Ontario Geological Survey, Map P.3047.

Sweeney, J. M. & Sutcliffe, R. H. (1986). Geology and platinum-group element mineralization of the Roby Zone, Lac des Iles Complex. In Ontario Geological Survey, Miscellaneous Paper 132, pp. 82–84.

Talkington, R. W. & Watkinson, D. H. (1984). Trends in the distribution of the precious metals in the Lac des Iles Complex, Northwestern Ontario. *Can. Mineral.*, **22**, 125–36.

Todd, S. G., Keith, D. W., Le Roy, L. W., Schissel, E. L., Mann, E. L. & Irvine, T. N. (1982). The J-M platinum-palladium Reef of the Stillwater Complex, Montana: I. Stratigraphy and petrology. *Econ. Geol.*, **77**, 1454–80.

Watkinson, D. H. (1987). Hydrothermal aspects of PGE mineralisation. In *Seminar on Platinum-Group Elements*, ed. R. Goldie, Geoscience Canada, in preparation.

Watkinson, D. H. & Dunning, G. R. (1979). Geology and platinum-group mineralisation, Lac des Iles complex, Northwestern Ontario. *Can. Mineral.*, **17**, 453–62.

Winfield, B. & Workman, A. (1976). Report on Lakeshore mapping, Lac des Iles, October–November 1975, Unpublished Report, Texasgulf Canada Ltd., 27 pp., with two appendices.

23

Platinum-Group Mineral Precipitation from Fluids in Pegmatitic Gabbro: Two Duck Lake Intrusion, Coldwell Complex, Ontario, Canada

DAVID H. WATKINSON & RICHARD DAHL

Ottawa-Carleton Geoscience Centre, Carleton University, Ottawa, Canada K1S 5B6

Platinum-group elements (Pt/Pt + Pd = 0·22) occur with very Cu-rich sulphide assemblages (Cu/Cu + Ni = 0·90) in a 46 million tonne deposit within coarse-grained to pegmatitic gabbro-monzonite of the Two Duck Lake Intrusion. This unit invades and has assimilated Archean metavolcanic rocks near its base and the layered succession of the Eastern Coldwell Complex Gabbro near its top. Assimilation and mixing of rheomorphic material is apparently responsible for the abundant xenoliths, granophyric patches, aplitic dykes, K-rich minerals and deuteric alteration that are spatially related to concentrations of the PGE in disseminated Cu-rich sulphides. Much of the disseminated sulphide is interstitial to and intergrown with biotite, plagioclase, amphibole and ilmenomagnetite. Platinum-group minerals (michenerite, mertieite-II, sperrylite, hollingworthite, stannopalladinite, zvyagintsevite, atokite, kotulskite, merenskyite, as well as electrum) occur mainly with chalcopyrite and sometimes bornite. Many grains of PGM (and pentlandite, argentian pentlandite and pyrite) are found at grain boundaries of chalcopyrite and ilmenomagnetite where the magnetite, but not ilmenite lamellae, is partly altered to chlorite or replaced by chalcopyrite. None has been identified in pyrrhotite, which has apparently been partly replaced by chalcopyrite. Late-stage fluid, perhaps of juvenile origin, but more likely from the breakdown of included xenoliths, mobilized elements from the intruded rocks, especially Cu, Ag, Pb, Pd, Pt and Rh. These elements were precipitated when this fluid reacted with pyrrhotite, and especially magnetite; this reaction is not unlike that often proposed for epigenetic Au precipitation in Fe-rich rocks.

24

Interelement Correlation, Stratigraphic Variation and Distribution of PGE in the Ultramafic Series of the Bird River Sill, Canada*

R. F. J. SCOATES, O. R. ECKSTRAND

*Geological Survey of Canada, 601 Booth Street,
Ottawa, Ontario, Canada K1A 0E8*

&

L. J. CABRI

*Canada Centre for Mineral and Energy Technology, 555 Booth Street,
Ottawa, Ontario, Canada K1A 0G1*

ABSTRACT

Anomalous concentrations of platinum-group elements (PGE) in the Bird River Sill (S.E. Manitoba, Canada) occur in peridotites and chromitites in the lower part of the Chromitiferous Zone. Harrisitic peridotites, commonly associated with anomalous PGE concentrations, disrupt chromitite layers with which they are in contact. The harrisites may have resulted from the ponding of upward migrating, volatile-rich, intercumulus liquid.

Factor analysis of PGE, Au, Ni, Cu, Co, Zn, Cr, V, S and Se determined in 66 hand specimens indicates two distinct associations. In the first, PGE and Cr (+ V + Zn) are strongly correlated and illustrate a clear affiliation between PGE and chromite. In the second, Ni (+ Cu + Co) are correlated with S and Se (±Au) indicating a nickel sulphide factor. There is no association between PGE and the nickel sulphide factor.

The PGE can be separated into two groups which show distinct geochemical tendencies. The Os-Ir-Ru group are strongly linked with chromite, whereas Rh, Pt and Pd display conflicting associations. On the one hand, mean abundances and PGE profiles seem to indicate that Rh, Pt and Pd are associated with a stratigraphic zone of sulphide mineralization. However, factor analysis shows no association of Rh, Pt and Pd with sulphide but rather a strong linkage with chromite.

All six PGE are accounted for mineralogically by discrete platinum group minerals (PGM): sperrylite, kotulskite, merenskyite, laurite, hollingworthite, irarsite, keithconnite, and mertieite II. The most common association of the PGM is with silicate gangue. However, a few PGM inclusions occur within chromite or are

*Geological Survey of Canada Contribution No. 18088.

associated with sulphide. The common occurrence of PGM in silicate gangue suggests small-scale remobilization. This remobilization may reflect either late magmatic stage ponding of intercumulus liquid and/or subsequent serpentinization. The strongly modified sulphide-silicate textures suggest that recrystallization related to greenschist facies metamorphism played an important role in the microscopic scale redistribution of PGM.

The strong statistical association of PGE and Cr suggests a genetic relationship between chromite crystallization and PGE concentration, but the mechanism is not understood.

BIRD RIVER SILL, SOUTHEASTERN MANITOBA

The Bird River Sill, a differentiated ultramafic to mafic layered intrusion of Archean age in southeastern Manitoba, is well known as the repository of Canada's most significant chromite resource. Several nickel-copper sulphide deposits are also associated with the sill of which the most important are the Dumbarton and Maskwa West deposits.

The sill is exposed discontinuously along both limbs of an eastward plunging anticline, the core of which is occupied by the Maskwa Lake batholith (Fig. 1). The sill has a strike length of at least 25 km along the south limb, and is exposed over a strike length of 4 km in two widely separated localities along the north limb of the structure (Fig. 1). The exposed width rarely exceeds 600 m and, since the dip is commonly near vertical, this reflects the true thickness of the body.

The sill intruded metabasalts of the Lamprey Falls Formation, the second oldest of six formations that constitute the Bird River greenstone belt (Trueman, 1980). The upper contact is interpreted to be a fault which separates anorthositic gabbro of the sill from a series of debris flows that contain inclusions of anorthositic gabbro in addition to inclusions of metabasalt. The intrusion was apparently unroofed subsequent to its emplacement (Trueman, 1980).

The sill is differentiated with a lower Ultramafic Series and an upper Mafic Series. Where the section has been measured, the Ultramafic Series is about 200 m thick and the Mafic Series is 400 m thick (Fig. 2). However, as the top of the sill has been eroded, the latter thickness is a minimum. A U-Pb zircon age of 2745 ± 5 Ma (Timmins *et al.*, 1985) was determined for the Mafic Series gabbro.

Scoates (1983) established a stratigraphy for the Ultramafic Series which is illustrated in Fig. 2. This is based largely on observations on the Chrome claims but would appear to apply also to other exposures which are about 4·5 km along strike to the east. At this time the lateral extent over which this stratigraphy applies is not known. The Chromitiferous Zone is further subdivided, and contains six groups of chromitite layers (Scoates, 1983). These have been designated in ascending order the Lower, Disrupted, Lower Main, Banded and Diffuse, Upper Main and Upper Paired Groups.

Three main rock types have been identified in the course of mapping the Ultramafic Series: peridotite, dunite and chromitite. Peridotites are typically olivine-chromite cumulates with conspicuous amounts of intercumulus material.

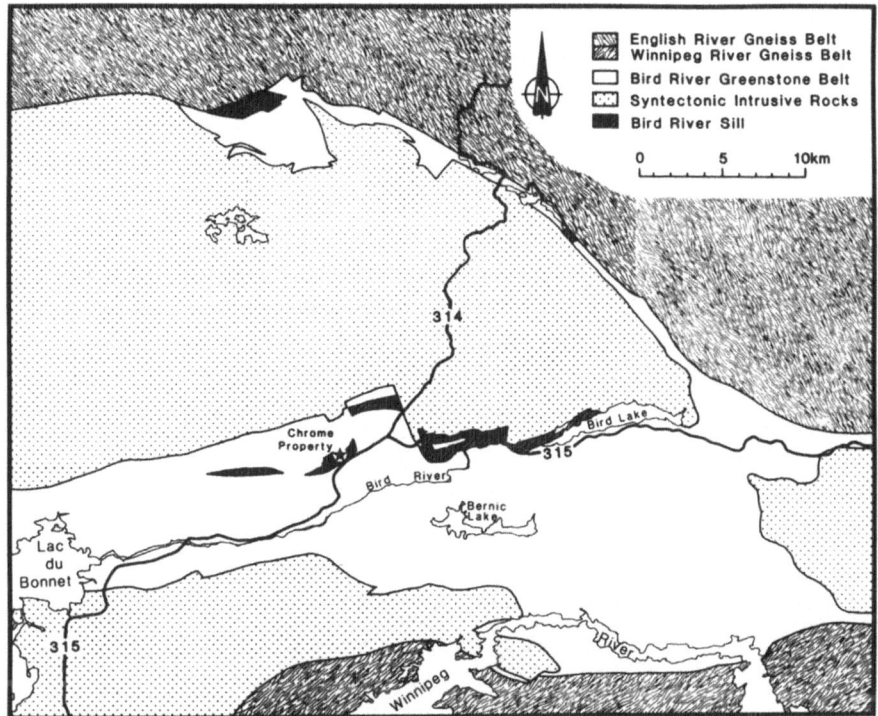

Fig. 1. Generalized geology of the Bird River greenstone belt, southeastern Manitoba, showing the disposition of the Bird River Sill and the location of the Chrome Property.

Olivine and chromite occur in their typical cotectic proportions: that is, chromite makes up from 1 to 5 modal %. Pervasive alteration makes the identification of primary intercumulus minerals uncertain. Some relict clinopyroxene has been observed but it is unclear how much orthopyroxene, hornblende or plagioclase may have been present.

PGE MINERALIZATION

Anomalous PGE concentrations in Bird River Sill Ultramafic Series rocks are defined as those greater than 100 ppb combined Pt, Pd and Au. Four intervals containing anomalous PGE concentrations occur in the Ultramafic Series. Two have been reported from the western part of the Chrome Property in Ultramafic Series rocks stratigraphically below the Chromitiferous Zone (Theyer, 1985). A third occurs within and immediately below Lower Group chromitite (Lower Group platinum-bearing unit, Fig. 3). The fourth interval is the Disrupted Group (Fig. 3) where sporadically distributed anomalous PGE concentrations occur. Of the four intervals, the Lower Group platinum-bearing unit is the most continuous and the most consistently anomalous.

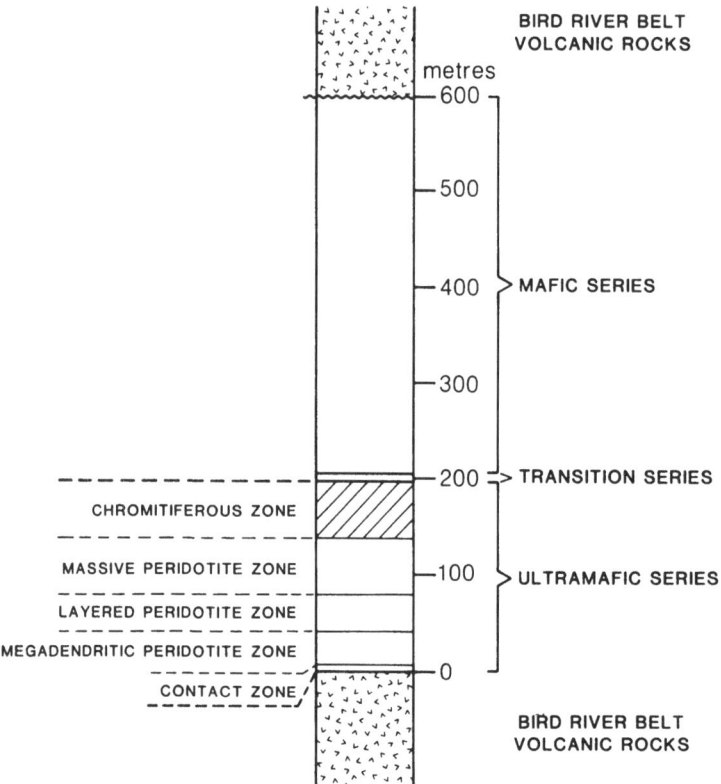

FIG. 2. Igneous stratigraphy of the Bird River sill.

Lower Group Platinum-Bearing Unit

This unit of anomalous PGE mineralization has been traced for 800 m along strike on the Chrome Property. The unit pinches and swells, ranging from several tens of centimetres up to 3 m thick. The mineralized unit includes the Lower Group chromitites and extends downward below the base of the Lower Group into peridotite of the underlying Massive Peridotite Zone.

This mineralized unit has been divided provisionally into three subunits (Fig. 3) on the basis of mineralogy and texture. Subunit 1, the stratigraphically lowest subunit, consists of disseminated-sulphide-bearing peridotite. The rocks are medium grained, original olivine crystals averaging 2 mm in size. Rounded to subangular chromitite inclusions are sporadically distributed throughout subunit 1 as are discontinuous, mm-scale chromitite layers. Coarser grained, harrisitic peridotite has been noted in places. Subunit 1 extends downward from 70 cm below to as much as 3 m below the base of the Lower Group. Subunit 2 overlies subunit 1 and represents the 70 cm interval from the top of subunit 1 to the base of the Lower

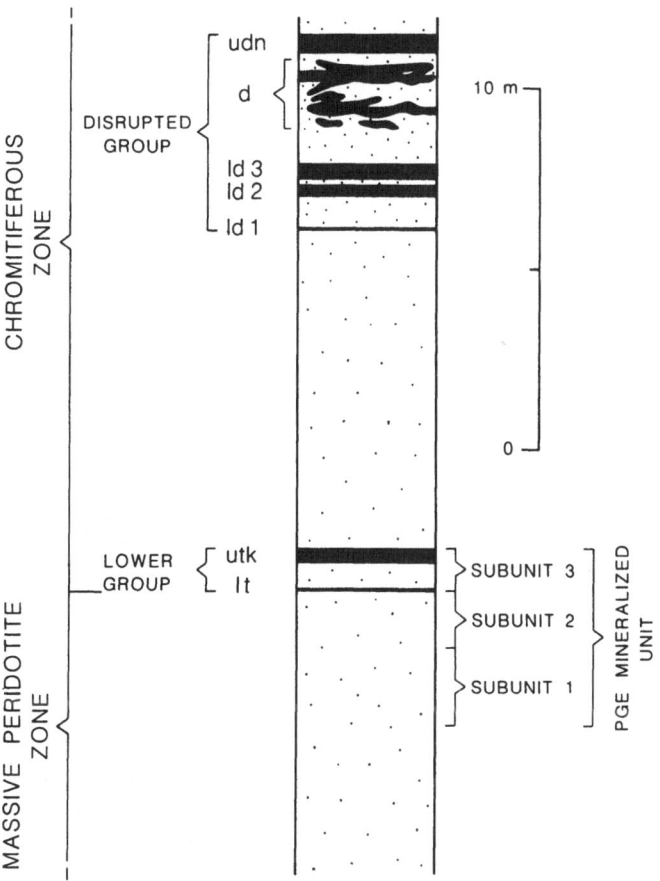

FIG. 3. Detailed stratigraphy of the contact between the Massive Peridotite Zone and the Chromitiferous Zone illustrating the subdivisions of the PGE mineralized unit.

Group. Subunit 2 consists of disseminated-sulphide-bearing peridotite, much of which is characterized by coarse grained olivine crystals (up to 2 cm), many of which are harrisitic. The volume of original postcumulus material increases from 5 to 10% in subunit 1, to 20% in subunit 2. Disseminated chromite, and disrupted fragments of massive chromitite are also more abundant than in subunit 1. Normal, fine grained peridotite is also present. Subunit 3 consists of the lower and upper chromitite members and the intervening olivine cumulate member of the Lower Group. The lower member is a 4–6 cm thick diffuse chromitite (chromite-olivine cumulate) layer that is disrupted in places and that locally bifurcates into two or more layers separated by peridotite. The chromitite is characterized by numerous peridotite inclusions up to 1 cm in size. The upper member ranges from diffuse to massive chromitite (10–25 cm thick) that commonly contains entrained peridotite inclusions. The chromitite locally bifurcates into two layers separated by up to

15 cm of peridotite. The olivine cumulate member that separates the lower and upper members ranges from fine grained peridotite (1–2 mm olivine) to coarse grained (1–2 cm olivine), harrisitic peridotite. The latter is commonly present where the lower member is disrupted. Disseminated sulphides are sporadically distributed throughout subunit 3.

The sulphides commonly occur as small clusters (a few millimetres in diameter) of tiny grains, suggestive of intercumulus magmatic droplets. However recrystallization (under upper greenschist to lower amphibolite facies metamorphism or during serpentinization) has produced fine grained sulphide-silicate intergrowth textures.

Platinum-Group Minerals

All six platinum-group elements are accounted for mineralogically by discrete platinum-group minerals (PGM). These include in approximate order of abundance, sperrylite ($PtAs_2$), kotulskite (PdTe), merenskyite ($PdTe_2$), laurite (RuS_2), hollingworthite (RhAsS), irarsite (IrAsS), keithconnite ($Pd_{3-x}Te$), and mertieite II (Pd_8Sb_3). The most common association of the PGM is as tiny grains (usually <5 μm) in silicate gangue (77% by volume, 91% by frequency) but some grains occur associated with sulphide (22% by volume, 3·6% by frequency) or included within chromite and magnetite (<1% by volume, 2·7% by frequency).

Platinum-Group Element Geochemistry

Pt ranges up to 700 ppb and Pd up to 1800 ppb in assays of 66 samples from Ultramafic Series rocks (Table 1). Geometric means for subunits 1–3 (Table 2) show that the highest average concentration for total PGE occurs in subunit 2, where Ni and Cu have the highest grades. Os, Ir and Ru have their highest concentrations in subunit 3 which contains the chromitite layers of the Lower Group.

Chondrite normalized profiles of PGE are grouped according to their association (Fig. 4). Peridotite-hosted sulphidic PGE mineralization displays a steep positive slope, similar to that of other PGE-bearing magmatic sulphide deposits. PGE-poor

TABLE 1
Bird River Sill—PGE assays (ppb, 66 samples)

	Low[a]	High[a]
Os	1	100
Ir	0·1	53
Ru	3	510
Rh	1	160
Pt	17	720
Pd	2	1 800
Au	4	520

[a]'Low' = minimum value obtained; 'high' = maximum value obtained.

TABLE 2

Bird River Sill—geometric means[a] of PGE in mineralized subunits

	Subunit 1 (*ppb*)	*Subunit 2* (*ppb*)	*Subunit 3* (*ppb*)
Os	4	6	15
Ir	2	5	10
Ru	18	35	91
Rh	6	19	19
Pt	75	165	147
Pd	110	370	187
Au	15	17	8
	230	617	477
	(*ppm*)	(*ppm*)	(*ppm*)
Ni	2 192	2 936	1 846
Cu	314	441	175
Cr	3 511	5 942	25 811

[a]The geometric mean (of Os, for example) = the antilog of the mean of the logarithms of all values of Os.

chromitite has a negative slope, similar to the negative slopes shown by ophiolitic chromitites in general (Barnes *et al.*, 1985). As in ophiolitic chromitites, this enrichment of Os, Ir and Ru in PGE-poor chromitites of the Bird River Sill expresses the tendency for laurite (RuS_2) and iridosmine (Os, Ir) to be the most common platinum-group minerals (Talkington *et al.*, 1983; Ohnenstetter *et al.*, 1986). The third group in Fig. 4, PGE-bearing chromitite has a shallow positive slope, intermediate between those of the other two groups. This profile appears to result from a combination of the high Os-Ir-Ru values of chromitite and the high Pt-Pd values of mineralized sulphidic peridotite.

Factor analysis of chemical data for 66 samples (Fig. 5) reveals a clear association of PGE and chromite. This is indicated by the fact that significant loadings of Cr, Zn and V (reflecting their co-occurrence in chromite) occur in the same factors as the heavy loadings of PGE (factor 2 in subunit 2, factor 1 in subunit 3). (Large positive loadings, i.e. values that approach +1, indicate that those elements tend to be strongly correlated with each other, and collectively serve to define that factor.) The association of Os, Ir and Ru with chromite is to be expected, as mentioned previously, but the fact that Rh, Pt and Pd are also associated with chromite is surprising. This association is based on analyses of hand samples, and consequently reflects spatial association of the elements at hand sample scale. It is present in both chromitite (subunit 3) and peridotite (subunit 2), even though only accessory amounts of chromite are present in the latter. In both subunits the other significant factor corresponds to base metal sulphide (Ni-Cu-S-Se). There is no suggestion in the factor analysis of any association of PGE with sulphide. This is despite the fact

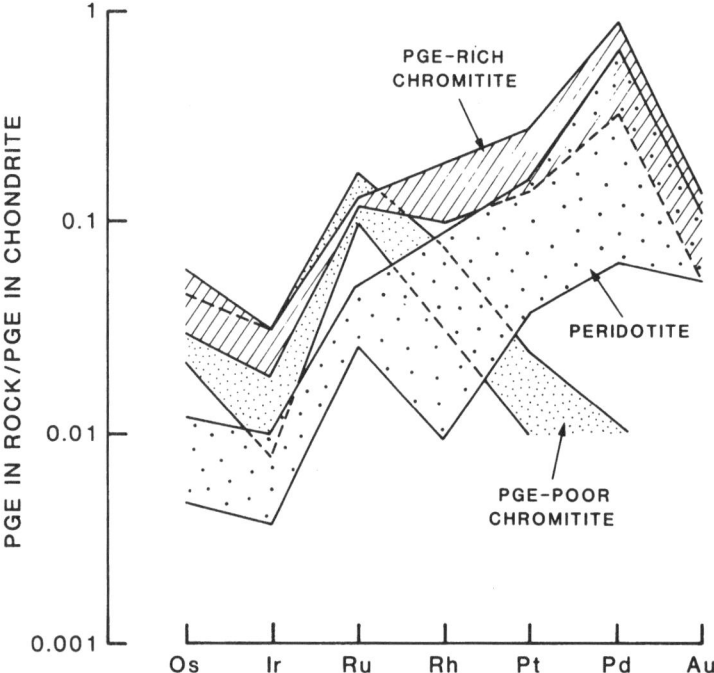

FIG. 4. PGE profiles for rocks of the Bird River Sill. Data were normalized to the chondrite values used by Naldrett and Duke (1980). The range shown for peridotite is based mainly on subunits 1 and 2; the range for PGE-rich chromitite is based on the Lower Group (= subunit 3) and some values from the Disrupted Group; the PGE-poor chromitite range is based on other chromitites (Talkington et al., 1983) in the sill.

that the PGE-enriched samples were derived from the mineralized unit, which is the only unit that is characteristically sulphide-bearing.

Au appears in the sulphide factor in subunits 2 and 3, but shows up in the PGE–chromite factor calculated for subunit 1 (not shown).

DISCUSSION AND CONCLUSIONS

Disseminated sulphides and anomalous PGE values in the Bird River Sill are mostly concentrated in a stratigraphic interval that includes the Lower Group chromitites. This crudely layer-like distribution of PGE, together with the textures of the sulphides is consistent with a magmatic genesis for both PGE and sulphides.

The soft-sediment-like, synmagmatic deformation of chromitite layers and the presence of harrisite immediately beneath massive chromitite layers in the Bird River Sill have been attributed to the upward migration of volatile-rich inter-cumulus liquid (Scoates et al., 1986). This liquid, derived through compaction of the underlying pile of cumulus crystals, ponded beneath massive chromitite layers

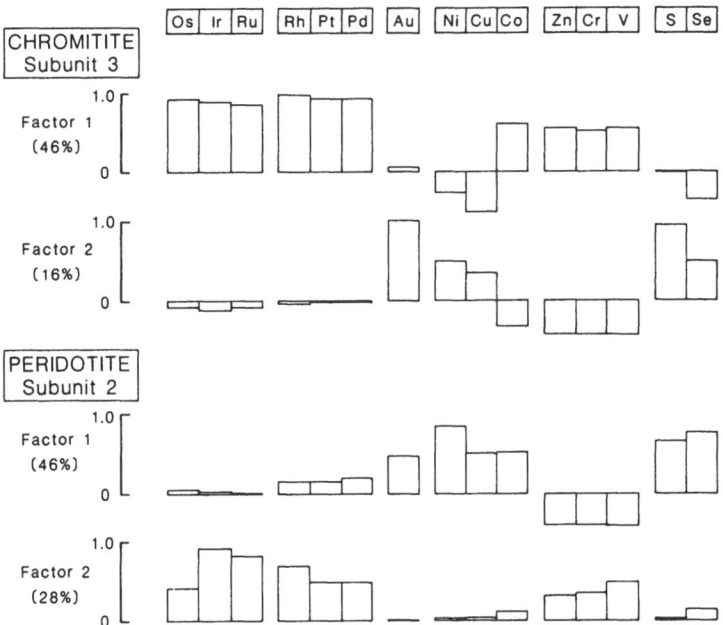

FIG. 5. Factor analysis of PGE and related elements in peridotite (subunit 2) and chromitite (subunit 3) of the Bird River sill. Data were log-transformed. Varimax-rotated factors were calculated using the SPSS mainframe program. The histograms represent the squares of the loadings for each element in the factors, and the percentage value shown represents the portion of total variance accounted for by each factor. Two-factor solutions appeared optimal for both subunits.

which are interpreted to have acted as semi-permeable membranes. Subsequent liquefaction of the underlying olivine cumulate mush produced the variety of soft-sediment-like deformational features observed in some chromitite layers. Harrisites and related pegmatites are thought to represent in-situ crystallization of volatile-rich intercumulus liquid under conditions of supercooling. Local disruption of Lower Group chromitites and the sporadic occurrence of coarse grained, harrisitic peridotite immediately beneath Lower Group chromitites illustrates the association of synmagmatic disruption of chromitites and the presence of harrisite with anomalous PGE concentrations.

The PGE can be separated into two groups which show distinct geochemical tendencies. The Os-Ir-Ru group (IPGE, Barnes *et al.*, 1985) are strongly linked with chromite. This linkage probably reflects an early high temperature co-crystallization of IPGE with chromite, as has been suggested previously by others (e.g. Talkington *et al.*, 1983).

The Rh-Pt-Pd group (PPGE, Barnes *et al.*, 1985) displays conflicting associations. On the one hand, mean abundances and PGE profiles seem to indicate that PPGE are associated with a stratigraphic zone of sulphide mineralization. However factor analysis based on hand sample chemistry shows no association of PPGE with sulphide but rather a strong linkage with chromite. This seeming inconsistency could be explained in two different ways. It could result if the PPGE were originally contained in intercumulus droplets of immiscible sulphide liquid associated with the harrisites. The PPGE could then have been mobilized out of the sulphides (during crystallization of the sulphides? late magma crystallization? regional metamorphism?) and crystallized with some kind of spatial affinity for chromite. Another possibility is that the chromite and immiscible sulphide droplets were originally accumulated together in the chromitite layers of the Lower Group. Then during the upward ponding of the harrisite-forming intercumulus melt, the dense sulphide droplets partially settled out of the chromitite and into the underlying peridotite, but maintained some kind spatial affinity for chromite.

The PPGE-chromite affinity is not obvious at microscopic scale; the PPGE-PGM are mostly enclosed in silicates. This distribution could result from a further, local scale late remobilization of PPGE away from chromite during serpentinization.

Continuing investigations involve detailed studies of mineralogy, texture and rock composition in relation to igneous stratigraphy and distribution of PGE in the various lithologic units. The objective of these studies is to understand the controls on PGE concentration and in particular the PPGE-chromite affinity.

ACKNOWLEDGEMENTS

We are grateful to Jennifer Shaw for performing the factor analysis, to Dogan Paktunç for producing the PGE profiles using his computer program, to Kim Nguyen and Gary Young for drafting the figures, and to John Bowles and Phil Potts for their valuable comments.

REFERENCES

Barnes, S.-J., Naldrett, A. J. & Gorton, M. P. (1985). The origin of the fractionation of platinum-group elements in terrestrial magmas. *Chem. Geol.*, **53**, 303–23.

Naldrett, A. J. & Duke, J. M. (1980). Platinum metals in magmatic sulfide ores. *Science*, **208**(4451), 1417–24.

Ohnenstetter, D., Watkinson, D. H., Jones, P. C. & Talkington, R. (1986). Cryptic compositional variation in laurite and enclosing chromite from the Bird River Sill, Manitoba. *Econ. Geol.*, **81**, 1159–68.

Scoates, R. F. J. (1983). A preliminary stratigraphic examination of the ultramafic zone of the Bird River Sill. Manitoba Department of Energy and Mines, Report of Field Activities 1983, pp. 70–83.

Scoates, R. F. J., Duke, J. M., Eckstrand, O. R. & Williamson, B. L. (1986). Layer disruption, PGE mineralization and the role of supercooling in the crystallization of the Bird River Sill, Manitoba. Current Activities Forum 1986 Program with Abstracts, Geological Survey of Canada Paper 86-8, p. 9.

Talkington, R., Watkinson, D. H., Whittaker, P. J. & Jones, P. C. (1983). Platinum-group mineral inclusions in chromite from the Bird River Sill, Manitoba. *Mineral. Deposita*, **18**, 245–55.

Theyer, P. (1985). Platinum-palladium distribution in ultramafic rocks of the Bird River complex, southeastern Manitoba. Manitoba Department of Energy and Mines Open File Report OF85-4.

Timmins, E. A., Turek, A. & Symons, D. T. A. (1985). U-Pb zircon geochronology and paleomagnetism of the Bird River Greenstone belt, Manitoba. Geological Association of Canada-Mineralogical Association of Canada Program with Abstracts, Vol. 10, p. A62.

Trueman, D. L. (1980). Stratigraphy, structure and metamorphic petrology of the Archean greenstone belt at Bird River, Manitoba. Unpublished Ph.D. thesis, The University of Manitoba, 150 pp.

25

Distribution of PGE in Sulphides of the Bay of Islands Ophiolite Complex, Newfoundland

J. W. LYDON & D. G. RICHARDSON

Geological Survey of Canada, 601 Booth Street, Ottawa, Ontario, Canada K1A 0E8

Sulphide concentrations of apparent magmatic affinity occur in feldspathic lithologies that presumably form part of the cumulate series of the Bay of Islands ophiolite complex. Reconnaissance-level studies have been carried out on three such occurrences in the Table Mountain massif, North Arm Mountain massif and Lewis Hills massif respectively, representing a 150 km spread along the ophiolite. In all three cases the primary sulphide assemblage is pyrrhotite-chalcopyrite-pentlandite with textures that vary from interstitial to massive sulphide. Cu:Ni ratios fall within the range of magmatic sulphides.

The Table Mountain occurrence is in a gabbro that has been tectonically dismembered from the main ophiolite and is now juxtaposed against pelitic sediments. The magmatic sulphides have mantle signatures for sulphur isotope and Se/S ratios. Interaction between sulphides and pore-waters derived from the sediments have resulted in the local sulphidization of pyrrhotite to pyrite with heavier sulphur isotopes. This relatively low temperature remobilization of chalcophile components may be responsible for perturbations shown by some samples to the otherwise flat chondrite-normalized PGE pattern.

The Lewis Hills occurrence is hosted by anorthosite, anorthositic gabbro and troctolite. Late stage hydrothermal(?) activity has mobilized a small amount of sulphide into tectonically juxtaposed dunite and caused local alteration of pentlandite to violarite. Sulphur isotope ratios have mantle values, but the sulphides tend to be weakly enriched in Se compared to mantle values. The average 100% sulphide chondrite-normalized PGE pattern is weakly positive with $Os = 0.08 \times$ chondrite and $Pd = 0.2 \times$ chondrite, though individual values for Pt, Pd and Au tend to be variable.

The North Arm Mountain occurrence is in anorthositic gabbro and, to a lesser extent, pyroxenite, and is by far the largest occurrence of sulphides in plutonic rocks of the Bay of Islands ophiolite. Sulphur isotope values are heavy (+7 per mil) and Se/S ratios, though variable, average within the Se depleted field of sedimentary sulphides, suggesting a magmatic system open to either contamination by crustal sulphur or devolatilization. The 100% sulphide chondrite-normalized PGE pattern is strongly positive, with $Os = 0.01 \times$ chondrite, $Pt = 0.1 \times$ chondrite and $Au = 1.0 \times$ chondrite. Pyroxenite appears to be enriched in PGE relative to anorthosite.

Collectively the results indicate variability within magmatic processes of oceanic crust and the strong effect of volatile fluids on the distribution of PGE at both magmatic and lower temperatures.

26

Chemical Evolution of Vapor during Crystallization of the Stillwater Complex

E. A. Mathez

Department of Mineral Sciences, American Museum of Natural History, New York, New York 10024, USA

V. J. Dietrich

Institut fur Mineralogie und Petrographie, ETH, CH-8092 Zurich, Switzerland

J. R. Holloway

Department of Chemistry, Arizona State University, Tempe, Arizona 85287, USA

&

A. E. Boudreau

Department of Geological Sciences, University of Washington, Seattle, Washington 98195, USA

The nature of the vapor evolved during the crystallization of the Stillwater Complex is deduced from data on C abundances, from compositional limits imposed by the condensed assemblage and from computation of equilibria in the C-O-H-Cl system.

Petrographic relations and the common presence of calcite and graphite indicate metamorphic redistribution of C. However, comparison of C abundances in kg-size samples of different lithologies suggests that redistribution was localized. Abundances of C in troctolites of OBZ 1, in which the Howland Reef is located, are 400–1100 ppm wt; those of troctolites from higher stratigraphic positions are generally <400 ppm. The C contents of norites and anorthosites are typically <500 ppm but also tend to be lower in samples from stratigraphically higher positions.

The C-O-H-Cl system has been studied for temperatures of 600-1200°C and pressures of 1–3 kb. The graphite surface extends from nearly pure HCl to CO_2-CO-H_2O-CH_4 mixtures in the Cl-free system. In the absence of alkalis, the important Cl species in fluids expected in nature are HCl and CH_3Cl. The addition of Cl to a C-O-H fluid leads to an essentially quantitative decrease in H_2O and increase in HCl plus CH_3Cl contents. In graphite-bearing systems, HCl is more abundant than H_2O for all fluids containing >10 at.% Cl.

For the probable evolution of fO_2 during cooling of the Stillwater Complex, the first fluid to separate from the magma was essentially a pure CO_2-CO-HCl mixture. With cooling, the fluid was driven to more H_2O- and CH_3Cl-rich compositions by precipitation of graphite, which was accompanied by rapid decrease in HCl/H_2O. The relative importance of alkali chlorides also increased as the fluid cooled. The hypothesized fluid evolution explains the variations in halogen contents of Stillwater and Bushveld apatites (Boudreau *et al.*, 1986, *J. Petrol.*, **27**, 967) as well as the Stillwater C data.

27

The Distribution of the Platinum–Group Elements in the Palisades Sill, New Jersey and New York

Raymond Talkington*

Geology Program-NAMS, Stockton College, Pomona, New Jersey 08240, USA

David Gottfried, Bruce Lipin, Norma Raitt

US Geological Survey, Reston, Virginia 22092, USA

John Puffer

Department of Geology, Rutgers University, Newark, New Jersey 07102, USA

&

David Shirley

Department of Geology and Geophysics, University of California, Berkeley, California 94720, USA

The Palisades sill is a high-titanium quartz-normative shallow diabase intrusion with a relatively simple geometry; it was emplaced during the Early Jurassic into Triassic strata of the Newark Supergroup. The sill is approximately 300 m thick and crops out along the western shore of the Hudson River. It has well-developed upper and lower chill margins, which are exposed in several areas along its strike length. Immediately above the lower chill is a cumulate zone (olivine and orthopyroxene), which is overlain by a diabase zone, above which is a ferrodiorite zone of late differentiates. An anomalous unit with very high Pd contents was found in the lower part of the ferrodiorite zone.

Whole-rock platinum (Pt) and palladium (Pd) analyses have been completed on 25 samples collected from several areas in the chilled margin, cumulate zone, diabase zone, and the ferrodiorite zone, including its anomalous unit. Table 1 gives a summary of the Pt and Pd contents and Pd/Pt ratios (n = number of samples analysed; $Mg\# = Mg/(Mg+Fe)$ in atomic proportions with all iron calculated as Fe^{2+}).

Pd and Pt contents and the Pd/Pt ratios for the chill margin of the Palisades sill are similar to those found in chilled margins of other high-TiO_2 quartz-normative Mesozoic sheets of eastern North America. The data further indicate that Pt is enriched relative to Pd in the cumulate zone, whereas Pd is enriched relative to Pt in the later differentiates in the upper part of the sill.

* Present address: Department of Earth Sciences, Carleton University, Ottawa, Ontario, Canada, K1S 5B6.

Raymond Talkington et al.

TABLE 1

	Mg#	Pd (ppb)	Pt (ppb)	Pd/Pt
Chilled margin ($n = 3$)	54–57	9·4–12	12	0·78–1
Cumulate zone ($n = 9$)	61–69	4·6–16	15–35	0·31–0·53
Diabase zone ($n = 5$)	48–61	0·7–12	5·7–18	0·36–1·9
Ferrodiorite zone ($n = 6$)	6–37	<0·5–14	<0·5–4·3	2·8–5·2
Anomalous unit ($n = 2$)	22–27	150–190	5·6–5·9	26·8–32·2

The decrease in Pd/Pt from the chilled margin upward into the cumulate zone followed by a regular increase in Pd/Pt upward through the diabase zone and into the later differentiates is, we believe, significant. First, the change in Pd/Pt is almost entirely the result of changing Pt concentrations; Pd remains relatively constant throughout. Second, Cu increases upward from the cumulate zone, indicating that silicate fractionation occurred without early sulphide liquid separation. Therefore, under conditions of *silicate* fractionation in the Palisades magma, the bulk distribution coefficient (crystals/liquid) of Pt is greater than 1 (i.e. Pt is associated with the earliest crystallizing minerals), whereas Pd has a bulk distribution coefficient close to 1.

The anomalous unit, which is located in the lower part of the later differentiates, is unusually high in Pd with Pt values within the range of the diabase. The authors speculate that this unit is the result of post- (or very late) magmatic hydrothermal processes.

28

Platinum-Group Element Mineralogy of the Pole Corral Podiform Chromite Deposit, Rattlesnake Creek Terrane, Northern California

BARRY C. MORING, NORMAN J PAGE & R. L. OSCARSON

US Geological Survey, 345 Middlefield Road, Menlo Park, California 94025, USA

The Pole Corral deposit (Red Mountain group) was selected for mineralogic studies because it contained some of the highest platinum-group element (PGE) contents found in a geochemical survey of 280 podiform chromite deposits in California and Oregon. The deposit is located about 11 airline km southwest from the town of Beegum, CA. It occurs in serpentinized dunite hosted by serpentinized harzburgite, part of a disrupted ophiolite, in the fault bounded Rattlesnake Creek terrane, which is a subdivision of the western Palaeozoic and Triassic belt of the Klamath Mountains province. Chromitite crops out on the walls of a prospect pit and consists of 25–50 mm, thick layers in dunite, disrupted by faulting and shearing, with an aggregate length of about 12 m. In 11 samples of chromitite, Pd, Pt, Rh, Ir and Ru, contents range from 1 to 15, 35 to 2530, 3 to 74, 70 to 2930, and 70 to 4930 ppb, respectively, and average 4·3, 271, 23, 999, and 1909 ppb, respectively. The mineralogy also indicates Os is present. Increased PGE contents are not associated with increased Cu (maximum 7 ppm) or Ni (maximum 1500 ppm) contents, which suggests that the chromitites are poor in base-metal sulphides.

A back-scattered scanning electron microscope study of seven polished sections of chromitite from Pole Corral identified 24 grains of PGE minerals that were analysed by qualitative methods. The grains ranged from about 2 to 20 μm in diameter and occur either as inclusions in chromite or at the boundary of chromite and serpentinized olivine crystals. Ten of the grains contained a single phase; the remainder contained multiple-phase grains (as many as four phases). Most of the grains consist of alloys. Ru–Fe alloy is the most common phase occurring as exsolution blebs and as single phase grains (found in 13 grains) in the chromitites. One grain of Ru–Fe alloy appears to be zoned with an Fe-rich core, and two grains have Os-rich rims. Alloys composed of various proportions of Os, Ir and Ru occur in six grains and, within several of these, are domains enriched in Cr and Fe or composed of Pt–Fe alloy. Four grains contain Pt–Fe alloys with variable contents of Ni and Cr(?). An Au–Pd alloy formed one grain. Laurite and compositions indicating the irarsite-osarsite-ruarsite group are the only sulphide or sulpharsenide PGE minerals identified.

29

PGE in Hawaiian Basalt: Implications of Hydrothermal Alteration on PGE Mobility in Volcanic Fluids

J. H. Crocket & A. Kabir

Department of Geology, McMaster University, Hamilton, Ontario, Canada L8S 4M1

Gold, palladium and iridium were determined by radiochemical neutron activation analysis in suites of fresh and hydrothermally altered basalt from Kilauea volcano, Hawaii. The objective was to evaluate the mobility of noble metals in basaltic rocks in response to hydrothermal exhalative activity. Most of the samples were collected from outcrops of post-1970 flows from Kilauea. The altered suites were taken from the Halemaumau crater area, and include some condensate material from active fumarole fields, the Mauna Ulu crater area and Sulfur Bank. Most of the altered basalts were characterized by a surface rind of haematite and/or sulphur-rich material.

The average noble metal contents of 22 unaltered basalts are: Au, 1·9; Pd, 3·2 and Ir, 0·38 ppb. A suite of 13 altered basalts averaged 5·75, 6·0 and 1·4 ppb Au, Pd and Ir respectively. The higher noble metal content of the altered rocks is thought to result from net addition of metals from hydrothermal fluids.

A more extensive study of Ir distribution in altered rocks was carried out as the metal is often considered relatively inert and immobile in many geological environments. Five separate localities (3 southeast of Halemaumau crater, Mauna Ulu crater and Sulfur Bank) were sampled. In two localities the average Ir content is statistically indistinguishable from the average for fresh basalt whereas in 3 localities altered rock suites are significantly higher in Ir than fresh rock. In the most Ir-rich area (east of Halemaumau crater rim, 1982 eruptive vent) Ir concentrations range from 5 to 29 ppb and average 16·3 ppb or 43 times the average of fresh basalts. These high Ir contents tentatively suggest that the metal may be readily concentrated and efficiently transported in volcanic processes.

30

Platinum-Group Minerals in Chromite-rich Horizons of the Niquelandia Complex (Central Goias, Brazil)

A. FERRARIO

Dipartimento di Scienze della Terra, Università di Milano, Via Botticelli 23, I-20133 Milano, Italy

&

G. GARUTI

Istituto di Mineralogia e Petrologia, Università di Modena, Via S. Eufemia 19, I-41100 Modena, Italy

ABSTRACT

Platinum-group minerals (PGM) are present as very small grains ($<10\,\mu m$) in chromitites and their dunite host in the Niquelandia layered intrusion. The chromite-hosted PGM occur as primary inclusions of sulphides (laurite, erlichmanite) and alloys (iridosomine, platinum–iron alloy). Laurite has a low Ir content (3–6 at%) and a Ru ratio ($Ru/(Ru + Os + Ir)$) in the range 0·90–0·53. The compositional field of published laurite analyses from dunite-hosted stratiform chromitites is extended towards Os-rich compositions by the data presented here. The Os content of laurite increases upsection with a corresponding increase in Ti in the host chromite, possibly as a result of the increased degree of differentiation of the parental magma.

The PGM in the dunite consist of Pt–Fe alloys included in orthopyroxene and Pd- and Pt-bearing tellurides, bismuthides and antimonides, some of these appear to be related to the magmatic segregation of an immiscible sulphide liquid, and some were possibly formed by the mobilization and re-deposition of Pt and Pd minerals during the serpentinization process.

INTRODUCTION

The Niquelandia complex in Central Goias (Brazil), also known as the Tocantins complex, has received much attention since the 1930s as a source of lateritic nickel (Morais, 1935; Leonardos, 1939; Pecora, 1944; Pecora & Barbosa, 1944). Fleischer and Routhier (1970) recognized the complex as a layered intrusion of Precambrian age, comparable with those of Stillwater and Bushveld. Only recently have Rivalenti *et al.* (1982) and Girardi *et al.* (1986) provided petrological support for this model. The complex forms a monocline dipping westwards at 40–60° and

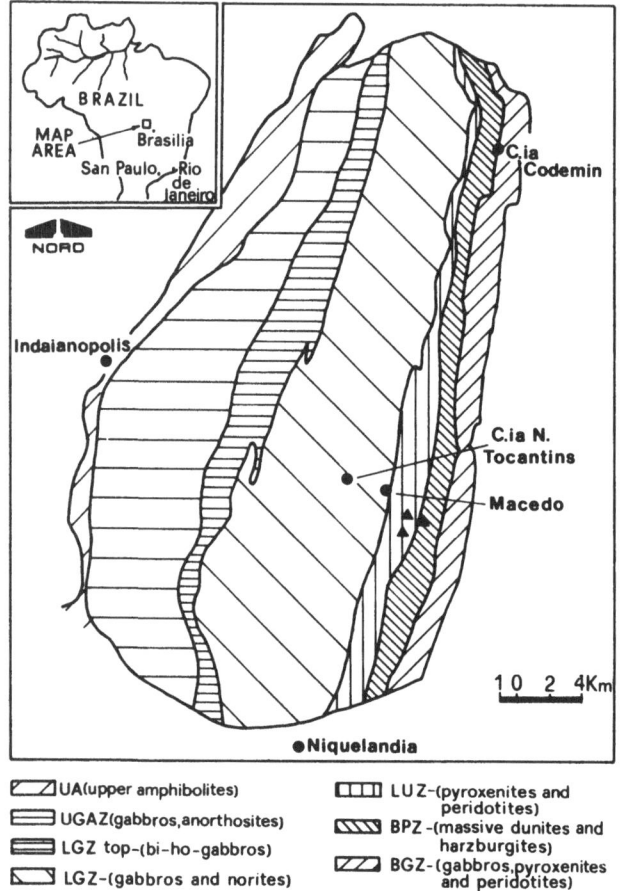

FIG. 1. Geological sketch map of the Niquelandia complex. BGZ = Basal Gabbro Zone; BPZ = Basal Peridotite Zone; LUZ = Layered Ultramafic Zone; LGZ = Lower Gabbro Zone; UGAZ = Upper Gabbro-Anorthosite Zone; UA = Upper Amphibolites; bi = biotite; ho = hornblende. Black triangles = sample localities.

having a total thickness of about 14 500 m. It can be divided into six principal zones . (Fig. 1). Chromite and chromium-rich spinel are known to occur as low-grade disseminations in most peridotites and some pyroxenites in the BPZ and LUZ ultramafic units. However, massive chromitites appear to be restricted to dunite layers well exposed southeast of Macedo.

The chromitites are grouped in two horizons set parallel to the main layering (Fig. 2). The lower horizon (CHR 1) consists of swarms of thin chromitites occurring in a 1 m wide tubular zone located close to the top of the Basal Peridotite Zone (BPZ). The higher chromitite horizon (CHR 2) consists of several chromite seams intercalated over a stratigraphic interval of about 20 m, and about 1000 m above the lower horizon (CHR 1), well inside the LUZ unit. Both horizons can be traced for more than 10 km.

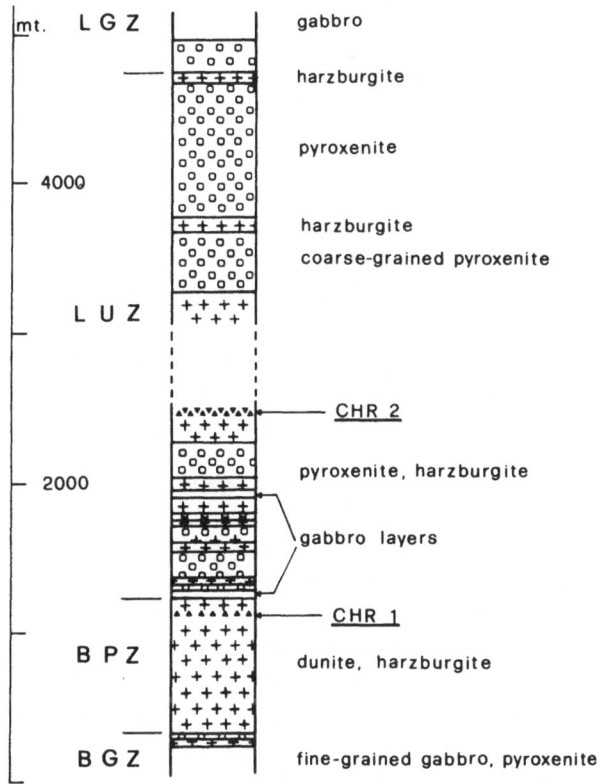

FIG. 2. Stratigraphic section through the ultramafic units BPZ and LUZ of the Niquelandia complex and location of the chromite-rich horizons CHR 1 and CHR 2.

White *et al.* (1971) noted platinum-group elements (Pt + Pd + Rh) up to 3·42 ppm in the chromitites, while Sighinolfi *et al.* (1983) found Ru + Pt + Pd anomalies of 170 ppb, sometimes accompanied by Au and Ag, in serpentinized peridotites at the same stratigraphic levels.

The present work reveals the occurrence of two suites of platinum-group minerals (PGM) in the chromitites and their dunite host, each characterized by different paragenetic associations. It also discusses some genetic aspects of the mineralization.

ANALYTICAL NOTES

Microprobe analyses of chromite and PGM were carried out on polished sections at the laboratories of Milano and Modena Universities using fully automated ARL-SEMQ instruments. On-line data reduction was performed by the use of a modified version of the MAGIC IV program.

The operating conditions for PGM analysis were as follows: 25 kV accelerating voltage, and 20 nA beam current, with a beam diameter of about 1 μm. The Kα analytical lines were used for Ni, Fe and S; Lα lines for As, Te, Sb, Pd, Pt, Rh, Ir, Ag and Au; Mα lines for Bi and Os. Corrections were made for the interferences Ag-Pd and Ru-Rh. The standards were synthetic NiAs, FeS$_2$, PdTe, PtTe, Ag$_2$Te, Bi$_2$Te$_3$, Sb$_2$S$_3$ and pure Ir, Os, Ru, Rh and Au.

Owing to the minute size ($<$2–10 μm) of the analysed PGM, fluorescence effects evidenced by the presence of varying amounts of Cr and Mg in chromite-hosted grains, or of S in PGM associated with pentlandite, were frequently observed. The analytical totals were often below 80 for Ru-Os sulphide grains less than 2 μm in diameter. The atomic proportions, however, were generally in good agreement with the theoretical stoichiometry so that the analyses could be accepted for discussion.

THE NIQUELANDIA CHROMITITES AND THEIR DUNITE HOST

In both horizons the chromitites occur as discontinuous layers, elongated lenses and less regular pods, varying in thickness from 1 to 30 cm (although some individual lenses more than 1 m thick have been observed). The chromitite bodies taper at their ends and may peter out into a sparse dissemination of chromite grains. The Cr$_2$O$_3$ content varies between 48 and 40 wt% in massive chromitites and grades towards chromian-aluminous spinel (Cr$_2$O$_3 < 35$ wt%) in the disseminated mineralization. Chromite compositions straddle the ophiolitic and stratiform fields when plotted in diagrams involving Mg-Fe^{2+} and Al-Cr-Fe^{3+} relationships but exhibit a clear stratiform affinity in the TiO$_2$ versus Cr$_2$O$_3$ diagram (Fig. 3). The TiO$_2$ content of the chromite increases from 0·25–0·36 in CHR1 to 0·30–0·52 wt% in CHR2, being correlated with its FeO/FeO + MgO ratio (Fig. 4). This trend of evolution has been considered a distinctive feature of stratiform as opposed to ophiolitic chromitites by Dickey (1975) and is consistent with the evolution of the parental magma for Niquelandia, which shows similar Ti-(Fe/Mg) relations with increasing degrees of fractionation (Girardi et al., 1986). Increasing Ti content in chromite upsection, reflecting the degree of differentiation of the magma, has also been observed in the Lower and Critical Zones of the Bushveld complex (Cameron, 1977; Hulbert & Von Gruenewaldt, 1985), although the TiO$_2$ variation range was much wider than at Niquelandia.

Olivine, orthopyroxene, amphibole and sulphides occur as primary inclusions in the massive chromite. The sulphides are typically Ni- and Cu-rich and, because of their drop-like morphology, are considered as derived from an immiscible sulphide melt trapped in the chromite at high temperature, during an early stage of crystallization.

In the lower horizon CHR 1, the host dunite is partly serpentinized (10–50%) and consists of olivine and minor orthopyroxene relics enclosed in a matrix of lizardite, chrysotile and talc (White et al., 1971). The secondary minerals pseudomorph the primary minerals but primary textures are preserved suggesting that serpenti-

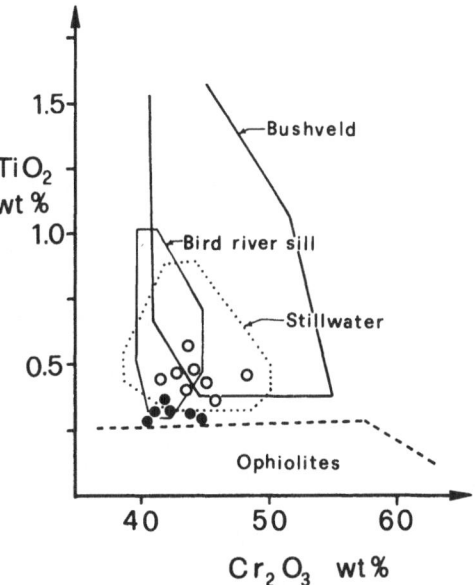

FIG. 3. Weight per cent TiO_2 versus Cr_2O_3 for the Niquelandia chromitites. CHR 1 = black dots, CHR 2 = open circles. Compositional fields of chromitites from three layered intrusions and several ophiolitic complexes are reported for comparison based on literature data and unpublished analyses available to the authors (see Mussalam *et al.*, 1981, for bibliography, and the following additional sources: Hatton & Von Gruenewaldt, 1985; Ohnenstetter *et al.*, 1986; Talkington *et al.*, 1986).

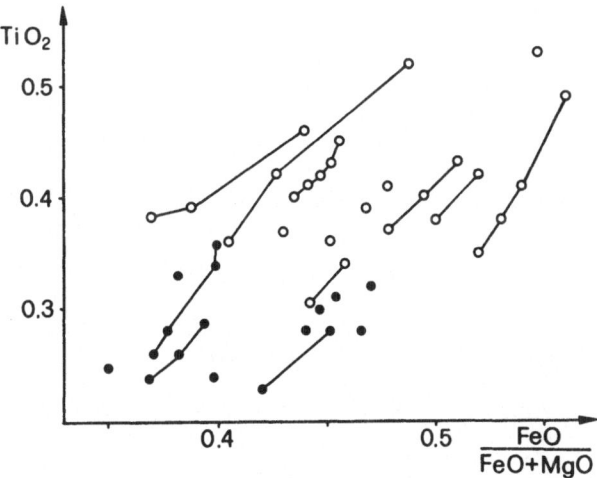

FIG. 4. Weight per cent TiO_2 versus FeO/FeO+MgO ratio for the Niquelandia chromitites (FeO is total ferrous iron). CHR 1 = black dots, CHR 2 = open circles. Lines connect compositions from the same chromitite seam.

nization must have occurred in a static tectonic environment (Rivalenti *et al.*, 1982). Disseminated chromite grains are frequently lined by secondary ferritchromite and chromian magnetite. Coarse grains of pentlandite exsolving abundant chalcopyrite lamellae occur interstitially to olivine. They are strongly veined by secondary magnetite, and contain heazlewoodite or awaruite possibly as a result of late alteration. However their texture and morphology are suggestive of an inter-cumulus magmatic origin (Garuti *et al.*, 1986). These magmatic sulphides can be distinguished from minute grains of Ni-sulphides including millerite, heazle-woodite, polydimite, and Co-pentlandite which occur dispersed in serpentine. These represent a by-product of the serpentinization process which is responsible for the movement of nickel from silicates to sulphide minerals (Eckstrand, 1975).

The dunite host in the upper CHR 2 horizon is strongly affected by in-situ lateritization, so that the samples were not suitable for reflected light study.

THE PLATINUM-GROUP MINERALS

Representative microprobe compositions of the chromite-hosted PGM are reported in Table 1. The PGM are found as small inclusions ($<10\,\mu$m) in unfractured chromite and mainly consist of euhedral, single-phase grains of laurite. Only one grain of laurite was found to be associated with interstitial sulphides in the dunite surrounding CHR 1. Laurite analyses show appreciable As (0·2–1·6 wt%) and Ni (0·09–0·15 wt%). Trace amounts of Pt and Pd were detected in some grains. No Rh was found to be present after correction for the Ru interference. Although we do not exclude that the minor elements may be related with the presence of fine inclusions, these were never observed. Therefore, we believe they represent a true composition. Ru, Os and Ir compositions of the Niquelandia laurites plot close to

TABLE 1

Representative compositions of platinum-group mineral inclusions in the Niquelandia chromitites (Brazil)

	1	2	3	4	5
Os	7·4	32·8	56·4	77·1	—
Ir	3·7	4·2	13·4	23·1	—
Ru	51·7	27·8	1·4	0·33	—
Rh	—	—	—	—	0·09
Pt	0·08	—	0·09	—	83·7
Pd	—	—	—	—	0·07
Ni	0·15	0·09	—	—	0·09
Fe	—	—	—	—	13·7
S	36·2	32·4	25·7	—	0·83
As	—	0·84	0·23	—	—
Total	98·63	98·13	97·22	100·53	98·48

1: Laurite; 2: Os-Laurite; 3: Erlichmanite; 4: Iridosmine; 5: Pt-Fe alloy. Composition (1) is from CHR 1, compositions (2), (3), (4) and (5) from CHR 2.

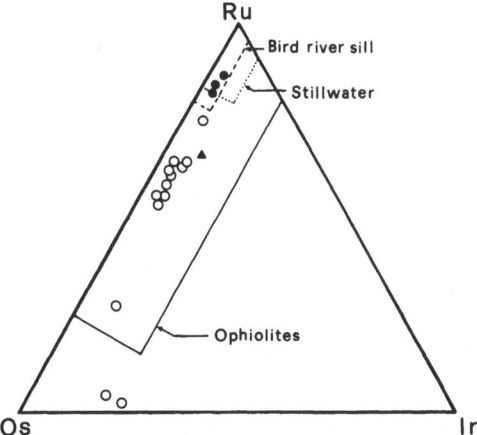

FIG. 5. Compositions of laurite and erlichmanite from Niquelandia plotted on the Ru-Os-Ir triangle (atomic proportions). CHR 1 = black dots, CHR 2 = open circles. The black triangle represents the laurite grain dispersed in the CHR 1 dunite. Compositional fields of Ru-Os-Ir sulphides from two layered intrusions and ophiolitic complexes are reported (data from Augé, 1985, 1986, 1988; Ohnenstetter *et al.*, 1986; Talkington & Lipin, 1986).

the Ru-Os join (Fig. 5), having as low an Ir content as laurite from other stratiform intrusions (Talkington & Lipin, 1986; Ohnenstetter *et al.*, 1986). However, it shows a more extensive range of Os-Ru substitution, comparable with laurites from ophiolitic complexes (Constantinides *et al.*, 1980; Augé, 1985, 1986, 1988). The laurite grain isolated in the CHR1 dunite has higher Ir and Os contents than the chromite-hosted laurites of the same horizon. If this analysis is ignored, two distinct compositional fields are apparent for laurites from CHR 1 and CHR 2, the Ru ratio (Ru/(Ru + Os + Ir)) being in the ranges 0·90–0·80 and 0·75–0·53, respectively. The increasing Os content of laurite upsection is correlated with the Ti increase of the chromite host (Fig. 6). In addition to Os-rich laurites, the CHR 2 chromitites also contain erlichmanite (OsS_2) and iridosmine (Os, Ir), both of which are usually euhedral and occur either as single or composite grains. One grain of Os-Ir-Pt sulphide, too small to give reliable microprobe analysis, was found associated with a Ni- and Cu-rich sulphide droplet in the CHR 1 chromitites. Two anhedral grains of Pt–Fe alloys were also identified as inclusions in the CHR 2 chromitites (analysis 5, Table 1), containing traces of Rh, Pd, Ni and S.

Many other PGM were identified as minute grains in the CHR 1 dunite. Representative microprobe compositions are listed in Table 2. One grain of Pt-Fe alloy with a composition approaching PtFe (analysis 5, Table 2), was found included in orthopyroxene. Several Pt- and Pd-bearing bismuthides, antimonides and tellurides, with rather complex compositions, were encountered in various textural positions. In the absence of X-ray data, they were provisionally attributed to known PGM species on the basis of their stoichiometries.

Two angular grains with kotulskite- and geversite-type stoichiometries but displaying complex substitutions (analyses 1 and 4, Table 2) were found within and

A. Ferrario, G. Garuti

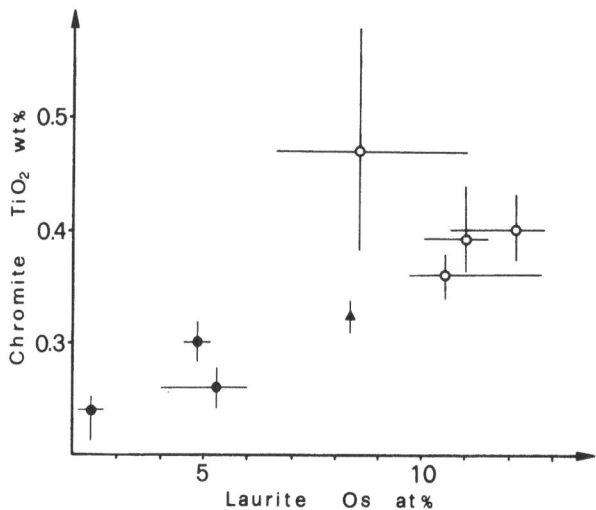

FIG. 6. Positive correlation between the Os content of laurite (as atomic percent) and the TiO_2 content of the chromite host. Symbols as in Fig. 5. Horizontal and vertical bars represent the observed variation ranges.

TABLE 2

Representative compositions of platinum-group minerals in the dunite host of the Niquelandia chromitites (Brazil)

	1	2	3	4	5
Ir	—	—	12·7	13·0	—
Rh	—	0·92	—	0·40	0·27
Pt	1·1	27·3	33·6	27·6	68·9
Pd	35·0	0·14	1·3	—	—
Au	—	2·4	3·4	—	—
Ag	—	—	1·6	—	—
Ni	2·1	5·8	2·9	1·8	—
Fe	—	—	—	—	27·6
Te	16·4	44·8	11·2	7·1	—
Sb	7·0	1·2	27·3	45·7	—
Bi	37·2	19·3	5·6	1·9	—
Total	98·8	101·9	99·6	97·5	96·8

1: (Pd, Ni, Pt)$_{1.01}$(Bi, Te, Sb)$_1$ associated with interstitial pentlandite; 2: (Pt, Ni, Rh, Au, Pd)$_{1.09}$(Te, Bi, Sb)$_{1.91}$; 3: (Pt, Ir, Ni, Au, Ag, Pd)$_{0.99}$(Sb, Te, Bi)$_{1.01}$ in ferritchromite; 4: (Pt, Ir, Ni, Rh)$_{1.07}$(Sb, Te, Bi)$_{1.93}$ in serpentine; 5: Pt–Fe alloy included in orthopyroxene. All PGM are from CHR 1.

adjacent to intercumulus pentlandite, respectively. Another composite grain consisting of moncheite and stumpflite-type compositions (analyses 2 and 3, Table 2) was included as a small nodule in porous ferritchromite lining a relic of intercumulus chromite. Both these phases contain appreciable Au and Ag substituting for platinum-group elements (PGE).

DISCUSSION

The chromite-included PGM in the Niquelandia layered intrusion are (Ru, Os, Ir)-sulphides along with (Os, Ir)- and (Pt, Fe)-alloys. Their mode of occurrence does not conflict with the conclusion drawn by many authors that most of the PGM inclusions in chromitites from stratiform and ophiolitic complexes are magmatic in origin (Constantinides *et al.*, 1980; Talkington *et al.*, 1984; Stockman & Hlava, 1984; Augé, 1985; Talkington & Watkinson, 1986; Prichard *et al.*, 1986), and are not the product of exsolution from the chromite host during cooling (Gijbels *et al.*, 1974). Further, there is no evidence that they were formed or affected by secondary processes during the serpentinization processes. We therefore assume that these minerals were formed at an early magmatic stage as part of the chromite precipitation event.

To explain the association of PGM sulphides and alloys Talkington *et al.* (1984) proposed that the PGE first separate from the silicate melt in the metallic state, then convert to metal sulphides at increasing sulphur fugacity and decreasing temperature and are finally trapped in the chromite. Owing to their different Gibbs-free-energies of formation (Westland, 1981), the PGM sulphides should form in the order RuS_2, PtS and OsS_2. If suitable sulphur fugacities are not reached before chromite crystallization, PGE alloys may persist in the final assemblage. This is possibly the case for the CHR 2 chromite-rich horizon of Niquelandia, where iridosmine (Os, Ir) was not completely converted to erlichmanite (OsS_2), the reaction requiring the highest sulphur fugacity. Pt-bearing PGM are very rare as inclusions in the Niquelandia chromitites. However, when present, they consist of Pt–Fe alloys, whereas Pt-sulphides which should become stable before Os-rich sulphides, are absent. This is possibly due to the high stability of the Pt–Fe alloys, which once formed cannot easily be converted to a cooperite–pyrrhotite assemblage.

The Os-enrichment of the PGM inclusions from the lower to the upper chromitite horizon of Niquelandia would appear to conform to the general increase in the Os/Ru ratio in the parental magma and is parallel to the increasing Ti content upsection in the chromite host. Although experimental data are lacking to support any conclusion, this correlation strongly suggests that the increase in the Os/Ru ratio is related to the increasing degree of differentiation of the parental magma. A similar feature has been described by Ohnenstetter *et al.* (1986), who suggested laurite to be sensitive to changes in the physical-chemical parameters of the parental magma. They proposed that the appearance of a fluid phase during chromite precipitation might be responsible for cryptic variations in the Ru/Os ratio in laurite, evolving to Os-rich compositions upsection.

Os, Ir, Pt-bearing PGM are associated with sulphide droplets included in the chromite, albeit infrequently. This suggests that slight sulphur saturation could have been reached in some cases during chromite precipitation and that the sulphide melt (particularly enriched in Cu) acted as a collector for PGE. This evolved to a PGM + base metal sulphide assemblage on cooling.

The primary deposition of PGM in the dunite host of the chromitites occurs: (1)

in association with the crystallization of orthopyroxene in the peridotite host (Pt–Fe alloys); (2) closely associated with the segregation of an immiscible sulphide liquid (Pd- and Pt-bearing bismuthides, tellurides and antimonides) occupying intercumulus positions in the rock. Laurite rarely forms at this stage. If present, it has higher Os and Ir contents than those of laurite included in adjacent chromitite, as expected for laurites crystallizing under relatively high sulphur fugacity (Stockman & Hlava, 1984).

The presence of Pd- and Pt-bearing tellurides and antimonides as free grains in serpentine or associated with secondary oxides, suggests that PGM could have been deposited also during the serpentinization process. Ag and Au anomalies appear to be related with strongly serpentinized peridotites (Sighinolfi et al., 1983) as a result of the circulation of possibly hydrothermal fluids. Significant amounts of these metals were also observed in some of the PGM indicating that these phases were deposited, or at least affected, by the action of circulating solutions. The primary PGM associated with interstitial sulphides or included in mafic silicates are likely to represent the source of Pt and Pd now occurring in the secondary assemblage.

Whatever the origin (primary or secondary) of the PGM now occurring in the dunite host of the chromitites, we believe that the two metals, Pt and Pd, must have been originally present in the mafic magma and that they were precipitated during the fractionation of the dunite, in contrast with Ru, Os and Ir, which were selectively concentrated in the chromitites.

CONCLUSIONS

The following inferences can be drawn from the preceding discussion:

(1) The compositional field of laurite from dunite-hosted stratiform chromitites is extended towards Os-rich compositions by the present data. The Os content is comparable with that of laurite from ophiolites, although the Ir content is significantly lower.

(2) Os-enrichment of laurite reflects Os-enrichment in the PGM assemblage and is accompanied by an increase in Ti in the host chromite. This can be ascribed to an increasing degree of fractionation of the parental magma. If confirmed, this conclusion would indicate that fractionation occurs between Ru and Os during magmatic differentiation processes.

(3) The Niquelandia occurrence confirms the general tendency of Pt and Pd to remain in the liquid phase during chromite precipitation, while the presence of an immiscible sulphide liquid appears to be critical for their deposition. However, some Pt may be deposited as alloys during the crystallization of both chromite and mafic silicates.

(4) The Niquelandia occurrence stresses the importance of secondary processes for the mobilization and the redistribution of PGE at low temperature. The chromite-rich horizons can be considered as source rocks for PGE placers which might be present in the extensive laterite deposits occurring in the area.

ACKNOWLEDGEMENTS

We would like to thank Professors L. Brigo (Università di Ferrara, Italy), V. A. V. Girardi (Universidade de Sao Paulo, Brazil) and G. Rivalenti (Università di Modena, Italy) for help in the field and for useful discussion basic to the final elaboration of this paper.

We are also grateful to C.ia Niquel Tocantins (particularly Drs D. M. Santos, J. A. Branquinho, Z. Rosado and J. N. Prado) for permission to carry out the field work and facilities provided.

Thanks are due to the Ministero della Pubblica Istruzione, and Consiglio Nazionale delle Ricerche (Rome, Italy) for financial support. The Consiglio Nazionale delle Ricerche is also acknowledged for financing the electron microprobe laboratories at Modena and Milano Universities.

REFERENCES

Augé, T. (1985). Platinum-group-mineral inclusions in ophiolitic chromitite from the Vourinos complex, Greece. *Can. Mineral.*, **23**, 163–71.

Augé, T. (1986). Platinum-group-mineral inclusions in chromitites from the Oman ophiolite. *Bull. Mineral.*, **109**, 301–4.

Augé, T. (1988). Platinum-group minerals in the Tiébaghi (New Caledonia) and Vourinos (Greece) ophiolites. In *Geo-platinum 87*, eds. H. M. Prichard, P. J. Potts, J. F. W. Bowles and S. J. Cribb, Elsevier Applied Science Publishers, London, p. 405.

Cameron, E. N. (1977). Chromite in the central sector of the eastern Bushveld Complex, South Africa. *Am. Mineral.*, **62**, 1082–96.

Constantinides, C. C., Kingston, G. A. & Fisher, P. C. (1980). The occurrence of platinum-group minerals in the chromitites of the Kokkinorotsos chrome mine, Cyprus. In *Ophiolites, Proceedings of International Ophiolite Symposium (Cyprus)*, ed. A. Panayiotou. Cyprus Ministry of Agriculture and Nat. Resources, Geol. Survey Dept., Nicosia, pp. 93–101.

Dickey, J. S. Jr (1975). A hypothesis of origin for podiform chromite deposits. *Geochim. Cosmochim. Acta*, **39**, 1061–74.

Eckstrand, O. R. (1975). The Dumont serpentinite: a model for control of nickeliferous opaque mineral assemblages by alteration reactions in ultramafic rocks. *Econ. Geol.*, **70**, 183–201.

Fleischer, R. & Routhier, P. (1970). Quelques grands themes de la geologie du Bresil. Miscellanes geologiques et metallogeniques sur le Planalto. *Sciences de la Terre*, **15**(1), 45–102.

Garuti, G., Rivalenti, G., Girardi, V. A. V., Brigo, L. & Fornoni Candia, M. A. (1986). Metalogenese dos sulfetos do complexo de Niquelandia, Goias. XXXIV Congresso Brasileiro de Geologia, Goiania-Goias 12 a 19 de Outubro de 1986, Resumos e Breves Comunicacoes. *Soc. Bras. de Geol.*, **1**, 233–4.

Gijbels, R. H., Millard, H. T., Desborough, G. A. & Bartel, A. J. (1974). Osmium, ruthenium, iridium and uranium in silicates and chromite from the eastern Bushveld complex, South Africa. *Geochim. Cosmochim. Acta*, **38**, 319–37.

Girardi, V. A. V., Rivalenti, G. & Sinigoi, S. (1986). The petrogenesis of the Niquelandia layered basic-ultrabasic complex, Central Goias, Brazil. *J. Petrol.*, **27**(3), 715–44.

Hatton, C. J. & Von Gruenewaldt, G. (1985). Chromite from the Swartkop chrome mine—an estimate of the effects of subsolidus reequilibration. *Econ. Geol.*, **80**, 911–24.

Hulbert, L. J. & Von Gruenewaldt, G. (1985). Textural and compositional features of chromite in the Lower and Critical Zones of the Bushveld complex south of Potgietersrus. *Econ. Geol.*, **80**, 872–95.

Leonardos, O. H. (1939). Os depositos niqueliferos de Goias. *Rev. Min. e. Met.*, **44**(19), 37–44.

Morais, L. J. (1935). Niquel no Brazil. *Bol. SFPM* (9), 1972.

Mussallam, K., Jung, D. & Burgath, K. (1981). Textural features and chemical characteristics of chromites in ultramafic rocks, Chalkidiki complex (Northeastern Greece). *TMPM*, **29**, 75–101.

Ohnenstetter, D., Watkinson, D. H., Jones, P. C. & Talkington, R. (1986). Cryptic compositional variation in laurite and enclosing chromite from the Bird River Sill, Manitoba. *Econ. Geol.*, **81**, 1159–68.

Pecora, W. T. (1944). Nickel-silicate and associated nickel-cobalt-manganese deposits near Sao Jose do Tocantins, Goias, Brazil. *U.S. Geol. Surv. Bull.*, **935E**, 247–305.

Pecora, W. T. & Barbosa, A. L. M. (1944). Jazidas de niquel e cobalto de S. Jose do Tocantins, Estado de Goias, Brazil. *Div. Fom. Prod. Mineral. Bol.*, **64**, 1–69.

Prichard, H. M., Neary, C. R. & Potts, P. J. (1986). Platinum-group minerals in the Shetland ophiolite. In *Metallogeny of the Basic and Ultrabasic Rocks, Edinburgh, 1985*, ed. M. J. Gallagher, R. A. Ixer, C. R. Neary & H. M. Prichard. Institution of Mining and Metallurgy, London, pp. 395–414.

Rivalenti, G., Girardi, V. A. V., Sinigoi, S., Rossi, A. & Siena, F. (1982). The Niquelandia mafic-ultramafic complex of Central Goias, Brazil: petrological considerations. *Rev. Bras. de Geocien.*, **12**(1–3), 380–91.

Sighinolfi, G. P., Girardi, V. A. V., Rivalenti, G., Sinigoi, S. & Rossi, A. (1983). PGE, Au and Ag distribution in the precambrian Niquelandia complex, Central Goias, Brazil. *Rev. Bras. de Geocien.*, **13**(1), 52–5.

Stockman, H. W. & Hlava, P. F. (1984). Platinum-group minerals in alpine chromitites from Southwestern Oregon. *Econ. Geol.*, **79**, 491–508.

Talkington, R. W. & Lipin, B. R. (1986). Platinum-group minerals in chromite seams of the Stillwater complex, Montana. *Econ. Geol.*, **83**, 1179–86.

Talkington, R. W. & Watkinson, D. H. (1986). Whole rock platinum-group element trends in chromite-rich rocks in ophiolitic and stratiform igneous complexes. In *Metallogeny of the Basic and Ultrabasic Rocks, Edinburgh, 1985*, ed. M. J. Gallagher, R. A. Ixer, C. R. Neary & H. M. Prichard. Institution of Mining and Metallurgy, London, pp. 427–40.

Talkington, R. W., Watkinson, D. H., Whittaker, P. J. & Jones, P. C. (1984). Platinum-group minerals and other solid inclusions in chromite of ophiolitic complexes: occurrence and petrological significance. *TMPM*, **32**, 285–301.

Talkington, R. W., Watkinson, D. H., Whittaker, P. J. & Jones, P. C. (1986). Platinum-group element-bearing minerals and other solid inclusions in chromite of mafic and ultramafic complexes: chemical compositions and comparisons. In *Metallogeny of Basic and Ultrabasic Rocks* (Regional Presentations), Theophrastus Publishers, Athens, pp. 223–49.

Westland, A. D. (1981). Inorganic chemistry of the platinum-group elements. In *Platinum-Group Elements: Mineralogy, Geology, Recovery*, ed. L. J. Cabri. CIM special, Vol. 23, pp. 5–18.

White, R. W., Motta, J. & De Araujo, V. A. (1971). Platiniferous chromitite in the Tocantins complex. Niquelandia, Goias, Brazil. In Geological Survey Research 1971, *U.S. Geol. Surv. Prof. Pap.* 750D, pp. D26–D33.

31

Further Studies of the Development of Platinum-Group Minerals in the Laterites of the Freetown Layered Complex, Sierra Leone

John F. W. Bowles

Department of Earth Sciences, The Open University, Walton Hall, Milton Keynes MK7 6AA, UK

ABSTRACT

The wide anorthosite outcrop of the western part of the Freetown Layered Complex, Sierra Leone, corresponds to the area from which platinum-group minerals (PGM) have been recovered. Published accounts of exploration for the source of the PGM have failed to reveal a bed-rock source. Comparison of the dip of the layering of the intrusion with the wide distribution of known occurrences of PGM does not indicate derivation from a single, narrow layer. The single unifying factor for all published occurrences of PGM from the Freetown Peninsula is that they were recovered from or near to stream channels. Study of the mineralogy of the PGM (Bowles, 1981; Bowles et al., 1983) reveals the large well-preserved and delicate crystal form of many of the samples. This indicates that there has been little or no attrition such as would have occurred during transport in a stream. The PGM are, therefore, found close to where they were formed. The geochemistry of the PGE at 25°C at high Eh and low pH in the presence of a sufficient concentration of anions such as chlorine, shows that the PGE may be taken into solution under very much the same conditions as exist in a lateritic soil (Bowles, 1986). A mechanism of solution of finely disseminated PGE from the anorthosite, transport in solution and deposition under conditions of lower Eh and higher pH has been proposed (Bowles, 1986). This paper attempts to take that conclusion further and relate it to the reported distribution of the PGM in the Freetown Peninsula. A simplistic model of the route of meteoric water through the soil is used to discuss the possible process. Meteoric water with intermediate Eh and pH and negligible dissolved species sinks into the laterite where these parameters are modified. The Eh rises and the pH decreases to the conditions typical of lateritic soils and the concentration of dissolved species increases. In this state the water is able to take PGE into solution from a finely disseminated form in the bed-rock as a part of the process of lateritisation. Soil water flow transports the PGE towards the streams. In the soil near the streams a transitional interface must exist between the laterite soil water and the stream water with lower Eh, neutral pH and low concentration of dissolved salts. At this interface, deposition of the PGE will occur and a deposit of large, well-formed PGM will be found. This proposed model provides an explanation of the distribution of PGM occurrences in the Freetown

Peninsula. The model has not been quantified and is speculative. It is intended to indicate a direction for future research and introduce a new parameter for consideration during exploration for the PGM in tropical countries.

PLATINUM IN THE FREETOWN PENINSULA, SIERRA LEONE

The Freetown Layered Complex was described by Wells (1962) and shown to consist of repeated sequences of troctolite, gabbro and anorthosite representing part of the eastern rim of a funnel shaped intrusion (Fig. 1). Four main Zones were recognised, each characterised by rhythmic layering in 30–200 m thick cycles, each cycle in turn characteristically displaying a fine layering on a centimetre scale. The intrusion consists of layers which lie parallel to the funnel shape of the floor of the intrusion and which are considered to have been formed (Wells, 1962; Wells & Bowles, 1981) by downdip growth of the layers as a result of the effects of the fractional crystallisation, multiple intrusion and double-diffusive convective processes envisaged by Irvine *et al.* (1983). The multiple intrusions prolonged the cooling of the intrusion, allowing extended subsolidus recrystallisation and permitting the development of cross-cutting schlieren of pyroxene in response to the stresses within the cooling intrusion (Wells & Bowles, 1981). Each cycle of rhythmic layering varies from troctolite at the base through olivine gabbro to anorthosite with a capping of pyroxene-rich pegmatite. Such a complete cycle is infrequently seen and most sections comprise repeated partial cycles showing the less extreme variation of the central part of that cycle. The older Zones lie inland to the east, while the later Zones are found closer to the west coast of the peninsula and Zone 3 is particularly notable for its greater development of anorthosite (Wells, 1962).

Platinum-group minerals (PGM) have been found only in those streams which drain the areas underlain by the main anorthositic region in the centre of Zone 3 (Junner, 1930; Pollett, 1931; Barber, 1962). Here the outcrop of the anorthosite is some 2–3 km across. Panning at a sequence of sites along the course of the streams shows clearly that the PGM are found where the stream crosses the anorthosite and downstream of that location, but that they abruptly cease to be evident as the upstream limit of the anorthosite is approached. Alluvial grains may be obtained from these streams, and well-preserved eluvial grains are found on the steep slopes above the streams in the areas underlain by anorthosite (see Bowles, 1981; Bowles *et al.*, 1983). Because of this association, attention has been directed to the anorthosite as a source rock for the PGM and an exploration programme (Barber, 1962) was instituted for the Sierra Leone State Development Company. This investigation covered prospect areas inland from Toke and York as well as the area around the rich eluvial deposits at Guma Water. Drilling of the anorthosite, examination of drill core obtained during investigation of a dam site at Guma Water and continuous sampling along 310 m of the wall of the tunnel used to draw water from the dam all failed to show significant platinum values. The investigation concluded that the platinum was derived from a source covering a wide area in which platinum is very sparsely disseminated. There is, of course, a considerable sampling problem

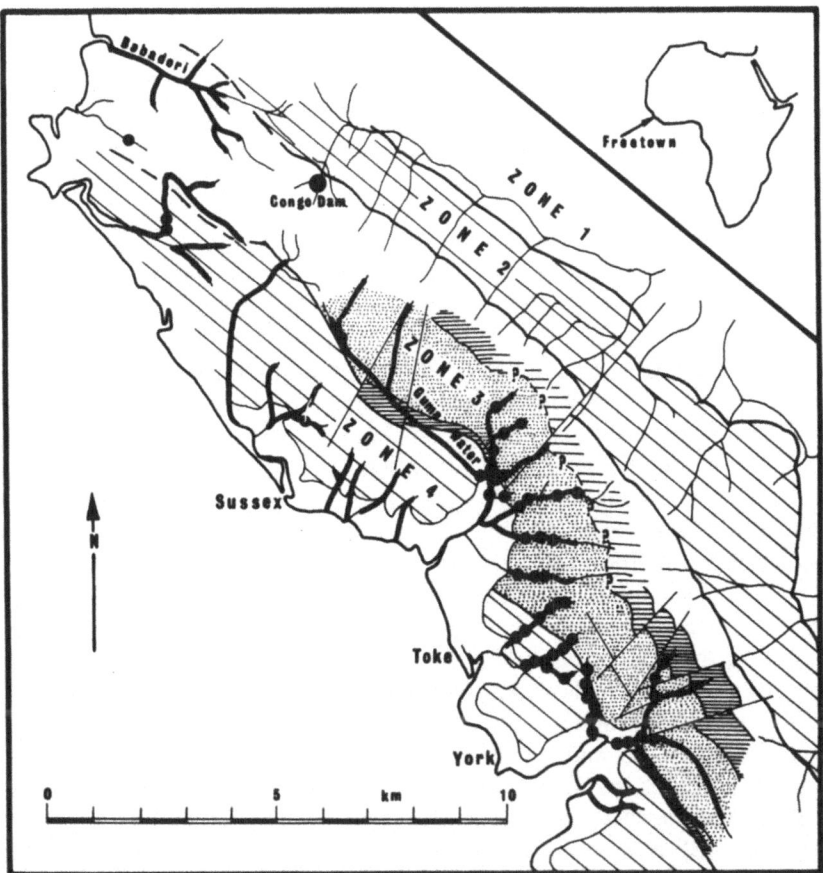

FIG. 1. The location and outline geology of the Freetown Intrusion. Along the drainage a thick line represents streams described as containing platiniferous gravels (after Junner, 1930) and solid circles indicate locations where platinum was recorded during prospecting (Barber, 1962). The principal elements of the geology of Zone 3 are given in outline (modified from Wells, 1962). Gabbros and troctolitic gabbros are shown shaded and the areas where anorthosite or anorthositic gabbro are the main rock types are shown stippled. The geological boundaries are best defined inland from Sussex and inland from York. The geology between the two areas has been inferred. Subsidiary layers of gabbro, troctolitic gabbro and impersistent pyroxenites occur within the anorthosites.

involved in this sort of investigation for platinum, so a cautious approach is required before accepting that conclusion. However, examination of the distribution of the PGE as outlined by the known occurrences in alluvial and eluvial deposits (Fig. 1) and the dip of the layering of the anorthosite (17–43°) make it appear unlikely that the principal source of the PGE is a single layer within the anorthosite.

Investigations in the Freetown Peninsula have concentrated on searching for a bed-rock source for the PGE beneath the laterite cover. Studies of the platinum content of the laterite were intended only as a convenient means of sampling. The sampling has not pointed consistently to any particular bed-rock source. Barber (1962) reports stream sampling of 127 pits, of which 54 showed platinum. In contrast 1260 samples of soil from a series of cut lines through the forest in the same area only showed platinum in 14 samples. Seven of these platiniferous samples were in streams and in the remaining seven the platinum content was low. In terms of reported occurrences, there seems to be a considerable tendency for platinum to occur in stream channels. Pollett (1931) indicates that much of the platinum in the streams is found directly on the surface of the bed-rock or embedded in its decomposed surface. Platiniferous samples that were not recovered from streams contain non-water worn crystalline growths of PGM and were recovered from eluvium near the streams or in small gullies on the steep slopes above the streams.

In order to explain the distinctive PGM of the Freetown Layered Intrusion, Sierra Leone, Bowles (1986) examined the possibility of low temperature solution of the PGE in the conditions which exist within lateritic soils. Given a sufficiently high concentration of chlorine it is possible for the PGE to enter into solution in ground water circulating through basic rocks during progressive lateritisation and for that water to transport the PGE in solution to another location where under changed conditions the PGE come out of solution to form large free-growing crystals. Chlorine is frequently mentioned in this context because sufficient data are available (Westland, 1981) to enable the process to be quantified. Other anions are likely to take the PGE into solution; sulphate and cyanide solutions and solutions rich in free carboxylic acid or complex soil organic matter have all been proposed but not quantified (Westland, 1981; Bowles, 1986; see also Mountain and Wood, this volume, pp. 57–82, and Plimer and Williams, this volume, pp. 83–92).

Figure 2 indicates the field of conditions of low pH and high Eh prevalent in iron-rich and leached lateritic soils which is to some extent coincident with the field in which the PGE may be taken into solution as chloride complexes. The fields of conditions of river and lake waters, meteoric water and lateritic soils are taken from the data of Baas Becking *et al.* (1960). It will be helpful to examine what occurs at the interfaces between these waters. Given water flow and circulation through a laterite, it is useful to consider several situations on an idealised profile from hilltop to stream in a tropical rain forest with a thick lateritic soil overlying basic rocks containing finely disseminated PGE. Mean annual rainfall in coastal Sierra Leone ranges from 450 to 500 mm. Some 100 ppm of dissolved chlorine are required for the field of PGE solubility to overlap significantly with that of lateritic soils.

Hilltop. Here there is a net rainfall input. The ranges of Eh and pH conditions of meteoric water, lake and river waters and lateritic soils, determined by Baas Becking *et al.* (1960), are indicated on Fig. 2. The field of meteoric waters lies intermediate between the conditions of lake and river waters and those determined for lateritic soils. In rainwater the concentration of chlorine and other anions will be very low so this water will enter the ground water and dilute it.

Laterite–bed-rock interface. Ground water at this level can be expected to be rich in dissolved species and to have the *Eh* and pH characteristics of the lateritic soil. This will enable the PGE to be taken into solution.

Intermittent waterlogged soil and hill slopes. These two situations are grouped

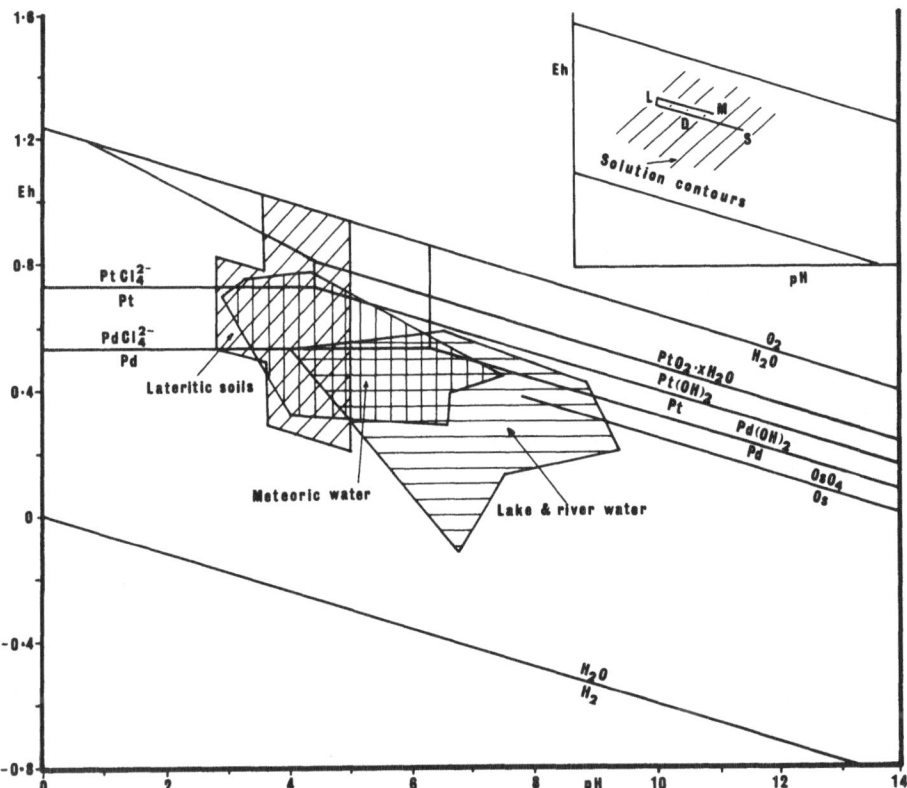

FIG. 2. The *Eh*–pH relations of solution reactions involving representative PGE compared with the *Eh*–pH conditions found in lateritic soils (diagonal ornament), meteoric waters (vertical shading), and lake and river waters (horizontal shading). Data from Westland, 1981; Baas Becking *et al.*, 1960; Bowles, 1986. The reactions of platinum and palladium with chlorine are shown for chlorine concentrations of 100 ppm. Inset shows changing water conditions from input as meteoric water (at M) to lateritic ground water (at L) and finally to stream water (at S). The slight decrease in Eh shown at L is intended to clarify the route and has no other significance. Deposition occurs (at D) where the water leaves the field of conditions in which the PGE are soluble. The 'solution contours' representing the combined change in *Eh*, pH and concentration of dissolved species are shown qualitatively across the water path.

together because both are characterised by varying conditions from net rainfall input to net evaporation. Within waterlogged soil the concentration of dissolved species increases. Local conditions and the degree of variation between net water input and evaporation will be important here. To a lesser extent the same is true of hill slopes. Rain water which provides run-off will not contribute to the ground water. Locally and occasionally, some rain water will enter and dilute the water flow but in areas characterised by net evaporation the chlorine concentration will probably remain sufficiently high for the PGE to remain in solution.

Stream bank. A range of conditions exist at this point varying from a net flow of water from the ground water into the stream via an equilibrium condition with no net flow, to a condition where water is lost from the stream (i.e. influent to effluent conditions). Whatever the flow regime, a transitional interface must exist between the soil waters and the stream water. The vital question is what happens at this point? To what extent are the PGE in solution carried into the stream and washed away? This may happen if inflow into the stream is rapid but it seems that the question of chlorine concentration has a key role since variation in the chlorine content causes large variation in the solubility of the PGE (Westland, 1981; Bowles, 1986).

SOLUTION CONTOURS

As a first approximation, the system varies between a high Eh, low pH, high solute concentration condition and a medium Eh, medium pH, low solute concentration situation. It should be possible to draw separate contours in the soil profile for Eh, pH and concentration but given the approximation to a bimodal system the situation can be simplified by using a single set of 'solution contours' normal to the

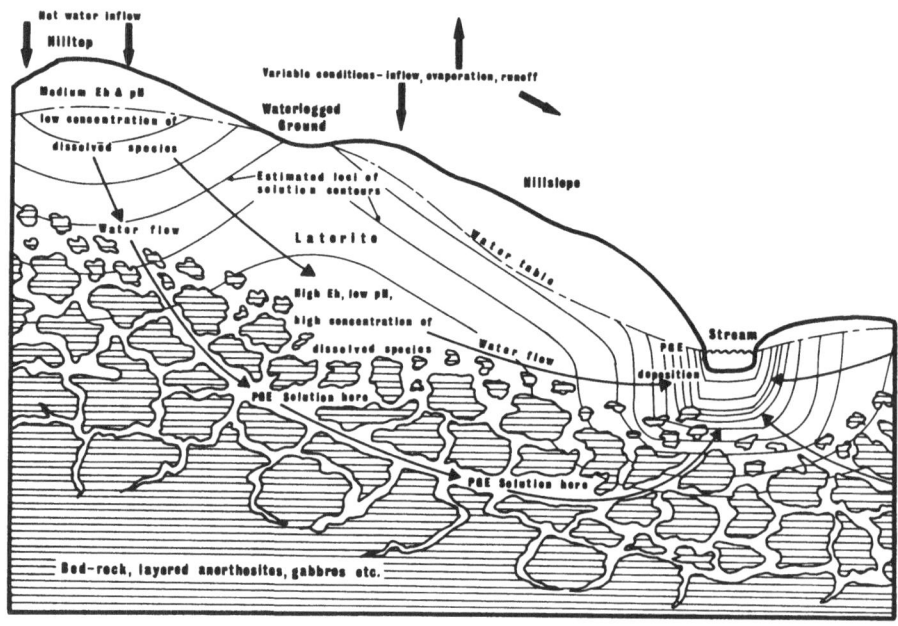

FIG. 3. Schematic representation of water flow through a laterite in a tropical region with high rainfall. Water entering the soil rises in Eh and decreases in Ph and is presumed to take the PGE into solution if the chlorine content is sufficiently high. Water flow transports the PGE towards the stream where interaction with the stream waters causes the Eh to fall, the pH to rise, and the concentration of dissolved anions to fall. It is predicted here that this will trigger precipitation of the PGE close to the stream.

ground water flow. Figure 3 represents the idealised soil profile discussed above with a single set of solution contours drawn giving the hilltop, laterite–bed-rock interface, stagnant pool and hillslope of the properties already described. No attempt has been made with the available data to quantify these contours.

PLATINUM DEPOSITION

Contours have been drawn at the stream bank normal to the direction of ground water flow through the soil below the water table. These contours indicate the transition from the high Eh, low pH, high concentration condition to the condition of the stream water. Given that the water is in equilibrium or is moving slowly through the soil, not pouring into the stream as at a waterfall or spring, then it seems likely that there will be a steady and progressive change in these parameters through the soil beneath the water table. In these circumstances there will be a transitional region in the soil between ground water conditions in which the PGE are soluble and conditions of Eh, pH and concentration where the PGE are insoluble. Where such a situation exists then deposition will occur. Thus this model predicts that PGM formation will occur in the vicinity of the stream channel at sites determined by the dynamic equilibrium which exists between the ground water and the water in the stream.

CONCLUSIONS

Previous studies have attempted to relate the distribution of PGM in the Freetown Peninsula directly to the geology of the underlying rocks. In general this has not been successful despite extensive shallow drilling and no satisfactory explanation for the wide distribution of the PGM has been established. Proposals for a low temperature solution mechanism have gained some cautious acceptance. This mechanism would take finely disseminated PGE into solution and precipitate them to form large well-formed crystals such as are found in the area. Whilst this offers a possible explanation for the size and surface morphology of the minerals it does not indicate where the precipitation would occur or explain their geographical distribution. Here the mechanism is further developed to show the parameters which could cause deposition and a likely site where those parameters may be expected to occur. Ground water flow into an area of lower Eh, higher pH and lower anion (particularly chlorine) concentration is indicated and it is likely that such a situation occurs where ground water interacts with the water in a stream. The PGM have all been found either in eluvium close to the streams or in the streams themselves in a condition indicative of a very close source. These observations are consistent with the mechanism of deposition proposed here which effectively concentrates the PGE close to the streams. The proposed mechanism is speculative and has not been quantified. It is intended to indicate a direction for future research and to suggest that in an area of high rainfall where a laterite has formed over PGE bearing basic rocks, exploration effort should be concentrated on the laterite in the immediate vicinity of existing or recent drainage channels.

ACKNOWLEDGEMENTS

Platinum-group minerals which have prompted this study are on loan from the British Geological Survey. This paper has benefited from constructive criticism from Drs H. M. Prichard and P. J. Potts and unknown reviewers who patiently reviewed a draft manuscript.

REFERENCES

Baas Becking, L. G. M., Kaplan, I. R. & Moore, D. (1960). Limits of the natural environment in terms of pH and oxidation-reduction potentials. *J. Geol.*, **68**, 243–84.

Barber, M. J. (1962). A report on the prospecting for gold, platinum and molybdenum carried out during 1961–1962 on behalf of The Sierra Leone Government. The Sierra Leone State Development Company Limited, Freetown, 34 pp.

Bowles, J. F. W. (1981). The distinctive suite of platinum-group minerals from Guma Water, Sierra Leone. *Bull. Minéral*, **104**, 478–83.

Bowles, J. F. W. (1986). The development of platinum-group minerals in laterites. *Econ. Geol.*, **81**, 1278–85.

Bowles, J. F. W., Atkin, D. Lambert, J. L. M., Deans, T. & Philips, R. (1983). The chemistry, reflectance, and cell size of the erlichmanite (OsS_2)-laurite (RuS_2) series. *Mineral. Mag.*, **47**, 465–71.

Irvine, T. N., Keith, D. W. & Todd, S. G. (1983). The J-M platinum-palladium reef of the Stillwater Complex, Montana: II. Origin by double-diffusive convective magma mixing and implications for the Bushveld Complex. *Econ. Geol.*, **78**, 1287–334.

Junner, N. R. (1930). Geology and mineral resources of Sierra Leone. *Mining Mag.*, **42**, 73–82.

Mountain, B. W. & Wood, S. A. Solubility and transport of platinum-group elements in hydrothermal solutions: thermodynamic and physical chemical constraints. In *Geo-platinum 87*, eds H. M. Prichard, P. J. Potts, J. F. W. Bowles and S. J. Cribb, Elsevier Applied Science Publishers, London, pp. 57–82.

Plimer, I. R. & Williams, P. A. New Mechanisms for the mobilization of the platinum-group elements in the supergene zone. In *Geo-platinum 87*, eds H. M. Prichard, P. J. Potts, J. F. W. Bowles and S. J. Cribb, Elsevier Applied Science Publishers, London, pp. 83–92.

Pollett, J. D. (1931). Platinum mining in Sierra Leone. *Engng. Mining World*, **2**, 747–8.

Wells, M. K. (1962). *Structure and Petrology of the Freetown Layered Basic Complex of Sierra Leone*. HMSO, London, Overseas Geology Mineral Resources, Bull. Supp. 4, 115 pp.

Wells, M. K. & Bowles, J. F. W. (1981). The textures and genesis of metamorphic pyroxene in the Freetown Intrusion. *Mineral. Mag.*, **44**, 245–55.

Westland, A. D. (1981). Inorganic chemistry of the platinum-group elements. In *Platinum-Group Elements: Mineralogy, Geology, Recovery*, ed. L. J. Cabri. Canadian Institute of Mining and Metallurgy, Special Volume 23, Montreal, pp. 5–18.

32

The Geology and Economic Potential of the PGE-rich Main Sulphide Zone of the Great Dyke, Zimbabwe

M. D. PRENDERGAST*

Union Carbide Geology Department, Box 384, Kwekwe, Zimbabwe

ABSTRACT

The Main Sulphide Zone (MSZ) of the Great Dyke is a major world PGE resource containing several hundred million tonnes of PGE-bearing potential ores with in situ unit dollar values which compare favourably with those of the Merensky Reef and UG-2 (Bushveld Complex, South Africa). There are no major technical obstacles to the exploitation of the MSZ, and the future of this resource depends on a sustained high Pt price, a low cost production system and the appropriate capital investment.

The MSZ is a laterally-continuous zone of disseminated sulphides hosted within a complexly-layered pyroxenite and stratigraphically located a few metres below the contact between the ultramafic and overlying mafic rock sequences. Two principal geological features of the MSZ are its highly characteristic and ubiquitous vertical metal distribution patterns and its systematic lateral variations in width and metal content between the margins and axis of the Great Dyke.

The stratigraphic location and form of the MSZ can be interpreted in terms of the fluid dynamic behaviour of, and temperature dependence of sulphur saturation in, mafic magmas. Lateral variations within the host pyroxenite, and within the MSZ itself, can be ascribed ultimately to the shape of the Great Dyke magma chamber(s) which created strong transverse gradients in temperature and magma composition, both together controlling such factors as fractionation, crystallisation and precipitation rates, fluid dynamic behaviour and partition coefficients.

INTRODUCTION

The Main Sulphide Zone of the Great Dyke, Zimbabwe, is one of three major, PGE-rich, stratiform sulphide zones in layered mafic-ultramafic intrusions. The other two, the Merensky Reef (Bushveld Complex, South Africa) and the J-M Reef (Stillwater Complex, USA), together with the PGE-rich UG-2 chromitite layer of

---

*Present address: Cluff Mineral Exploration (Zimbabwe) Ltd, Box 1795, Harare, Zimbabwe.

the Bushveld Complex, are important current sources of world PGE supply and have been fully described in the scientific and industrial literature (Cousins, 1969; Newman, 1973; von Gruenewaldt, 1977; Conn, 1979; Brynard *et al.*, 1976; Schwellnus *et al.*, 1976; Vermaak & Hendriks, 1976; Bow *et al.*, 1982; McLaren & de Villiers, 1982; Todd *et al.*, 1982; Gain, 1985; Hiemstra, 1985, 1986). By contrast, although known (Lightfoot, 1926; Mennell & Frost, 1926; Wagner, 1929) and intermittently explored for more than 60 years, the Main Sulphide Zone (MSZ) has not been developed to the production stage and remains virtually unknown to science and the minerals industry at large.

The geological setting, nature and origin of the MSZ are now being examined in the light of recent models of PGE mineralisation in layered intrusions. This paper summarises the principal geological features, possible origin, and economic potential of this important PGE resource, as a prelude to forthcoming contributions on specific geological aspects of the MSZ and its host rocks.

GEOLOGICAL BACKGROUND

Regional Setting, Stratigraphy and Structure

The Great Dyke (530 km long, up to 11 km wide) is an early Proterozoic (2·47 Ga; Wilson, 1982) layered intrusion set in a rifted cratonic environment (Fig. 1) (Wilson, 1981; Wilson *et al.*, 1987; Podmore & Wilson, 1987). In its present plane of erosion, it comprises a slightly sinuous, and locally faulted, line of four, narrow, layered mafic-ultramafic complexes (Musengezi, Hartley, Selukwe and Wedza; Worst, 1960).

Each complex consists of two major stratigraphic portions: a lower Ultramafic Sequence and an upper Mafic Sequence (Wilson & Wilson, 1981). Both sequences are broadly similar in all four complexes. The Ultramafic Sequence as a whole, however, displays differences in total thickness and stratigraphic detail both between each complex (see Worst, 1960), and between the southern and northern parts of the Hartley Complex (Hughes, 1970*a*). Figure 2 summarises the principal stratigraphy in the northern half of the Hartley Complex, the best known part of the Great Dyke (Wilson, 1976, 1982; Wilson & Wilson, 1981; Podmore & Wilson, 1987; Wilson & Prendergast, in press).

Important features of the Ultramafic Sequence throughout the Great Dyke include the well-developed cyclic layering (each cyclic unit ideally comprising, from the base up, chromitite—dunite—harzburgite—olivine bronzitite—bronzitite), and the predominance of dunites and chromitites, and bronzitites and harzburgites, in the lower and upper portions respectively. In addition, the uppermost ultramafic Unit 1 is markedly similar wherever it occurs, with its most characteristic stratigraphic components (chromitites Clc and Cld, the olivine-bronzitite footwall of Cld, and the Pyroxenite No. 1 Layer) readily recognisable at equivalent stratigraphic levels in all four complexes (Worst, 1960; Prendergast, 1987). Unlike the lower pyroxenites, the Pyroxenite No. 1 (P1) Layer comprises both a thick bronzitite (Bronzitite No. 1) and a relatively thin overlying layer of websterite (Main

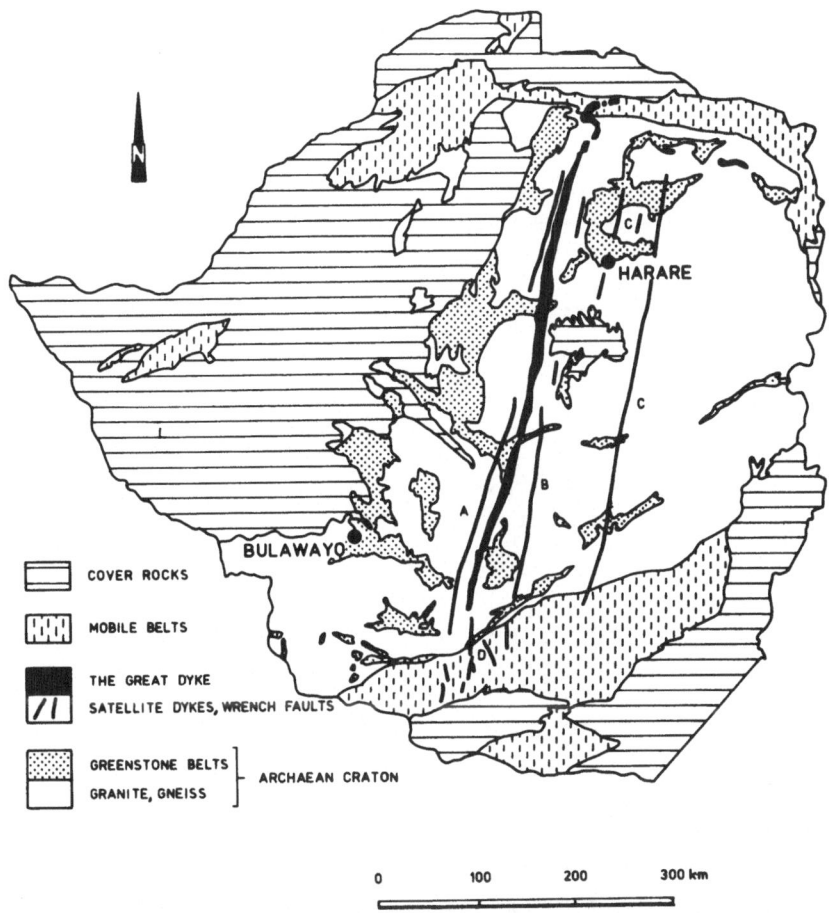

FIG. 1. Geological map of Zimbabwe showing the regional setting of the Great Dyke. Other members of the Great Dyke fracture set shown here are the Umvimeela Dyke (A), East Dyke (B), sinistral wrench faults (C), and the Southern Satellite Dykes (D). (For details see Wilson & Prendergast, in press.)

Websterite), with a persistent stratiform zone of sulphide mineralisation (the Main Sulphide Zone) situated at or near the top of Bronzitite No. 1. An important developing theme of Great Dyke geology is the presence of significant primary lateral variations within many layers, particularly of Unit 1 (Prendergast, 1987; Wilson & Prendergast, in press; Prendergast & Wilson, in press; Prendergast & Keays, in press).

The present form of each of the four complexes is essentially a shallow, boat-like, doubly-plunging synclinal structure (see Worst, 1960, map sheets 1–9). Longitudinally, the layers pitch very gently but unevenly towards the centre where a mafic remnant is preserved above the ultramafic rocks. Transversely, the layers dip moderately inwards from the margins towards the axis. The transverse synclinal

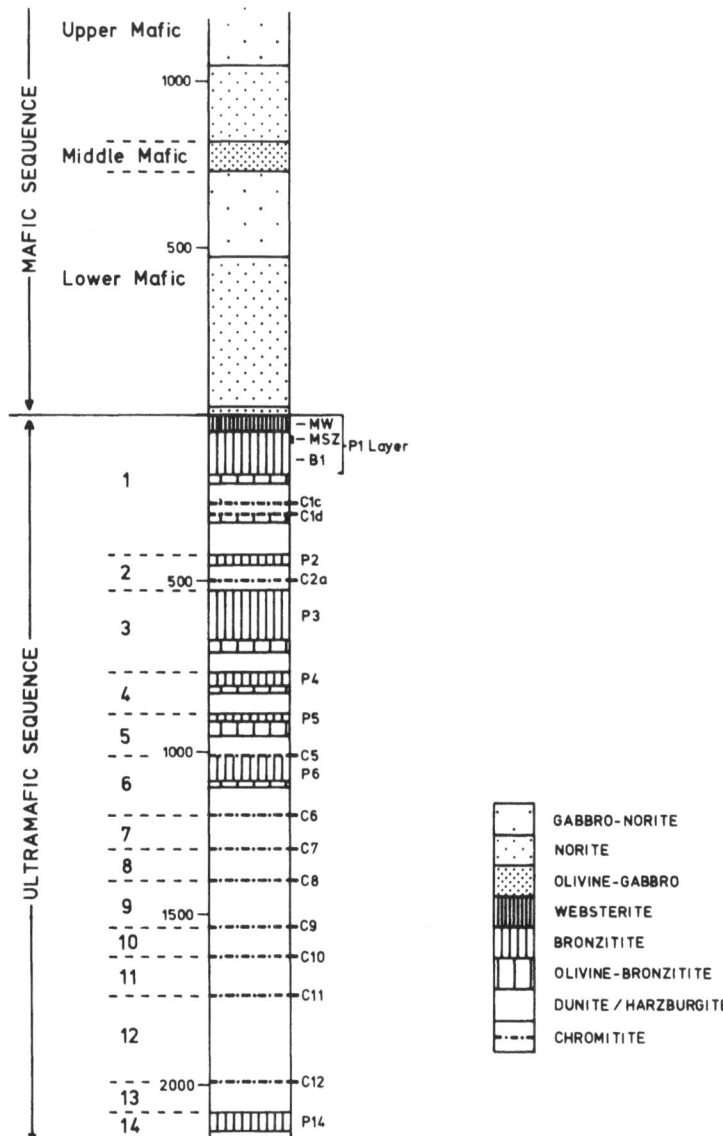

FIG. 2. Summary of the stratigraphy of the Mafic and Ultramafic Sequences of the Great Dyke in the northern half of the Hartley Complex (after Podmore & Wilson, 1987; Wilson, 1982; Wilson & Prendergast, in press). Cyclic units, and pyroxenite and chromitite layers, in the Ultramafic Sequence, are numbered from the top downwards. Note that some cyclic units lack either a lower chromitite or an upper pyroxenite. Subunits defined by minor chromitites (e.g. Prendergast & Wilson, in press) are not shown. Main Websterite: MW; Main Sulphide Zone: MSZ; and Bronzitite No. 1: B1. Stratigraphic heights above and below the mafic-ultramafic contact are given in metres.

structure is considered to be, in part, primary (Wilson & Prendergast, in press), and, in part, the result of several secondary factors including continuous compaction of axial cumulates under the weight of overlying crystal layers and magma (e.g. Irvine, 1980), and later crustal loading which caused the dense ultramafic rocks to subside between the lighter and more rigid wall rocks (Prendergast, 1987; Worst, 1960).

Gravity measurements (Podmore & Wilson, 1987) show that the Great Dyke has a deep V- or Y-shaped transverse structure with a pronounced, upwards-flaring, trumpet shape up to 12 km wide, at least at the level of the Ultramafic Sequence, and a 1 km wide axial feeder dyke along most of its length at a depth of 4–10 km. Together, the geophysical, stratigraphic, structural and petrological data (Prendergast, 1987; Podmore & Wilson, 1987; Wilson & Prendergast, in press) suggest that the Great Dyke is the remains of two major, initially-composite, magma chambers (North and South) both with a very high length:width ratio, and separated at Lalapanzi between the Hartley and Selukwe Complexes (Fig. 3; for a recently proposed structural subdivision of the Great Dyke see Wilson & Prendergast, in press).

Crystallisation of the Ultramafic Sequence

In the lower portion of the Ultramafic Sequence, orthopyroxene and chromite compositions show initial upward enrichments of MgO and Cr_2O_3 respectively (Prendergast, 1987; Wilson, 1982). This phenomenon is variously interpreted as the result of early supercooling near the base which delayed the onset of equilibrium crystallisation (Jackson, 1970; Wilson, 1982) or replenishment at the base of compositionally-stratified magma (Wilson & Prendergast, in press; Wilson & Engell-Sorensen, 1986). Higher in the Ultramafic Sequence, the major minerals follow a normal Fe-enrichment trend which continues upwards into the overlying mafic rocks.

Mineral compositional data also indicate that the cyclic layering in the Ultramafic Sequence resulted from a series of major replenishments with new hot, dense, parental magma, probably a high-Mg basalt with ∼15% MgO, followed by slow mixing with cooler, more evolved magma (Wilson, 1982). The stratigraphic and mineral compositional data are also consistent with the possibility of several other minor replenishments with the same magma type (Podmore & Wilson, 1987; Wilson & Prendergast, in press; Prendergast & Keays, in press; Wilson, 1982; A. H. Wilson, personal communication, 1986).

Crystallisation in separate compartments of varying size would account for the different cumulate successions in the lower Ultramafic Sequence. The compartments later became fully interconnected throughout each magma chamber at the level of ultramafic Unit 1 (e.g. Podmore & Wilson, 1987) and possibly between the North and South magma chambers as well (Wager & Brown, 1968; Hughes, 1970a, b) perhaps following the injection of particularly large volumes of magma, thus breaching the former barrier at Lalapanzi.

The narrow elongate shape of each magma chamber led to a strong transverse heat gradient. It is envisaged that longitudinal variations in the volumes of magma

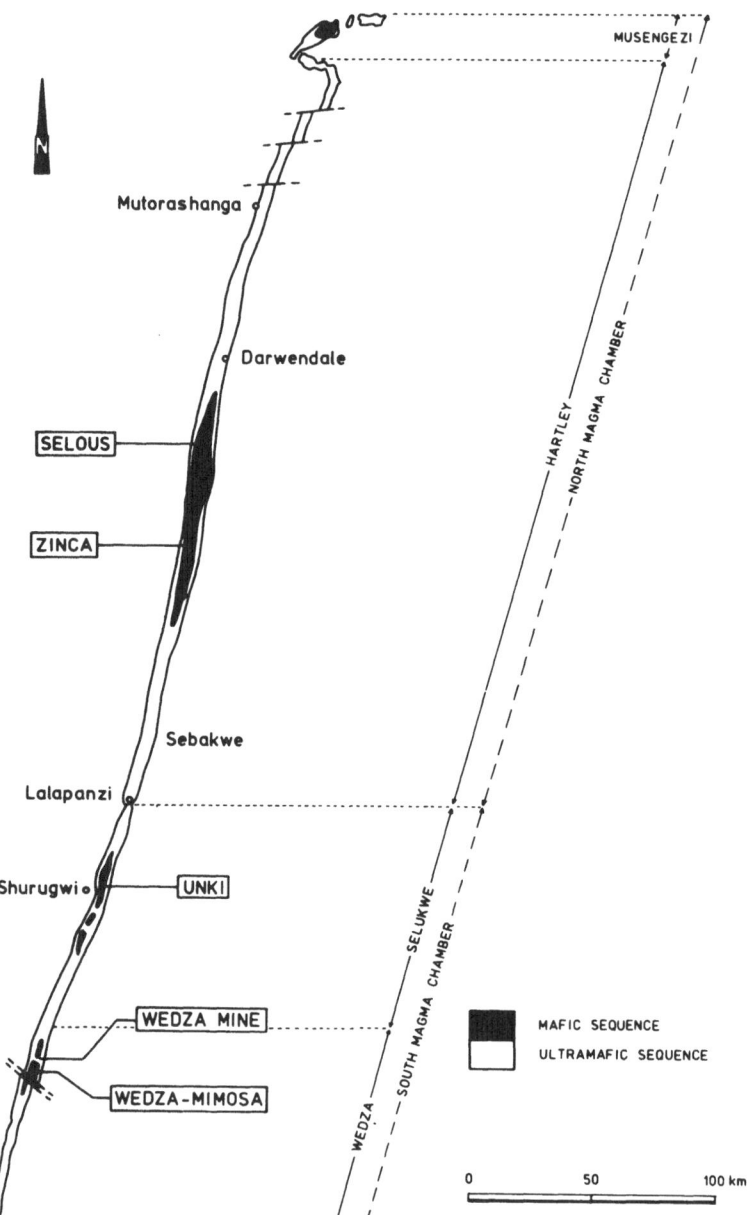

FIG. 3. Geological map of the Great Dyke showing the Musengezi, Hartley, Selukwe and Wedza Complexes, and the proposed North and South magma chambers. For details, see Wilson & Prendergast, in press. The MSZ occurs in the centre of each complex, directly beneath the mafic remnants. Also shown (boxed) are the locations of the principal MSZ mining prospects, 1926–83.

replenishments and degrees of mixing between old and new magmas, coupled, in the lower Ultramafic Sequence, with compartmentalisation of the magma chambers, probably produced hybrid magmas of slightly different compositions within each compartment. These factors, together with variations in the height: width aspect ratio, are considered to be the principal cause of lateral variations within individual layers.

The Pyroxenite No. 1 Layer
The P1 Layer is now preserved directly beneath the central mafic remnant in each complex (Fig. 3), with the stratiform MSZ situated in Bronzitite No. 1 up to several metres below the base of the Main Websterite. As a result of intensive exploration of the MSZ, the petrology and stratigraphy of the upper portions of the P1 Layer are now relatively well known in the Wedza Complex and in large parts of the Hartley Complex. In the Selukwe and Musengezi Complexes however, available information is confined to reconnaissance mapping (Worst, 1960) and to a few boreholes; consequently, the stratigraphic successions presented here are not necessarily typical of the P1 Layer in these two complexes.

The thicknesses of Bronzitite No. 1 and the Main Websterite, as well as the stratigraphic location of the MSZ, vary both between each complex, and within especially the Hartley Complex (Fig. 4). In the relatively wide Hartley Complex (11 km), the main layers appear to become significantly thinner towards the margins (Wilson & Prendergast, in press); thinning of the main layers is much less evident in the narrower Wedza Complex (6 km; Prendergast & Keays, in press).

In many areas, an irregular semi-concordant pegmatoid with coarse PGE-poor sulphide mineralisation, and several metres thick in places, is located at the top of the Main Websterite immediately below the mafic rocks. Locally, smaller and less persistent barren pegmatoids are also present elsewhere in the P1 Layer.

In places, the stratigraphy of the P1 Layer is highly complex, with both the main bronzitite and websterite layers being intercalated with lenses of other cumulate types. For example, olivine-bronzitite lenses occur in the bronzitites of the Hartley (A. H. Wilson, personal communication, 1986) and Musengezi Complexes. In the Wedza Complex, the bronzitites are often interlayered with minor lenses of websterite and transitional bronzitite (a distinctive textural variant of bronzitite forming lateral and vertical transition zones between bronzitite and websterite in which postcumulus augite occurs as either very small oikocrysts or interstitial grains).

In the Wedza and Hartley Complexes, layering subunits have been defined within the P1 Layer principally on the basis of cumulate lithology and texture, sulphide distribution, and reversals in orthopyroxene compositions (Wilson, 1976; A. H. Wilson, personal communication, 1986; Prendergast & Keays, in press). Subunit 1 of the Wedza Complex, loosely defined here as the Main Websterite plus the upper MSZ-bearing portion of Bronzitite No. 1, can be tentatively correlated in all four complexes (e.g. Subunit 1a of the Hartley Complex; Wilson & Prendergast, in press); subunits 2 and 3 of the Wedza Complex can also be recognised, on the

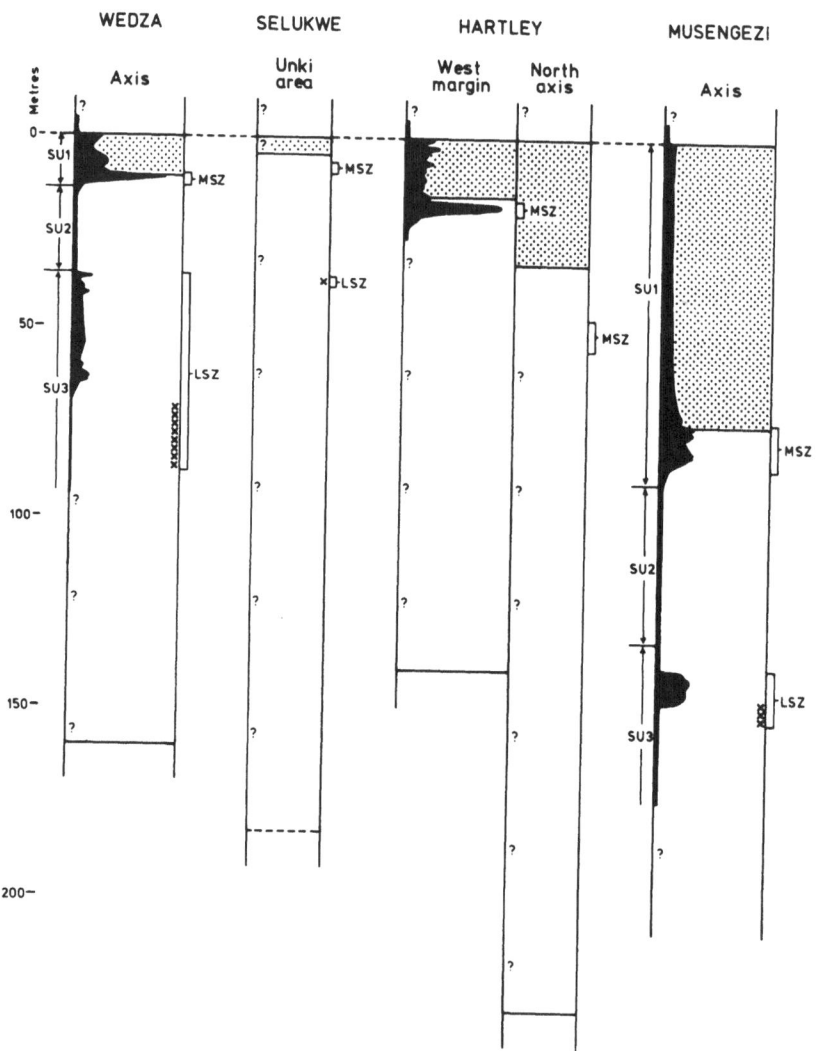

FIG. 4. Stratigraphy of the Pyroxenite No. 1 (P1) Layer in parts of the Wedza, Selukwe, Hartley and Musengezi Complexes, showing Bronzitite No. 1 (blank), Main Websterite (stippled), and layering subunits (SU1–3) for the Wedza and Musengezi Complexes, as defined in text; minor layering features are not shown. Other features portrayed are: vertical variations in sulphide distribution (black, with relative sulphide volume on horizontal axis) based, where available, on sulphide Cu, Ni assays of continuous borehole core; the Main Sulphide Zone (MSZ), the Lower Sulphide Zone (LSZ) and the basal PGE enrichments of the LSZ (crosses), where known; see text for definitions. There is no detailed information on the LSZ in the Selukwe and Hartley Complexes. The successions in all four complexes have been arbitrarily levelled to the base of the overlying Mafic Sequence. Scale at left shows depth in metres below the mafic contact. The base of the P1 Layer in the Selukwe and Musengezi Complexes is not well constrained.

basis of cumulate lithology and sulphide distribution, in the Musengezi Complex (Fig. 4).

Detailed stratigraphic analysis of the P1 Layer in the Wedza Complex has revealed strong lateral variations in layering, and in cumulus mineralogy, mode, texture and fabric (Prendergast & Keays, in press). The most striking variations occur along the east margin where, in places, the normally massive, medium-grained Main Websterite contains lenses of fine-grained laminated bronzitite and transitional bronzitite or is substituted by normal bronzitite and bronzite-phyric augitite. Elsewhere along the east margin are zones within which the Main Websterite interdigitates with lenses of the overlying mafic rocks, which die out towards the axis, and is cut, with or without the MSZ and upper bronzitites, by large erosional depressions now filled with fine-grained mafic rocks. Similar erosional features at the base of the Mafic Sequence have been observed in the Selukwe (Mennell & Frost, 1926) and Hartley (A. H. Wilson, personal communication, 1986) Complexes.

In both the Hartley and Wedza Complexes, the most consistent lateral variations within the P1 Layer occur between the margins and the axis. Towards the east margin of the Wedza Complex there are systematic decreases in bronzite Mg and augite Cr contents, and increases in both cumulus augite:bronzite modal ratio and trapped liquid content (Prendergast & Keays, in press). Away from the axis, transitional bronzitites and then websterites become increasingly common within Bronzitite No. 1 and transitional bronzitites and bronzite-phyric augitites begin to appear in the Main Websterite. Very similar transverse variations in trapped liquid content and orthopyroxene compositions are observed in the Hartley Complex (Wilson, 1976; A. H. Wilson, personal communication, 1986). In the Wedza Complex, the transverse variations within the P1 Layer are ascribed principally to the narrow width of the magma chamber and the high heat gradient between the margins and the axis (Prendergast & Keays, in press).

Sulphide and PGE Mineralisation in the P1 Layer
Subunit 1 of the Wedza Complex, and its correlative subunits in the three other complexes, each contain a broad zone of disseminated sulphide mineralisation (Fig. 4). The PGE-rich Main Sulphide Zone (MSZ) is geologically defined as the strong sulphide concentration at the base of this broad mineralised zone. Another broad sulphide-bearing zone, also with basal PGE-enrichment, is situated in the upper parts of subunit 3 in the Wedza and Musengezi Complexes. Sulphide and PGE mineralisation is also present at a broadly equivalent stratigraphic level in the Selukwe Complex (J. M. Clutten, personal communication, 1986; Mennell & Frost, 1926). This lower mineralised zone, including the basal PGE enrichment, is poorly known, variably developed and economically unimportant by comparison with the MSZ, and is loosely termed here the Lower Sulphide Zone.

Subunit 2, as recognised in the Wedza and Musengezi Complexes, is very weakly mineralised in comparison with subunits 1 and 3 (Fig. 4). On present evidence, there is no significant sulphide mineralisation in the lower portions of Bronzitite No.1.

The broad sulphide zones of subunits 1 and 3 (Wedza and Musengezi Complexes) have several important common features, suggesting a similar origin for both:

1. stratigraphic locations within and immediately beneath a websterite layer (subunit 1), or in a bronzitite which, in the Wedza Complex, grades laterally into websterite (subunit 3);
2. basal PGE-enrichments with similar vertical metal distributions, and
3. lateral variations in the width and intensity of both sulphide and PGE mineralisation.

The Main Sulphide Zone

The broad lateral continuity of the MSZ has been proved in outcrop, and by trenching and drilling, throughout the P1 Layer remnants in the Wedza Complex and in most of the Hartley and Selukwe Complexes. In the Musengezi Complex, its presence has been confirmed by four closely-spaced boreholes drilled in the axis, and by analogy with the other complexes, it is assumed to persist towards the margins as well. On a smaller scale, the MSZ is absent only where it has been removed by magmatic erosion. Such instances are rare, and affect only a minute proportion of the MSZ. There is no conclusive evidence for Merensky-type 'pot-hole' structures.

The pristine host rock is generally a medium-grained bronzitite (or transitional bronzitite) comprising cumulus bronzite plus 10–40% trapped liquid (postcumulus augite, plagioclase, biotite, potassic feldspar and quartz). Sulphide mineralisation varies from finely-disseminated grains to almost net-textured concentrations. In general, sulphide content is proportional to the trapped liquid content of the host rock which is highest near the margins. This effect is more pronounced in the wide Hartley Complex for example, than in the narrower Wedza Complex. In places, the sulphides and silicates have been secondarily coarsened by volatile-rich fluids (e.g. Lightfoot, 1926), but such pegmatoid zones are generally rare. The MSZ is affected by varying degrees of late magmatic-'hydrothermal' alteration. In the Wedza Complex, alteration is most intense near the margins where primary textures are often completely replaced by an intergrown assemblage of sulphides, hydro-silicates, quartz, carbonate and chromian spinel, together with remnant pyroxene and plagioclase. Alteration is minimal in the axis where cumulus textures are often well preserved.

Examination of a large number of routine borehole MSZ assay profiles from different parts of the Great Dyke shows that, with only minor exceptions, the following features are characteristic of the MSZ in all areas (Fig.5):

1. The MSZ comprises two main subzones, a lower PGE subzone rich in Pt,Pd and other precious metals, and an upper Base Metal (BM) subzone with very low precious metal contents.
2. The PGE subzone itself comprises two main portions (upper and lower), defined on the basis of bulk Cu,Ni and Pd,Pt contents.
3. Within the PGE subzone as a whole, and within each of its component portions, bulk base and precious metals contents increase upwards, whereas Pd:Pt ratios and Pd + Pt contents per unit sulphide (as bulk Cu + Ni content)

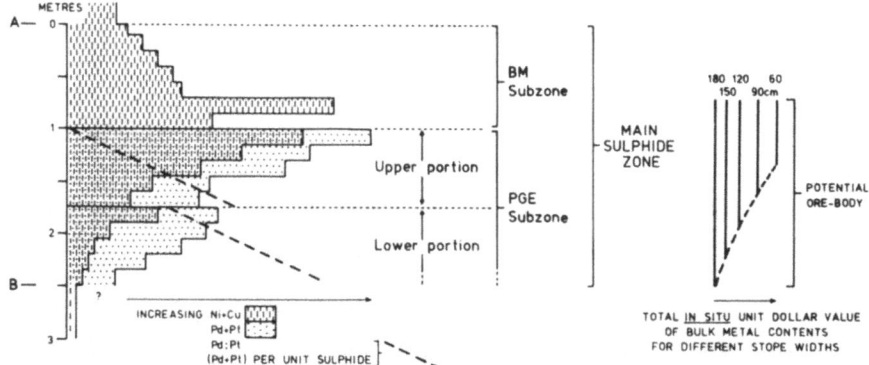

FIG. 5. Model MSZ assay profile, showing its main components as defined by vertical variations of Ni + Cu and Pd + Pt contents and Pd:Pt ratios, and Pd + Pt per unit sulphide in consecutive 15 cm samples. Points A and B are, respectively, the upper and lower limits of the MSZ arbitrarily defined by the marked changes in slope of Cu + Ni content; significant Pd and Pt contents (detectable by standard fire assay with AA finish) persist for up to 1 m below B in many real profiles. Total *in situ* unit dollar values of the potential MSZ orebody are also shown for different stope widths. This model is a summary of MSZ profiles in the Wedza Complex and along the west margin of the Hartley Complex. MSZ profiles elsewhere essentially differ only in total thickness and average metal content, as explained in the text. Ore grades are proprietary company information; absolute grades are therefore not shown in Figs 5–8.

increase downwards, so that the highest metal contents and the lowest Pd:Pt ratios and Pd + Pt contents per unit sulphide occur at the top of the PGE subzone.

In general, this complex vertical MSZ assay profile is remarkably uniform over wide areas. In the Wedza Complex, it persists unchanged in both altered and unaltered MSZ and is considered to be entirely magmatic in origin. There is no evidence for significant remobilisation of PGE by late fluids.

In potentially economic areas, optimum MSZ stope widths are designed to include both the bulk of the PGE subzone and the lower part of the overlying BM subzone. The *in situ* unit dollar value of the ore decreases as the stope width increases below a given hanging-wall cut off (Fig. 5).

The thicknesses of the MSZ and its component subzones vary significantly in different areas (Fig. 6). In some areas, especially in the Wedza Complex and towards the west margin of the Hartley Complex, the MSZ is 2–3 m thick, the PGE subzone being about 1·5 m thick with well-defined upper and lower portions (Fig. 6A,B). In others, especially in and near the axis of the Hartley and Musengezi Complexes, the typical axial MSZ profile is highly attenuated (Fig. 6C,D,F), with relatively low grade mineralisation distributed through a much greater thickness (up to 20 m in some cases). In certain relatively small areas, the attenuated axial MSZ is highly segmented, the barren intervals coinciding with intercalated lenses of fine-grained laminated bronzitite (Fig. 6E).

There are significant lateral variations in bulk metal contents, intermetallic

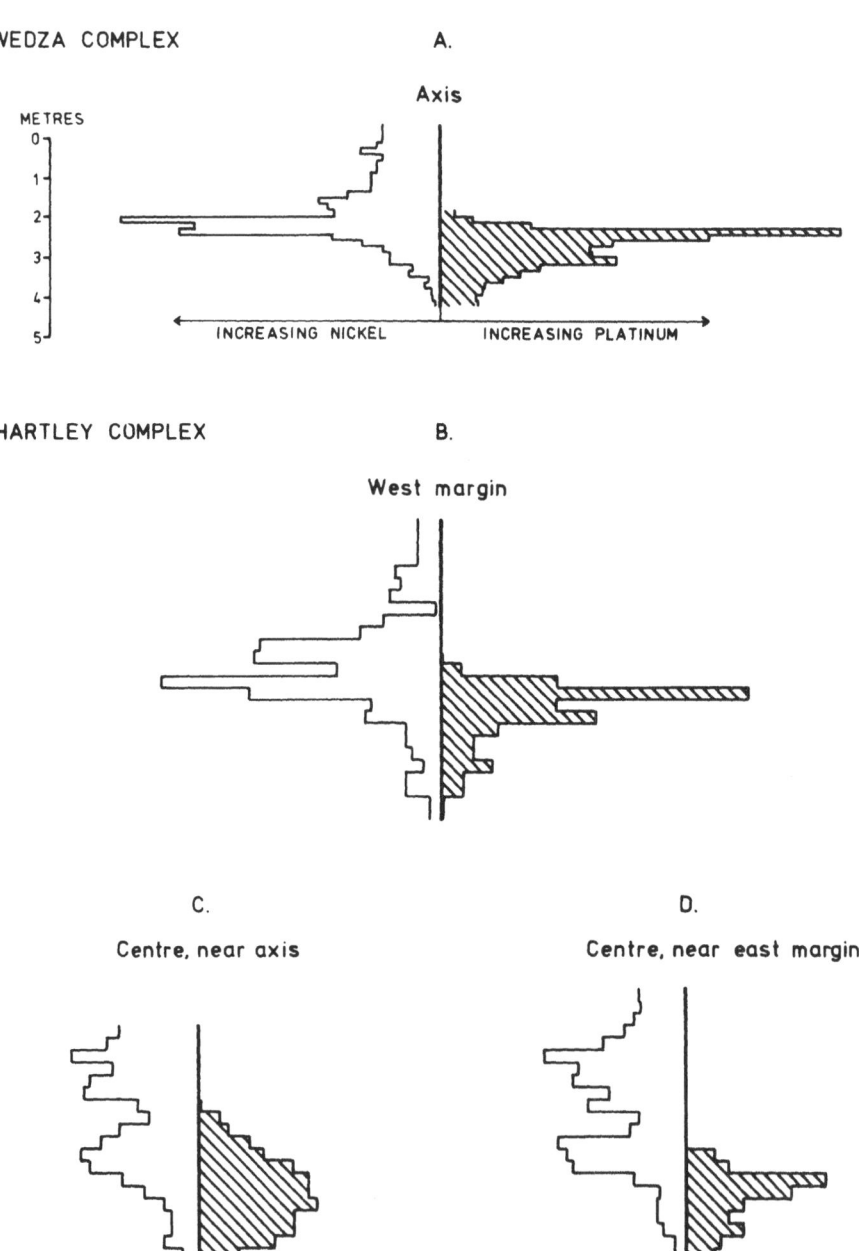

WEDZA COMPLEX

A.

Axis

METRES

HARTLEY COMPLEX

B.

West margin

INCREASING NICKEL INCREASING PLATINUM

C.

Centre, near axis

D.

Centre, near east margin

FIG. 6. Examples of MSZ Ni (blank) and Pt (hatched) assay profiles from boreholes in different parts of the Great Dyke. All profiles are drawn to the same vertical and horizontal scale as in A. Individual sample widths are 30 cm except for A and F (15 cm in both).

E.   F.

HARTLEY COMPLEX   MUSENGEZI COMPLEX

North, axis   Axis

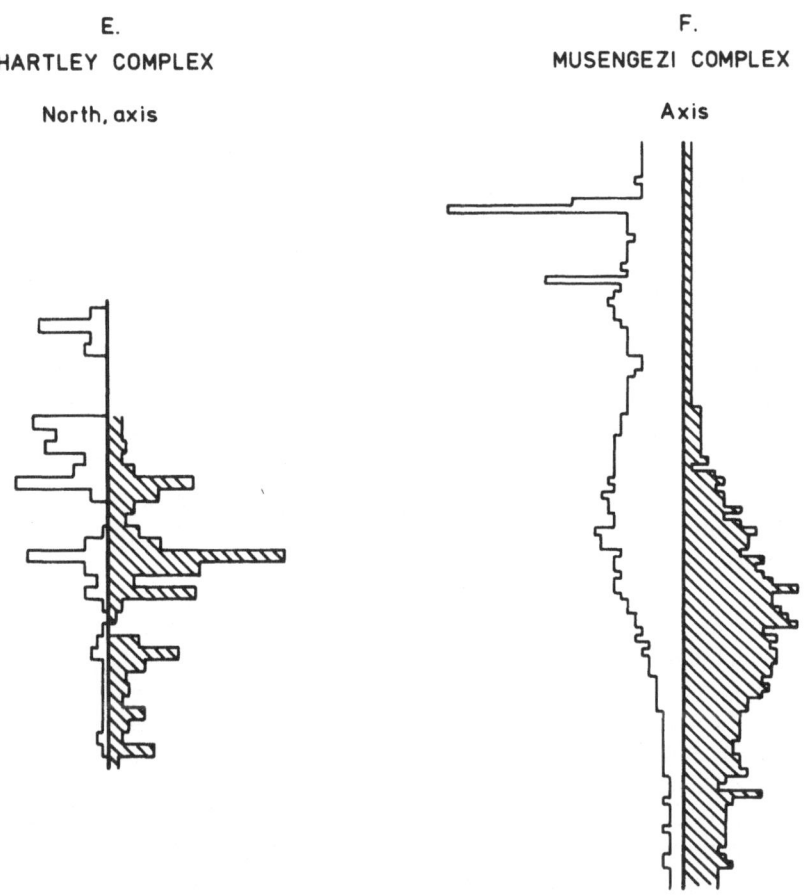

Fig. 6.—contd.

ratios, and Pd + Pt contents per unit sulphide within the PGE subzone. Along the
east margin of the Wedza Complex, for example, there are marked antipathetic
trends in bulk Pd + Pt contents and Pd:Pt ratios, with a zone of magmatic erosion
and interdigitating mafic contacts coinciding with low bulk Pd + Pt contents and
high Pd:Pt ratios. The principal lateral variations in MSZ bulk metal contents and
intermetallic ratios occur in a transverse direction. In the Wedza Complex, for
instance, Cu:Ni and Pd:Pt ratios increase whereas bulk Pd + Pt content per unit
sulphide decreases from the axis towards the margins. In the Hartley Complex, the
MSZ is relatively weakly mineralised in the axis by comparison with the margins.
By contrast, in the much narrower Wedza Complex, bulk sulphide content is only
very slightly lower in the axis than in the margins. In the axis of the Hartley
Complex, sulphide content is too low to compensate for the high Pd + Pt content
per unit sulphide. Given that Pt content is the critical revenue factor, this in part
explains why potentially economic zones in the Hartley Complex are confined to

the west margin, whereas in the narrower Wedza Complex, the whole deposit is potentially economic from margins to axis. An important implication of this phenomenon is that MSZ bulk metal contents may improve toward the margins of the Musengezi Complex. To date, this area has been drilled only in the axis, where MSZ bulk metal contents and assay profiles are broadly similar to the poorly-mineralised axial areas of the Hartley Complex.

Special grade control problems are associated with mining the MSZ in certain potentially economic areas. In the Selukwe Complex for instance, the PGE subzone commonly has relatively high bulk PGE contents, but very poor sulphide mineralisation, whereas, in addition, the overlying BM subzone is unusually rich in base metals. Effectively, there are two discrete Ni- and Pt-rich zones which are individually uneconomic and too far apart to be stoped together (J. M. Clutten, personal communication, 1986). A different instance occurs near the east margin of the Wedza Complex where there are relatively small areas in which the BM subzone is absent. These phenomena are ascribed to extreme lateral variations in sulphide contents and Pd + Pt content per unit sulphide, and, in the latter case, to small-scale current action which either prevented sulphides accumulating or removed pre-existing accumulations.

The Lower Sulphide Zone

The Lower Sulphide Zone is presently well known only from a few deep boreholes in the Wedza and Musengezi Complexes. Generally it is poorly mineralised and highly attenuated compared to the MSZ, with much lower bulk PGE contents. Relatively strong PGE enrichments however occur in at least two areas. In the axis of the Musengezi Complex, the basal portion of the Lower Sulphide Zone is very strongly enriched in PGE, with assay profiles strikingly similar in size and shape to those of the MSZ in the same area (Fig. 7). By analogy with the MSZ, the metal content of the Lower Sulphide Zone in the Musengezi Complex may also increase towards the margins. In the Selukwe Complex, a 'lower platinum horizon' was found 30 m below the MSZ, with comparable mineralised widths and bulk Pt contents (Mennell & Frost, 1926). This mineralisation may be a PGE-rich lateral variant of the 'base metal reef' now known to be present below the MSZ elsewhere in the Selukwe Complex (J. M. Clutten, personal communication, 1986). On present evidence, both these mineralised zones are provisionally correlated with the Lower Sulphide Zone as defined in the Wedza and Musengezi Complexes.

The Origin of the MSZ: Current Ideas

The MSZ is located within a layered pyroxenite that can be interpreted in terms of bottom crystallisation from a system of liquid layers set up largely by a major influx of hot, dense, parental magma at the level of chromitite Clc (Huppert & Sparks, 1980; Sparks & Huppert, 1984; Turner & Campbell, 1986; Prendergast & Keays, in press). The cumulate succession and orthopyroxene compositional variations are consistent with overall fractionation from chromitite Clc through to the overlying mafic rocks, together with periodic overturns between basal liquid layers and one or more minor replenishments. Sulphur saturation and sulphide precipitation

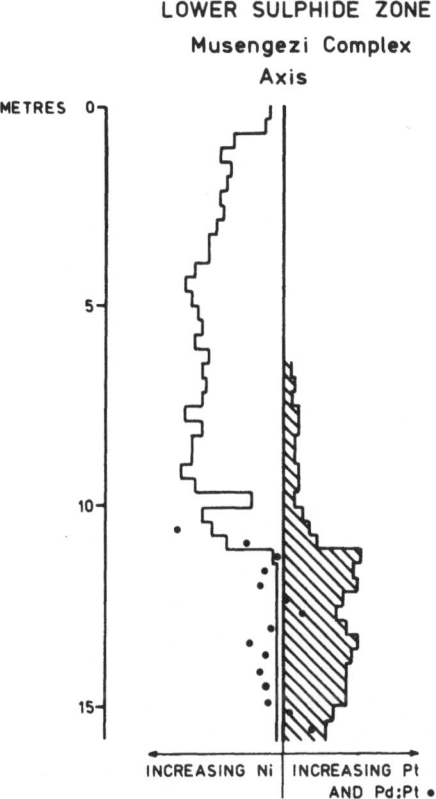

LOWER SULPHIDE ZONE
Musengezi Complex
Axis

FIG. 7. Ni (blank) and Pt (hatched) assay profiles of the Lower Sulphide Zone in the axis of the Musengezi Complex. The Pd:Pt ratio profile is also shown. Individual sample widths are 30 cm; the vertical and horizontal scales are the same as in Fig. 6. Note, as in the MSZ, the well-defined BM and PGE subzones and the two portions of the PGE subzone (defined here by the Pt and Pd:Pt profiles).

probably occurred largely as a result of overall cooling of the magma (Haughton *et al.*, 1974; Buchanan *et al.*, 1983) both within the P1 Layer in general, and within each liquid layer in particular, with the largest volumes of sulphide precipitating near the cooler margins (cf. Campbell, 1977; Keays & Campbell, 1981). The characteristic form of the vertical MSZ metal profile can be explained in terms of at least two successive overturns between relatively thin convecting basal liquid layers in a double diffusive regime (Prendergast & Keays, in press). It is envisioned that each overturn was followed by rapid sulphide precipitation at the floor of the magma chamber with efficient stripping of precious metals (high R value; Campbell & Naldrett, 1979; Campbell *et al.*, 1983) in the order of their relative partition coefficients from the overlying magma convecting within each successive basal liquid layer.

Lateral variations within the MSZ are considered to be the result of several factors: first, variations in magma composition and metal content set up by variations in magma mixing, fractionation and distance from sites of magma replenishment; second, the effect of transverse gradients in temperature, composition and fO_2 on rates of crystallisation and sulphide precipitation, and on metal partition coefficients, and third, the relative efficiency of the PGE-enrichment process in different areas. In addition, magma currents, which caused the redistribution of precipitating sulphides, and 'cold spots' which affected metal partition coefficients, and other processes, may have been important local factors along the margins.

The Lower Sulphide Zone probably originated in broadly similar fashion, but with major differences in the relative rate and efficiency of the processes which controlled the formation of the MSZ.

ECONOMIC POTENTIAL

Exploration and Resource Development

The MSZ was discovered in 1925 soon after the Merensky Reef in the Bushveld Complex, and was soon traced throughout most areas of the Great Dyke. The 'potato reef', a hanging-wall marker of characteristically-weathered gossanous pyroxenite, proved an effective guide to early MSZ prospectors (Lightfoot, 1926; Mennell & Frost, 1926; Wagner, 1929). Initial interest focused on the near surface, oxidised zone. This zone extends up to 250 m from outcrop, and in the Wedza area, with its relatively good exposure, flat to moderate dips and subdued topography, the oxidised MSZ was especially suitable to shallow open-cast mining with a low waste to ore ratio. At the Wedza Mine (1926–28), the ore was mined from long strike trenches, crushed and the pulp passed over a series of long, gently sloping riffled cement strakes. Average head grade was claimed to be 4·3 g/t Pt, most of the relatively mobile Pd having been leached from the weathered zone. Despite considerable experimentation however, recoveries were rarely better than 50% and the mine closed in 1928 (Golding, 1932; Wagner, 1929).

The main Pt-bearing mineral, now known to be sperrylite (Wagner, 1929) is too fine-grained for efficient concentration by either wet gravity or froth flotation methods. Metallurgical testwork in the 1950s and 1960s (Technical files, Union Carbide Zimbabwe) however, showed that the valuable metals could be efficiently recovered by smelting the oxidised ore in an electric arc furnace followed by normal treatment of the ferronickel or nickel matte product. The oxidised MSZ ore represents a significant resource in its own right. The high cost of electric smelting could, in part, be offset by cheap surface mining and by savings in crushing the relatively soft, friable ore without the need for subsequent grinding.

Between the late 1960s and early 1980s the deeper sulphide ore was investigated in parts of all four complexes by means of several hundred boreholes. Trial mining and plant-scale metallurgical extraction projects were set up at Wedza-Mimosa, in the Wedza Complex, and at Selous and Zinca, near the west margin of the Hartley Complex, where initial exploration indicated the presence of large tonnages of potential ore; in addition an exploration shaft was sunk at Unki in the Selukwe

Complex (Fig. 3). Together, these pilot projects investigated and proved the technical feasibility of appropriate mining schemes and extraction processes through to the sale of refined metals: first, various forms of Bushveld-type mining systems (Newman, 1973), utilising scrapers and/or trackless vehicles loading to footwall haulages, are suitable for the shallow dips and relatively narrow stope widths of the orebody; second, approximately 70% of the base- and precious-metal content can be recovered by normal flotation, arc smelting and converting, followed by adaptations of INCO- and Outukumpu-type refining processes (e.g. Queneau, 1961; Newman, 1973; Liddell *et al.*, in press). In certain areas, poor ground conditions and the presence of excessive talc remain potential problems, but in general, the principal technical aspects of exploiting MSZ sulphide ore have been resolved.

Valuation and Grade Control

In February, 1987, metal prices, the total *in situ* unit dollar value of the MSZ was ~14% higher, at a 90 cm stope width, than average Merensky Reef and UG-2 chromitite valuations, but ~22% lower at a 180 cm stope width (Fig. 8).

The potential revenue distribution is broadly similar to those of the Bushveld orebodies although Au and Ni account for significantly greater proportions of potential MSZ revenue. At present valuation, the MSZ is effectively a five-metal orebody (major Pt plus important Pd, Rh, Au and Ni) which may provide slightly improved flexibility relative to the four-metal Bushveld ores.

An analysis of the revenue–cost equation of underground mining shows that the most economic production rate is likely to be (very approximately) 5000 t/day from 180 cm stopes, the lower unit dollar value of the ore being offset by lower mining costs at this width (P. Batty, personal communication, 1987). The long production lead time and substantial capital costs required for such an underground operation may be significantly reduced in parts of the Wedza Complex where geological conditions may permit the economic open cast mining of the sulphide ore.

Because of the relatively marginal dollar value at 180 cm, efficient underground grade control will be essential on any future producing mine. There are no well-defined visual markers within the MSZ which, at first sight, presents as a super-ficially-uniform, dark grey-green pyroxenite with sparsely-disseminated sulphides. With experience however, a careful observer can identify the orebody with good precision, and sulphide-sensitive chemical paints and self-potentiometry have proved useful additional aids for this purpose. Nevertheless, economic hanging-wall cutoffs can only be accurately determined by assay. In practice only Ni cutoffs are used for reasons of time and cost, but Ni content is not a reliable guide to PGE content. Consequently, Ni face profiles must be carefully interpreted by close reference to model assay profiles and to visual inspections of the face by trained observers.

Resource Estimates

Because of the broad lateral variations in grade and thickness, and the lack of information for certain areas, the total tonnage and mean grade of the MSZ cannot be accurately computed. Assuming an average 180 cm width however, the MSZ

M. D. Prendergast

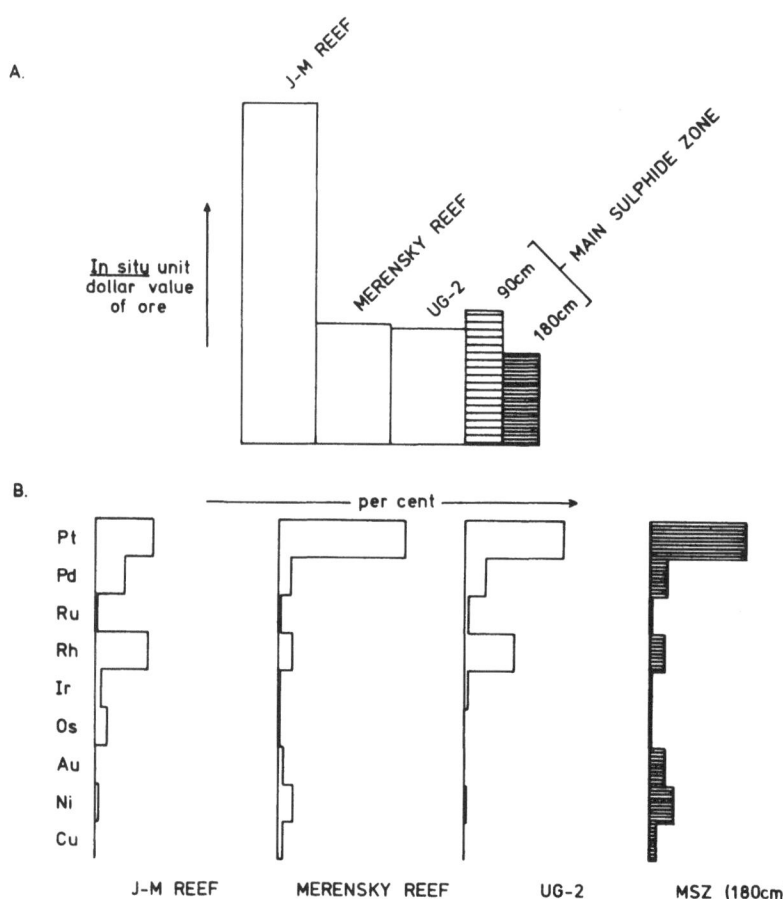

FIG. 8. Relative *in situ* unit dollar values (A), and potential revenue distribution for nine metals (B), for the four major world stratiform PGE-rich mineralised zones in layered intrusions. These data are based on published 'average' ore grades, with the exception of the MSZ valuation which is based on one of the potential ore blocks referred to in the text, and on metal prices as at February, 1987.

contains a total resource of approximately 4·4 billion tonnes (Table 1). About 9% is near-surface, oxidised MSZ, some of which may be amenable to open-cast mining and direct electric smelting. Much of the sulphide resource, especially in the axis of the Hartley Complex, is too deep (up to 1300 m vertical) and too low grade to be considered for mining under foreseeable circumstances. Certain other MSZ resources are also ruled out for development in the near future because of their remote, rugged location (e.g. Musengezi Complex) and special geological features (e.g. parts of the Selukwe Complex). However, a number of large blocks each containing up to several hundred million tonnes of potential ore at relatively shallow depth have been established by close drilling in the central Wedza area and along the west side of the Hartley Complex.

TABLE 1

Resource estimates (in million tonnes) for the Main Sulphide Zone of the Great Dyke

	Resource		
Complex	*Sulphide ore*	*Oxidised ore*	*Total*
Musengezi	166	24	190
Hartley	3 288	186	3 475
Selukwe	389	122	511
Wedza	170	56	226
Total	4 012	389	4 401

The Future

To date, no economically-viable producing mine has been established on the MSZ, despite considerable, technically-successful, exploration and evaluation effort, and, since 1978, a more than fivefold increase in local dollar ore value. All the Great Dyke PGE prospects remain dormant, and there is no active local industrial interest in the MSZ, although government and foreign interest has recently increased.

The failure to develop the PGE resources of the Great Dyke is a major continuing disappointment for Zimbabwe's government and mining industry. Principal constraints on future development are the marginal grade of the potential ore which can only be offset by highly cost-efficient mining and extraction, the sharp escalation in capital costs required to build a mine of the optimum economic size, and the poor market perception of the local investment climate. Attractive features are the large proven resources of potential ore in an essentially continuous mineralised zone, with relatively uniform grades over wide areas, and the absence of major technical obstacles.

Other factors in favour of the MSZ as a potential future source of PGE are the continuing rise in demand for, and the strategic status of, PGE for industrial use, as well as special problems which may affect future supplies from each of the current major producers (Buchanan, 1979; Robson 1986); for example, Canada's present production is closely-linked to the nickel market and cannot be increased for its own sake; Russian sales to the West have declined in recent years and cannot be considered a reliable source; and in the medium term, South African production is under threat from sanctions and social unrest. Additionally, recently commenced production from the J-M Reef is likely to remain small relative to total demand, and, with the exception of the Lac des Iles deposit in Ontario (Macdonald, this volume pp. 215–36.), current PGE exploration in Canada, Australia, Finland and elsewhere has yet to make a major significant find.

In summary, future development of the MSZ could be potentially attractive given a sustained high platinum price, the necessary capital investment, and cost efficient production.

ACKNOWLEDGEMENTS

This paper is published by permission of the management of Union Carbide Zimbabwe Limited, whose support and encouragement over many years are gratefully acknowledged. Additional information was kindly supplied by Rio Tinto Zimbabwe Limited and Anglo American Corporation. The author would also like to thank the following for their helpful critical comments: P. Batty, F. C. Bohmke, L. G. Kimble, A. H. Wilson, J. F. W. Bowles, P. J. Potts and two anonymous reviewers. Mrs L. Bismark-Pettit typed the manuscript.

REFERENCES

Barnes, S.-J. & Naldrett, A. J. (1985). Geochemistry of the J-M (Howland) reef of the Stillwater Complex, Minneapolis adit area: I. Sulfide chemistry and sulfide-olivine equilibrium. *Econ. Geol.*, **80**, 627–45.

Brynard, H. J., de Villiers, J. P. R. & Viljoen, E. A. (1976). A mineralogical investigation of the Merensky Reef at the Western Platinum Mine, near Marikana, South Africa. *Econ. Geol.*, **71**, 1299–1307.

Bow, C., Wolfgram, D., Turner, A., Barnes, S., Evans, J., Zdepski, M. & Boudreau, A. (1982). Investigations of the Howland Reef of the Stillwater Complex, Minneapolis Adit area: Stratigraphy, structure and mineralisation. *Econ. Geol.*, **77**, 1481–92.

Buchanan, D. L. (1979). Platinum-group metal production from the Bushveld Complex and its relationship to world markets. Bureau for Minerals Studies Report 4, 31 pp. University of the Witwatersrand, Johannesburg.

Buchanan, D. L., Nolan, J., Wilkinson, N. & de Villiers, J. P. R. (1983). An experimental investigation of sulfur solubility as a function of temperature in synthetic silicate melts. *Spec. Publ. Geol. Soc. S.Afr.*, **7**, 383–91.

Campbell, I. H. (1977). A study of macro-rhythmic layering and cumulate processes in the Jimberlana Intrusion, Western Australia. Part 1: The upper layered series. *J. Petrol.*, **18**, 183–215.

Campbell, I. H. & Naldrett, A. J. (1979). The influence of silicate:sulfide ratios on the geochemistry of magmatic sulfides. *Econ. Geol.*, **76**, 1503–6.

Campbell, I. H., Naldrett, A. J. & Barnes, S.-J. (1983). A model for the origin of the platinum-rich sulfide horizons in the Bushveld and Stillwater Complexes. *J. Petrol.*, **24**, 133–65.

Conn, H. K. (1979). The Johns-Manville platinum-palladium prospect, Stillwater Complex, Montana, USA. *Can. Mineral.*, **17**, 463–8.

Cousins, C. A. (1969). The Merensky Reef of the Bushveld igneous complex. In *Magmatic Ore Deposits*, ed. H. D. B. Wilson. Economic Geology Monograph, Vol. 4, pp. 239–51.

Gain, S. B. (1985). The geologic setting of the platiniferous UG-2 chromitite layer on the farm Maandagshoek, eastern Bushveld Complex. *Econ. Geol.*, **80**, 925–43.

Golding, E. (1932). Notes on the Wedza Platinum Mine, Southern Rhodesia. *J. Chem. Metall. Min. Soc. S. Afr.*, December, 4.

Haughton, D. R., Roeder, P. L. & Skinner, B. J. (1974). Solubility of sulfur in mafic magmas. *Econ. Geol.*, **69**, 451–67.

Hiemstra, S. A. (1985). The distribution of some platinum-group elements in the UG-2 chromitite layer of the Bushveld Complex. *Econ. Geol.*, **80**, 944–57.

Hiemstra, S. A. (1986). The distribution of chalcophile and platinum-group elements in the UG-2 chromitite layer of the Bushveld Complex. *Econ. Geol.*, **81**, 1080–6.

Hughes, C. J. (1970*a*). Major rhythmic layering in ultramafic rocks of the Great Dyke of

Rhodesia, with particular reference to the Sebakwe area. *Geol. Soc. S.Afr. Spec. Publ.*, **1.**, 594–609.

Hughes, C. J. (1970*b*). Lateral cryptic variation in the Great Dyke of Rhodesia. *Geol. Mag.*, **107**, 319–25.

Huppert, H. E. & Sparks, R. S. J. (1980). The fluid dynamics of a magma chamber replenished by influx of hot, dense, ultrabasic magma. *Contrib. Mineral. Petrol.*, **75**, 279–89.

Irvine, T. N. (1980). Infiltration metasomatism, adcumulate growth, and double-diffusive fractional crystallisation in the Muskox intrusion and other layered intrusions. In *Physics of Magmatic Processes*, ed. R. B. Hargraves. Princeton University Press, Princeton, N.J., pp. 325–83.

Jackson, E. D. (1970). The cyclic unit in layered intrusions—a comparison of repetitive stratigraphy in the ultramafic parts of the Stillwater, Muskox, Great Dyke and Bushveld Complexes. *Geol. Soc. S.Afr. Spec. Publ.*, **1**, 391–424.

Keays, R. R. & Campbell, I. H. (1981). Precious metals in the Jimberlana Intrusion, Western Australia: Implications for the genesis of platiniferous ores in layered intrusions. *Econ. Geol.*, **76**, 1118–41.

Liddell, K. S., McRae, L. B. & Dunne, R. C. (in press). Process routes for the beneficiation of noble metals from Merensky and UG-2 ores. *Extraction Metallurgy '85*, London.

Lightfoot, B. (1926). Platinum in Southern Rhodesia. Southern Rhodesia Geological Survey, Short Report, 19, 13 pp.

Macdonald, A. J. (1988). Platinum-group element mineralisation and the relative importance of magmatic and deuteric processes: Field evidence from the Lac de Iles deposit, Ontario, Canada. In *Geo-platinum 87*, eds H. M. Prichard, P. J. Potts, J. F. W. Bowles and S. J. Cribb. Elsevier Applied Science Publishers, London, pp. 215–36.

McLaren, C. H. & de Villiers, J. P. R. (1982). The platinum-group chemistry and mineralogy of the UG-2 chromitite layer of the Bushveld Complex. *Econ. Geol.*, **77**, 1348–66.

Mennell, F. P. & Frost, A. (1926). Notes on the occurrence of platinum in the Great Dyke with special reference to Belingwe and Selukwe. *Proc. Rhod. Scientific Assoc.*, **25**, 1–8.

Newman, S. C. (1973). Platinum. *Trans. Inst. Min. Metall.*, **82**, A52–A68.

Podmore, F. & Wilson, A. H. (1987). A reappraisal of the structure and emplacement of the Great Dyke, Zimbabwe. In *Mafic Dyke Swarms*, eds H. C. Halls & W. F. Fahrig. Geological Association of Canada Special Paper 34.

Prendergast, M. D. (1987). The chromite orefield of the Great Dyke, Zimbabwe. In *Evolution of Chromium Ore Fields*, ed. C. W. Stowe. Van Nostrand Reinhold, New York, 340 pp.

Prendergast, M. D. & Keays, R. R. (In press). Controls of platinum-group element mineralisation and the origin of the PGE-rich Main Sulphide Zone in the Wedza Subchamber of the Great Dyke, Zimbabwe: Implications for the genesis of, and exploration for, stratiform PGE mineralisation in layered intrusions. In *5th Magmatic Sulphides Field Conference, Harare, Zimbabwe*, eds M. D. Prendergast & M. J. Jones. Institution of Mining and Metallurgy, London.

Prendergast, M. D. & Wilson, A. H. (In press). The Great Dyke of Zimbabwe: II. Mineralisation and mineral deposits. In *5th Magmatic Sulphides Field Conference, Harare, Zimbabwe*, eds M. D. Prendergast & M. J. Jones. Institution of Mining and Metallurgy, London.

Queneau, P. (Ed.) (1961). *Extractive Metallurgy of Copper, Nickel, and Cobalt*. Interscience, New York, London, 647 pp.

Robson, G. G. (1986). *Platinum 1986*, Johnson Matthey, London, 52 pp.

Schwellnus, J. S. I., Hiemstra, S. A. & Gasparrini, E. (1976). The Merensky Reef at the Atok platinum mine and its environs. *Econ. Geol.*, **71**, 249–60.

Sparks, R. S. J. & Huppert, H. E. (1984). Density changes during the fractional crystallisation of basaltic magmas: fluid dynamic implications. *Contrib. Mineral. Petrol.*, **85**, 300–9.

Todd, S. G., Keith, D. W., Le Roy, L. W., Schissel, D. J., Mann, E. L. & Irvine, T. N. (1982). The J-M platinum-palladium reef of the Stillwater Complex, Montana: 1. Stratigraphy and petrology. *Econ. Geol.*, **77**, 1454–80.

Turner, J. S. & Campbell, I. H. (1986). Convection and mixing in magma chambers. *Earth Science Reviews*, **23**, 255–352.

Vermaak, C. F. & Hendriks, L. P. (1976). A review of the mineralogy of the Merensky Reef, with specific reference to new data on the precious metal mineralogy. *Econ. Geol.*, **71**, 1244–69.

Von Gruenewaldt, G. (1977). The mineral resources of the Bushveld Complex. *Minerals Sci. Engng.*, **9**, 83–95.

Wager, L. R. & Brown, G. M. (1968). *Layered Igneous Rocks.* W. H. Freeman, San Francisco, California, 578 pp.

Wagner, P. A. (1929). Platinum deposits and mines of South Africa. C. Struik (Pty) Ltd, Cape Town, 338 pp.

Wilson, A. H. (1976). The petrology and structure of the Hartley Complex of the Great Dyke, Rhodesia. D.Phil. thesis, University of Rhodesia. Salisbury.

Wilson, A. H. (1982). The geology of the Great Dyke, Zimbabwe: the ultramafic rocks. *J. Petrol.*, **23**, 240–92.

Wilson, A. H. & Prendergast, M. D. (In press). The Great Dyke of Zimbabwe: I. Tectonic setting, stratigraphy, petrology, structure, emplacement and crystallisation. In *5th Magmatic Sulphides Field Conference, Harare, Zimbabwe*, eds M. D. Prendergast & M. J. Jones. Institution of Mining and Metallurgy, London.

Wilson, A. H. & Wilson, J. F. (1981). The Great Dyke. In *Precambrian of the Southern Hemisphere*, ed. D. R. Hunter. Elsevier, Amsterdam, pp. 572–8.

Wilson, J. F. (1981). The granite-gneiss greenstone shield, Zimbabwe. In *Precambrian of the Southern Hemisphere*, ed. D. R. Hunter. Elsevier, Amsterdam, pp. 454–88.

Wilson, J. F., Jones, D. L. & Kramers, J. D. (1987). Mafic dyke swarms in Zimbabwe. In *Mafic Dyke Swarms*, eds H. C. Halls & W. F. Fahrig. Geological Association of Canada, Special Paper 34, 433–44.

Wilson, J. R. & Engell-Sorensen, O. (1986). Basal reversals in layered intrusions are evidence for emplacement of compositionally-stratified magma. *Nature*, **323**, 616–18.

Worst, B. G. (1960). The Great Dyke of Southern Rhodesia. *Southern Rhodesia Geological Survey Bulletin*, 47, 234 pp.

33

PGE- and Au-Distribution in Rift-related Volcanics, Sediments and Stratabound Cu/Ag Ores of Middle Proterozoic Age in Central SWA/Namibia

G. Borg,[a]* M. Tredoux,[b] K. J. Maiden,[c] J. P. F. Sellschop[b]
& O. F. D. Wayward[a]

[a]Department of Geology, [b]Wits-CSIR Schonland Research Centre for Nuclear Sciences, University of Witwatersrand, Johannesburg, Wits 2050, South Africa
[c]33 Cobran Road, Cheltenham, NSW 2119, Australia

ABSTRACT

A failed continental rift system, consisting of several partly discontinuous grabens, developed during Middle Proterozoic times in SWA/Namibia and Botswana, possibly in response to a mantle plume. The basins contain thick successions of bimodal volcanics, and continental and marine sediments. High-potassium acid volcanics developed as a result of partial melting of mainly sedimentary continental crust. The basal acid volcanics are enriched in PGE and Au which might have been supplied by mantle derived fluids, released during the initial phase of crustal melting and fracturing of the lithosphere. After this anomalous phase of acid volcanism, 'more normal' acid volcanics were extruded. Mantle derived tholeiitic basalts, which were extruded along deep reaching fractures, show typically magmatic PGE and Au distribution patterns. Overlying coarse continental red beds, derived from the denudation of the older basement and the reworking of uplifted and blockfaulted acid volcanics, are generally barren of PGE and Au. Laterally extensive, sediment-hosted, stratabound Cu/Ag mineralization at the base of a dark pyritic shale unit is slightly enriched in Pt (0·04–0·12 g/t), Au (0·04–0·06 g/t), Th and Ni. Generally, the bimodal volcanics display typical tholeiitic chondrite normalised PGE patterns. Altered volcanic rocks, as well as the ore samples, show PGE patterns which may indicate hydrothermal reworking. The authors suggest a model wherein the PGE and Au are mantle derived, possibly from the mantle plume which triggered the development of the continental rift system. Selective remobilisation leached Cu, Zn and Co from the basalts and Pt and Au from the red beds. The metals precipitated at the redox-cline between the pyritic sediments and the red beds, whereas overlying sediments, also pyritic, remained unmineralised.

*Present address: Federal Institute for Geosciences and Natural Resources, PO Box 510153, D-3000 Hannover 51, Federal Republic of Germany.

INTRODUCTION

Regional Geological Setting

A belt of late Middle Proterozoic rocks stretches in an intermittent, arcuate shape from central SWA/Namibia northeast into Botswana and south and southeast into the northern parts of South Africa (Fig. 1). Evidence suggests that these volcano-sedimentary basins belong to a failed continental rift system which developed approximately parallel to the margin of the Kaapvaal Craton (Borg, 1988). This study deals exclusively with the basins which are aligned along the northeastern branch of the Koras-Sinclair-Ghanzi Rift (Borg, 1988). The areas under investigation are the Klein Aub Basin, the Dordabis/Witvlei Basins southeast and east of Windhoek and the Oologsende Porphyry near the SWA/Namibian border to Botswana.

Stratigraphy

These Middle Proterozoic basins generally display similar stratigraphic sequences, which have been described in detail by Borg & Maiden (1986*b*,*c*) and Borg (1988). The rift-fill sequences rest unconformably on a 1600–2000 Ma old basement (SACS, 1980) (Fig. 1), which comprises granitic and granodioritic rocks and highly metamorphosed sediments, with acid volcanics and minor basalts.

The basal unit of the rift-fill sequence, comprising the Nückopf and Grauwater Formations and, as a lateral equivalent, the Oologsende Porphyry, consists mainly of acid volcanics. These formations contain increasing intercalations of coarse clastic sediments towards the top of the unit (Fig. 1). Contemporaneous high level granitic intrusions, e.g. the Gamsberg Granite Suite, were emplaced, locally intruding the rift-fill sequence (SACS, 1980).

Overlying the basal acid volcanic unit is a succession of up to 3000 m of coarse clastic sediments with a varying portion of basalt close to the base (Fig. 1). The sediments of the Doornpoort Formation and equivalents consist of reddish conglomerate, arkose (locally evaporitic), protoquartzite and minor shale derived from the underlying basement and reworked acid volcanics. The sediments have been deposited under oxidising continental conditions during a period of intense, periodic block faulting (Borg & Maiden, 1986*a*). The overall thickness of the tholeiitic volcanics varies in the different basins from a few hundred metres in the Klein Aub area to approximately 2500 m in the vicinity of Dordabis and Witvlei.

Overlying, and locally interfingering with, the continental red beds are fine-clastic, dark grey, pyritic sediments and minor carbonate (Borg & Maiden, 1986*a*,*c*) (Fig. 1). The sediments are of shallow marine origin (Borg *et al.*, in preparation), but possibly locally lacustrine (Ruxton & Clemmey, 1986). The marine transgression occurred during a phase of thermal subsidence (Borg, 1988), after volcanic and tectonic activity had ceased.

The rift succession has formed between 1250 Ma and approximately 950 Ma and is unconformably overlain by sediments of the Late Proterozoic Damara Sequence and the Cambrian Nama Group (SACS, 1980).

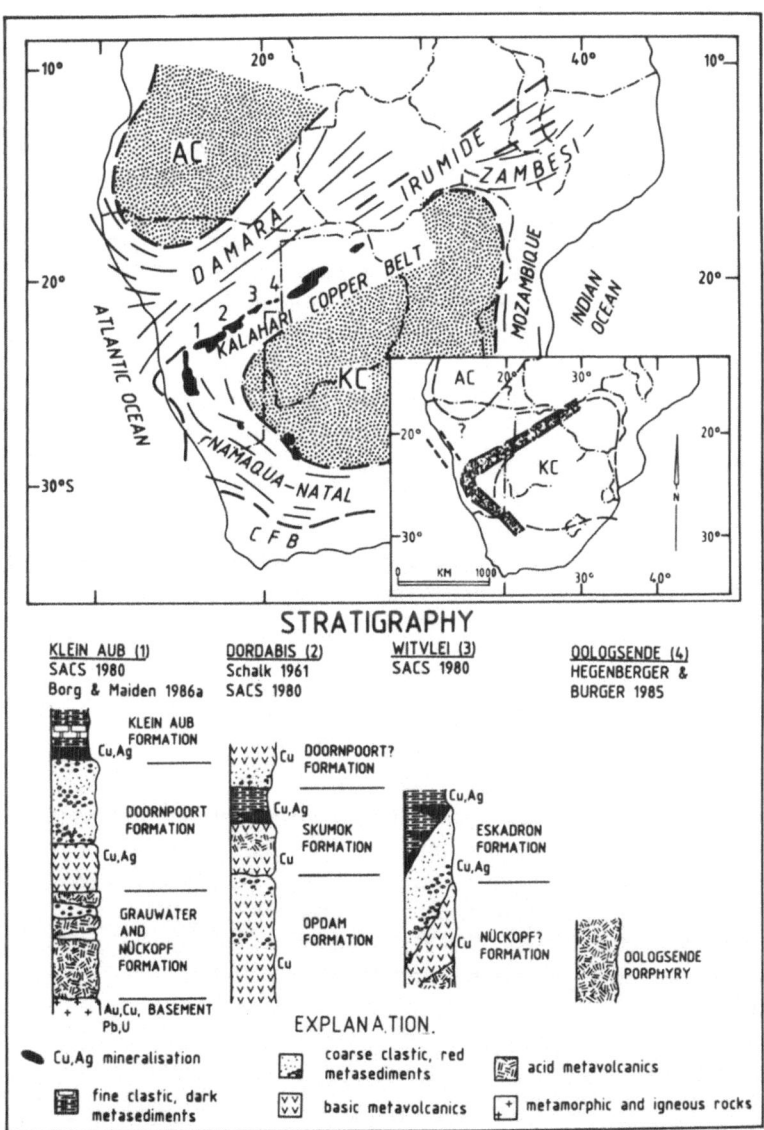

FIG. 1. Location map of Middle Proterozoic basins in southern Africa and their relation to cratonic regions and mobile belts. AC = Angola Craton; KC = Kalahari Craton; CFB = Cape Fold Belt. Insert map shows the orientation of the proposed continental rift system.

The rocks have been thrusted, faulted and moderately folded during the Damaran Orogenesis. Metamorphism of lower greenschist grade (with temperatures not exceeding 350°C) has affected the succession (Ahrendt et al., 1978).

Mineralisation

The >1600 Ma old metamorphic and igneous complexes, which form the local basement to the rift sequence, show numerous small Au, Cu, Zn and Pb occurrences (De Kock, 1934). Au commonly occurs in quartz veins and shear zones. Pb is found in many veins and together with Cu and Zn as massive sulphides (e.g. Kobos; Brewitz, 1974).

The 1300–950 Ma volcano-sedimentary basins are host to laterally extensive, sediment-hosted, stratabound Cu/Ag mineralisation (Ruxton, 1981, 1986; Ruxton & Clemmey, 1986; Borg & Maiden, 1986c). The mineralisation generally occurs at the interface between red beds and dark pyritic sediments and is partly diagenetic, partly epigenetic (Borg & Maiden, 1986b,c): at Klein Aub mine, diagenetic mineralisation has been upgraded during regional metamorphism in the vicinity of a wrench fault system (Borg et al., 1987). The association of Cu/Ag-mineralised dark shale with underlying altered, metal depleted basalts is remarkably consistent throughout the basins.

Unpublished exploration reports mention single occurrences of Au, associated with the stratabound Cu/Ag mineralisation, from exploration bore holes (Borg & Maiden, 1986c), but generally Au and Pt have not been determined during exploration.

GEOCHEMISTRY

Analytical Methods

(a) Neutron activation analyses (NAA): the analytical method used for the determination of the platinum-group elements (PGE) and Au was NiS preconcentration plus NAA, as specified in Davies & Tredoux (1985). Blanks were run with each sample and standard batch. Data are listed in Tables 1A and 1B.

(b) X-ray fluorescence analyses (XRF): the major and trace elements (Tables 1A and 1B) were analysed by using standard XRF facilities at the Geological Survey of South Africa.

Analytical Results

Basement

One sample of gneissic metasediments from a fuchsitic shear zone has been analysed. It contains insignificant amounts of base metals but PGE contents are relatively high with an Au value of 10 ppb (Table 1B and Fig. 2). The chondrite normalised concentrations (using the values given for C1 chondrite by Taylor & McLennan (1985)) display a relatively shallow curve (Fig. 3) with absolute concentrations that are two orders of magnitude higher than concentrations reported from granite analysed by Crocket (1981) and even an order of magnitude higher than sills

TABLE 1A

Major- and trace-element analyses (XRF) and PGE values (NAA) of samples from acid volcanic rocks

	Acid volcanics										Basal acid volcanics				
	NAA 3	NAA 6	NAA 5	NAA 11	NAA 14	NAA 15	NAA 16	NAA 19	NAA 20	NAA 21	NAA 24	NAA 27	NAA 2	NAA 6	NAA 7
Major elements (%)															
SiO₂	74·23	76·43	77·44	67·65	67·44	69·72	72·72	71·50	69·64	67·85	72·06	62·47	77·04	69·42	75·09
TiO₂	0·07	0·05	0·11	0·45	0·33	0·36	0·33	0·28	0·34	0·33	0·27	0·64	0·05	0·40	0·10
Al₂O₃	13·09	11·99	11·02	15·98	13·20	12·29	10·98	10·93	12·28	12·90	11·47	16·52	12·07	13·97	12·18
Fe₂O₃	1·68	1·84	3·41	4·08	6·73	6·89	5·73	6·52	6·75	7·01	5·67	8·59	1·58	4·29	2·35
FeO	0·25	0·25	0·25	1·50	0·25	0·25	1·37	1·18	1·01	2·78	1·10	0·80	0·25	0·25	0·74
MnO	0·14	0·11	0·07	0·11	0·08	0·17	0·11	0·08	0·16	0·11	0·12	0·11	b.d.l	0·15	0·13
MgO	0·05	b.d.l.	b.d.l.	0·83	0·16	0·27	0·20	0·12	0·26	0·29	0·37	1·22	0·24	0·80	0·08
CaO	0·53	0·53	0·48	1·77	0·55	0·35	2·85	2·26	1·94	2·34	1·73	7·08	b.d.l.	2·02	0·28
Na₂O	3·56	7·64	3·13	2·74	4·08	3·25	3·43	2·48	2·53	2·97	2·39	0·40	1·44	5·82	3·25
K₂O	6·41	1·27	4·60	5·37	6·22	5·62	3·08	4·79	5·72	4·98	5·28	1·99	7·05	3·56	6·12
P₂O₅	0·20	0·19	0·21	0·30	0·25	0·26	0·25	0·24	0·35	0·27	0·24	0·33	0·16	0·32	0·17
S	0·00	0·00	0·01	0·00	0·00	0·04	0·01	0·01	0·02	0·01	0·00	0·01	0·00	0·02	0·00
H₂O⁺	0·13	0·15	0·35	0·38	0·14	0·45	0·28	0·14	0·28	0·36	0·26	0·06	0·32	0·12	0·19
H₂O⁻	0·02	0·03	0·03	0·01	0·02	0·02	0·06	0·04	0·03	0·01	0·01	0·01	0·05	0·02	0·01
CO₂	0·34	0·20	0·17	0·12	0·07	0·15	0·18	0·32	0·29	0·75	0·19	4·14	0·07	0·41	0·09
XRF data (ppm)															
Cu	5	13	29	180	13	6	30	-5	44	10	-5	40	12	100	9
Y	66	41	53	54	120	97	88	104	120	115	125	52	83	43	64
Pb	34	28	40	24	25	24	35	33	36	26	34	29	33	30	29
Zn	24	-5	26	49	-5	45	48	-5	37	107	31	-5	53	41	32
Zr	186	183	260	360	615	577	624	608	693	676	693	417	159	301	212
Th	29	26	18	-5	22	18	24	17	23	19	17	27	21	18	25
Co	-5	-5	-5	10	17	17	13	17	18	17	12	-5	-5	9	-5
Rb	274	40	156	164	228	216	116	211	235	211	229	28	349	87	204
Nb	20	23	17	13	41	38	40	37	43	41	44	-5	19	17	27
U	-5	-5	-5	-5	-5	-5	-5	-5	-5	-5	-5	-5	-5	-5	-5
Ni	8	-5	12	-5	10	8	10	6	14	7	9	36	22	13	13
Sr	54	125	135	319	-5	14	152	45	34	34	42	33	5	321	51
NAA data (ppb)															
Os	b.d.l.	b.d.l.	1	b.d.l.	b.d.l.	b.d.l.	b.d.l.	b.d.l.	b.d.l.	b.d.l.	b.d.l.	b.d.l.	b.d.l.	13	b.d.l.
Ir	0·17	0·64	0·41	0·17	0·13	0·12	0·34	0·17	0·90	1·46	0·32	0·33	b.d.l.	9·01	1·49
Ru	<10	8	7	<10	<10	<10	<10	5	<10	<10	<10	<10	b.d.l.	24	<10
Pt	5	15	12	12	17	7	7	<10	24	22	17	10	<10	168	44
Pd	3	7	12	<10	<10	3	5	<10	<10	4	4	5	192	145	121
Au	0·8	3·1	3·2	2·3	2·2	1·4	3·6	1·2	7·1	1·3	6·3	4·2	10·1	9·1	6·3

b.d.l.: Below the lower limit of detection. In case of the NAA data, high background counts occasionally obscured the true detection limit; concentrations were then indicated as <10 ppb.

N.B.: Elements indicated as <10 in plotting the graphs of Fig. 3.

TABLE 1B

Major- and trace-element analyses (XRF) and PGE values (NAA) of samples from basement rocks, sedimentary rocks, basalts and Cu/Ag ores

	Green beds				Ore samples			Red beds				Basalt	Basement		
	GC 8	GC 8A	GC 9	NAA 25	KA 65	KA 52	NAA 13	D1	D2	D3	DP 56	DP 59	NAA 8	NAA 26	NAA 1
Major elements (%)															
SiO_2	71·06	63·23	63·31	50·46	53·74	44·34	52·13	76·91	78·06	78·67	51·14	48·26	47·95	50·31	90·2
TiO_2	0·45	0·55	0·50	0·46	0·34	0·56	0·48	0·37	0·23	0·20	1·15	1·53	1·57	1·40	b.d.
Al_2O_3	14·14	16·72	15·55	18·20	16·34	19·98	17·79	9·20	8·68	9·92	16·37	15·06	15·28	14·11	3·7
Fe_2O_3	5·36	8·92	6·40	6·82	6·94	9·26	8·52	3·95	3·29	2·76	10·90	15·56	16·04	14·27	1·5
FeO	3·26	3·53	5·38	0·80	5·56	7·76	5·41	b.d.l.	b.d.l.	b.d.l.	0·35	0·79	0·25	5·17	0·2
MnO	0·17	0·22	0·31	0·22	0·26	0·18	0·22	0·14	0·13	0·15	0·21	0·31	0·19	0·19	0·(
MgO	1·59	2·71	3·32	3·64	3·64	4·42	4·26	2·29	0·42	0·20	2·39	7·92	1·68	5·62	0·1
CaO	1·03	1·23	3·25	2·18	1·56	1·08	3	2·29	2·88	0·41	16·45	6·71	16·05	9·55	0·1
Na_2O	1·99	1·88	3·69	3·14	4·77	3·33	7·10	1·15	2·05	2·98	0·77	3·24	0·18	3·39	0·2
K_2O	3·68	4·27	1·88	13·20	2·73	5·22	4·94	5·14	3·19	4·59	0·10	b.d.l.	0·10	0·35	2·(
P_2O_5	0·40	0·40	0·45	0·70	0·66	0·64	0·64	0·27	0·26	0·22	0·44	0·45	0·44	0·31	0·1
S	n.d.a.	n.d.a.	n.d.a.	1·27	n.d.a.	n.d.a.	1·64	n.d.a.	n.d.a.	n.d.a.	n.d.a.	n.d.a.	0·01	0·01	0·(
H_2O^+	1·74	2·41	2·29	1·93	2·52	2·86	2·20	0·76	0·40	0·38	1·05	2·82	0·59	1·02	0·1
H_2O^-	0·05	0·09	b.d.l.	0·04	0·07	0·22	0·01	b.d.l.	b.d.l.	b.d.l.	0·20	b.d.l.	0·02	0·03	0·0
CO_2	0·25	0·27	1·87	1·40	0·69	0·13	1·68	2	1·93	0·07	0·11	0·11	0·48	0·41	0·1
XRF data (ppm)															
Cu	15	174	102	20 616	25 000	30 000	33 275	b.d.l.	50	40	780	342	16	321	27
Y	51	59	46	57	44	57	44	49	31	40	19	21	19	26	19
Pb	16	29	61	70	16	695	20	23	20	25	13	12	16	17	39
Zn	78	135	100	32	93	87	100	31	8	12	16	105	13	118	195
Zr	230	272	283	203	155	155	157	295	165	213	98	127	106	130	40
Th	15	16	16	15	19	42	22	10	8	16	11	7	-5	-5	-5
Co	25	44	30	28	35	43	35	13	11	9	42	93	63	75	-5
Rb	129	156	70	220	86	152	72	167	97	130	b.d.l.	b.d.l.	-5	12	96
Nb	13	13	14	17	10	14	13	10	b.d.l.	13	b.d.l.	b.d.l.	6	8	-5
U	b.d.l.	6	10	-5	14	0	-5	b.d.l.	b.d.l.	b.d.l.	b.d.l.	b.d.l.	-5	-5	-5
Ni	27	40	30	44	52	78	67	14	7	b.d.l.	60	184	179	118	21
Sr	54	76	122	-5	19	21	38	92	107	79	3 494	1 300	1 346	260	-5
NAA data (ppb)															
Os	b.d.l.	b.d.l.	b.d.l.	b.d.l.	b.d.l.	b.d.l.	b.d.l.	b.d.l.	b.d.l.	b.d.l.	b.d.l.	b.d.l.	4	2	b.d.l.
Ir	0·44	0·04	0·20	1·49	0·15	0·23	1·03	0·09	0·06	0·07	0·23	0·44	4·14	2·88	1·6
Ru	n.d.a.	n.d.a.	n.d.a.	<10	n.d.a.	n.d.a.	16	n.d.a.	n.d.a.	n.d.a.		<10	14	11	14
Pt	14	b.d.l.	7	57	47	34	120	3	b.d.l.	b.d.l.	55	26	134	117	52
Pd	6	b.d.l.	5	18	8	<10	n.d.a.	b.d.l.	b.d.l.	b.d.l.	5	8	37	45	19
Au	0·9	0·4	0·9	2·9	46	36	7·2	0·3	0·6	0·4	0·5	0·5	8·7	14·3	9·8

n.d.a.: No data currently available.

b.d.l.: Below the lower limit of detection. In case of the NAA data, high background counts occasionally obscured the true detection limi concentrations were then indicated as <10 ppb.

N.B.: Elements indicated as <10 in plotting the graphs of Fig. 3.

Two additional basalts were analysed for PGE only. The values are:

	Os	Ir	Ru	Pt	Pd
DP51	b.d.l.	0·28	<10	19	4
DP54	b.d.l.	0·20	n.d.a.	8	4

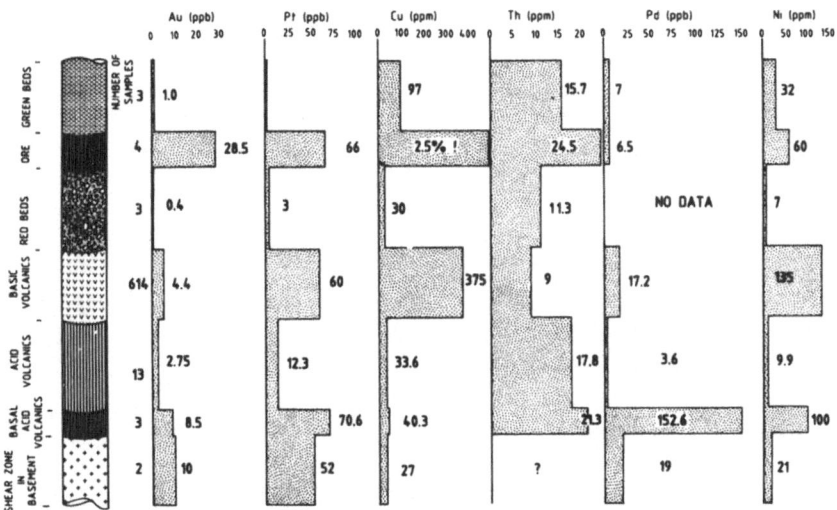

FIG. 2. Simplified stratigraphic column and distribution of selected elements in the different formations.

from the Bushveld Complex (Davies & Tredoux, 1985). Since presently no other analyses from these basement rocks are available it can not be determined whether the high values are related to the shear zone or represent generally high PGE and Au contents of the local basement.

Basal Acid Volcanics

Samples from the base of the acid volcanic pile, in the western part of the Klein Aub Basin, differ distinctly in their chemical composition from those higher up in the succession (Table 1A, Figs 2,3). They are corundum normative, more potassic and have lower Sr contents than the overlying acid extrusive rocks. The chondrite normalised concentrations show a pattern with Pd values markedly higher than Pt values, a trend which distinguishes these samples from most other samples in the suite. Sample NAA 6 from this unit shows mineralogical and chemical signs of alteration. The samples are considerably enriched in elements such as Au, PGE, Ni, Th and Rb, compared to the acid volcanics higher in the sequence.

Acid Volcanics

Samples taken from the other acid volcanics can be divided into three groups, depending on the sample location: (1) Samples NAA 3, 4, 5 and 11 overlie the basal acid volcanics from the western part of the Klein Aub Basin. Compared to the latter, they show decreasing Rb concentrations and K_2O/Na_2O ratios but increasing Sr contents. Au, Ni, Th and Rb contents are lower than in the basal acid volcanics and represent more normal values for this type of volcanic rocks. The PGE are still nearly an order of magnitude higher than would be expected. (2) Samples NAA 14, 15, 16, 19, 20, 21 and 24 are from the Oologsende Porphyry (Fig.

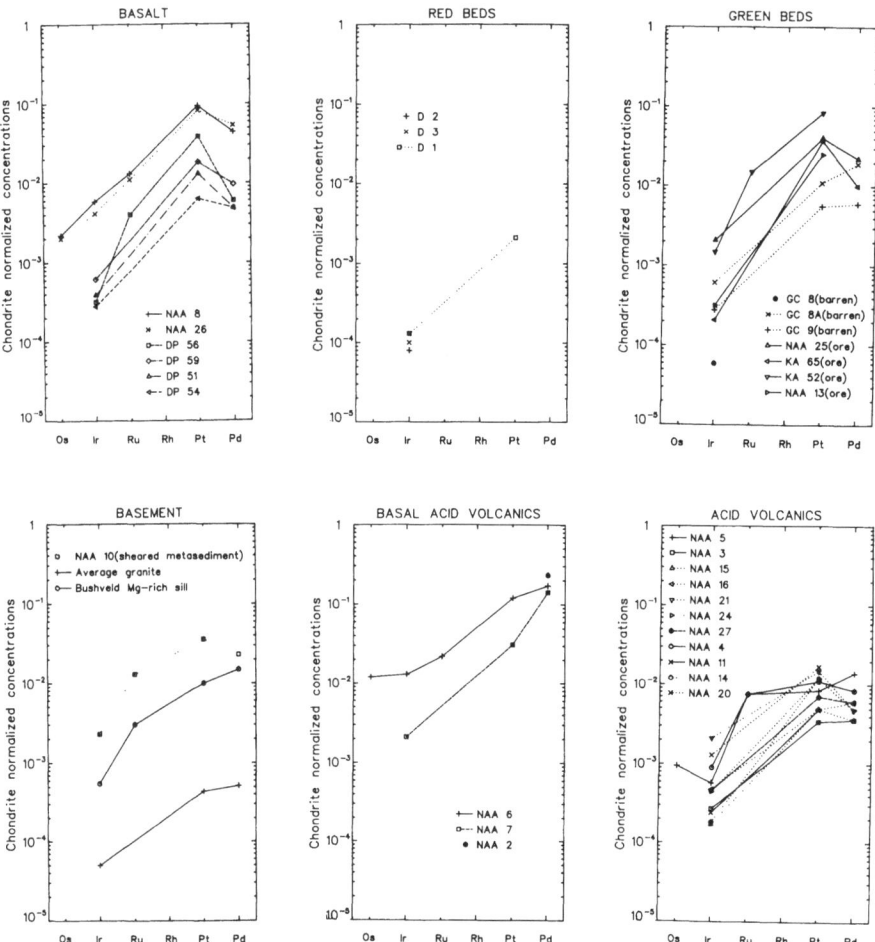

Fig. 3. Chondrite normalised concentrations of PGE in the different rock types from the rift sequence. C1 chondrite values used for normalisation are those of Taylor and McLennan (1985). Note that no points are plotted for elements indicated as <10 in Table 1. As Ru data are sparse (and Rh data not available yet for this data set), the curves are often extrapolated from Ir to Pt (or Pd), as an estimation of the full PGE patterns.

1) and show Rb, Sr values and K_2O/Na_2O ratios similar to those of the basal acid volcanics from the Klein Aub Basin, but with PGE patterns more like those of group (1). This group seems to have a slightly higher Pt/Pd ratio relative to the other acid volcanics. (3) Sample NAA 27, from the Witvlei area, appears to be altered and shows relatively erratic Sr, Rb and K_2O contents. Its PGE pattern is subparallel to those of group (1) above. The chondrite normalised PGE concentrations of all three groups have a constant positive trend, with a spread in absolute values of up to one order of magnitude (e.g. Ir in Fig. 3), and generally higher Pt/Pd ratios than the basal volcanics.

Basalt

The chondrite normalised patterns of the basic volcanics (Fig. 3) all have steep positive trends typical of some tholeiitic magmas. Two groups of basalts can be distinguished: Samples DP 51, 54, 56 and 59 which show concentrations nearly one order of magnitude lower than those of samples NAA 8 and 26 but follow a similar trend. The former are from the central part of the Klein Aub Basin and the latter come from the eastern part of the same basin and from the Witvlei Basin. The patterns of these rocks are very similar to those of the basaltic sills of the Bushveld Complex (Davies & Tredoux, 1985), although the absolute concentrations of the basalts NAA 8 and 26 are 3–10 times higher. On a worldwide basis, however, tholeiites tend to have Pd >Pt in their PGE trends. It is not known whether the positive Pt anomaly of these basalts were an original feature of the magmas, or whether it developed due to the relatively higher mobility of Pd during subsequent leaching.

Red Beds

The oxidised, coarse clastic red beds from the Klein Aub Basin contain extremely low concentrations of PGE (Pt = 3 ppb, Ir = 0·08 ppb) and Au (0·4 ppb). The relatively high Th values (average 11·3 ppm) are probably due to zircons contained in the granitic basement clasts.

Green Beds

Three samples of unmineralised, pyritic, fine sediments from the Klein Aub Basin contain PGE and Au concentrations far below the contents of ores hosted by the same sediments. However, PGE and Au contents of the unmineralised, pyritic sediments are distinctly higher than those of the red beds (Figs 2, 3). The Ir concentrations are higher than average marine sediments (Crocket & Kuo, 1979).

Stratabound Sediment-hosted Cu-ore

The PGE (with the exception of Pd), Au, Th and Ni are enriched in the Cu/Ag ores from Klein Aub Mine and Eskadron, an exploration shaft. The mineralisation occurs at the interface between the underlying, oxidised red beds and dark, pyritic sediments. The ores from Klein Aub Mine contain an average of 2% Cu and 7 g/t Ag with high grade zones of up to 8% Cu and 170 g/t Ag. To date, no Pt and

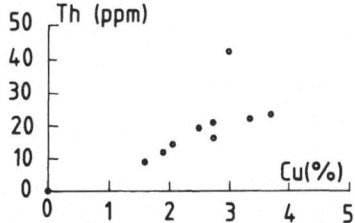

FIG. 4. A plot of the correlation between Cu and Th in ore samples from Klein Aub mine, SWA/Namibia.

only subordinate Au have been recovered from Klein Aub ores, but both elements occur as a significant potential by-product in the ore. Pt values are between 0·036 and 0·12 g/t and Au values are as high as 0·066 g/t. A sample of chalcocite concentrate from Klein Aub Mine shows high concentrations of Pt (155 ppb) and Au (118 ppb), Pd concentrations are relatively low, compared with unmineralised samples (Fig. 2).

Ore samples show a positive correlation between Cu and Th (Fig. 4). The ores from these localities are very similar with regard to their chemical composition (Table 1B).

INTERPRETATION

The high K_2O/Na_2O ratio (2·2) of the acid volcanics, including the basal acid volcanics, indicates that they originated from partial melting of a lower continental crust of mainly metasedimentary composition. The evolution of the magma is documented by both the decreasing Rb and increasing Sr contents, with stratigraphic height, from the basal acid volcanics to the higher parts of the acid volcanic pile.

Although the exact stratigraphic position of the Oologsende Porphyry is uncertain, since neither an upper nor lower geological contact is exposed, Rb and Sr contents are similar to those of the basal acid volcanics in the Klein Aub Basin. This might imply that the Oologsende Porphyry also represents an early product of partial crustal melting.

Mantle derived fluids (or gases?) might have contaminated the basal acid volcanics, locally introducing Pt, Pd and Ni (e.g. NAA 6). This event might document a phase of high heat flow, initial crustal melting and initial fracturing of the continental crust, caused by an underlying mantle plume. The effects of fluids, derived from the mantle plume, would show most strongly in the first melts and decrease with continuing crustal melting and more fully developed acid volcanic activity. Overall concentrations of PGE and Au in the acid volcanics (and in the basement shear sample) are comparable to concentrations of the overlying basalts and thus unusually high for such rock types.

The tholeiitic, mantle derived basalts were probably extruded along deeply penetrating faults (Borg & Maiden, 1986*b*). PGE (except perhaps Pd) in these rocks have remained immobile during diagenesis and metamorphism, although locally the basalts were strongly altered. Base metals (Cu, Zn and Co) and other constituents such as MgO and NaO have been leached strongly during both events (Borg & Maiden, 1986*b*). The immobility of the PGE might be due to the refractory nature of PGE minerals and their association with phases like chromite, which were unaffected by the alteration. The composition of the altering fluids must have resulted in the leaching of a very limited suite of elements, but the present data are insufficient to allow an in-depth discussion of the fluid composition.

The low PGE and Au concentrations in the red beds are rather surprising considering the source of the clasts, which were derived from basement rocks and reworked material of the acid volcanics. Both of these rock suites contain PGE and

Au in considerably higher amounts than found in the red beds. It can be assumed that the red beds have lost most of their PGE and Au at some stage, either during erosion, transport, diagenesis or metamorphism. These elements subsequently most likely remained within the closed system of relatively narrow, fault-bounded rift grabens.

The highest concentrations of Pt, Au, Th and Ni occur in stratabound sediment-hosted Cu/Ag ores at the redox-cline interface between red beds and pyritic sediments. The Cu/Ag mineralisation (and probably the associated concentrations of PGE and Au) is of diagenetic and partly epigenetic (syn-metamorphic) origin (Borg & Maiden, 1986*b*,*c*). Although the higher parts of this unit also contain pyrite and do not differ lithologically from the basal part, ore concentrations are restricted to the base. The sulphur supply (from pyrite), the chemically reducing environment (and perhaps organic carbon), probably caused the precipitation of Cu, Ag, Au, Pt, Ni and Th at the base of the dark pyritic sediments. The basal enrichment suggests a general direction of migration for the mineralising fluids from below, approximately normal to the red bed/dark pyritic sediment interface.

A MODEL

Sources of PGE and Au

The model favoured by the authors relates the origin of PGE and perhaps Au, to the underlying mantle plume and process of continental rifting. The initial rift phase was probably characterised by partial crustal melting, fracturing of the continental lithosphere and high heat flow. Mantle derived fluids (or gases), released during this first phase of fracturing and expelled contemporaneously together with extrusion of the basal acid volcanics, might have caused the PGE and Au enrichment of these lavas, locally altering them to anomalous chemical compositions (e.g. NAA 6). Such a model might explain the typically magmatic PGE chondrite normalised patterns, which would have been unlikely to originate from partial crustal melting. After the release of an initial 'pulse' of contaminating fluids (gases), the chemical character of the later acid volcanics reflect 'more normal' composition for partial crustal melts. PGE concentrations are lower than in the basal volcanics, but still relatively high for such rock types, perhaps indicating a prolonged, although diminished influence of the mantle plume. The mantle plume also supplied the tholeiitic, within-plate basalts with their typical magmatic PGE distribution patterns similar to those of the acid volcanics.

The model of an elongated mantle plume underlying the rift might explain the similar chemical character and PGE and Au contents of all the volcanic rocks along strike, although the samples come from localities up to 400 km apart.

Ore Genesis
Placer Origin

The Cu/Ag mineralisation at the base of the green beds has been described as having a syndepositional, modified placer origin (Ruxton, 1981, 1986). Since most of the mineralisation is controlled by zones of either diagenetic or tectonic permeability (Borg & Maiden, 1986c) this would seem improbable. Data for all six PGE in placer deposits are scarce. Although some placers (those derived from 'Alaskan type' ultramafics) are dominated by Pt alloys (Cabri, 1981); Os–Ir–Ru alloys and minerals are always present and very often constitute the major component (Cabri, 1981). Pd usually is the least abundant of the PGE in placers. PGE patterns of such deposits can be expected to display a negative slope from Os (Ir) to Pd, with a positive Pt anomaly. The PGE trends of the Cu/Ag mineralised zones do show a positive Pt anomaly, but also a positive slope from Ir (Os could not be detected in any) to Pd. Such patterns do not support a placer origin for the PGE in these ores.

Diagenetic and/or Epigenetic Origin

Redistribution of base metals (Cu, Zn and Co), PGE and Au has occurred in the different volcanic and sedimentary formations of the rift system. Leaching mechanisms were probably active both during basin dewatering as a result of sediment compaction and during metamorphism (compare Borg & Maiden, 1986c). Based on work by Borg & Maiden (1986b) and the present study, the following ore genesis model is proposed.

Percolating fluids selectively leached Cu, Zn, Co and maybe some Au, from permeable zones of the tholeiitic basalts while PGE (except Pd?) remained immobile. This accords with previous work (e.g. Keays *et al.*, 1982) which indicates that the PGE are not very mobile during alteration of mafic rocks. The base metal bearing fluids migrated through the overlying red beds where oxidising conditions and the absence of sulphur prevented any precipitation of these elements. PGE, Au and minor amounts of base metals (Ni, Co) were leached from the red beds, during erosion of the source rocks or diagenesis and/or metamorphism, and added to the circulating fluids, which might have been Cl-rich basinal brines (evaporites in red beds!) at low temperatures, low pH and high *Eh*. The assumption, implicit in the model presented here, that sedimentary strata might be more readily stripped of PGE than volcanic rocks, is probably a reasonable one. It must be noted that the extent to which the PGE can be mobilised in low temperature processes is currently the subject of serious debate (e.g. Bowles, 1986), and that very little is known about the nature of the mobilising fluids.

The fluids, enriched in base metals, Au and presumably the PGE, migrated upwards until they encountered the dark, pyritic (reduced) sediments of the Klein Aub Formation and equivalents. The sulphur (supplied by abundant diagenetic pyrite) and the low *Eh* caused the precipitation of Cu, Ag, Au, Pt, Th and minor Ni. Zn, Co and, apparently, Pd were either not precipitated from the fluid to the same extent or, alternatively, leached from the ore at a later stage. The authors favour the first option since the ore textures do not indicate any remobilisation of Zn or Co sulphides. The mineralised samples have higher Pt/Ir ratios than the basalts and acid volcanics (see Fig. 3). This would support the theory (Keays *et al.*, 1982) that Pt (and Pd) is more readily redistributed than Ir.

Presently it can not be decided whether the more impermeable character of the host rock (cap rock) or its function as a chemical trap prevented the mineralisation of the overlying pyritic (potential host-) rock. Textural studies (Borg & Maiden, 1986c) have shown that a portion of the mineralisation was emplaced during diagenesis, prior to destruction of sedimentary permeability. During metamorphism, the diagenetic mineralisation has been upgraded in zones of tectonically induced permeability. The present data base is too small to decide whether the emplacement of PGE and Au should be attributed to one or both Cu-mineralising events.

CONCLUSIONS

Samples from Middle Proterozoic rift-related (volcanic and sedimentary) formations in SWA/Namibia have revealed consistently similar distribution patterns of PGE and Au over 400 km strike length. PGE and Au were enriched during the initial pulse of potassic, acid volcanism of the Nückopf Formation and equivalents, which was derived by partial melting of (mainly sedimentary) continental crust. Mantle derived tholeiitic basalts of the Doornpoort formation and equivalents were extruded and subsequently covered by coarse clastic red beds, supplied by the denudation of basement rocks and reworked acid volcanics. Although the source rocks contain considerable amounts of PGE and Au, the red beds are nearly barren of PGE and Au. Sediment-hosted stratabound Cu-Ag mineralisation, associated with previously unrecognised contents of Pt and Au, occurs at the base of the overlying dark pyritic (reduced) sediments. The pyritic sediments further up show only low concentrations of PGE and Au.

It is suggested that PGE and Au in the volcanic rocks were supplied from a mantle plume associated with the formation of the continental rift. Base metals (Cu, Zn, Co) were leached from permeable, altered zones of the basaltic flows, whereas Pt, Au and probably more Cu were derived from the red beds. Most PGE in the basalts remained immobile. The metals were transported in acid solutions with a high *Eh*. A reversal in redox conditions and an abundance of sulphur caused the precipitation of metals at the base of the pyritic sediments, producing Cu/Ag ores with up to 0·12 g/t Pt and 0·66 g/t Au. A multi-phase model of metal emplacement is envisaged with an early phase of precipitation during diagenesis and the upgrading of this mineralisation in zones of tectonically induced permeability during deformation and regional metamorphism.

ACKNOWLEDGEMENTS

This study forms part of a research project funded by the Geological Survey of SWA/Namibia on regional controls on the localisation of stratabound copper deposits in SWA/Namibia and Botswana. The work has benefited from discussions with T. S. McCarthy, R. G. Cawthorn and G. N. Phillips, Department of Geology, and M. J. de Wit, Bernard Price Institute for Geophysics, University of the

Witwatersrand. We also thank J. F. W. Bowles and an unknown reviewer for their valuable comments. The Geological Survey of South Africa provided some of the geochemical analyses.

REFERENCES

Ahrendt, H., Hunziker, J. C. & Weber, K. (1978). Age and degree of metamorphism and time of nappe emplacement along the southern margin of the Damara Orogen/Namibia (SW-Africa). *Geol. Rundsch.*, **67**, 716–42.

Borg, G. (1988) The Koras-Sinclair-Ghanzi Rift in southern Africa – volcanism, sedimentation, age relationships and geophysical signature of a late Middle Proterozoic rift system. *PreCambrian Res.*, **38**, 75–90.

Borg, G. & Maiden, K. J. (1986a). A preliminary appraisal of the tectonic and sediment-ological environment of the Sinclair Sequence in the Klein Aub Area, SWA/Namibia. *Communs geol. Surv. S. W. Africa/Namibia*, **2**, 65–73.

Borg, G. & Maiden, K. J. (1986b). Stratabound copper-silver-gold mineralization of Late Proterozoic age along the margin of the Kalahari Craton in SWA/Namibia and Botswana. *Can. Mineral.*, **24**, 178.

Borg, G. & Maiden, K. J. (1986c). The Kalahari Copper Belt of SWA/Namibia and Botswana. *Geol. Ass. Canada, Spec. Paper* (in press).

Borg, G. & Maiden, K. J. (1987). Alteration of late Middle Proterozoic volcanics and its relation to stratabound copper-silver-gold mineralization along the margin of the Kalahari Craton in SWA/Namibia and Botswana. In *Geochemistry and Mineralization of Proterozoic Volcanic Suites*, ed. T. C. Pharaoh, R. D. Beckinsale & D. T. Rickard. Geol. Soc. Spec. Publ., pp. 347–54.

Borg, G., Graf, N. & Maiden, K. J. (1987). The Klein Aub Fault Zone—a wrench fault system in Middle Proterozoic Metasediments in central SWA/Namibia. *Communs. geol. Surv. S. W. Africa/Namibia*, **3**.

Borg, G., Stanistreet, I. G. & Maiden, K. J. (in preparation). Evidence of Middle Proterozoic rift-related fluvial and marine sedimentation, Sinclair Sequence, Central SWA/Namibia.

Bowles, J. F. W. (1986). The development of platinum-group minerals in laterites. *Econ. Geol.*, **81**, 1278–85.

Brewitz, H. W. (1974). Montangeologische Erkundung und Genese der metamorphen, exhalativsedimentären Zn-Cu Lagerstätte Kobos im Altkristallin des Nauchas Hoch-landes, Südwest-Afrika. *Clausth. geol. Abh.*, **18**, 1–128.

Cabri, L. J. (1981). Relationships of mineralogy to the reocvery of platinum-group elements from ores. In *Platinum-Group Elements: Mineralogy, Geology, Recovery*, ed. L. J. Cabri. CIM Special Volume, 23, pp. 233–50.

Crocket, J. H. (1981). Geochemistry of the platinum-group elements. In *Platinum-Group Elements: Mineralogy, Geology, Recovery*, ed. L. J. Cabri. CIM Special Volume 23, pp. 47–64.

Crocket, J. H. & Kuo, H. Y. (1979). Sources for gold, palladium and iridium in deep sea sediments. *Geochim. Cosmochim. Acta*, **43**, 831–42.

Davies, G. & Tredoux, M. (1985). The platinum-group element and gold contents of the marginal rocks and sills of the Bushveld Complex. *Econ. Geol.*, **80**, 838–48.

De Kock, W. P. (1934). The geology of the western Rehoboth. *Mem. Dep. Mines, S. W. Afr.*, **1**, 148 pp.

Keays, R. R., Nickel, E. H., Groves, D. I. & McGoldrick, P. J. (1982). Iridium and palladium as discriminants of volcanic-exhalative, hydrothermal and magmatic sulphide mineralization. *Econ. Geol.*, **77**, 1535–47.

Ruxton, P. A. (1981). The sedimentology and diagenesis of copper-bearing rocks on the southern margin of the Damara orogenic belt, Namibia and Botswana, Ph.D. thesis, University of Leeds, 241 p. (unpublished).

Ruxton, P. A. (1986). Sedimentology, isotopic signature and ore genesis of the Klein Aub Copper Mine, South West Africa/Namibia. In *Mineral Deposits of Southern Africa, II,* eds C. R. Anhaeusser & S. Maske. Geological Society of South Africa, pp. 1725–38.

Ruxton, P. A. & Clemmey, H. (1986). Late Proterozoic stratabound red bed-copper deposits of the Witvlei area, South West Africa/Namibia. In *Mineral Deposits of Southern Africa, II,* eds C. R. Anhaeusser & S. Maske. Geological Society of South Africa, pp. 1739–54.

South African Committee for Stratigraphy (SACS) (1980). Stratigraphy of South Africa. Part 1 (Comp L. E. Kent). Lithostratigraphy of the Republic of South Africa, South West Africa/Namibia, and the Republics of Bophuthatswana, Transkei and Venda. *Handb. geol. Surv. S. Afr.,* **8.**

Taylor, S. R. & McLennan, S. M. (1985). *The Continental Crust: its Composition and Evolution: an Examination of the Geologic Record Preserved in Sedimentary Rocks.* Blackwell, Oxford, 385 pp.

34

PGE in the 3·5 Ga Jamestown Ophiolite Complex, Barberton Greenstone Belt, with Implications for PGE Distribution in Simatic Lithosphere

MAARTEN J. DE WIT[a] & MARIAN TREDOUX[b]

[a]BPI Geophysical Research and [b]Schonland Research Centre for Nuclear Sciences, School of Earth Sciences, University of the Witwatersrand, Wits 2050, Johannesburg, South Africa

ABSTRACT

The 3·5 Ga mafic-ultramafic rocks of the Barberton greenstone belt, South Africa, form a pseudostratigraphy comparable to that of Phanerozoic ophiolites. This Archaean complex, referred to here as the Jamestown Ophiolite Complex, consists of high temperature tectonometamorphic peridotites overlain by an intrusive-extrusive igneous section, which in turn is capped by a chert-shale sequence. Elevated PGE concentrations with distinctly different chondrite normalized patterns have been found at three pseudostratigraphic levels of this ophiolite. From the top down, high PGE values are first encountered in an undeformed chromitite, interlayered with meta-pyroxenites. The PGE abundances and normalized patterns are similar to those reported from PGE mineralized zones in large mafic layered complexes. Chromite-magnetite pods occur stratigraphically below the chromitite. Their PGE patterns resemble those reported from chromite pods in Phanerozoic ophiolites and those of the overlying chromitite. The host rocks of the pods (deformed talc-carbonate rocks) are interpreted as a mixture of deformed high-temperature cumulates and depleted upper mantle residue, within a ductile shear zone. In all the above occurrences, the PGE were concentrated by igneous upper mantle-crustal fractionation processes of mantle derived liquids. The third type of PGE concentration is found in an unusual, sulphur-poor NiFe-silicate-spinel pod (called the Bon Accord-[BA]-deposit) and an associated hematite-magnetite-chromite (HMC) pod, both of which occur in coarse-grained meta-dunites at the lowest levels in the ophiolite. These host rocks are interpreted as high-temperature tectonites (strongly depleted in REE); they probably represent depleted upper mantle residue from which the overlying rocks were derived by partial melting. BA can be divided into two groups. Group A has flat, chondrite-like PGE patterns. Group B has a lower concentration of all the PGE, but the relative depletion is most severe for Os and Ir (which are the most refractory); thus the patterns have a positive slope. Both BA groups have negative Pt anomalies in their PGE patterns (more pronounced for Group A), while the HMC-pod, which is depleted in the other PGE, relative to BA, has a strong positive Pt anomaly. It is difficult to explain these fractionations in terms of known upper mantle-crustal processes, especially in the absence of sulphur. The PGE differentiation of BA is interpreted as an indication of chemical fractionation which started in the lower mantle. We conclude that the body represents a metal-silicate sample derived from an almost chondrite-like lower mantle where it resided as a left-over from inefficient core formation. This siderophile-enriched mantle fractionated during ascent and some of its residue was incorporated into the lower portion of the lithosphere during formation of the Archaean oceanic crust. Tectonic and chemical data suggest that the thick and old 'lithospheric keel' beneath the South African craton has a large component of Archaean residual sima and is therefore likely to be relatively enriched in PGE.

INTRODUCTION

Early Archaean greenstone belts (>3·0 Ga) are commonly assumed to contain remnants of ancient oceanic crust (see Windley, 1976, 1984; Condie, 1981 for overviews). Therefore, the mafic and ultramafic units, with their associated cherts, have been implicitly equated to Phanerozoic ophiolite complexes. Ophiolites, however, are very complex sequences of mafic-ultramafic igneous and tectono-metamorphic units (Coleman, 1977; Gass *et al.*, 1984; Karson, 1984; Nicolas, 1986 for overviews), whereas, with exceptions, the mafic-ultramafic units of the greenstone belts have been reported as geologically simple (Viljoen & Viljoen, 1969; Emslie, 1972; Naldrett, 1972; Anhaeusser, 1973, 1983; Condie, 1981). This difference has raised doubts about the validity of equating Archaean greenstone sequences with ophiolites (McCall, 1981; Nisbet, 1982).

The 3·3–3·5 Ga Archaean Barberton greenstone belt (see Fig. 1) offers an excellent opportunity to examine the planet's oldest (and very well-preserved) record of mafic-ultramafic rock associations. These rocks have been interpreted as part of a thick (~17 km) continuous volcanic sequence (the Onverwacht Group; Viljoen & Viljoen, 1969; Anhaeusser, 1983 for overview), dominated by ultramafic lavas in a lower unit (Sandspruit, Theespruit and Komati Formations; now known as the Tjakastad Subgroup, SACS, 1980), and mafic to felsic lavas in an upper unit (Hooggenoeg, Kromberg and Swartkoppie Formations now known as the Geluk Subgroup; SACS, 1980) respectively.

More recent work, however, has emphasized extreme complexities within and between the above mentioned formations (Williams & Furnell, 1979; de Wit, 1983; de Wit *et al.*, 1987a,b; Armstrong *et al.*, 1988). Specifically, de Wit *et al.* (1987a) have documented the intrusive nature of mafic-ultramafic units from the Komati and Kromberg Formations into overlying tholeiitic pillow lavas and sediments. They have also presented evidence of multiple intrusive events within the igneous portions, including cross-cutting intrusives, multiple injection of magma in the Komati section, and sheeted intrusions in the Kromberg section. In addition, they have shown that associated mafic-ultramafic complexes, which were previously interpreted as layered intrusions (Viljoen & Viljoen, 1969; Anhaeusser, 1976, 1985), are, at least in some cases, peridotitic tectonites, similar to alpine peridotites

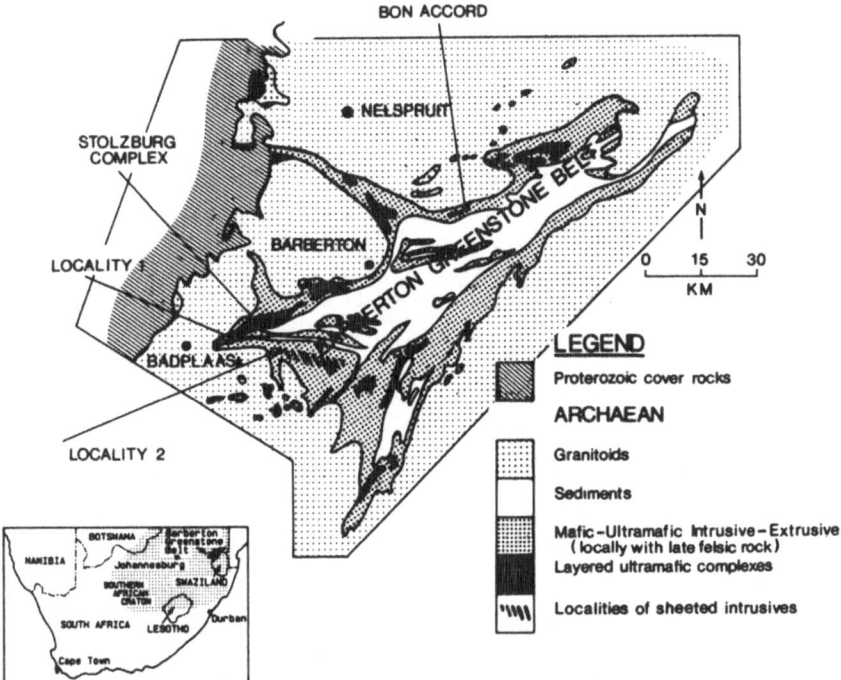

FIG. 1. Schematic diagram and location of the Barberton greenstone belt, South Africa, showing the location of the Bon Accord [BA] occurrence, and the location of the Stolzburg Complex within the mafic-ultramafic sequences of the Onverwacht Group. For a more complete distribution of the simatic rocks of this greenstone belt see maps in Anhaeusser (1983) and de Wit *et al.* (1987a).

which form the upper mantle base of the oceanic lithosphere (cf. Dewey & Kidd, 1977; Christensen & Salisbury, 1979; Girardeau & Nicolas, 1981; Casey *et al.*, 1983; Nicolas, 1986).

The observations of de Wit *et al.* (1987a,b) undermine the simplicity of the widely proposed simple 'layer cake' stratigraphy of the igneous rocks of this greenstone belt: their work reveals a complex intrusive-extrusive simatic section overlying a high-temperature tectonometamorphic base, and capped by a chert sequence which it locally intrudes. Thus, when reconstructed, an integral section through the greenstones of the Barberton belt consists of a lower (REE depleted) peridotitic tectonite zone, overlain by a zone which consists of a complex array of magma chambers and conduits, which in turn intrude and are covered by a substantial carapace of pillow lavas and thin cherts. Such a sequence can be classified as an ophiolite *senso stricto* according to the 'Steinmann Trinity' and the Penrose Conference definition in which ophiolites conform to a specific pseudostratigraphy (Anon., 1972).

The REE and oxygen isotope profiles across the Barberton ophiolitic pseudo-stratigraphy are similar to those of some Phanerozoic ophiolites (Fig. 2; Hoffman *et al.*, 1986; de Wit *et al.*, 1987a). Geochemical data on the hydrothermal meta-

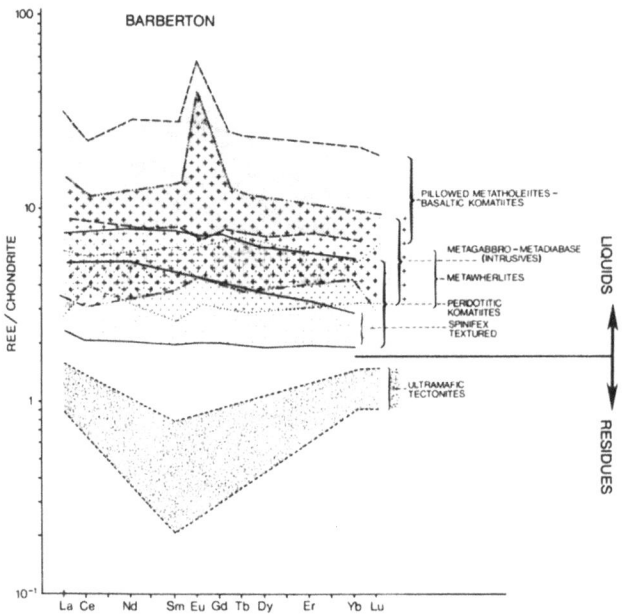

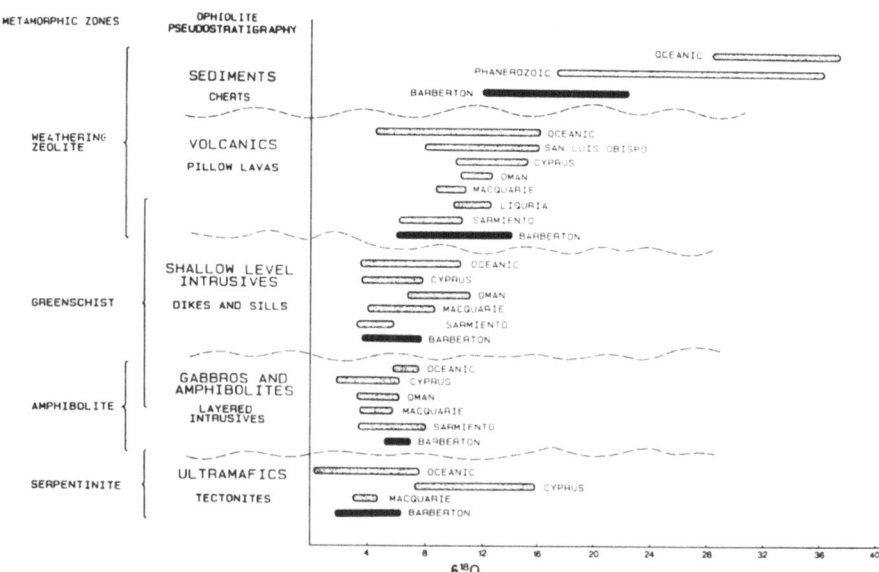

FIG. 2. (a) REE chondrite-normalized summary plot of the mafic-ultramafic rocks of the Onverwacht Group, plotted in their relative pseudostratigraphic position, as reconstructed as the Jamestown Ophiolite Complex. This plot is similar to that of some Phanerozoic Ophiolites (from de Wit *et al.* 1987*a*). (b) Schematic representation of the relationship between $\delta^{18}O$ and the pseudostratigraphy oceanic crust, of Phanerozoic ophiolites and the Jamestown Ophiolite. The overall configuration of the $\delta^{18}O$ of the Barberton rocks, when reconstructed as the Jamestown Ophiolite Complex, is comparable to that of all ophiolites analysed for $\delta^{18}O$ so far (from de Wit *et al.*, 1987*a*).

morphism and alteration of these greenstone rocks can also be correlated with ocean floor rocks and Phanerozoic ophiolites. De Wit *et al.* (1987*a*) refer to this Archaean ophiolite as the Jamestown Ophiolite Complex.

Geochronological studies indicate that the mafic-ultramafic rocks were formed and metamorphosed (pervasively hydrated) between 3·45 and 3·49 Ga (de Wit *et al.*, 1987*a,b*; Armstrong *et al.*, 1988). This can be interpreted in a framework of simatic crust formation and hydrothermal metamorphism, as expected at ocean ridge spreading centres (de Wit *et al.*, 1987*a*).

In this paper we examine the platinum-group element (PGE) geochemistry at various levels of the Jamestown Ophiolite. The purpose of this is three-fold:

(1) Do the PGE concentrations and trends show similarities to those of Phanerozoic ophiolites, and hence verify the interpretations of the mafic-ultramafic rocks of the greenstone belt as being ophiolitic in nature?

(2) Can PGE geochemistry help to distinguish between deformed ultramafic cumulates and deformed residual ultramafic rocks; thus revealing more about the nature of the simatic crust–mantle transition, which is often only cryptically exposed in Phanerozoic ophiolites?

(3) What does the PGE geochemistry of the Jamestown Ophiolite Complex reveal about simatic lithosphere formation and mantle compositions in the Archaean? This, in turn, might allow inferences to be made about Archaean spreading and partial melting processes (i.e. lithospheric/asthenospheric processes leading to the formation of Archaean simatic lithosphere).

DISTRIBUTION OF PGE-ENRICHMENTS IN THE OPHIOLITE

Stratigraphy

Elevated PGE concentrations, with distinctly different chondrite-normalized patterns have been found at three different pseudostratigraphic levels in the Jamestown Ophiolite (Fig. 3). From the top downwards, high PGE values are first encountered below komatiites in an *undeformed* chromitite interlayered with meta-pyroxenites of the Stolzburg ultramafic complex (Fig. 1). Stratigraphically below the chromitite, PGE enrichment occurs in chromite-magnetite pods; the host rocks of the pods are *deformed* layered metaperidotites (now talc-carbonate rocks). A third type of PGE concentration is found in a sulphur-poor NiFe pod and an associated hematite-magnetite-chromite pod, both of which occur in coarse grained metadunites at the lowest exposed levels of the ophiolite. The relative positions of these three different occurrences are shown schematically in Fig. 3.

Analytical Methods

A suite of 24 samples from the Jamestown ophiolite were analysed for 17 elements. These included mineralized horizons, as well as unmineralized (host) rocks. The REE, Cr, Fe, Co, Ni, As and Sb were determined by routine INAA as described by Erasmus *et al.* (1977). For the PGE and Au, the samples were preconcentrated by NiS fire-assay before the NAA. A flow-chart of the procedure is given in Fig. 4.

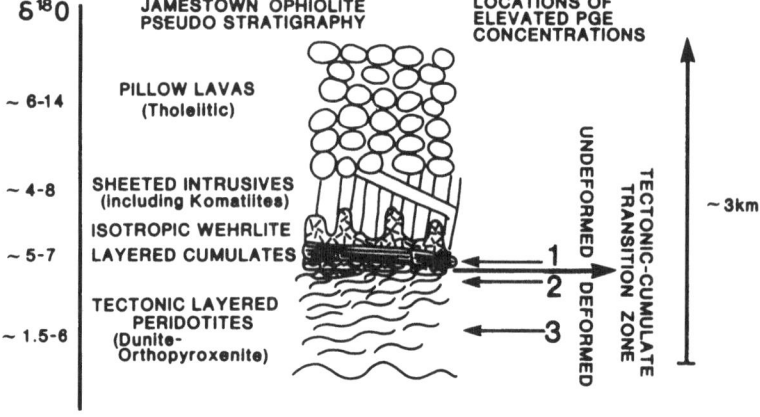

1. IN CHROMITE LAYER

2. IN CHROMITE PODS

3. IN Ni-Fe POD

FIG. 3. Schematic section through the Jamestown Ophiolite Complex, showing the positions of the three occurrences of PGE enrichment, discussed in this paper. Note that this ophiolite is relatively thin, with an isotropic section of wehrlite (and gabbro) and apparently only a thin section of cumulate ultramafics.

The data are listed in Table 1 and chondrite normalized values are plotted in Fig. 5. Values used for normalization are those preferred for C1 chrondrites by Taylor and McLennan (1985): Os = 1049; Ir = 710; Ru = 1071; Rh = 201; Pt = 1430; Pd = 836; in ppb.

Nature and Geochemistry of the PGE Concentrations

Chromitite-layer

A 20–30 cm thick chromitite occurs interlayered with metapyroxenites at the easternmost extremity of the Stolzburg layered ultramafic complex (locality 1, Fig. 1). The chromitite is undeformed, and can be traced along strike for more than 10 m. It is presumed to be of cumulate origin. The layer consists of a massive aggregate of well-packed chromite; the chromite grains have lobate rims of magnetite. Ilmenite occurs in small quantities, typically as a cement between chromite euhedra. Poikiloblastic pyroxene megacrysts, now chlorite pseudomorphs, give the rock a porphyritic texture.

The chromitite generally has low concentration of the moderately siderophile (Ni, Co) and chalcophile (As) trace elements, but Sb values are unusually high (see Table 1, Fig. 5). It is at present still unknown whether this is a primary feature of the layer or a result of secondary Sb enrichment. The PGE abundances and normalized patterns of this rock (Table 1, Figs 5 and 6(a)) are similar to those reported from PGE mineralized zones in large mafic layered complexes (viz. the Bushveld and

TABLE 1

Chemical data for samples from the Jamestown Ophiolite Complex. Base metal and REE data in ppm, unless otherwise indicated. PGE and Au data in ppb

	Cr (%)	Fe (%)	Co	Ni (%)	As	Sb	La	Ce	Sm	Eu	Os	Ir	Ru	Rh	Pt	Pd	Au	Description
PBT-1	nda	9·7	nda	nda	nda	nda	nda	nda	nda	nda	bdl	0·1	bdl	0·5	15	12	2·4	Pillowed tholeiite
PBT-4	nda	7·4	nda	nda	nda	nda	nda	nda	nda	nda	bdl	0·8	2	0·4	7	6	1·6	Spinifex komatiite
STD-2	0·130	6·7	130	0·27	0·3	2·2	0·03	bdl	0·12	0·14	10·5	6·9	9	4	21	<10	2·3	Coarse-grained metadunite
DBS-1	18·10	18·6	400	0·07	0·4	7 800	nda	0·06	nda	nda	nda	72	316	220	nda	1 546	24	Undeformed chromitite
SCR-1A	14·75	15·1	299	0·08	0·4	0·29	0·79	bdl	0·05	bdl	60	31	76	8	212	<10	11	Chromite pod, Stolzburg
SCR-1B	18·18	18·2	371	0·07	0·5	0·38	1·84	bdl	0·06	bdl	74	67	342	28	201	<10	39	Chromite pod, Stolzburg
SCR-3A	0·067	40·2	222	0·35	23·0	0·52	0·65	bdl	0·04	bdl	bdl	0·3	2	bdl	74	22	12	Magnetite pod, Stolzburg
SCR-3B	0·044	43·9	346	0·36	201·9	0·55	bdl	bdl	0·03	bdl	bdl	0·3	bdl	bdl	112	<10	15	Magnetite pod, Stolzburg
SCR-3C	0·046	46·1	304	0·42	156	0·66	bdl	bdl	0·04	bdl	bdl	1·3	bdl	13	172	<10	29	Magnetite pod, Stolzburg
SCR-3D	0·260	25·7	232	0·28	39·9	0·87	bdl	bdl	0·04	bdl	bdl	0·4	12	6	94	13	7	Magnetite pod, Stolzburg
SCR-3E	0·099	57·3	194	0·31	84·5	0·59	0·26	bdl	0·03	bdl	5·1	1·3	bdl	5	nda	<10	13	Magnetite pod, Stolzburg
SCR-3F	0·035	17·1	156	0·36	5·4	1·47	bdl	bdl	0·04	bdl	1·3	1·3	5	7	186	61	18	Magnetite pod, Stolzburg
SCR-3G	0·068	41·1	285	0·37	65·2	3·23	bdl	bdl	0·04	bdl	0·3	0·3	3	3	120	42	11	Magnetite pod, Stolzburg
L-4.5	0·011	33·8	4 100	27·7	40	300	0·64	2·24	0·68	bdl	277	384	1 123	nda	nda	1 561	122	Bon Accord—group A
R-4.5	0·020	32·3	3 900	30·5	60	1 600	0·28	1·19	0·48	0·08	627	560	1 830	478	1 529	1 754	170	Bon Accord—group A
DBS-3	0·017	34·5	3 800	31·3	60	2 300	0·26	1·08	0·49	0·05	583	501	1 771	493	1 478	1 845	187	Bon Accord—group A
BA-83.1	0·006	33·3	3 700	27·7	40	4 200	0·44	1·45	0·49	0·05	550	932	1 314	nda	1 794	1 794	nda	Bon Accord—group A
BA-83.B	0·005	32·3	3 600	29·6	70	5 100	0·36	1·47	0·61	0·11	1 113	730	1 829	515	1 764	1 537	252	Bon Accord—group A
BA-84.3	0·002	33·9	4 100	27·9	30	2 400	0·38	1·49	0·58	0·08	441	932	1 722	426	1 694	1 533	341	Bon Accord—group A
D-4.5	nda	30·7	3 300	23·8	50	180	12·58	2·46	0·62	bdl	65	97	849	242	1583	461	197	Bon Accord—group B
BA-84.1	0·008	30·1	3 300	18·4	90	15	9·8	1·11	0·46	0·19	117	120	618	221	1135	720	178	Bon Accord—group B
BA-84.2	0·011	18·9	4 000	16·8	400	24	2·3	7·92	2·44	bdl	73	72	617	225	1078	833	212	Bon Accord—group B
BAE-1	0·707	nda	673	0·57	1 723	nda	bdl	bdl	bdl	bdl	119	204	130	nda	15 530	<10	248	Hematite-magnetite-chromite body
BAE-2	0·781	nda	386	0·63	1 217	nda	bdl	bdl	bdl	bdl	129	247	201	nda	bdl	<10	158	Hematite-magnetite-chromite body

bdl: below the limit of detection.
nda: no data available.

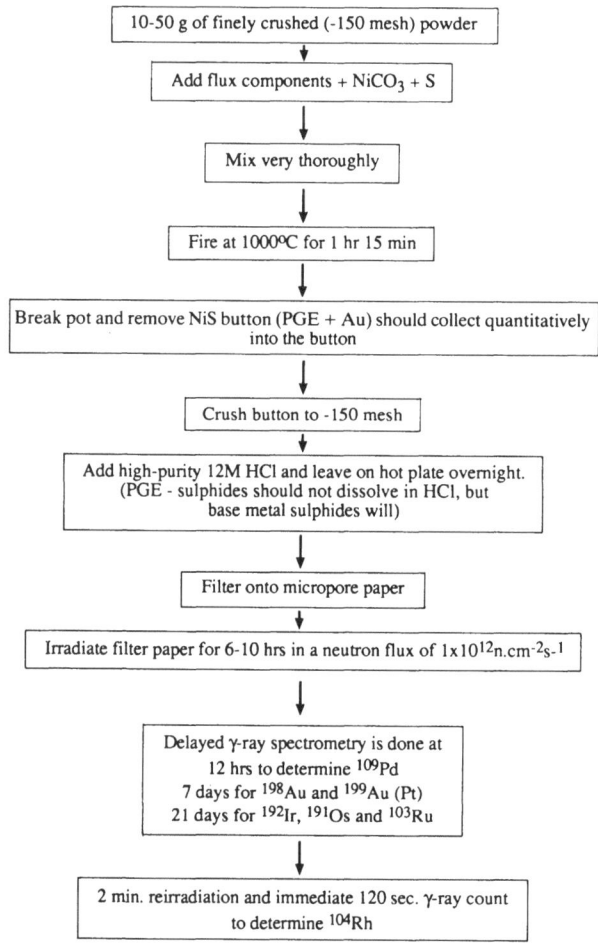

FIG. 4. Flow diagram of the procedure used to determine the PGE and Au. The international standard SARM 7 and reagent blanks were processed with each batch.

Stillwater complexes; see Fig. 6(a)) and in the magmatic cumulates of the Zimbales ophiolite (Bacuta *et al.*, 1987 and this volume). This is thus consistent with the interpretation that (this) part of the Stolzburg ultramafic complex has a cumulate origin.

Chromite-magnetite pods

A large number of small (10–50 cm in diameter) chromite-magnetite pods and boudinaged layers occur throughout a distinctly layered (cm-dm scale) thin sequence of talc-carbonate rocks, some 20–30 m *below* the chromitite described above (Locality 1, Fig. 1). These rocks were probably ultramafics. Westward, over 50–100 m, these talc-carbonates grade into the metadunite-orthopyroxenite massif which constitutes most of the Stolzburg complex (de Wit *et al.*, 1987*a*). These

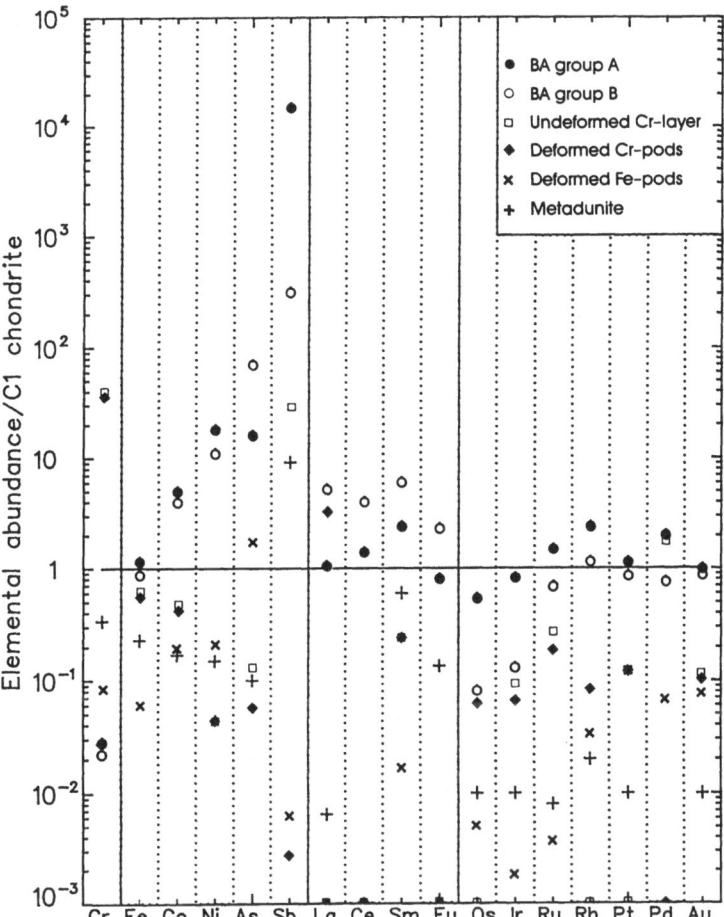

FIG. 5. Multi-element diagram of the chromitite layer, the chromitite-magnetite pods, BA and its host peridotite (STD-2). All values normalized to C1 chondrite, using values of Taylor & McLennan (1985).

coarse-grained, ultramafics display complex, high-temperature tectonometamorphic textures and are markedly depleted in REE (Fig. 2(a); de Wit *et al.*, 1987*a*). The orthopyroxenite units display isoclinal folding on a km scale. These peridotites have been interpreted as suboceanic depleted (residual) mantle, similar to those reported from the lowermost sections of Phanerozoic ophiolites (Moores, 1969; Menzies *et al.*, 1977; Casey *et al.*, 1983; Nicolas & Prinzhofer, 1983; Prinzhofer & Allegre, 1985; Nicolas, 1986).

The magnetite-chromite bearing talc-carbonate rocks are strongly deformed; folding of the layering is common and ductile shear zones are ubiquitous. It is not possible to tell if the layering is igneous (cumulate) or tectonic in nature, but at least some of it compares with cumulate layering in small ultramafic (now also talc-carbonate) plutons nearby (Fig. 7). The talc-carbonate unit is probably separated

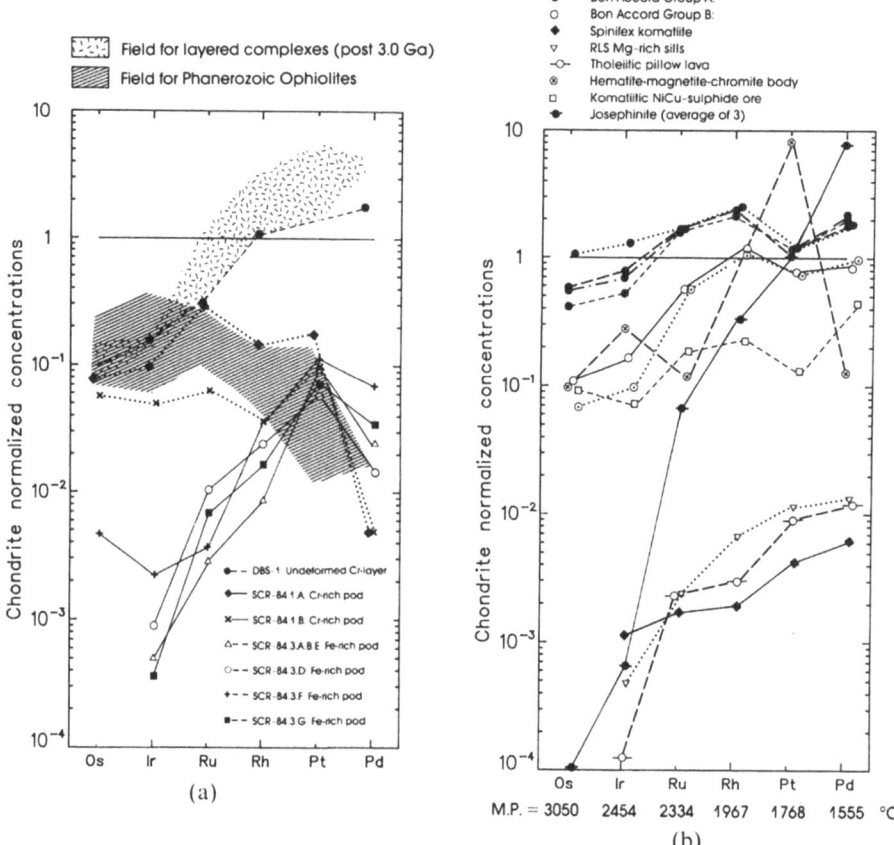

FIG. 6. Chondrite-normalized PGE plots, using the same C1 values as in Fig. 5. The PGE are plotted in order of decreasing melting temperatures from left to right. (a) The undeformed chromite layer and deformed chromite-magnetite pods from the Stolzburg Complex. Also shown are the fields for ores from the Bushveld and Stillwater Complexes (Naldrett, 1981; Davies & Tredoux, 1985) and podiform chromitites from Phanerozoic ophiolites (Page & Talkington, 1984; Page *et al.*, 1982, 1984). (b) The BA samples and BAE-1 (from the HMC body). Also plotted are (i) a pillowed tholeiite (PBT-1); (ii) a spinifex komatiite (PBT-4); (iii) a chill zone from an early sill associated with the Rustenburg Layered suite of the Bushveld Complex; (iv) the average of three analyses of josephinite samples from the Josephine peridotite, Oregon; (v) the average trend of two komatiitic Ni-Cu sulphide ores (Kambalda, Australia and Langmuir, Canada; Naldrett, 1981).

from the overlying cumulated rocks by a shear zone, but this is not well exposed. These talc-carbonate rocks, with their chromite-magnetite pods, thus occupy a deformed transition zone between ultramafic cumulates above and upper mantle-like ultramafic residuals below, and may contain deformed components of both these environments.

The PGE concentrations of the pods and boudins show two distinct trends (Figs 5 and 6a):

(i) PGE patterns which resemble those of the overlying chromitite in trends, but

with lower concentrations. These pods are much less Cr-rich than the undeformed layer (Table 1, Fig. 5). Sb values are also much lower, but Ni and As
are higher than the undeformed layer.

(ii) PGE patterns—with distinct negative Rh to Pd slopes—and concentrations
which closely resemble those reported from podiform chromitite in
Phanerozoic ophiolites. These pods are, in terms of Cr, Fe, Co, Ni and As
concentrations, very similar to the undeformed layer, but do not show the Sb
enrichment. The PGE patterns of these pods are very different from those of
the undeformed layer and the Fe-rich pods. In fact, the two sets of patterns
are complementary (see Fig. 6(a)).

Although Phanerozoic podiform chromitites are often regarded as magma-
chamber cumulates (see Coleman, 1977), others clearly indicate that such chromitites are part of the residual (restite) ultramafics (Leblanc *et al.*, 1980; Nicolas &
Prinzhofer, 1983; Boudier & Nicolas, 1985; Nicolas, 1986), but which may at times
be carried into magma chamber (Cox, 1987). It is often difficult from field and
petrographic interpretation to distinguish between these possibilities. At the transition zone between the residual mantle rocks and the overlying crustal cumulate
ultramafics of most ophiolites, both ultramafic sequences have frequently been
deformed together at high temperatures: even in the most classical and well-
exposed ophiolites, the original contact between these two rock sequences is cryptic
(viz. the Bay of Island ophiolites, J. Malpas and D. Elton, personal communi-
cations, 1986).

The PGE patterns of the chromitites might be a useful criterion to distinguish
between residual and cumulate (melt) rocks. Basaltic melts are always enriched in
Pd and Pt relative to Os, Ir and Ru (Davies & Tredoux, 1985; Barnes *et al.*, 1985)
and their PGE patterns have positive slopes (see Fig. 6(b)). Chromite cumulates
forming from such a melt will either have similar slopes to that of the melt (e.g.
SCR-84.3 A, B, C, D, E, and G; or somewhat flatter slopes because of the
association of Os, Ir and Ru with some (early?) chromites (SCR-84.3 F?), but
complete inversion of PGE patterns between a magma and its cumulate chromitites
has, to the authors' knowledge, not been observed. Cumulate chromite can therefore be recognized by a positive PGE trend. Residual mantle, from which basalt has
been extracted (as during the formation of oceanic crust), would be left with PGE
patterns with negative slopes. Chromitites which form part of these restite ultramafics can then be distinguished from the magma chamber chromitites by negative PGE trends (e.g. SCR-84.1A and B and most podiform chromitites of
Phanerozoic ophiolites, Fig. 6(a)).

Thus, the PGE results corroborate our field observations that the chromite-
magnetite pods and boudins may have been derived from two sources; a cumulate,
and a depleted residual one. As a corollary, the talc-carbonate rocks may thus
represent a zone of mixed ultramafic rocks in a ductile shear environment, in which
material from the magmatic cumulates *and* the upper mantle residues were brought
tectonically in juxtaposition. This zone could thus represent the true petrological
moho, separating simatic crust from mantle.

FIG. 7. Remnant cumulate layering with original igneous phase grading in talc-carbonate pluton, southwest of the Stolzburg Complex (see Fig. 1, locality 2). The layering is slightly folded due to a late tectonic overprint. The layering is similar to that seen in the southern parts of Stolzburg complex. See text for further explanation. The outcrop is about 1·5 m across.

NiFe-spinel pod

An unusual Fe-Ni-Co rich pod ($\sim 6 \times 3 \times 0.35$ m; 20–25 t) occurs in an allochthon of depleted, high-temperature metaperidotitic tectonites, along the northwestern margin of the Barberton greenstone belt (Fig. 1). This pod (called the Bon Accord—[BA]-deposit), together with an associated hematite-magnetite-chromite (HMC) pod, occurs in coarse-grained metadunites similar to those of the depleted residual ultramafics at the base of the Stolzburg complex. These pods are therefore inferred to occur in the lowermost exposed levels of the Jamestown ophiolite, namely in the depleted upper mantle residue from which the overlying rocks were derived by partial melting (de Wit *et al.*, 1987*a*; Tredoux *et al.*, 1988).

The BA deposit has an extremely high Ni content (up to 31% Ni) and has an unusual mineralogy containing *inter alia* minerals such as bunsenite (NiO), liebenbergite (Ni-olivine), trevorite (nickeliferous spinel), nichromite and cochromite (Ni- and Co-spinels, respectively), and chromite. The chromites are similar in composition to those from the undeformed chromitite layer of the Stolzburg complex (Fig. 8). Unusual nickeliferous serpentine, talc, chlorite, borate and carbonate minerals have replaced much of the inferred original mineralogy of BA (Tredoux *et al.*, 1988).

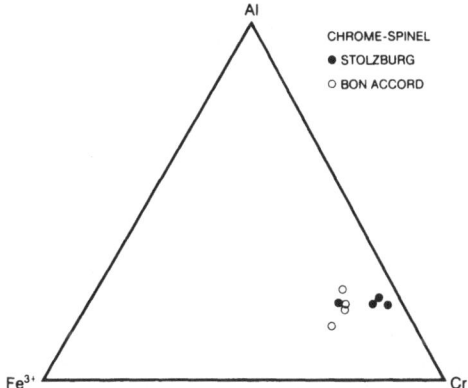

F<small>IG</small>. 8. Chromian-spinel compositions from the Bon Accord deposit compared with those from chromitite layer of the Stolzburg complex.

The whole rock geochemistry of BA has been described in detail elsewhere (de Waal, 1978, 1979; Tredoux *et al.*, 1988). Here we summarize the PGE geochemistry (as reported in Tredoux *et al.*, 1988); the data are listed in Table 1 and the chondrite-normalized PGE concentrations are plotted in Fig. 6(b).

In their overall shape the PGE patterns of BA display a sweep towards a Rh maximum, *a Pt dip*, and a final Pd rise. BA can be divided into the two distinct groups (as reflected in the major and trace element chemistry): (1) group A samples, characterized by relatively higher Ni and Sb concentrations, have relatively enriched PGE patterns with the highest Os and Ir values (0·5–1·3 × Cl), and with definite positive Rh and negative Pt anomalies; (2) group B samples (enriched in LREE and radiogenic Pb), which have more depleted PGE trends, especially for Os and Ir (0·06–0·11 × Cl) with Rh values peaking near Cl and a weaker Pt trough, lying consistently below the group A values. This bimodality is mineralogically controlled (Tredoux *et al.*, 1988): the group A samples are dominated by bunsenite and trevorite and the group B samples by the Ni-silicates.

The mineralogy of the HMC body is similar to that of BA, except that the minerals are Fe-rich and contain very little Ni. However, although the concentrations of Ir, Ru and Os are comparable to the group B samples of BA, parts of the body (e.g. BAE-1) are *enriched* in Pt relative to all BA samples. The chondrite-normalized PGE pattern of BAE-1 is flat, with a very pronounced *positive* Pt anomaly (Fig. 6(b)), which complements the negative Pt anomaly seen in all the BA samples.

A meteorite origin has previously been suggested for BA (de Waal, 1978), but de Wit *et al.* (1986) and Tredoux *et al.* (1988) have now shown that the field observations and geochemistry are incompatible with such an interpretation. The PGE geochemistry clearly substantiates this conclusion.

Some other possible models for the genesis of BA are: (i) precipitation from a pool of immiscible sulphide liquid in the mantle; (ii) a massive sulphide body such as is associated with many komatiites (e.g. Kambalda, Australia; Langmuir, Canada); (iii) a hydrothermal/volcanic-exhalative sulphide deposit; (iv) metasomatic/metamorphic remobilization and concentration; and (v) an altered chromitite pod. The merits and demerits of these models are discussed in detail by Tredoux *et al.* (1988) and will not be elaborated on here. Suffice to say that the major problem with scenarios (i)–(iii) is that they explicitly demand sulphides to be a major component of BA, which is not the case; BA contains only 100 ppm S (de Waal, 1978) and no petrographic evidence of a replaced sulphide morphology has been found. Major desulphurization of a pre-existing sulphide mineralogy therefore seems unlikely. Moreover, since BA occurs in depleted ultramafic rocks, a volcanic association (viz. (ii), (iii)) seems incompatible with its geologic setting. Secondary remobilization and alteration could perhaps be the cause of the unusually high Ni and PGE enrichment, although such a process has not yet been documented. In fact, the PGE (with the possible exception of Pd, and sometimes

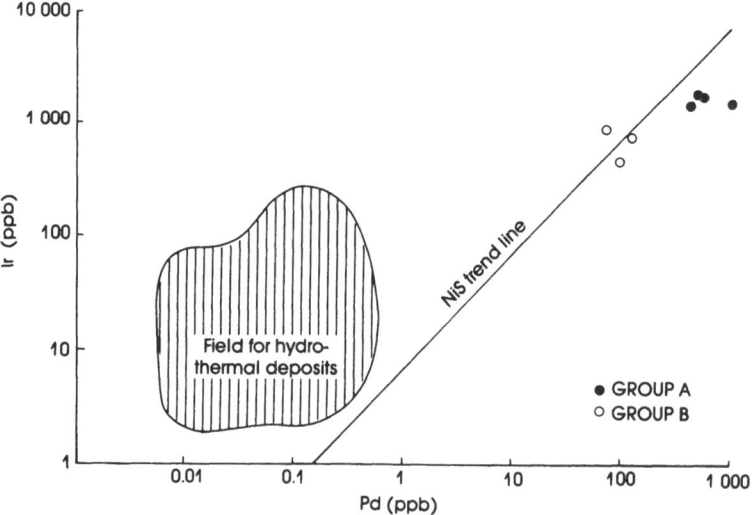

FIG. 9. Comparison of the Ir/Pd ratios of BA with the NiS trend line of Keays *et al.* (1982) and the distribution of this ratio in hydrothermal Ni-ores from the Yilgarn, Australia.

Pt) are generally considered to be immobile in the crustal alteration environments (Keays *et al.*, 1982; Prichard *et al.*, 1986; Oshin & Crocket, 1986). The slope of the PGE patterns of BA (which lack a strong Pd enrichment) and the Pd/Ir ratios (see Fig. 9) also do not support a secondary origin.

Because none of the above models appear to account satisfactorily for the unusual geology, geochemistry and mineralogy of BA, Tredoux *et al.* (1987) suggest that another possible explanation of BA might be that it represents a residual siderophile-enriched mantle heterogeneity that developed within up-welling asthenosphere during the formation of 3·5 Ga Archaean oceanic lithosphere.

DISCUSSION

The fact that the uppermost *magmatic* phases of the Jamestown ophiolite have similar PGE patterns (i.e. the spinifex komatiites, and pillowed tholeiites; Table 1 and Fig. 6(b)), but do *not* show the marked enrichment in the noble metals as seen in the chromitite layer, suggests that enrichment in the cumulate rocks was due to crystallization processes. This has also been suggested for the PGE ores associated with chromitite in the Bushveld Complex (Davies & Tredoux, 1985); high PGE contents in these chromite-rich horizons have been attributed to crystallization processes during multiple magma injection and mixing in magma chambers (Campbell *et al.*, 1983; Davies & Tredoux, 1985). Such mixing processes have now been well documented in the Bushveld Complex (Kruger & Marsh, 1982; Kruger & Mitchell, 1985). Magma mixing is even more likely to occur in the evolving 'open system' magma chambers of ophiolites where magma replenishment is a more continuous process. Perhaps the smaller size of mid-oceanic ridge magma chambers (compared to continental layered intrusions) and their transient nature (i.e. Dewey & Kidd, 1977; Nicolas, 1986; Detrick *et al.*, 1987) prevents the extensive development of the type of ore bodies seen in their 'closed system' continental counterparts.

As asthenospheric mantle rises towards the earth's surface at spreading centres to form oceanic lithosphere, it undergoes continuous fractionation due to small degrees of melt segregation (cf. McKenzie, 1984; Takahashi & Scarfe, 1985). The starting depth and extent of this partial melting of hot rising asthenosphere are ill-constrained and not well understood (Herzberg, 1983; McKenzie, 1984; Herzberg & O'Hara, 1985; Takahashi & Scarfe, 1985; Walker, 1986). During the formation of oceanic crust (as fossilized in ophiolites), the most extensive partial melting (10–30%) of the asthenosphere apparently takes place between about 30–60 km and the depth of the Moho (~6 km), depending on the spreading rates (McKenzie, 1985; Nicolas, 1986). In the region below this zone of extensive melting, the degree of partial melting is likely to be much smaller (<1%) but extracted from large volumes of material, most of which never reach to shallow enough depths to undergo extensive melting (McKenzie, 1985). From the shallow environment of extensive melting, the silicate melt fraction rises to form the simatic crustal carapace, leaving behind a harzburgitic-dunitic to lherzolitic residue, depleted in lithophile trace elements.

Data for olivines from komatiites indicate that this mineral is enriched in Ir[Os] relative to Pd[Pt] (Keays et al., 1982; see Fig. 6(b)). Furthermore, Mitchell & Keays (1981) suggest that, in mantle nodules, nearly all Ir and Os appear to be present as metallic alloys, which would remain in the residue during partial melting of the suboceanic mantle or coprecipitates with early formed olivine and/or chromite (Keays et al., 1982; Davies & Tredoux, 1985). Cox (1987) has shown how some of the residue may sometimes be carried up with the partial melts into the magma chambers and be emplaced in the cumulate section. The residual material would be relatively depleted in Pt-Pd, because the latter preferentially participate (with sulphur?) in the partial melts (Barnes et al., 1985; Davies & Tredoux, 1985). Such a shallow level differentiation of the noble metals in the extensive melting environment below spreading centres, can adequately account for the relative enrichment of Pt-Pd in the chromitite layer of the Stolzburg complex, as opposed to their distinct depletion in some of the underlying chromitite pods. This interpretation suggests that the PGE patterns of podiform chromitites reported from ophiolites (i.e. Page et al., 1982, 1984; Page & Talkington, 1984), which are similar to those of the Stolzburg Cr-rich pods, may indicate a residual, rather than a cumulate, origin for such chromitites.

It has now also been clearly shown that PGE concentrations of some chromitite-pods from the Phanerozoic Zambales ophiolite (Philippines; Bacuta et al., 1987, and this volume) are similar to those of layered complexes. These are interpreted as magmatic crustal cumulates (Bacuta et al., 1987, and this volume). The distinctly fractionated PGE patterns of the Cliff chromitites from the Shetland ophiolite complex (Prichard et al., 1986) could also support such an interpretation. The growing body of evidence that ophiolites contain chromitites of both cumulate and residual origins, again emphasizes the need to be able to distinguish between these possibilities. The usefulness of the PGE patterns is very apparent.

BA is clearly different to any of the other PGE-enriched zones in the Jamestown ophiolite. The absolute concentrations of Os and Ir are exceptionally high—the only other rocks which display similar Os-Ir concentrations are those described by Prichard et al. (1986) from the Cliff and Harold's Grave localities in the Shetland ophiolite. It would therefore seem that the mantle might contain sporadic enrichment of the PGE: a chemical condition that could have resulted from disequilibrium core-mantle segregation (i.e. inefficient core formation; Ringwood, 1966; Stevenson, 1981, 1983).

Inefficient core formation is envisaged as leaving a lower-mantle enriched in siderophile elements (Ringwood, 1966; Stevenson, 1981, 1983; Jones & Drake, 1986). If one considers chemical fractionation during the rise of a siderophile-rich silicate diapir from this region of the mantle prior to 3·0 Ga, as part of a deep thermal plume (cf. Olsen et al., 1987; Fig. 10) it is clear that the highly siderophile nature of the PGE would cause them to partition with any small Ni-Fe-Co metallic fraction. The rate and efficiency of such partitioning would depend largely on the ambient temperature. As such a rising diapir moved through 2500°C isotherm

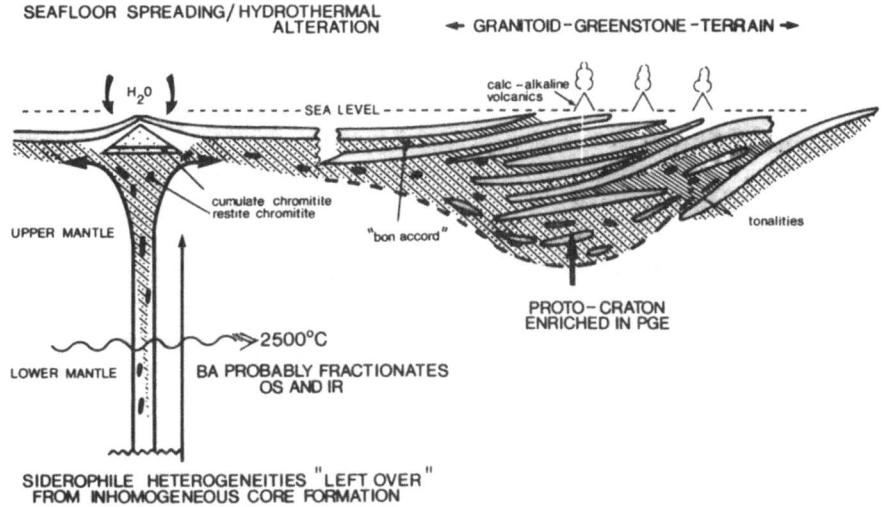

FIG. 10. Schematic model showing the source, evolution and final emplacement of PGE during the formation of Archaean oceanic crust and proto-craton (granite-greenstone terrain and depleted ultra-mafic keel).

(below 2000 km; Stacey, 1977), some of the Os and Ir would preferentially remain in the solid residue. Some Ru would be immobilized in a similar way at somewhat shallower depths. Due to their lower crystallization temperatures, Pt and Pd (and probably Rh) would continue to equilibrate until liquid-crystal fractionation in the upper-mantle was terminated with simatic lithosphere formation, as discussed earlier (see also Fig. 10).

During the ascent through the mantle, any metallic material would eventually be oxidized. Fe-oxides will form first and the metallic portion will be progressively enriched in Ni. Experimental studies (Blum *et al.*, 1988) indicate that the PGE will follow the Ni away from the oxidation front; but Pt, which has a greater affinity for Fe in geological systems than the other PGE (Kinloch, 1982), might have been retained to some extent in the Fe-oxides. Such a sequence of events could be used to explain the association between BA (Ni-rich, with a negative Pt anomaly) and the HMC pod (Fe-rich, with a positive Pt anomaly), i.e. that they represent parts of the same body that were tectonically separated from each other.

In the above scenario, BA could represent a dynamically emplaced hetero-geneity which developed, at least in part, due to very small degrees of partial melting during ascent of a 3·5 Ga lower-mantle diapir. Subsequently, a part of the heterogeneity remained embedded in the residual upper-mantle section during the formation of the Jamestown ophiolite complex. As a corollary, and unlike today, partial melting of the upwelling Archaean asthenosphere destined for spreading centres may have started deep in the mantle. If this is correct, spreading centres at >3·5 Ga were perhaps fed from a one-mantle reservoir. This would be consistent with more vigorous early mantle convection related to greater radiogenic heat production in the Archaean mantle.

Bird & Basset (1980) envisage a similar deep-mantle origin for Os–Ir–Ru alloys found with josephinite (a Pd-rich NiFe alloy) in the Mesozoic Josephine Ophiolite Complex (western USA). However, the PGE pattern of josephinite (see Fig. 6(b)) is different to any of the patterns in the Jamestown Ophiolite. This could be a reflection of the longer residence in the upper mantle of the younger rocks, which might have allowed BA-like material to 'sweat out' an Os–Ir–Ru component. The high PGE concentrations in the Shetland ophiolite are also interesting, as they may, in part, have been derived from previously depleted mantle; as indeed the supra-subduction model for this ophiolite implies (Prichard & Lord, this volume, p. 162).

Finally, it has been suggested (de Wit *et al.*, 1987*b*; Tredoux *et al.*, 1987) that, during mid-Archaean tectonism, the Archaean simatic lithosphere of southern Africa formed intraoceanic thrust stacks and continuously underplated granite-greenstone terrains to form mafic-ultramafic 'keels' to Archaean continental lithosphere. Analyses of 3·3–3·2 Ga diamonds and the geochemistry of associated ultramafic xenoliths found in Mesozoic kimberlites yield independent evidence that, in southern Africa, such Archaean 'keels' reached depths of at least 120 km and retained this minimal thickness throughout further development of the tectosphere which underlies the southern African craton (Jordan, 1981; Richardson *et al.*, 1984; Nickel & Green, 1985; Haggerty, 1986; MacGregor & Manton, 1986). By inference from our work on the Jamestown Ophiolite, this thick lower (mantle) lithosphere is a potential PGE 'store' and contributor to post 3·0 Ga igneous rocks, should their parent magmas pass through or originate within this region. Phanerozoic kimberlitic magmas have indeed penetrated this keel episodically between 600 and 60 Ma; we note that some associated diamonds may contain up to 160 ppm of iridium (no other PGE analysis has been quantified: Sellschop, 1979), which is consistent with our postulated Archaean, oceanic derived, PGE-enriched mantle lithosphere (Fig. 10).

CONCLUSION

Elevated PGE concentrations occur in three distinct rock units of the circa 3·5 Ga Jamestown Ophiolite Complex:

(1) in thin cumulative chromitites;
(2) in deformed chromitite-magnetite pods;
(3) in a large Fe-Ni pod.

The chromitite layers and pods were derived from shallow level partial-melting and magma chamber fractionation processes directly beneath oceanic-like spreading centres. The layers and pods represent melts and residual products of this differentiation process. The process is chiefly responsible for Rh-Pt-Pd enrichment in the melt fraction relative to the residue.

The Fe-Ni pod, which might have been derived from a chondrite-like lower mantle source, apparently underwent continuous fractionation in an environment of small degrees of partial melting at elevated temperatures. If this conclusion is correct, it must constrain models of a postulated, deeply buried Archaean magma ocean (cf. Nisbet & Walker, 1982); Archaean spreading-centres may have been operative during single-mantle-reservoir convection. Alternatively, could fractionation in a deep mantle magma ocean (or a poroelastic region, cf. Nicolaysen (1985)) have had a significant effect on upper-mantle Os-Ir concentrations?

To return to the questions posed in the Introduction section of this paper, it would seem that they can all be answered in the affirmative: (1) The PGE concentrations and patterns of some rocks in the Jamestown Ophiolite are very similar to what has been described in Phanerozoic ophiolites; (2) the PGE patterns are useful to distinguish between magma chamber (cumulate) chromite and residual (restite) chromite, especially where both are deformed together; and (3) significant new information about the Archaean simatic lithosphere and its formation can be obtained from the PGE geochemistry of mafic-ultramafic rocks in old greenstone belts.

Because the Archaean simatic lithosphere may have had significant amounts of PGE-enriched heterogeneities embedded within it, and because this lithosphere may have been a significant 'building block' of Archaean cratons (in particular of their lower [mantle] 'keels'), Archaean mantle lithosphere may be a significant PGE 'store'. Figure 10 summarizes these models.

ACKNOWLEDGEMENTS

This work was funded by a CSIR-FRD grant, and the University of the Witwatersrand: we thank D. Mthembu and D. Maseko for typing and O. Corner for help with some of the diagrams. We gratefully acknowledge the encouragement and support of J. P. F. Sellschop, R. A. Hart and R. J. Hart. We thank unknown referees for constructive reviews of the paper.

REFERENCES

Anhaeusser, C. R. (1973). The evolution of early Precambrian crust of southern Africa. *Phil. Trans. R. Soc., Ser. A*, **273**, 359–88.

Anhaeusser, C. R. (1976). The nature of crysotile asbestos occurrences in southern Africa: A review. *Econ. Geol.*, **71**, 96–116.

Anhaeusser, C. R. (1983). Contribution to the geology of the Barberton Mountain Land. Special Publication of the Geological Society of South Africa.

Anhaeusser, C. R. (1985). Archaean layered ultramafic complexes in Barberton Mountain Land, South Africa. In *Evolution of Archaean Supracrustal Sequences*, ed. L. B. Ayers, P. C. Thurston, K. T. Kard & W. Weber. Geological Association of Canada Special Paper 28, pp. 281–320.

Anon. (1972). Ophiolites. *Geotimes*, **17**, 24–5.

Armstrong, R. A., Compston, W. & de Wit, M. J. (1988). Re-examination of the 3·5–3·3 Ga Barberton Greenstone belt stratigraphy: a zircon ion-microprobe experiment. (Under review.)

Bacuta, G. C. Jr, Lipin, B. R., Gibbs, A. K. & Kay, R. W. (1987). PGE abundances in chromite deposits of the Acoje ophiolite block, Zambales ophiolite complex, Philippines (abstract in this volume, pp. 381–3).

Barnes, S.-J., Naldrett, A. J. & Gorten, M. P. (1985). The origin of the fractionation of platinum-group elements in terrestrial magmas. *Chem. Geol.*, **53**, 303–23.

Bird, J. M. & Basset, W. A. (1980). Evidence of a deep mantle history in terrestrial osmium-iridium-ruthenium alloys. *J. Geophys. Res.*, **85**, 5461–70.

Blum, J. D., Wasserburg, G. J., Hutchean, I. D., Beckett, J. R. & Stolper, E. M. (1988). 'Domestic' origin of opaque assemblages in refractory inclusions in meteorites, *Nature*, **331**, 405–9.

Boudier, F. & Nicolas, A. (1985). Harzburgite and lherzolite subtypes in ophiolites and oceanic environments. *Earth Planet Sci. Lett.*, **76**, 84–92.

Campbell, I. H., Naldrett, A. J. & Barnes, S.-J. (1983). A model for the origin of the Platinum-rich sulfide horizons in the Bushveld and Stillwater Complexes. *J. Petrol.*, **24**, 133–65.

Casey, J. F., Carson, S. A., Elthon, P., Rosencrantz, D. & Citus, M. (1983). Reconstruction of the geometry of accretion during the formation of the Bay of Islands Ophiolite Complex. *Tectonics*, **2**, 509–28.

Christensen, N. I. & Salisbury, M. H. (1979). Seismic anistrophy in the Oceanic upper mantle: Evidence from the Bay of Islands ophiolite complex. *J. Geophys. Res.*, **84**, 4601–10.

Coleman, R. G. (1977). *Ophiolites—Ancient Oceanic Lithosphere?* Springer, New York.

Condie, K. C. (1981). *Archaean greenstone belts. Developments in Precambrian Geology 3.* Elsevier, Amsterdam.

Cox, K. G. (1987). Postulated restite fragments from Karoo picrite basalts: their bearing on magma segregation and mantle deformation. *J. Geol. Soc. London*, **144**, 275–80.

Davies, G. & Tredoux, M. (1985). The platinum-group element and gold contents of the marginal rocks and sills of the Bushveld complex. *Econ. Geol.*, **81**, 838–48.

Detrick, R. S., Buhl, P., Vera, E., Muter, J., Orcutt, J., Madsen, J. & Brocher, T. (1987). Multichannel seismic imaging of a crustal magma chamber along the East Pacific Rise. *Nature*, **326**, 35–41.

De Waal, S. A. (1978). The nickel deposit at Bon Accord, Barberton, South Africa—a proposed paleometeorite. In *Mineralization in metamorphic terraines. Spec. Publ. Geol. Soc. S. Afr.*, **4**, pp. 87–98.

De Waal, S. A. (1979). The metamorphism of the Bon Accord nickel deposit by the Nelspruit granite. *Trans. Geol. Soc. S. Afr.*, **82**, 335–42.

Dewey, J. F. & Kidd, W. S. F. (1977). Geometry of plate accretion. *Bull. Geol. Soc. Amer.*, **88**, 960–8.

De Wit, M. J. (1983). Notes on a preliminary 1:25 000 geological map of the southern part of the Barberton Greenstone Belt. *Spec. Publ. Geol. Soc. S. Afr.*, **9**, 185–7.

De Wit, M. J., Tredoux, M., Hart, R. J., Armstrong, R. A., Lindsay, N. M. & Sellschop, J. P. F. (1986). Platinoids in a 3·6 Ga nickel-iron occurrence: implications for early terrestrial evolution and iridium anomalies. *Lunar Planet. Sci.*, **XVII**, 184–5.

De Wit, M. J., Hart, R. A. & Hart, R. J. (1987*a*). The Jamestown ophiolite complex, Barberton Mountain Belt: a section through 3·5 Ga simatic crust. *J. African Earth Sciences*, **6**, 681–730.

De Wit, M. J., Armstrong, R. A., Hart, R. A. & Wilson, A. H. (1987*b*). Felsic igneous rocks within the 3·3–3·5 Ga Barberton greenstone belt: High crustal level equivalents of the surrounding tonalite-trondhjemite terrain, emplaced during thrusting. *Tectonics*, **6**, 529–49.

Emslie, R. F. (1972). Oceanic crust and the identification of ancient oceanic crust on the continents: a summary. In *The Ancient Oceanic Lithosphere*, ed. E. Irving. Dept. of Energy, Mines and Resources, Earth Physics Branch. Spec. Publ. 42, pp. 153–6.

Erasmus, C. S., Fesq, H. W., Kable, E. J. D., Rasmussen, S. E. & Sellschop, J. P. F. (1977). The NIMROC samples are reference materials for neutron activation analysis. *J. Radioanal. Nucl. Chem.*, **39**, 323–34.

Gass, I. G., Lippard, S. J. & Shelton, A. W. (1984) (eds). *Ophiolites and oceanic lithosphere*. Geological Society Special Publication No. 13, Blackwell, Oxford.

Girardeau, J. & Nicolas, A. (1981). The structure of two ophiolite massifs, Bay-of-Islands, Newfoundland: a model for the oceanic crust and upper mantle. *Tectonophysics*, **77**, 1–34.

Haggerty, S. E. (1986). Diamond genesis in a multiply constrained model. *Nature*, **320**, 34–8.

Herzberg, C. T. (1983). Solidus and liquidus temperatures and mineralogies for anhydrous garnet lherzolite to 15 GPa. *Phys. Earth Planet. Int.*, **32**, 193–202.

Herzberg, C. T. & O'Hara, M. J. (1985). Origin of mantle peridotite and komatiite by partial melting. *Geophys. Res. Lett.*, **12**, 541–4.

Hoffman, S. E., Wilson, M. & Stakes, D. S. (1986). An inferred oxygen isotope of Archaean oceanic crust, Onverwacht Group, South Africa. *Nature*, **321**, 55–8.

Jones, J. H. & Drake, M. J. (1986). Geochemical constraints on core formation in the Earth. *Nature*, **322**, 212–28.

Jordan, T. H. (1981). Continents as a chemical boundary layer. *Phil. Trans. Roy. Soc. Lond.*, **A301**, 359–73.

Karson, J. A. (1984). Variations in structure and petrology in the Coastal Complex, Newfoundland: anatomy of an oceanic fracture zone. In *Ophiolites and Oceanic Lithosphere*, ed. I. G. Gass, S. J. Lippard & A. W. Shelton. Geological Society of London Special Publication No. 13, pp. 131–44.

Keays, R. R. (1982). Palladium and iridium in komatiites and associated rocks: application to petrogenic problems. In *Komatiites*, ed. N. I. Arndt & E. G. Nisbet. Allen & Unwin, London, pp. 435–57.

Keays, R. R., Nickel, E. H., Groves, D. I. & McGodrick, P. J. (1982). Iridium and palladium as discriminants of volcanic-exhalative, hydrothermal and magmatic nickel sulphide mineralization. *Econ. Geol.*, **77**, 1535–47.

Kinloch, E. D. (1982). Regional trends in the platinum-group mineralogy of the critical zone of the Bushveld complex. *Econ. Geol.*, **77**, 1328–47.

Kruger, F. J. & Marsh, J. S. (1982). The significance of $^{87}Sr/^{86}Sr$ ratios in the Merensky cyclic unit of the Bushveld Complex. *Nature*, **289**, 53–5.

Kruger, F. J. & Mitchell, A. A. (1985). Discontinuities and variations of Sr-isotope systematics in the Main Zone of the Bushveld complex, and their relevance to PGE mineralization. *Can. Mineral.*, **23**, 306.

Leblanc, M., Dupuy, C., Cassard, D., Moutte, J., Nicolas, A., Prinzhofer, A., Rabinovitch, M. & Routhier, P. (1980). Essai sur la genese des corps podiformes de chromitite dans les peridotites ophiolitiques: Étude des chromites de Nouvelle-Caledonie et comparison avec celle Méditerranée orientale. In *Ophiolites*, ed. A. Panayiotou. Cyprus Geological Survey Dept. Special Publication, pp. 691–701.

MacGregor, I. D. & Manton, W. I. (1986). Roberts Victor eclogites: Ancient Oceanic Crust. *J. Geophys. Res.*, **91**, 14 063–79.

McCall, G. J. H. (1981). Progress in research into the early history of the earth: a review 1970–1980. In *Archaean Geology*, ed. J. E. Glover & D. I. Grooves. Geological Society of Australia, pp. 3–18.

McKenzie, D. (1984). The generation and compaction of partially melted rock. *J. Petrol.*, **25**, 713–65.

McKenzie, D. (1985). The extraction of magma from crust and mantle. *Earth Planet Sci. Lett.*, **74**, 81–91.

Menzies, M., Blanchard, J., Brannon, J. & Koroter, R. (1977). REE geochemistry of fused ophiolitic and alpine lherzolites II. Beni Bouchera, Ronda and Lanzo. *Contrib. Mineral. Petrol.*, **64**, 53–74.

Mitchell, R. H. & Keays, R. R. (1981). Abundance and distribution of gold, palladium and iridium in some spinel and garnet lherzolites: implications for the nature and origin of precious metal, sulphur-rich intergranular components in the upper mantle. *Geochim. Cosmochim. Acta*, **45**, 2425–42.

Moores, E. M. (1969). Petrology and structure of the Vourinos ophiolite complex of northern Greece. Geological Society of America Special Paper No. 118, pp. 1–71.

Naldrett, A. J. (1972). Archaean ultramafic rocks. In *The Ancient Oceanic Lithosphere*, ed. E. Irving, Dept. of Energy, Mines and Resources, Earth Physics Branch Special Publication 42, pp. 141–51.

Naldrett, A. J. (1981). Platinum-group element deposits. In *Platinum-Group Elements: Mineralogy, Geology, Recovery*, ed. L. J. Cabri, CIM Special Volume 23, pp. 197–231.

Nickel, K. G. & Green, D. H. (1985). Empirical geothermobarometry for garnet peridotites and implications for the nature of the lithosphere, kimberlites and diamonds. *Earth Planet. Sci. Lett.*, **73**, 158–70.

Nicolas, A. (1986). Structure and petrology of peridotites: clues to their geodynamic environment. *Rev. Geophys.*, **24**, 875–95.

Nicolas, A. & Prinzhofer, A. (1983). Cumulative or residual origin for the transition zone in ophiolites: structural evidences. *J. Petrol.*, **24**, 188–206.

Nicolaysen, L. O. (1985). On the physical basis for the extended Wilson Cycle, in which most continents coalesce and then disperse again. *Trans. Geol. Soc. S. Afr.*, **88**, 561–80.

Nisbet, E. G. (1982). The tectonic setting and petrogenesis of komatiites. In *Komatiites*, ed. N. T. Arndt & E. G. Nisbet. Allen & Unwin, London, pp. 501–21.

Nisbet, E. G. & Walker, D. (1982). Komatiites and their structure of the Archaean mantle. *Earth Planet. Sci. Lett.*, **60**, 105–13.

Olsen, P., Schubert, G. & Anderson, C. (1987). Plume formation in the D″-layer and the roughness of the core-mantle boundary. *Nature*, **327**, 409–12.

Oshin, I. O. & Crocket, J. H. (1986). Noble metals in the Thetford Mines Ophiolites, Quebec, Canada. Part II: Distribution of gold, silver, iridium, platinum, and palladium in the Lac de l'Est Volcano—sedimentary section. *Econ. Geol.*, **81**, 931–45.

Page, N. J & Talkington, R. W. (1984). Pd, Pt, Rh, Ru and Ir in peridotites and chromitites from ophiolite complexes in Newfoundland. *Can. Mineral.*, **22**, 137–49.

Page, N. J, Pallister, J. S., Brown, M. A., Smewing, J. D. & Haffty, J. (1982). Pd, Pt, Ph, Ir and Ru in chromite-rich rocks from the Samail Ophiolite, Oman. *Can. Mineral.*, **20**, 537–84.

Page, N. J, Engin, T., Singer, D. A. & Haffty, A. (1984). Distribution of platinum-group elements in the Bati Kef chromite deposit, Guleman-Elazic area, eastern Turkey. *Econ. Geol.*, **79**, 177–84.

Prichard, H. M. & Lord, R. A. (1988). The Shetland ophiolite: evidence for a supra-subduction zone origin and implications for platinum-group element mineralization. Abstract in this volume p. 161. Also in *Mineral Deposits in the European Commutiny*, eds J. Boissonnas and P. Omenetto, Springer Verlag (in press).

Prichard, H. M., Neary, C. R. & Potts, P. J. (1986). Platinum-group minerals in the Shetland ophiolite. In *Metallogeny of the Basic and Ultrabasic Rocks*, ed. M. J. Gallagher, R. A. Ixer, C. R. Neary & H. M. Prichard. Institution of Mining and Metallurgy, pp. 395–414.

Prinzhoffer, A. & Allegre, C. J. (1985). Residual peridotites and the mechanisms of partial melting. *Earth Planet Sci. Lett.*, **74**, 251–65.

Richardson, S. H., Gurney, J. J., Erlank, A. J. & Harris, J. W. (1984). Origin of diamonds in old enriched mantle. *Nature*, **30**, 198–200.

Ringwood, A. E. (1966). Chemical evolution of the terrestrial planets. *Geochim. Cosmochim. Acta*, **30**, 41–104.

SACS (1980). South African Committee for Stratigraphy. Stratigraphy of South Africa, Part I (compiler, L. E. Kent). *Handbk Geol. Surv. S. Afr.*, **8**.

Sellschop, J. P. F. (1979). Nuclear probes in physical and geochemical studies of natural diamonds. In *The Properties of Diamond*, ed. J. E. Field. Academic Press, London, pp. 106–63.

Stacey, F. D. (1977). A thermal model for the earth. *Phys. Earth Planet. Int.*, **15**, 341–8.

Stevenson, D. J. (1981). Models for the Earth's core. *Science*, **211**, 611–19.

Stevenson, D. J. (1983). The nature of the earth prior to the oldest known rock record: The Hadean Earth. In *Earth's Earliest Biosphere*, ed. J. W. Schopf. Princeton University Press, N.J., pp. 32–40.

Takahashi, E. & Scarfe, C. M. (1985). Melting of peridotite to 14 GPa and the genesis of komatiite. *Nature*, **351**, 566–8.

Taylor, S. R. & McLennan, S. M. (1985). *The Continental Crust: its Composition and Evolution*. Blackwell, Oxford, 328 pp.

Tredoux, M., De Wit, M. J., Hart, R. J., Armstrong, R. A., Lindsay, N. M. & Sellschop, J. P. F. (1988). Evidence of a deep mantle origin in an Archaean nickel-iron-platinum occurrence. *J. Geophys. Res.* (in press).

Viljoen, M. J. & Viljoen, R. P. (1969). An introduction to the geology of the Barberton granite-greenstone terrain. *Trans. Geol. Soc. S. Afr.*, Special Publication 2, pp. 9–28.

Walker, D. (1986). Melting equilibria in multicomponent systems and liquidus/solidus convergence in mantle peridotite. *Contrib. Mineral. Petrol.*, **92**, 303–7.

Williams, D. & Furnell, R. G. (1979). Reassessment of part of the Barberton type area, South Africa. *Precamb. Res.*, **9**, 325–47.

Windley, B. F. (1976). *The Early History of the Earth*. Wiley-Interscience, London.

Windley, B. F. (1984). *The Evolving Continents*, 2nd edn. John Wiley, New York.

NOTE ADDED IN PROOF

It has been brought to our attention that the C-1 chondrite value used for the normalization of Os (viz 1049 ppb) is in error. A more correct value would be 507·5 (the average of 10 analyses of the C-1 chrondrite Orgueil), adjusted by a factor of 1·5 (for volatile loss during accretion) = 761 ppb. The result of this correction is to change the slopes from Os to Ir in Figs 5 and 6 slightly. This has no effect on the arguments or conclusions presented in this paper.

35

Evidence for Fluid in the Footwall Beneath Potholes in the Merensky Reef of the Bushveld Complex

R. Grant Cawthorn & Kelly L. Poulton

Department of Geology, University of the Witwatersrand, Wits 2050, Johannesburg, South Africa

ABSTRACT

Platinum mineralization in the Merensky Reef has recently been attributed to the ingress of fluids, possibly via depressions in the Reef referred to as potholes. In this study samples from immediately below three potholes and from the same horizon under normal Reef have been analysed to investigate this possibility.

In terms of lithology, magmatic texture and mineral compositions, sample profiles from the two environments are identical. The rocks range from norite to leuconorite, containing orthopyroxene in the range En_{74-82} and plagioclase in the range An_{74-82}. However, samples from below potholes display minor retrogressive mineral alteration to assemblages stable over a wide range of temperatures.

Incompatible, immobile elements, such as phosphorus, display no difference in the two environments ($P_2O_5 = 0.024 \pm 0.006\%$ under potholes, and $0.023 \pm 0.004\%$ under normal Reef), arguing against a greater concentration of trapped residual liquid in pothole environments. However, incompatible, mobile elements, such as potassium, rubidium and copper are enriched and far more variable in concentration under potholes than under normal Reef ($K_2O = 0.29 \pm 0.26\%$ under potholes, and $0.16 \pm 0.05\%$ under normal Reef). This contrast in behaviour between mobile and immobile elements may be the result of reaction between the cumulates and a fluid phase.

While this study presents evidence supporting the preferential concentration of fluids beneath potholes it is not possible to comment on the possible transport or remobilization of the platinum-group elements (PGE) by these fluids from these data.

GEOLOGICAL RELATIONS

The Merensky Reef occurs close to the top of the Critical Zone of the Bushveld Complex (Vermaak, 1976), in a series of incompletely developed cyclic units, comprising pyroxenite, norite and anorthosite with occasional bands of chromite. Most layers within this sequence are traceable over considerable distances, although variations in thickness occur and some layers may be eliminated, possibly

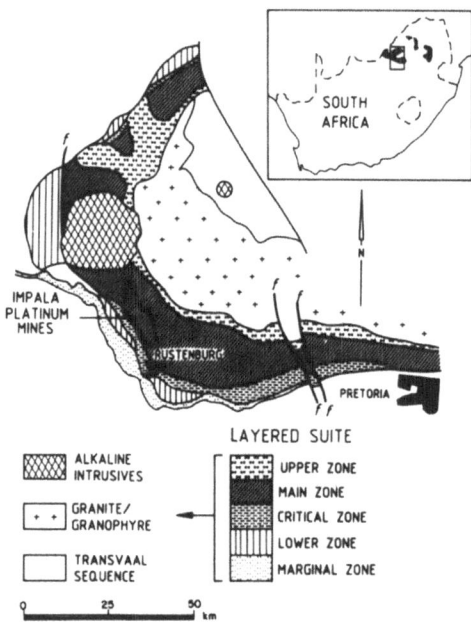

FIG. 1. Map of the setting of Impala Platinum Mines within the western lobe of the Bushveld Complex.
The inset shows the outcrop of mafic rocks of the Bushveld Complex in South Africa.

by erosion (Viljoen *et al.*, 1986). The basal contact of the Merensky Reef is
irregular and, in places, occurs in potholes which have been described in detail by
Schmidt (1952), Ferguson & Botha (1963), Schwellnus *et al.* (1976), Leeb-Du Toit
(1986), Viljoen *et al.* (1986) and Farquhar (1986). They range in size from a few
metres to 0·5 km in diameter and may plunge from 1 to 100 m below the normal
plane of the Reef. They may have sloping to steep, or even overhanging, margins,
and the rocks immediately underlying the normal Reef have been totally removed
in these potholes.

The samples which have been studied came from Impala Platinum Mines, which
are situated 10 km north-north-west of Rustenburg in the western lobe of the
Bushveld Complex (Fig. 1). The detailed geology and mineralization has been
documented by Leeb-Du Toit (1986). The typical morphology and relation of
potholes to the Merensky Reef are described in that publication and the typical
cross-section in Fig. 2a is based on that work. Three potholes were accessed in the
mine and boreholes sampled. The potholes plunge down to the top of a leuconorite
unit, referred to as Footwall 3 in the mine. This unit is usually 5 m in thickness, but
in the three boreholes studied it is only 4 m thick, and it appears that up to 1 m has
been removed from the top of the Footwall 3 unit by the pothole-forming event.
The underlying unit, Footwall 4, contains two 5 cm thick anorthosite layers which
are very distinctive and these were taken as the datum level for sampling. All
heights quoted are measured upwards from the top of these anorthosites.

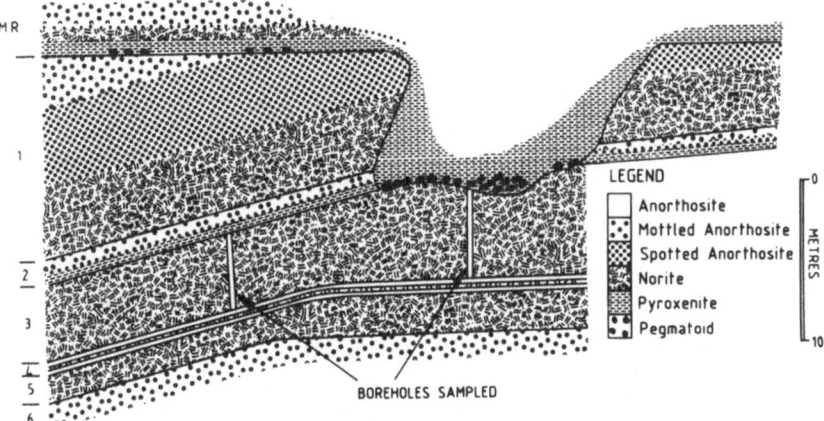

FIG. 2(a). Generalized cross-section of a pothole in the Merensky Reef showing the sequence of rocks in the footwall (from Leeb-Du Toit, 1986). Horizontal scale is compressed relative to vertical scale. MR refers to the Merensky Reef unit. Numbers refer to different footwall horizons recognized by mine geologists.

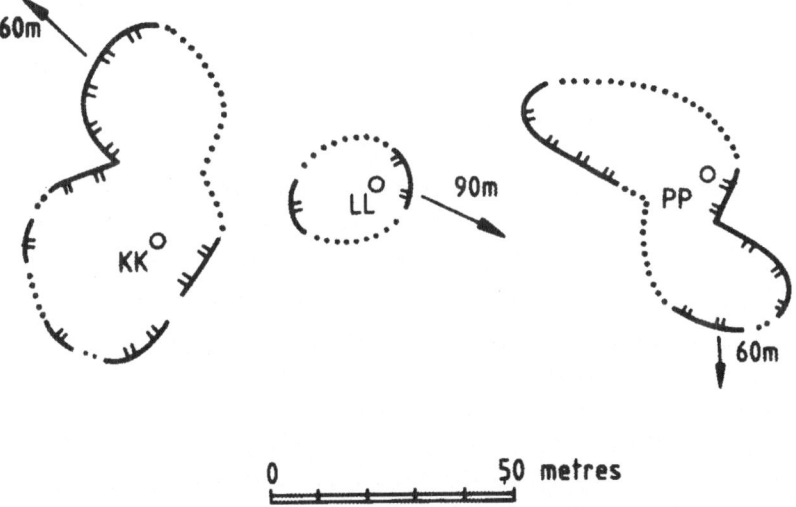

FIG. 2(b). Geometry of potholes studied. The edge of the pothole is shown with a solid line where it has been accurately delineated by mining. The dotted edges are inferred. The position of the boreholes sampled is indicated. The direction and distance to the profile sampled under normal Reef is shown by the arrow.

Mining development in the footwall below the Reef allowed the Footwall 3 and 4 horizons to be sampled under normal Reef. The three profiles under normal Reef were between 65 and 80 m from the edge of the pothole where a profile had been taken, and from mine plans it appears that there are no nearer potholes to the location where these normal footwall profiles were taken. The pothole profiles were drilled as near as possible to the centre of the potholes which ranged in diameter from 25 m (where profile LL was taken) to 70 m by 30 m (profiles KK and PP). These elliptical potholes may owe their shape to the partial overlap of two closely-spaced potholes as suggested by the limits of mining, shown in Fig. 2b.

LITHOLOGY

Footwall 3 is a leuconorite, bounded by pure anorthosite below (Footwall 4) and a pyroxenite above (the base of Footwall 2). This unit is atypical in that the proportion of pyroxene increases upwards. At the base there are anhedral to poikilitic orthopyroxene grains up to 1 cm in size separated by almost pure anorthosite. Upwards the rock assumes a more uniform texture with evenly distributed 2 mm orthopyroxene grains. Under potholes the top of Footwall 3 is frequently missing, but in the normal succession the uppermost rocks are true norite, using the criterion of Eales *et al.* (1986) that the boundary between norite and leuconorite should be drawn at 62·5% plagioclase by weight. This change in the proportion of plagioclase is shown in Fig. 3. These mineral proportions are based on the normative composition of the rock. The lowest rocks are very heterogeneous with large mottles of poikilitic orthopyroxene, which makes modal analysis unreliable. Towards the top of the sequence where the rock is more homogeneous, point-counting was undertaken on a few samples and the results included in Fig. 3. They show close agreement with the normative data except for one sample. Three separate thin sections were prepared of the lowest sample from borehole KK and point-counted. This has large, irregular, poikilitic orthopyroxene grains and the variation in modal orthopyroxene in these three sections ranged from 11 to 21%, based on 1000 points over an area of 3 cm by 2 cm.

No systematic differences exist in the mineral proportions of the rocks at any specific height in Footwall 3 between profiles under normal Reef and under potholes. However, the more pyroxene-rich samples observed in the normal Reef sections are missing together with Footwall 1 and 2 under potholes. Campbell (1986) suggested that potholes terminated downwards against a pyroxene-rich layer, because the remelting of pyroxene-rich material produces a dense liquid that would collect in the bottom of the pothole and hinder further erosion. In the present instance, it appears that about 1 m of pyroxene-rich material has been removed and that the erosion has been terminated at a leuconorite horizon.

Plagioclase always occurs as euhedral to subhedral cumulus grains up to 3 mm in size, while orthopyroxene changes systematically from a poikilitic to interstitial phase at the bottom of the unit to having a subhedral texture towards the top. Even at the top, the orthopyroxene grains are up to 6 mm in size and often contain

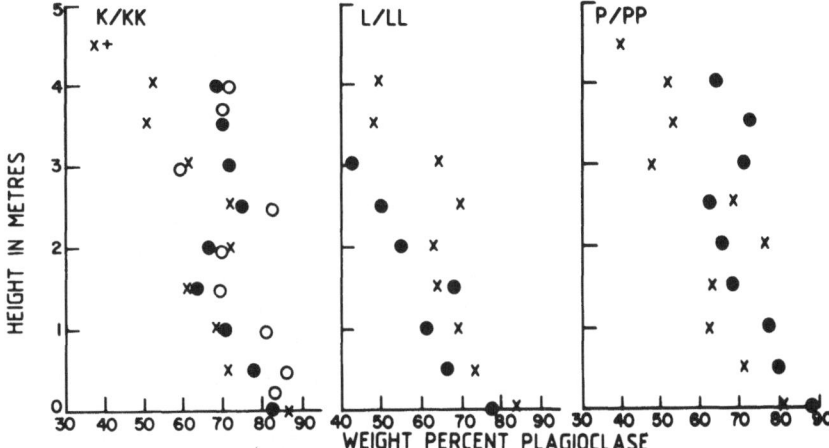

FIG. 3. Plot of the feldspar content of samples from the Footwall 3 horizon as a function of height showing the general upward decrease in abundance in all six profiles. Symbols: crosses and pluses refer to samples below normal Reef (determined by norm and mode respectively); solid and open circles below potholes. Each pothole and adjacent normal section is shown as a separate profile.

inclusions of subhedral plagioclase grains, significantly smaller (1 mm) than those elsewhere in the rock. The orthopyroxene also shows extensive overgrowth of both orthopyroxene and clinopyroxene between plagioclase grains destroying its original more euhedral texture. This upward change in morphology is paralleled by an increase in orthopyroxene abundance from 20 to 60%. Clinopyroxene is a minor, interstitial phase, and there are also infrequent flakes of biotite. Both minerals are associated with the orthopyroxene, but constitute more than 2% in only one sample. No chromite and only one subhedral grain of olivine was found in all the samples studied. Samples from below normal Reef and potholes are similar in their mineralogy and magmatic texture.

In contrast, the extent of post-magmatic alteration of the rocks does vary in the two settings. Under normal Reef, alteration is very minor. Occasionally, the orthopyroxene contains irregular veinlets of serpentine and the plagioclase has cloudy areas reflecting the formation of clay or zeolite minerals. Under potholes the alteration is more variable, although many samples are indistinguishable from those under normal Reef. In others, individual grains of orthopyroxene may be totally replaced by an extremely fine-grained alteration product identified as serpentine and chlorite by X-ray diffraction. Other grains in the same slide are unaltered. Clinopyroxene may show variable replacement by tremolite and chlorite and the plagioclase may be replaced by clay minerals and paragonite identified by X-ray diffraction. Local patches of epidote have formed from the feldspar, whereas other grains are perfectly fresh. The alteration is extremely localized and does not represent a regionally extensive retrograde metamorphism. The existence of fresh and altered grains in the same slide demonstrates that equilibrium has not been reached even on the scale of a thin section.

GEOCHEMISTRY

From the 6 profiles, 54 samples were taken at half metre intervals and analysed for major and trace elements. The important geochemical data are presented in Figs 3–7 and specific average values are given in Table 1. The compositions of the orthopyroxene and plagioclase as determined from the norm are shown in Fig. 4. Electron microprobe analyses of some samples give comparable results. No zoning in terms of Mg/Fe can be detected in the electron microprobe analyses of orthopyroxene, although variations of up to 1·5 mol% En are randomly detected within single grains. The Cr_2O_3 content showed a considerable range, typically from 0·49 to 0·30% within one sample, but again no systematic zonation could be detected. The normative En content of pyroxene and An content of plagioclase both range from 74 to 82%, but no regular change in composition with height is discernible. For example, in profile LL the An content shows a decrease with increasing height whereas in PP the uppermost samples have the highest An content. There is generally less variation in the En content of the orthopyroxene, but in all profiles there is an upwards increase in this value over the bottom 1·5 m which parallels the increasing proportion of orthopyroxene in the whole rock (Fig. 3).

The average An and En compositions of all the profiles are shown in Table 1. The average An content in plagioclase from all the profiles under normal Reef is 1% lower than from under potholes whereas for the En in orthopyroxene samples from under normal Reef are 1% higher. There does not appear to be evidence for a difference in the degree of fractionation in these two environments.

In Fig. 5 the Cr and Ni contents of the whole rock are plotted against the MgO content. Orthopyroxene is the only important magnesian phase and so the MgO axis represents the proportion of orthopyroxene. As there is an extremely good correlation between these elements it indicates that there are no other important minerals such as chromite or sulphide which contain any Cr and Ni in the rock. Samples from both the pothole and normal Reef environment display identical trends for these elements.

There are very few accessory minerals. The rocks are predominantly plagioclase and orthopyroxene and texturally these rocks are adcumulates (Wager et al., 1960). This can also be substantiated using chemical analyses. The incompatible trace elements, such as Rb, Zr, Nb and Y, are present in concentrations close to their detection limits by X-ray fluorescence spectrometry and so it is difficult to use these data quantitatively. However, their extremely low abundance demonstrates that all these rocks are adcumulates.

Phosphorus is another incompatible element, but whose concentration is sufficiently high for meaningful interpretation (Henderson, 1968). The average for 25 samples below potholes is $0·024 \pm 0·004\%$ P_2O_5, while for 29 samples below normal Reef the average is $0·023 \pm 0·006\%$. This constant P_2O_5 content demonstrates that there is no enrichment of trapped liquid in the section beneath potholes, as suggested by von Gruenewaldt (1979). If there had been the partial entrapment of some residual magma which did not crystallize apatite, the P_2O_5 would have been the same as actually observed. However, there would have been a decrease in the

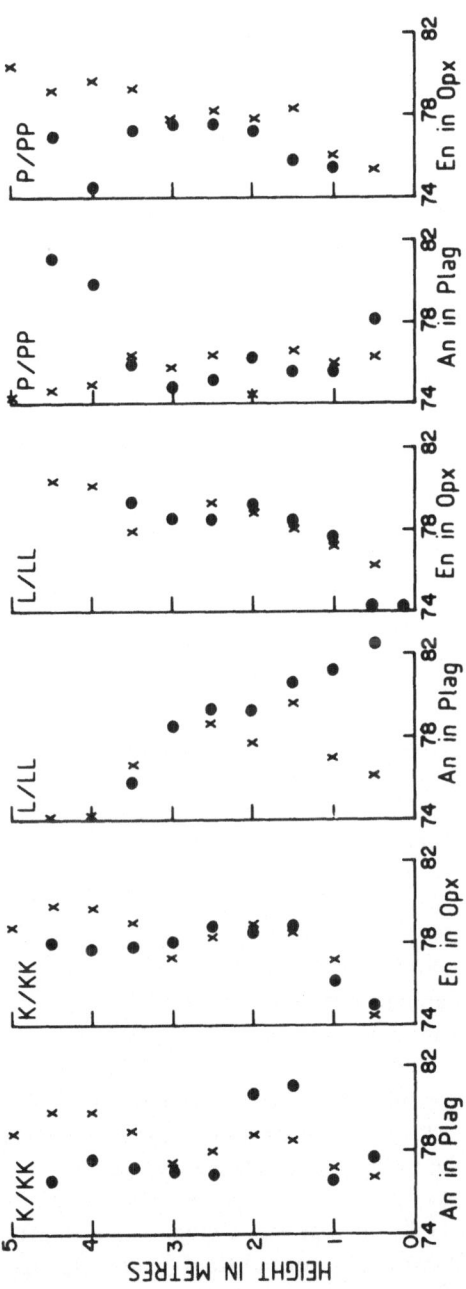

FIG. 4. Plot of normative composition of An and En contents of plagioclase and orthopyroxene as a function of height, showing no obvious differentiation pattern.

TABLE 1

Average concentrations of certain elements and mineral compositions in samples below potholes and normal Merensky Reef

		K_2O	P_2O_5	Rb	Cu	Ni	Sr	An in Plag	En in Opx	No. of samples
Normal Reef	K	0·13	0·023	0·5	21	280	317	77·2	78·2	10
		0·02	0·009	0·6	5	123	67			
	L	0·19	0·024	1·4	24	251	323	76·7	78·5	9
		0·07	0·004	0·7	4	62	51			
	P	0·15	0·021	1·6	22	266	301	75·6	78·2	10
		0·02	0·003	0·8	8	96	63			
	Average	0·16	0·023	1·4	22	266	313	76·5	78·3	29
		0·05	0·006	1·4	6	99	62			
Pothole Reef	KK	0·18	0·022	0·6[a]	54	219	353	77·9	77·6	9
		0·07	0·003	0·5[a]	33	66	29			
	LL	0·32[a]	0·027	5·0	54	278	319	79·5	77·9	7
		0·09[a]	0·005	0·8	18	102	71			
	PP	0·27	0·024	3·2	39	266	359	77·0	75·9	9
		0·09	0·002	2·3	25	182	40			
	Average	0·29[a]	0·024	4·1[a]	48	253	345	78·0	77·1	25
		0·26[a]	0·004	2·4[a]	28	131	51			

Second row below each average refers to 1 standard deviation.
K_2O and P_2O_5 are reported in weight %, other elements in ppm.
[a] Indicates that one very high value has been excluded.

normative An and En contents of the plagioclase and pyroxene. It has been demonstrated from Fig. 4 and Table 1 that there is no such change in mineral compositions beneath potholes and normal Reef. The P_2O_5 and mineral data therefore argue against the relative enrichment of trapped residual liquid beneath potholes.

In contrast, K_2O under potholes has a concentration of 0·291%, while under normal Reef it is 0·160%. There is no difference in the proportion of plagioclase in the two environments and so this difference cannot be attributed to its very slight substitution into feldspar. The different contents of K_2O and P_2O_5 in these rocks is shown in Fig. 6. A linear correlation would be predicted if their abundances were controlled by variations in the mesostasis content. However, many of the samples from below potholes are enriched in K_2O regardless of their P_2O_5 content, which suggests that enrichment of K_2O under potholes must be due to a non-magmatic process; presumably the influence of volatiles. While Rb contents for individual samples cannot be used quantitatively because of the large relative error and low concentrations, it is significant that the average value for Rb is 4·1 ppm under potholes and only 1·4 ppm under normal Reef (Table 1), suggesting that Rb has also been concentrated in zones under potholes.

Copper and Ni also display different behaviour in the two environments. The average values for Ni in the footwall succession below potholes is 253 ppm and below normal Reef is 266 ppm, demonstrating that no systematic differences exist.

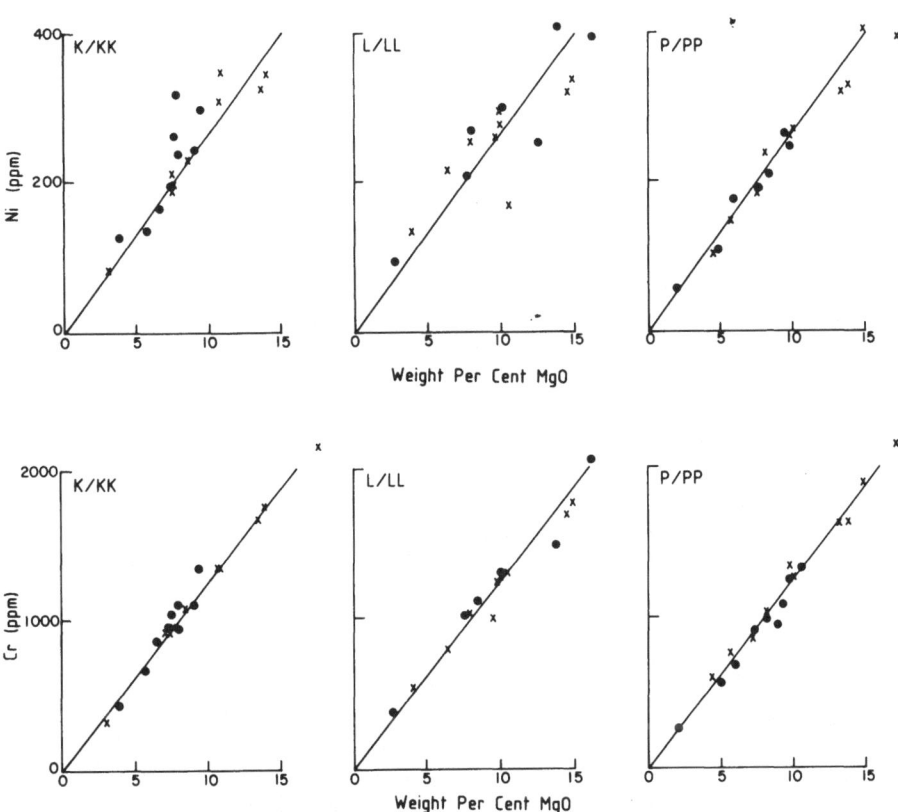

FIG. 5. Plot of Ni and Cr in ppm versus MgO in Footwall 3 horizon below potholes (solid circles) and normal Reef (crosses), showing a control on Ni and Cr by mafic mineral content. The lines in each diagram for Cr and Ni are identical indicating the comparable trends in all sections regardless of whether they are from below pothole or normal Reef.

However, for Cu the average value beneath potholes is 48 ± 28 ppm, compared to 22 ± 8 ppm under normal Reef (Table 1). These differences are mainly due to the enrichment in Cu at the top of the pothole profiles in the immediate footwall to the Merensky Reef. This relative enrichment in Cu compared to Ni is shown in Fig. 7. It was argued on the basis of the MgO versus Ni plot (Fig. 5) that there were no nickel-bearing sulphides present in these samples, and so the enrichment in Cu cannot be attributed to the presence of Merensky Reef-type sulphide accumulation. It is more probable that this enrichment is due to the introduction or redistribution of Cu by hydrothermal fluids.

MODELS FOR PLATINUM MINERALIZATION AND POTHOLE FORMATION

By far the most important source of platinum is in layered intrusive complexes, where it is usually associated with sulphides (Naldrett, 1981) and possibly also with

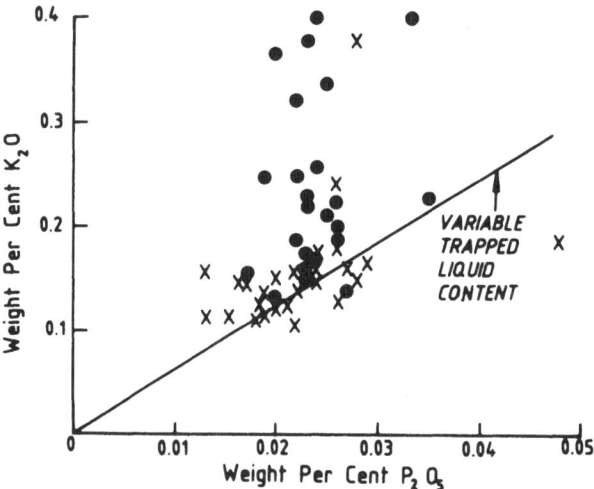

FIG. 6. Plot of K_2O versus P_2O_5 showing that many of the samples from below potholes (solid circles) are enriched in K_2O relative to P_2O_5 compared to samples below normal Reef (crosses).

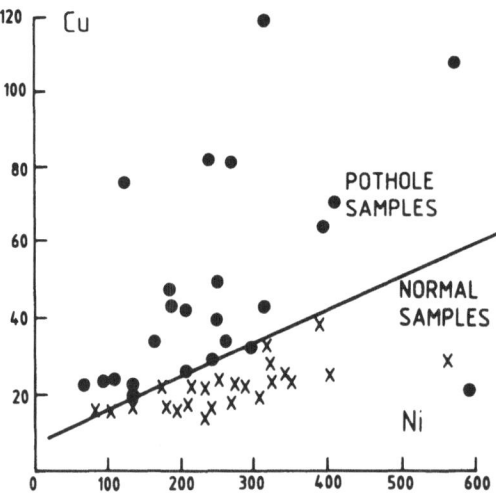

FIG. 7. Plot of Ni versus Cu in samples below potholes (solid circles) and normal Reef (crosses), illustrating the highly variable, but enriched values of Cu below potholes.

chromite (Hiemstra, 1986). Two fundamentally different models have been developed for the anomalously high concentrations of platinum and related elements in such deposits as the Merensky Reef of the Bushveld Complex (Vermaak, 1976) and the Stillwater Complex (Todd *et al.*, 1982). One hypothesis suggests that platinum is scavenged from a large reservoir of magma by an immiscible sulphide liquid, possibly as a result of magma addition and mixing (Campbell *et al.*, 1983; Irvine *et al.*, 1983). The other model involves a complex path of melt-fluid evolution, in the course of which the PGE are transported and deposited in mineralogically and texturally determined locations (Ballhaus & Stumpfl, 1985; Boudreau *et al.*, 1986). A further model which bridges the magmatic-hydrothermal debate has been presented, whereby fluids percolate through the cumulates until they breach the cumulate–magma interface (Elliot *et al.*, 1982). The reaction of fluid and magma causes sulphide precipitation. These fluids are claimed to cause the potholing of the originally planar interface. A variation of this theme, which regards the fluid as a more passive participant in the sulphide-forming and potholing processes, has also been discussed. The mixing of two magmas of suitably different composition could produce a hybrid magma which is superheated with respect to the cumulus layer with which it is in contact (Campbell, 1986). Thermochemical erosion may result. Campbell (1986) noted that in order to produce preferential erosion and pothole formation a pre-existing or latent zone of weakness must exist, and suggested that these could occur at sites of hydrothermal or fumarolic activity in the floor of the cumulus pile. The two fundamental aspects of this model are that the structure of the potholes is developed contemporaneously with the surrounding cumulates and a discrete fluid existed at that same time.

DISCUSSION

From the geochemical data, it appears that hydrothermal fluids have influenced the footwall rocks below potholes to the Merensky Reef compared to the same horizon elsewhere enriching them in K_2O, Rb and Cu. Some connection between fluids and potholes therefore seems probable. From this study it is impossible to discuss in quantitative terms what the composition of this fluid was. However, some comparisons with postulated compositions may be made. Boudreau *et al.* (1986) suggested that they are enriched in alkalis and iron. The K_2O content of samples beneath potholes is higher by almost a factor of two than under normal Reef (Table 1). The Na_2O content of these rocks is reflected in their normative albite content, and it can be seen in Table 1 that there is no significant difference in the plagioclase composition in the two different environments. In fact, the samples from below potholes may be slightly more calcic. However, in view of the high modal content of the plagioclase, the addition of extremely small amounts of Na_2O may not have a significant influence on the bulk plagioclase composition. Furthermore, if the fluids have infiltrated through the upper critical zone which is dominated by plagioclase-bearing rocks (Cameron, 1982), the fluid may have already reached equilibrium with plagioclase of composition An_{75-80}. Thus changes in composition of plagioclase in Footwall 3 would be minimal. There is a very small difference between the En content of orthopyroxenes in the two settings of 1% (Table 1) which may be attributed to the percolating fluid, although the same arguments may apply to the iron as to the Na_2O. Hence, these results do not contradict the suggestions of Boudreau *et al.* (1986), but suggest that there is little evidence for reaction between fluid and cumulates in terms of major elements.

The mineralogical changes reported here do not represent a retrogression to an equilibrium assemblage. The replacement of pyroxene by tremolite represents a reaction at higher temperature than the formation of chlorite and serpentine from the pyroxenes and zeolites and paragonite from the plagioclase. Instead, these assemblages indicate a continuous, but incomplete, reaction with decreasing temperature. These reactions are slightly more common, but not totally restricted to samples beneath potholes. It would be an oversimplification to suggest that the fluid reaction occurred at a single temperature. Prolonged existence of these fluids is implied. Thus, the hypotheses of Elliot *et al.* (1982) who suggested synchronous magmatic and fluid processes; Ballhaus & Stumpfl (1986) who estimated trapping temperatures of 730°C of fluid inclusions; Boudreau *et al.* (1986) who calculated equilibrium temperatures for the olivine–phlogopite pair of 545–700°C, and Ballhaus & Stumpfl (1985) who determined comparable temperatures based on graphite-bearing equilibria, need not be mutually exclusive. Once generated, these channelways of preferential fluid migration may have survived or been reactivated during the cooling of the intrusion.

There are several other features which may be attributed to these fluids. It has been demonstrated that the oxygen fugacity was lower in potholes than elsewhere in the Merensky Reef as reflected in the chromite composition at the base of the Reef (Elliot *et al.*, 1982). However, it is impossible to establish at what temperature this lower oxygen fugacity was imposed. Elliot *et al.* (1982) suggest that it is a magmatic phenomenon, but lower temperature fluids percolating through the footwall could produce the same effect.

The platinum-group minerals are distinctly different in potholes compared to normal Reef. In normal Reef there is a far greater proportion of Pt–Pd sulphides to Pt–Fe intergrowths, whereas the converse is true in potholed areas (Kinloch, 1982). The base metal sulphide mineralogy also changes from predominantly pyrrhotite, pentlandite, chalcopyrite in normal Reef to containing mackinawite and cubanite in addition to the previously mentioned phases in potholes. These two minerals are not produced at the high temperatures of magmatic sulphide formation but may be produced by low temperature reaction. Thus various geochemical and mineralogical observations are consistent with the ingress of fluids into the Merensky Reef via channelways which are concentrated under potholes.

ACKNOWLEDGEMENTS

The research was made possible by Gencor and Impala Platinum Mines whose logistical support and permission to publish is gratefully acknowledged. In particular, we thank Mr A. J. Craig, Mr K. G. Wulff, Mr U. F. Hahn, Mr C. D. Schulze and Mr S. R. Davey. Technical support was provided by Mrs S. Hall, Mrs A. Davies, Miss M. Krausey, Mrs P. King and Mr H. Hudson. The CSIR and Gencor provided financial support. Dr E. A. Mathez and an anonymous referee are thanked for their valuable comments on the manuscript, and Professor H. V. Eales for useful discussions.

REFERENCES

Ballhaus, C. G. & Stumpfl, E. F. (1985). Occurrence and petrological significance of graphite in the upper Critical Zone, western Bushveld Complex, South Africa. *Earth Planet. Sci. Lett.*, **74**, 58–68

Ballhaus, C. G. & Stumpfl, E. F. (1986). Sulfide and platinum mineralization in the Merensky Reef: evidence from hydrous silicates and fluid inclusions. *Contrib. Mineral. Petrol.*, **94**, 193–204.

Boudreau, A. E., Mathez, E. A. & McCallum, I. S. (1986). Halogen geochemistry of the Stillwater and Bushveld Complexes: evidence for transport of the platinum-group elements by Cl-rich fluids. *J. Petrol.*, **27**, 967–87.

Cameron, E. N. (1982). The upper critical zone of the eastern Bushveld Complex—precursor of the Merensky Reef. *Econ. Geol.*, **77**, 1307–27.

Campbell, I. H. (1986). A fluid dynamic model for the potholes of the Merensky Reef. *Econ. Geol.*, **81**, 1118–25.

Campbell, I. H., Naldrett, A. J. & Barnes, S.-J. (1983). A model for the origin of the platinum-rich sulfide horizons in the Bushveld and Stillwater Complexes. *J. Petrol.*, **24**, 133–65.

Eales, H. V., Marsh, J. S., Mitchell, A. A., De Klerk, W. J., Kruger, F. J. & Field, M. (1986). Some geochemical constraints upon models for the crystallization of the upper Critical Zone–Main Zone interval, northwestern Bushveld Complex. *Mineral. Mag.*, **50**, 567–82.

Elliot, W. C., Grandstaff, D. E., Ulmer, G. C., Buntin, T. & Gold, D. P. (1982). An intrinsic oxygen fugacity study of platinum-carbon association in layered complexes. *Econ. Geol.*, **77**, 1493–510.

Farquhar, J. (1986). The Western Platinum Mine. In *Mineral Deposits of Southern Africa, II*, ed. C. R. Anhaeusser & S. Maske. Geological Society of South Africa, Johannesburg, pp. 1135–42.

Ferguson, J. & Botha, E. (1963). Some aspects of igneous layering in the basic zones of the Bushveld Complex. *Trans. Geol. Soc. S. Afr.*, **66**, 259–78.

Henderson, P. (1968). The distribution of phosphorus in the early and middle stages of fractionation of some basic layered intrusions. *Geochim. Cosmochim. Acta*, **32**, 897–911.

Hiemstra, S. A. (1986). The distribution of chalcophile and platinum-group elements in the UG-2 chromitite layer of the Bushveld Complex. *Econ. Geol.*, **81**, 1080–6.

Irvine, T. N., Keith, D. W. & Todd, S. G. (1983). The J-M platinum-palladium reef of the Stillwater Complex, Montana. II. Origin by double-diffusive convective magma mixing and implications for the Bushveld Complex. *Econ. Geol.*, **78**, 1287–334.

Kinloch, E. D. (1982). Regional trends in the platinum-group mineralogy of the critical zone of the Bushveld Complex, South Africa. *Econ. Geol.*, **77**, 1328–47.

Leeb-Du Toit, A. (1986). The Impala Platinum Mines. In *Mineral Deposits of Southern Africa II*, ed. C. R. Anhaeusser & S. Maske. Geological Society of South Africa, Johannesburg, pp. 1091–106.

Naldrett, A. J. (1981). Platinum-group element deposits. In *Platinum-group Elements: Mineralogy, Geology and Recovery*, ed. L. J. Cabri. Canadian Institute Mining and

Metallurgy, Special Volume 23, pp. 197–232.

Schmidt, E. R. (1952). The structure and composition of the Merensky Reef and associated rocks on the Rustenburg Platinum Mine. *Trans. Geol. Soc. S. Afr.*, **55**, 233–79.

Schwellnus, J. S. I., Hiemstra, S. A. & Gasparrini, E. (1976). The Merensky Reef at the Atok platinum mine and its environs. *Econ. Geol.*, **71**, 249–60.

Todd, S. G., Keith, D. W., Schissel, D. J., Leroy, L. L., Mann, E. L. & Irvine, T. N. (1982). The J-M platinum-palladium reef of the Stillwater Complex, Montana: I. Stratigraphy and petrology. *Econ. Geol.*, **77**, 1454–80.

Vermaak, C. F. (1976). The Merensky Reef—thoughts on its environment and genesis. *Econ. Geol.*, **71**, 1270–98.

Viljoen, M. J., De Klerk, W. J., Coetzer, P. M., Hatch, N. P., Kinloch, E. & Peyerl, W. (1986). The Union Section of Rustenburg Platinum Mines, with reference to the Merensky Reef. In *Mineral Deposits of Southern Africa, II*, ed. C. R. Anhaeusser & S. Maske. Geological Society of South Africa, Johannesburg, pp. 1061–90.

Von Gruenewaldt, G. (1979). A review of some recent concepts of the Bushveld Complex, with particular reference to sulfide mineralization. *Can. Mineral.*, **17**, 233–56.

Wager, L. R., Brown, G. M. & Wadsworth, W. J. (1960). Types of igneous cumulates. *J. Petrol.*, **1**, 73–85.

36

Geochemistry of the Mount Ayliff Intrusion and the Origin of Associated Sulphide Mineralization

R. GRANT CAWTHORN

Department of Geology, University of the Witwatersrand, Wits 2050, Johannesburg, South Africa

The Mount Ayliff Intrusion (Insizwa Complex) is one of the thickest of all Karoo intrusive sills in southern Africa. Several aspects make it distinct from all other sills. It contains near-liquidus ilmenite grains with up to 10% MgO; it contains chromite with high Cr_2O_3 and TiO_2 concentrations; and it has nickel-copper sulphide mineralization at its base. Ilmenite with similar composition has been observed as phenocrysts in the picrite basalts in the Karoo and hence on the basis of the ilmenite and chromite compositions it is argued that the lower part of the intrusion crystallized from a highly magnesian magma.

In one vertical profile through the picrite at the base of the Tonti lobe of the intrusion, the relative changes in nickel content of the olivine provide evidence of two periods of separation of immiscible sulphide liquid which can be related to horizons where magma addition and mixing may have occurred. Incompatible trace element ratios such as Zr/P and K/P provide further evidence for magma mixing at these horizons. Through the adjacent Ingeli lobe of the intrusion a distinctly different profile for the nickel content of olivine exists which suggests continuous sulphur saturation of the magma rather than pulses of excessive sulphide separation.

An intermittent stratabound, pegmatoidal zone has been identified in the gabbro overlying the picrite in one area of the intrusion. It is capped by a medium-grained gabbro which may have acted as a barrier to residual magma and/or fluid. This pegmatoidal horizon contains minor, but higher, concentrations of platinum compared to surrounding rocks.

A study of the oxygen fugacity in the picrites has been attempted by use of the chromite ferrous:ferric ratios. The oxygen fugacity is invariably low in samples which show a significant depletion of nickel content in the olivine, suggesting a close relationship between oxygen fugacity and the presence of sulphides. This is inferred to be related to local subsolidus reequilibration rather than a magmatic effect. This demonstrates that the volatile species were buffered by the mineral assemblage of the host rock with which it was in intimate association, and that there was no major flow of fluids through this pile which imposed their own volatile fugacities on the rocks.

37

The Effects of Metasomatising Fluids on the PGE-content of the UG-1 Chromitite Layer

ROLAND K. W. MERKLE

Institute for Geological Research on the Bushveld Complex, University of Pretoria, Pretoria 0002, South Africa

Research in recent years has indicated the importance of late magmatic and hydrothermal processes for the formation and modification of platiniferous rocks in the Bushveld Complex. However, very little is known about the exact nature of these processes and their effects (i.e., mobilisation or addition of PGE) on the platinum-group minerals or on whole rock PGE patterns.

For this study 26 samples of the UG-1 chromitite layer in the vicinity of a magnetite-pipe in the eastern Bushveld were selected, where thin veins of magnetite-rich spinel crosscut unmetasomatised chromitite. Metasomatic changes of chromite composition and silicates are very localised and are only detectable in close proximity to these veins. Addition of water and Fe changed the assemblage of chromite plus orthopyroxene ($mg^{\#}$ 82) with minor clinopyroxene, hornblende and plagioclase in unmetasomatised UG-1 to predominantly clinopyroxene ($mg^{\#}$ 80 to 73) and an Fe-enriched chromite with increasing metasomatism. In the veins the assemblage consists of a magnetite-rich spinel with small amounts of olivine ($mg^{\#}$ 54). The increase of the Fe-content in the spinel allows samples to be arranged in order of increasing degree of metasomatism.

Some 2000 microprobe analyses of spinel in unaltered and metasomatised samples show that metasomatism not only led to an increase of Fe, but also of Ti, V, and Mn and a decrease of Mg, Al, and Cr. For Zn, Ni, and Co no systematic behaviour is indicated. Only in profiles across the contact between chromite and vein-spinel can it be demonstrated that Ni and Co contents increase and Zn content in the spinel decreases.

PGM in unmetasomatised UG-1 samples consist predominantly of laurite enclosed by chromite, occasionally together with native Ru, native Os, or Pt–Pd–Rh sulphides and base metal sulphides. In metasomatised samples laurite is also the dominant PGM. However, Pt–Fe alloys become more abundant and Pt–Rh–Fe and Pt–Pd–Fe alloys were observed.

Despite changes in the platinum-group element mineralogy, whole rock PGE analyses indicate that metasomatism of the UG-1 chromitite layer did in most cases not effect the bulk PGE in any way that would exceed analytical uncertainty. Only in some of the samples from the channelways for the metasomatising fluids, a decline in Pt content is indicated.

Preferential mobilisation of Pt from chromitites and silicate rocks in the critical zone could account for the Pt dominance in the hortonolitic dunite pipes and other hydrothermal PGE deposits of the Bushveld Complex.

38

Platinum-Group Element Mineralization in the Ultramafic Sequence of the Acoje Ophiolite Block, Zambales, Philippines

B. Orberger[a], G. Friedrich[a] & E. Woermann[b]

[a]Institut für Mineralogie und Lagerstättenlehre, [b]Institut für Kristallographie,
RWTH Aachen, Wullnerstrasse 2, D-5100 Aachen, Federal Republic of Germany

ABSTRACT

Platinum-group minerals (PGM) are related to the ultramafic sequence of the Acoje ophiolite block, which was recently mined for metallurgical chromite.
Two groups of PGM have been observed:

1. Os–Ir–Ru sulphides and alloys as subeuhedral inclusions in both chromitite in dunite enveloped lenses and chromitite in the transition zone at the mantle/crust boundary.

Laurites are Os-enriched, heterogeneous in their chemical composition and sometimes rimmed by Ni-sulphide. They are syngenetic with chromitite.

2. Pt–Pd–(Ni–Cu)–Te–Bi–S–As minerals in black dunite lenses, related to a disseminated Ni–Cu sulphide mineralization. They occur in oxidized parts of the sulphides, close to the sulphides in the serpentine matrix, as inclusions in magnetite or as replacements at the margin of magnetized sulphides.

Pt–Pd–(Ni–Cu)–Te–Bi–S–As minerals show irregular shapes. They are also heterogeneous in their chemical composition. It is suggested that primary Pt and Pd are remobilized from either distinct PGM or from another primary mineral phase and reprecipitated during or postserpentinization as low-temperature PGM-phases. Pt and Pd in Ni–Cu sulphides are rarely above the detection limit of the proton microprobe.

INTRODUCTION

Platinum-group element (PGE) mineralization, hosted in ophiolites, has been investigated by Prichard & Neary (1981), Prichard et al. (1986); Legèndre & Augé (1986), Talkington et al. (1983, 1984, 1986); Burgath & Mohr (1986) and others.

While these studies were mainly restricted to chromitites, our investigations deal with the PGM mineralization in the ultramafic sequence as a whole, based on detailed SEM-analysis, EDS and β-autoradiography. The purpose of the present study is to locate PGMs within the lower sequence of the Acoje ophiolite block and to determine their chemical compositions. The formation of PGMs is discussed with respect to their textures in comparison to PGMs, which have been investigated in layered intrusions and ophiolites.

THE ACOJE OPHIOLITE BLOCK

The Acoje/Barlo ophiolite block as well as the Coto/Eastern block belong to the Massinloc Massif, the northern part of the Zambales ophiolite complex in NW-Luzon (Fig. 1). The Acoje and Coto blocks are considered to be derived from an island arc- and back arc-magmatism, respectively (Hawkins & Evans, 1983). Bacuta (1978) and Hawkins & Evans (1983) distinguish from west to east: peridotites, including tectonized harzburgite, serpentinized dunites, websterites, pyroxenites and lherzolites, overlain by layered and massive gabbro. The upper parts of the sequence comprise basaltic and diabasic rocks, and the most eastern part tertiary volcanics.

The Zambales range has been translated northward along the E-Luzon shear (Fuller *et al.*, 1983). The emplacement process of the ophiolite, whether it is an obduction of lithospheric plates on Miocene sediments or a repeated east-dipping subduction of lithosphere was discussed by several authors (Violette, 1980; Hawkins & Evans, 1983; Karig, 1983; Bischke (personal communication)).

Podiform chromitite within the boundary harzburgite/serpentinized dunite is mined in the Acoje and Coto blocks. Coto hosts refractory (Al-rich) chromite, while Acoje chromite is of metallurgical grade (Cr-rich). Disseminated Ni-sulphides of economic value have been found only in black dunite lenses in Acoje and their chemistry has been investigated by Abrajano (1984). Platinum has been mined as a byproduct (Hulin, 1950). PGE distribution, particularly in chromitites has been studied by Bacuta *et al.* (1988). A few PGM have been reported in mineral separates or in the serpentine matrix by Cabri (1982, written communication), Abrajano (1984) and Orberger *et al.* (1985).

PETROGRAPHY OF THE ULTRAMAFICS IN ACOJE

A 2·5 km long profile was mapped and sampled in the Acoje ophiolite block in underground levels 1520f, 1250f, 1000f and on surface of the Acoje mine. Samples were taken every 4·5 m on level 1250f and every third sample has been analysed by XRF. These data are represented as a stratigraphic column (Fig. 2). The profile is based mainly on samples from level 1250f, 1520f and the surface.

Ultramafics of the Upper Mantle

The upper mantle consists of tectonized harzburgite, which contains dunite enveloped chromitite-lenses. At the mantle-crust boundary folded massive chromitite

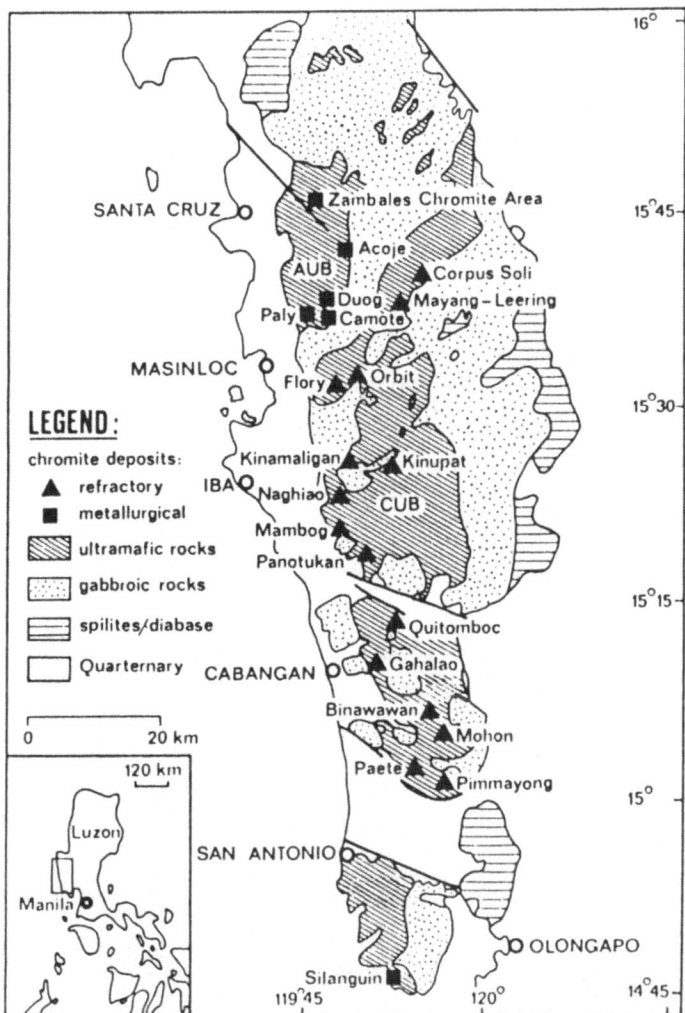

FIG. 1. Geological map and structural units of the Zambales range with localities of chromite deposits: AUB-Acoje ophiolite block. (Map: Philippine Bureau of Mines, modified by Hock & Friedrich (1986).)

occurs. Ru–Os–Ir sulphides and alloys are found as inclusions both in Cr-spinels of these chromitite-lenses and in massive chromitite of the transition zone (Fig. 2).

Ultramafic Cumulates—Lower Oceanic Crust

Upwards the ore grades from massive to banded and finally to disseminated chromitite in serpentinized dunite. The ultramafic sequence of the Acoje block consists mainly of four dunitic units (I–IV), which are highly serpentinized. The

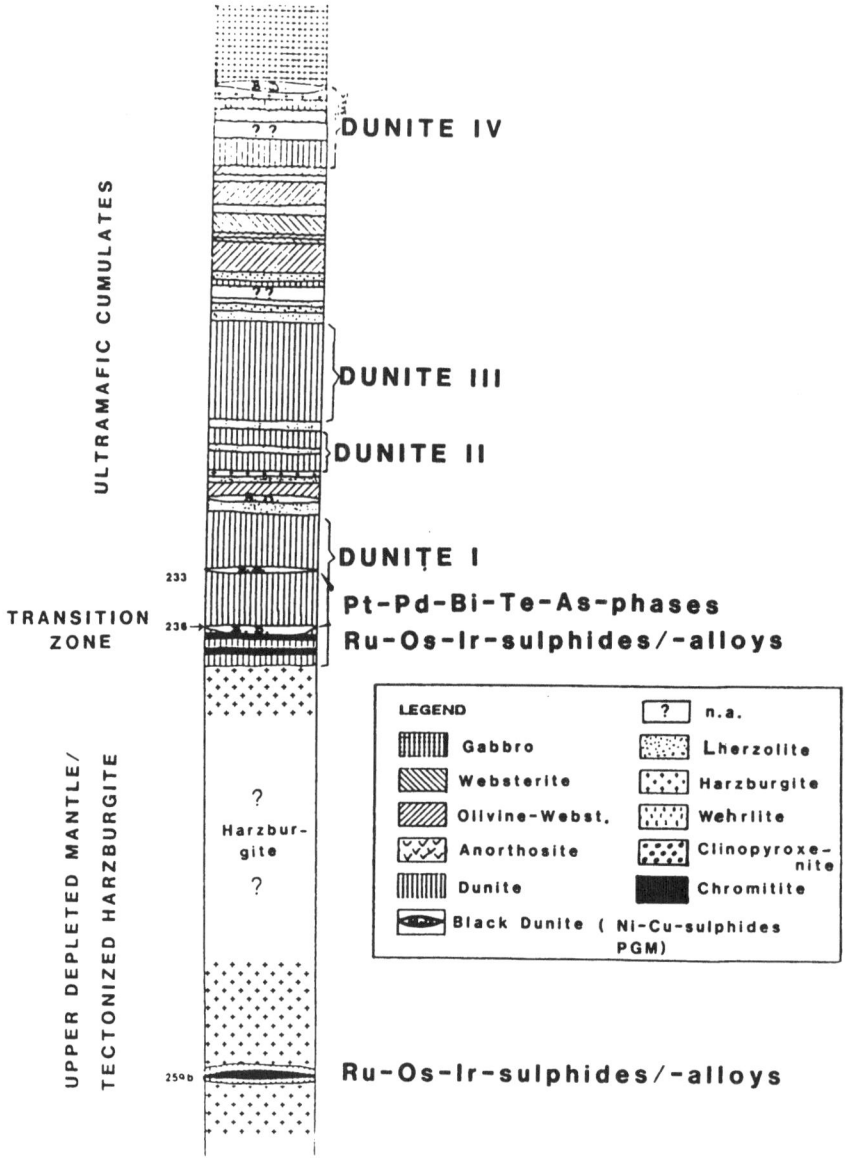

FIG. 2. Profile of level 1250f (Acoje Mine). Upper part, above the harzburgite is based on sampling (every 4·5 m) and XRF-analyses of every third sample, until now. The lower part (harzburgite) is reconstructed from surface sampling and XRF analyses of chromitite, enveloped by dunite (sample no. of analysed rocks are marked left of the profile). M/Si = Mg + Fe + Mn + Ni/Si. n.a. = not analysed.

sequence contains several 'black dunite' lenses. No tectonic contact is observed between green dunite and black dunite lenses, but there is a sharp, marked colour change from dark green to black. Underground this colour change is difficult to observe. Disseminated Ni- and Cu-sulphides of different generations often characterize these lenses.

Dunite I ($\sim$300 m) is the first unit of the ultramafic cumulates. At the base it contains chromitite and at least two black dunite lenses with disseminated Ni- and Cu-sulphides, associated with Pt–Pd–Bi–Te–As minerals. Dunite I and Dunite II ($\sim$100 m) are separated by an alternation of lherzolite and olivine-websterite. Lower lherzolite is overlain by a black dunite lens, which is poor in sulphide. The olivine-websterite is cpx-dominated. Between lherzolite and Dunite II ($\sim$200 m) a harzburgitic unit is intercalated.

Dunite II contains a pyroxene-rich unit and is separated from Dunite III also by lherzolite. Towards the top of the ultramafics pyroxene becomes more abundant. Dunite III is overlain by alternating olivine-websterites, websterites and clino-pyroxenites. Wehrlite is only present in a small band above Dunite III.

Close to the gabbro sequence, at the seismic Moho, a further dunitic unit IV ($\sim$50 m), harzburgite and a sulphide poor black dunite lens occur.

The entire ultramafic sequence, intersected by clinopyroxenite, gabbro and anorthosite dykes, is highly sheared. Petrographical studies on a larger scale profile, concerning the whole ophiolite sequence have been carried out by Hawkins & Evans (1983), Hock (1983) and Hock & Friedrich (1986).

Serpentinization affected mainly the dunitic units, while lherzolite, clino-pyroxenite and websterites are less serpentinized. The $Mg + Fe + Mn + Ni/Si$ ratio $> 1 \cdot 97$ in serpentinized dunites suggests an original composition of almost pure olivine.

ANALYTICAL TECHNIQUES

Scanning electron microscope (SEM) and energy dispersive spectroscopy (EDS) analyses have been carried out in the Max-Planck-Institut für Chemie/Mainz and the Gemeinschaftslabor für Elektronenmikroskopie of the RWTH, Aachen.

Platinum group minerals have been observed only by SEM because of the very small grain size ($1–10\ \mu$m). The EDS has been used for quantitative analyses of the PGE with pure element standards. The accelerating voltage was 20 kV and the sample current 100 pA.

The β-autoradiography technique is described in Potts (1984). Irradiation of polished thin sections on Sb-free glasses have been investigated for Ir-, Sb-, As-, Au- and Co-bearing phases. This technique is very useful for detecting laurites or Ir–Os–Ru alloys as inclusions in Cr-spinels. About 80% of the irradiated dunite samples showed black dots corresponding to Co–Fe phases in oxidized parts of Ni–Cu sulphides. In many cases investigation of these sections by SEM led to a detection of minute native Pt grains ($1/2\ \mu$m) in the vicinity of the Co–Fe phases.

PLATINUM-GROUP MINERALS AND GOLD

Two groups of platinum-group minerals can be distinguished in the Acoje ophiolite block:

1. Laurites and Ir–Os–Ru alloys:
   —in chromitite, enveloped by dunite within harzburgite;
   —in massive chromitite within the transition zone.
2. Pt–Pd–Bi–Te–As–(Cu–Fe) minerals in black dunite:
   —in the serpentine matrix close to Ni- or Cu-sulphides;
   —on oxidized parts of sulphides.

Locations of the PGM in the different ore bodies are marked in Fig. 3.

Laurite and Ru–Os–Ir Alloys

Laurite and Ru–Os–Ir alloys occur as euhedral inclusions in Cr-spinels of chromitite in dunite lenses within tectonized harzburgite and in chromitite within the transition zone between upper depleted mantle and lower oceanic crust.

The grain diameter is $2–10\ \mu m$. Different shapes of laurites are presented in Fig. 4(a–d). In the investigated samples laurites and Ru–Ir–Os alloys often coexist. Figure 4(d) shows a two-phase grain: the lower bright part has laurite composition while the upper dark part consists of Ru-, Os-, Ir-bearing millerite. Pt-, Pd- or Rh-contents are below the detection limit of EDS. Figure 4(b) shows a scanning picture of an inhomogeneous laurite-inclusion with a Ni–Fe sulphide rim. Os-enriched parts are dark, Ru–Ir dominated areas are light grey.

The results of quantitative EDS analyses are listed in Table 1 (305, 259b, 195). The analysed chromium probably presents a contribution from the surrounding Cr-spinels, due to the small size of laurites in relation to the electron beam diameter. Figure 5 shows the Acoje laurites in chromitite in harzburgite and in massive chromitite from the transition zone. The compositions of laurite inclusions in Cr-spinels from the Shetland ophiolite (Prichard *et al.*, 1986), Troodos, Vourinos and Tiébaghi (Legèndre & Augé, 1986) and the Meratus-Bobaris ophiolite zone/SE-Kalimantan (Burgath & Mohr, 1986) are shown for comparison.

Laurite inclusions from chromitite in harzburgite are slightly more Os- and Ir-enriched than laurite inclusions from chromitite within the transition zone. Others contain high amounts of Os ($\sim$11 at.%) and Ru ($\sim$20 at.%) only.

Pt–Pd–Bi–Te–As Minerals

Pt–Pd–Bi–Te–As minerals occur in black dunite lenses. They are always associated with Ni- and Cu-sulphides. Tiny grains of $2–10\ \mu m$ have been located:

(a) in oxidized parts of Cu-sulphides (Figs 6(c, d), 7(a, b)); or
(b) as inclusions in secondary magnetite; or
(c) in the serpentine matrix in the vicinity of oxidized Cu–Ni sulphides (Figs 6(a, b), 7(c, d)).

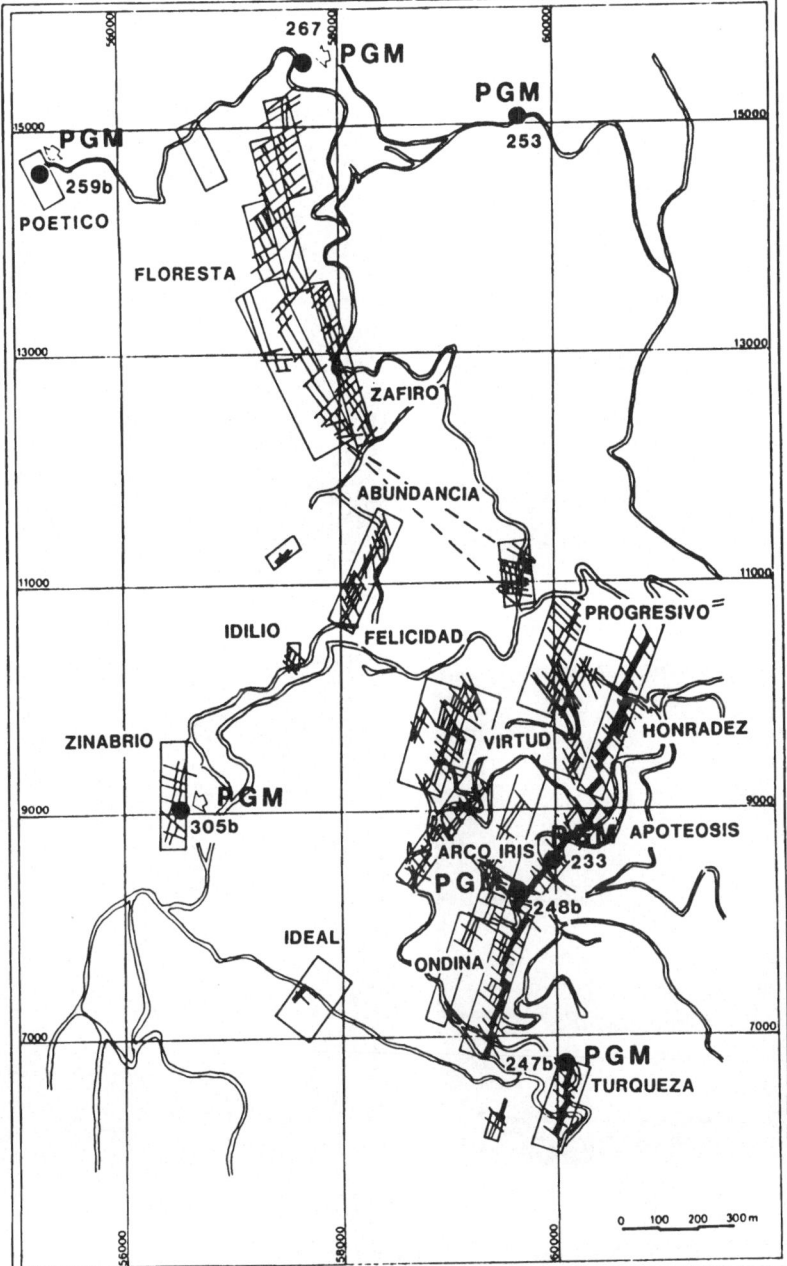

FIG. 3. Location of the chromite and sulphide ore bodies, projected to the surface. ☐ chromitite; ■ sulphide ore; PGM: locations of platinum group minerals reported here (Acoje Mine). (Mapping of the ore bodies: Ignacio, 1979.)

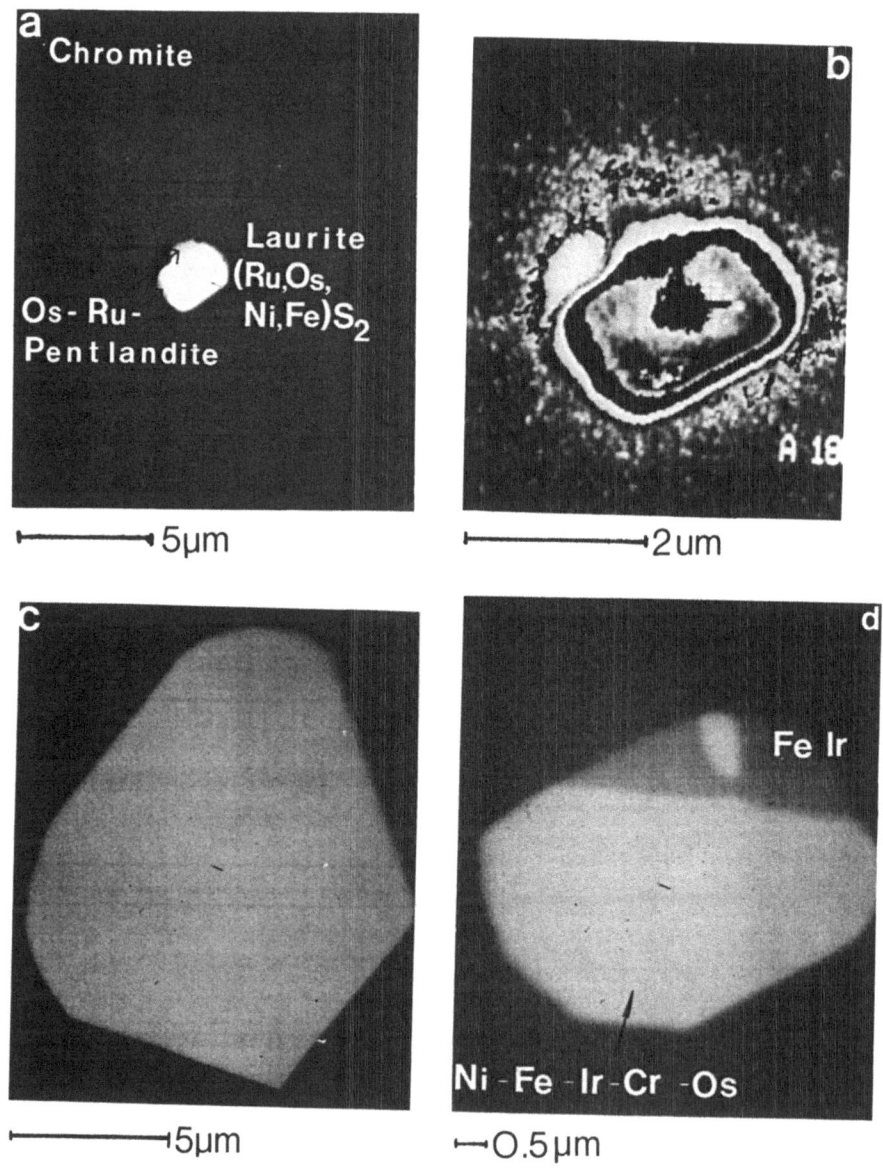

FIG. 4. (a) Subeuhedral laurite inclusion in Cr-spinel from dunite lense in harzburgite (259). (b) Scanning picture of (a): Os-Ru bearing Ni-sulphide forms a zone around laurite. (c) Subeuhedral laurite-inclusion in chromite within the transition zone: mantle harzburgite/lower cumulates (195). (d) Euhedral bi-phase-inclusion in chromite. Bright part: Ni-Fe-Ir-Cu-Os bearing laurite, upper dark part is Fe-richer and Ir-poorer (305b).

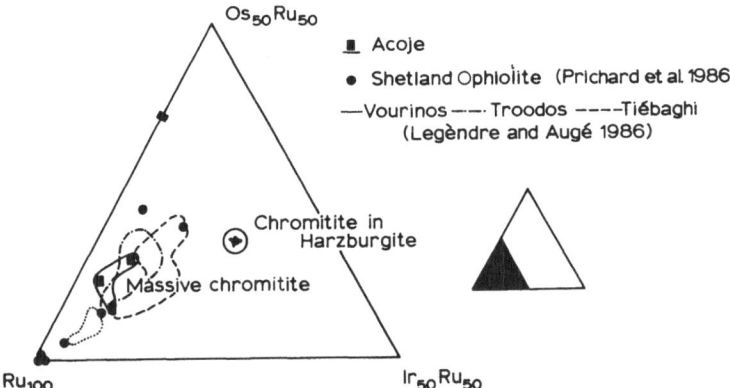

FIG. 5. Compositions of laurite inclusions in Cr-spinels from chromitites, Acoje Mine, compared to laurites from chromitites: Shetland (Prichard *et al.*, 1986), Vourinos-, Troodos- and Tiébiaghi-ophiolite complexes (Legèndre & Augé, 1986).

Four main types of Pt–Pd compounds can be distinguished (Table 2):

1. Tellurobismuthides   (233-2, 233c)
2. Arsenides   (233 a1,a2,h, 2333,3a)
3. Cu–Ni–Fe–Pt–Pd alloys   (233b,c,d,e,f)
4. Pt–Pd–Ag alloys

EDS investigations have been carried out on small PGM grains and also on exsolved or heterogeneous parts (Figs 6(b), 7(d)). High quality analyses could not be obtained because of the tiny grain size of 2–5 μm or even smaller. It was, however, important to get information about the major changes in the chemical composition. These analyses, normalized to 100, are presented in Table 2.

Pt–Pd Bearing Tellurobismuthides
Pt, Pd, Bi, Te minerals occur either as inclusions in magnetite, trapped during magnetite precipitation or in oxidized parts of Cu-sulphides (Fig. 6(c, d)). Pt, Pd, Bi and Te are inhomogeneously distributed. Quantitative analyses suggest a moncheite to merenskyite composition (233g, Table 2).

Pt–Pd Bearing Arsenides
Arsenides occur also in magnetite (Fig. 7(a, b)) or in serpentine matrix in contact with Ni-sulphide and magnetite. The composition of the major part of the grain in Fig. 6(a, b) (233h, Table 2) is ±Ni$_3$As, lighter grey patches consist of Cu–Pt alloy containing small amounts of Fe and Ni. Bright reflecting parts are particularly Pt–Fe-rich.

    The two-phase grain in Fig. 7(a, b) consists of a Cu–Pd–Pt rim with a Ni–As–Cu–Fe–Pt–Pd core. This arsenide is located at the margin of magnetite in an already altered part of magnetite.

B. Orberger, G. Friedrich, E. Woermann

TABLE 1

Composition of laurites and Ir–Os–Ni–Fe alloy in chromitites within harzburgite and from the transition zone of the Acoje ophiolite block.

Element	305 (wt.%)	305 (wt.%)	305 (wt.%)	305 (wt.%)	259b (wt.% (Rim))	259b (wt.%)	195 (wt.%)	195 (wt.%)	195 (wt.%)
Ru	35·42	36·49	37·69	37·32	9·48	25·09	41·52	42·64	43·14
Os	15·93	16·74	16·03	17·10	6·07	27·15	14·15	14·63	14·87
Ir	8·57	8·78	9·06	8·53	—	—	6·26	5·99	6·53
Fe	0·74	0·68	0·74	0·60	17·06	2·75	0·87	0·90	0·85
Ni	0·18	0·29	0·26	0·31	20·97	2·64	—	—	—
Cr	2·07	1·45	1·52	1·45	—	—	1·74	1·83	1·63
S	32·12	33·27	33·69	33·91	19·25	23·76	33·78	34·15	34·44
	95·03	97·70	98·99	99·22	72·83	81·39	98·32	100·14	101·46
				(at.% (normalized to 100))					
Ru	22·80	22·89	23·28	23·00	6·75	20·25	25·36	25·63	25·71
Os	5·45	5·58	5·26	5·60	2·30	11·64	4·59	4·67	4·71
Ir	2·90	2·90	2·94	2·77	—	—	2·01	1·89	2·05
Fe	0·86	0·77	0·83	0·66	22·00	4·02	0·96	0·98	0·91
Ni	0·20	0·31	0·28	0·33	25·72	3·66	—	—	—
Cr	2·59	1·76	1·83	1·74	—	—	2·07	2·14	1·89
S	65·19	65·79	65·58	65·90	43·23	60·43	65·02	64·69	64·72

Element	195 (wt.%) dark	195 (wt.%) grey	195 (wt.%)	195 (wt.%)	195 (wt.%)	195 (wt.%)
Ru	15·10	4·48	—	—	—	—
Os	3·72	0·68	0·96	1·22	0·43	0·96
Ir	1·13	0·72	6·78	13·39	18·40	6·78
Fe	14·09	19·80	17·63	8·66	7·79	17·63
Ni	24·78	35·69	73·75	70·87	71·52	73·75
Cr	3·75	7·18	—	—	—	—
S	29·82	28·29	—	—	—	—
	92·39	96·84	99·83	96·43	100·70	99·83
			(at.% (normalized to 100))			
Ru	8·07	2·18	—	—	—	—
Os	1·06	0·18	0·31	0·43	0·15	0·31
Ir	0·32	0·18	2·17	4·73	6·40	2·17
Fe	13·62	17·43	19·45	10·51	9·33	19·45
Ni	22·80	29·88	77·38	81·89	81·43	77·38
Cr	3·90	6·78	—	—	—	—
S	50·24	43·37	—	—	—	—

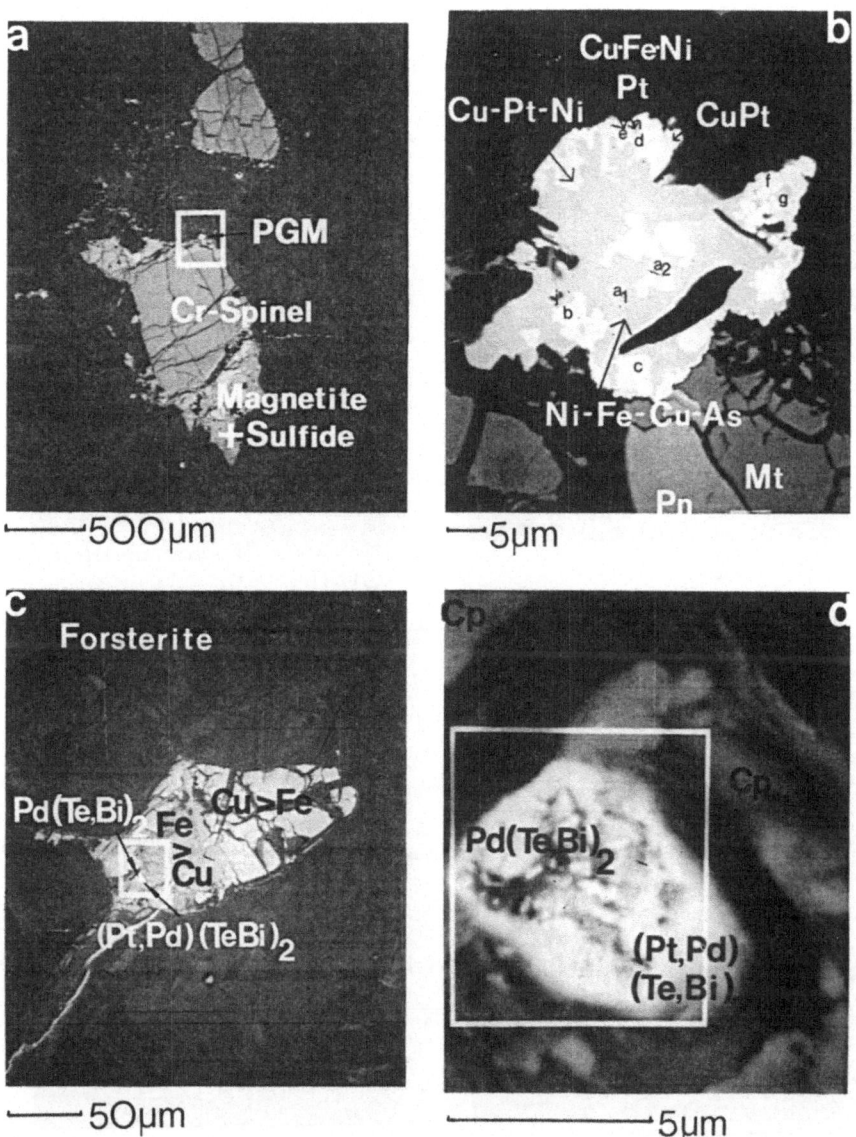

FIG. 6. (a) PGM indicated in contact with Ni-sulphide and magnetite surrounding an accessory Cr-spinel in black dunite (233). (b) Enlargement of PGM in Fig. 4(a). Ni-arsenides (with Cu and Fe) show exsolutions of various Pt alloys (233a,h, Table 2). (c) Altered Cu-sulphide with (Pt,Pd) (Te,Bi)₂(247b). (d) Enlargement of the lower part of the Cu-sulphide of (c): Inhomogeneous distributions of Pt-Pd-Bi-Te are characteristic. The upper part does not contain Pt (247b).

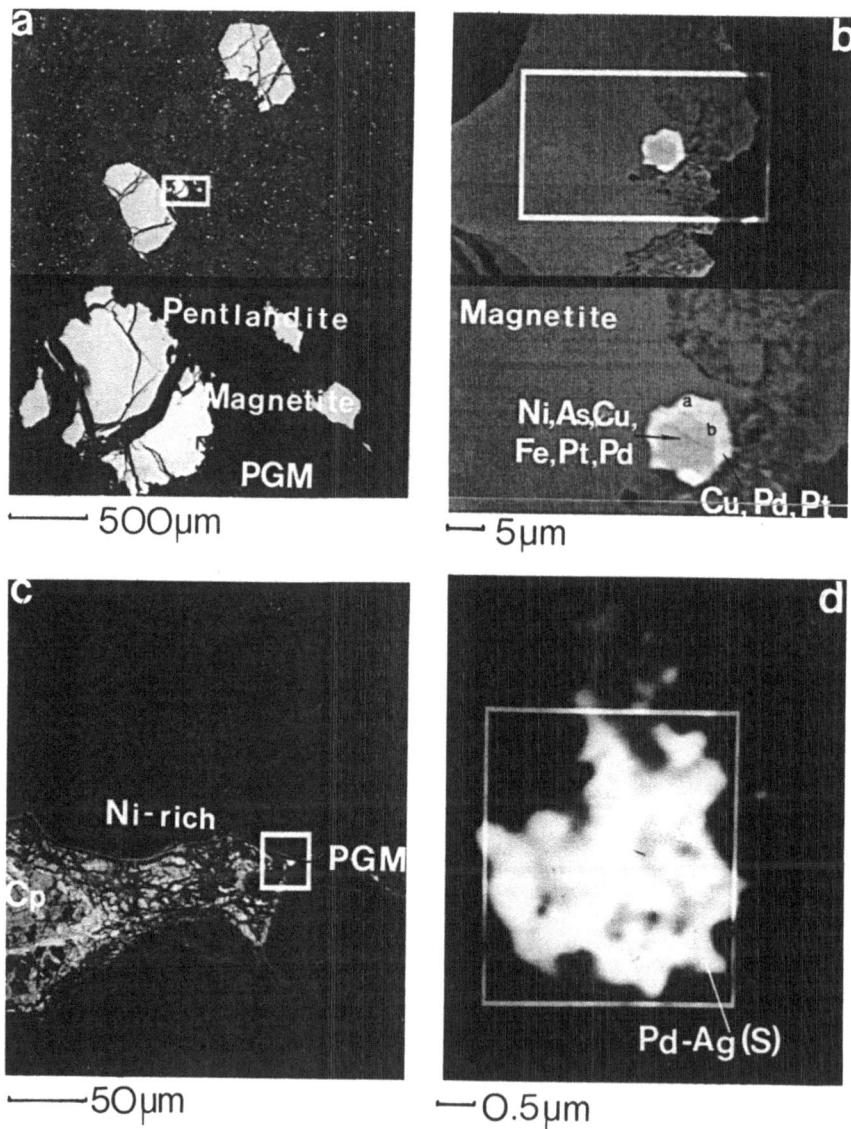

FIG. 7. (a) Accessory Cr-spinel and oxidized pentlandite in black dunite. PGM is located in magnetite (247b). (b) Enlargement of (a) showing the PGM: dark core consists of Ni–As–Fe, minor Pt,Pd; bright rim is a Cu–Pd–Pt intermetallic phase (247b). (c) Late formed Pt-Pd tellurobismuthide at the border of oxidized Cu-sulphide. The surrounding rim is Ni-rich (248b). (d) Diffuse aggregate of Pd–Ag–(S?) in magnetized part sulphide (253).

Cu–Ni–Fe–Pt–Pd Alloys
Metal alloys form exsolutions or replacements in Ni-arsenides (Fig. 6(a,b)). Chemical compositions of different alloys are given in Table 2 (233b–233f).

Pd–Pt–Ag Aggregates
Pd–Pt–Ag aggregates (Fig. 7(d)) occur on oxidized parts of Cu-sulphides or in the serpentine matrix, close to an altered sulphide. The grain size varies between 0·5 and 5–10 μm. Irregular shapes and heterogeneous compositions are characteristic.

Gold
Native gold grains have been detected within the serpentine matrix of black dunites in contact with chromitite. The grain size varies from 5 to 300 μm. Heterogeneous composition is characteristic. Figure 8(a,b) shows a Au-grain with Cu-rich areas at the ends and higher Ag content in the centre.

DISCUSSION

Intrusive rocks show a large range in PGE fractionation: from Pt–Pd depletion in ophiolitic chromitites to a Pd–Pt enrichment in, for example, the J-M Reef of the Stillwater Complex, Montana (Page *et al.*, 1982).

PGE distribution is caused by three major processes; partial melting, fractional crystallization and alteration (Barnes *et al.*, 1985). Chromitites from various ophiolites are characterized by strong enrichment of Os–Ir–Ru and depletion of Pd–Pt–Rh (Page *et al.*, 1982). This mineralogical study also confirms a PGE-fractionation in Acoje: laurites and Os–Ru–Ir alloys are restricted to inclusions in Cr-spinels, Pt–Pd intermetallic phases to black serpentinites associated with Ni–Cu sulphides.

Some atypical PGE patterns of ophiolites, with an enrichment of Pt–Pd are also known from ophiolitic chromitites, e.g. Shetland, where Pt–Pd–Rh minerals have been found in the serpentine matrix of chromite-rich samples (Prichard & Neary, 1981; Prichard *et al.*, 1986). Bacuta *et al.* (1987) also reported an enrichment of Pt and Pd in the upper level chromitites in Acoje. Those chromitites however are in contact with Ni-Cu sulphide bearing black dunite lenses and PGM occur in the corresponding serpentine matrix.

Formation of Laurites and Os–Ru–Ir Alloys
Different theories have been adopted for the formation of laurite and Os–Ru–Ir alloys.

An exsolution of PGE from chromite was proposed by Gijbels *et al.* (1974), whereas Talkington *et al.* (1983, 1984) and Constantinidis *et al.* (1980) favoured precipitation of PGE-bearing sulphides from liquids, which have been trapped during chromite crystallization.

Another possibility for the formation of laurite and alloys is the preconcentration of PGE in Fe-droplets (Hiemstra, 1979) or in Ni-droplets (Legèndre, 1982). Stockman and Hlava (1984), Prichard & Neary (1981) and Augé (1985) agree that

TABLE 2

Composition of Ni₃As, Pt–Cu alloys, Pt–Fe tellurides and Cu–Bi tellurides from black dunites of the Acoje ophiolite block

Element	233a1 (wt.%)	233a2 (wt.%)	233b (wt.%)	233c (wt.%)	233d (wt.%)	233e (wt.%)	233f (wt.%)	233g (wt.%)
Fe	0·86	0·93	1·43	1·70	15·83	18·12	0·96	1·04
Ni	66·91	66·65	2·18	2·30	5·85	6·10	1·66	59·45
Cu	1·79	2·14	57·86	55·46	5·89	6·01	53·93	3·44
Pt	—	—	35·46	32·90	60·53	67·78	28·23	—
As	30·68	31·19	—	—	—	—	—	27·36
	100·24	100·91	96·93	92·36	88·10	98·01	84·78	91·29
				(at.% (normalized to 100))				
Fe	0·97	1·04	2·22	2·75	36·06	37·27	1·65	1·28
Ni	71·55	70·87	3·21	3·52	12·68	11·94	2·72	69·81
Cu	1·77	2·10	78·83	78·55	11·79	10·87	81·70	3·73
Pt	—	—	15·74	15·18	39·48	39·92	13·93	—
As	25·71	25·99	—	—	—	—	—	25·18
	100·00	100·00	100·00	100·00	100·00	100·00	100·00	100·00

Element	233h (wt.%)	2332 (wt.%)	2332c (wt.%)	2332e (wt.%)	2332f (wt.%)	2333 (wt.%)	2333a (wt.%)	2481 (wt.%)
Fe	2·97	17·00	19·65	1·09	16·45	1·06	2·08	—
Ni	50·73	1·00	9·65	30·29	14·70	55·95	7·06	—
Cu	3·74	—	1·26	—	—	10·65	48·21	—
Pt	8·65	24·85	65·49	—	57·86	3·10	10·54	37·83
As	24·31	—	—	—	—	25·48	0·90	—
Pd	—	—	—	—	—	1·55	15·68	—
Bi	—	—	—	16·14	—	—	—	7·62
Te	—	43·50	4·04	50·87	10·99	—	—	54·55
S	—	—	—	1·61	—	—	—	—
	90·40	86·35	100·09	100·00	100·00	97·79	84·47	100·00
				(at.% (normalized to 100))				
Fe	3·95	38·55	39·02	1·84	31·76	1·25	3·30	—
Ni	64·25	2·15	18·05	48·60	26·99	63·11	10·64	—
Cu	4·38	—	2·20	—	—	11·10	67·17	—
Pt	3·30	16·13	37·22	—	31·76	1·05	4·78	29·48
As	24·12	—	—	—	—	22·52	1·07	—
Pd	—	—	—	—	—	0·96	13·05	—
Bi	—	—	—	7·27	—	—	—	5·54
Te	—	43·17	3·51	37·56	9·28	—	—	64·98
S	—	—	—	4·72	—	—	—	—
	100·00	100·00	100·00	100·00	100·00	100·00	100·00	100·00

TABLE 2—contd.

Element	2483	2482
Fe	9·23	6·26
Ni	—	1·21
Cu	17·44	15·03
Pt	4·54	10·78
Pd	8·46	3·55
Bi	16·63	4·09
Te	34·24	51·04
S	9·48	8·03
	100·00	100·00
	(at.%)	
Fe	13·93	9·94
Ni	—	1·82
Cu	23·15	20·97
Pt	1·95	4·90
Pd	6·71	2·96
Bi	6·71	1·73
Te	22·63	35·46
S	24·92	22·21
	100·00	100·00

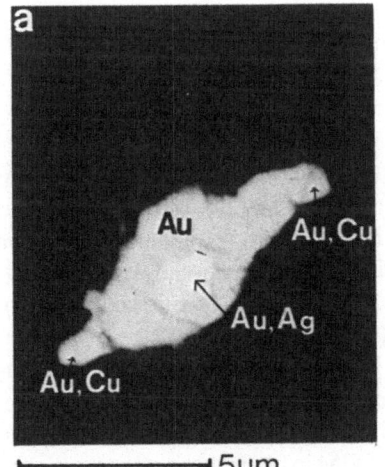
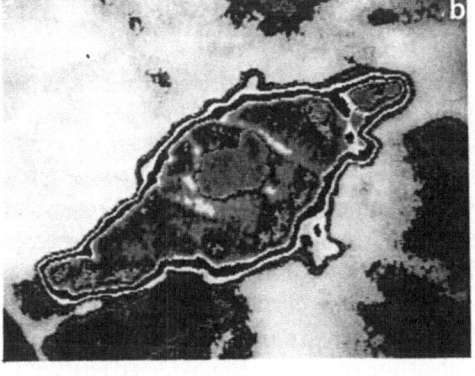

5µm

FIG. 8. (a) Native gold, showing inhomogeneous distribution of Cu and Ag. Cu is enriched in the edges, Ag in the central part of the grain (236). (b) Scanning picture of (a): Dark grey parts show enrichments of Au–Cu, Au–Ag respectively.

PGM form as a simple trapping of already formed sulphides or alloys depending on fS_2. Entrapment of immiscible PGE-bearing liquids during chromite precipitation, followed by fS_2 increase leading to a sulphurization of PGE-bearing alloys to precipitate sulphides was supported by Legèndre & Johan (1981).

Acoje laurites and alloys form euhedral inclusions, excluding an exsolution of PGE from chromitites. Preliminary proton-microprobe investigations on ruthenium in Cr-spinels indicate that the concentration of this metal is not above the detection limit of 14 ppm.

It is assumed that fractional crystallization leads to a precipitation of PGMs in the magma, followed by a nucleation of chromite. The formation of the Ni–Fe sulphide rim (Fig. 4(b)) could be due to prior crystallization of laurite at high temperatures and a precipitation of Ni–Fe sulphide at sufficiently low temperatures ($<600°C$). A sulphurization after entrapment of Ru–Os–Ir–Ni–Fe alloys by chromite seems to be excluded by the lack of fissures around the laurite crystal at least in the plane of this section.

Multiple phase inclusions of laurite + pentlandite + bornite, exsolutions during temperature decrease, have also been reported from the Meratus-Bobaris ophiolite zone in Kalimantan (Burgath & Mohr, 1986) and from the Shetland ophiolite (Prichard *et al.*, 1986). The coexistence of sulphides and alloys in the same sample is remarkable.

Formation of Pt–Pd Intermetallic Phases

Irregular, tiny Pt–Pd tellurides or bismuthides and arsenides, as found in Acoje have not yet been reported from other ophiolites. It must be kept in mind, however, that previous investigations in most of the ophiolites have been limited to chromitite layers. Sperrylite, Pd-antimonides, Pt-arsenides, potarite, etc., are known from chromitites in the Shetland ophiolite (Prichard *et al.*, 1986).

Many similarities exist between the J-M Reef, Stillwater complex and the Ni–Cu sulphide and Pt–Pd bearing zone in Acoje: Volborth *et al.* (1986) described a similar mineral association, i.e. sulphides, particularly pentlandite, rimmed by chalcopyrite, low temperature sulphides such as violarite and mackinawite, PGE-tellurides, arsenides, sulphides and alloys from the J-M Reef, Stillwater complex. The association of PGM and magnetite, as a product of serpentinization as well as oxidation of sulphides, and serpentine is also characteristic. Irregular composition of those minerals is typical. Au–Ag alloys are also present and are interpreted to be of secondary origin. High carbon contents and the presence of chlorine are additional features common in both environments (Orberger *et al.*, 1987).

The PGE are certainly of a common source. Ru, Os and Ir are primary, whereas Pt, Pd and Rh are fractionated into dunites. Textures of Pt–Pd minerals in Acoje, such as the occurrence in oxidized parts of sulphides, etc., heterogeneous compositions, irregular shapes and the very small size of PGM in black dunite lenses point to a secondary origin involving a later remobilization of PGEs. Mobilization of Pt, Pd and Rh from sulphides could be one possibility.

Experimental investigations on PGE solubilities (Pt, Pd, Rh, Ru) in sulphides and alloys, carried out by Mackowicky *et al.* (1986), report an incorporation of those elements in pyrrhotite at 900°C and moderate to high fS_2. However pyrrhotite is rather rare in Acoje, while Cu- and Ni-sulphides, chalcopyrite, pentlandite and their low temperature alteration products are dominant. Mackowicky *et al.* (1986) did not find any appreciable solubility of Pt, Pd, Rh and Ru in chalcopyrite, but pentlandite dissolves up to 12·5 wt.% Pd and up to 12·4 wt.% Rh, but no Pt.

Proton-microprobe investigations, carried out on sulphides from the Stillwater complex, Montana and Sudbury, Canada by Cabri *et al.* (1984) resulted in the detection of 1900–13 600 ppm Pd, 26–110 ppm Rh and 14–80 ppm Ru in Stillwater pentlandites, whereas in Sudbury sulphides Pd and Rh are below the detection limits of 1·2–3 ppm, Pt is below the detection limit of 50–60 ppm. Preliminary qualitative proton-microprobe studies on Acoje sulphides indicate no PGE in the Cu- and Ni-sulphides above these detection limits. Only one pentlandite grain contained low amounts of Pd. Ag has been found in many Cu- and Ni-sulphide grains.

The previous hypothesis, that PGEs are liberated from the actually present Cu–Ni sulphide structures has not been confirmed by proton-microprobe analyses, perhaps small amounts of palladium could have been derived from pentlandites. The source of the other PGM-forming metals such as Bi and Te cannot be answered at present. No evidence for their solid solution in sulphides is indicated by proton microprobe.

On the basis of these results, the question about the secondary origin of the Pt and Pd in the Acoje ophiolite block still remains unsolved.

Anomalous high halogen concentrations (Orberger *et al.*, in preparation) suggest that Pt and Pd could have been transported as, for example, chloride complex. Remobilization of Pt and Pd and their transport as halogen complexes, especially complexing with chlorine will be discussed in Orberger *et al.* (in preparation) with regard to the calculations by Mountain & Wood (1988).

The occurrence of Pt, Pd, Bi, Te and As in Acoje serpentinites and the observation of their intermetallic phases agree with the suggestion made by Mountain & Wood (1987) that the presence of Te, Bi and As in hydrothermal solutions reduces the solubility of PGE.

SUMMARY AND CONCLUSION

The Acoje block, part of the Zambales ophiolite complex in NW Luzon, Philippines, hosts metallurgical chromitite in the transition zone between upper depleted mantle and lower oceanic crust. Chromitite lenses, enveloped by dunite, occur within the tectonized harzburgite close to the mantle/crust boundary. Disseminated Ni–Cu sulphide mineralization occurs in black dunite lenses within green dunite. The sequence of ultramafic cumulates has a thickness of about 650 m and consists of four major, highly serpentinized dunitic layers. The $(Mg + Fe + Mn + Ni)/Si$ ratio $> 1·97$ in serpentinites indicates an original composition of almost pure olivine. Serpentinized dunites alternate with layers of less intensive serpentinized olivine-websterite, websterite, clinopyroxenite and rarely lherzolite; wehrlite is presently observed only as a 20 m thick layer in the upper part of the ultramafic cumulates.

Two groups of platinum-group element bearing minerals within the ultramafic sequence can be distinguished:

1. Os–Ir–Ru sulphides and alloys in chromitite lenses;
2. Pt–Pd–(Ni–Cu)–Te–Bi–S–As minerals in black dunite lenses, associated with Ni–Cu sulphides.

Laurites are Os-enriched and form euhedral inclusions (5–10 μm) with Cr-spinels. Some laurites are characterized by an Os-rich core and a Ni-sulphide rim. The zoning is probably due to T- and fO_2 variation during crystallization. Laurites and Os–Ru–Ir alloys are formed during chromite precipitation. Pt–Pd–(Ni–Cu)–Te–Bi–S–As minerals (2–5 μm) on the other hand are located in oxidized parts of Ni–Cu sulphides or as inclusions in magnetite and replacements at the margin of magnetized sulphides in serpentinites.

It is suggested that the formation of the PGM in black dunites is due to syn- or postserpentinization processes. Platinum and palladium are redistributed, the presence of tellurium, bismuth and arsenic reduced their solubility and result in a precipitation of low-temperature intermetallic phases.

Preliminary proton-microprobe investigations on Cu- and Ni-sulphides indicate that they contain Pt, Pd, Rh, Ru, Te, Bi and As below the detection limit. Single pentlandite grains show low Pd concentrations.

ACKNOWLEDGEMENTS

We wish to thank the 'Deutsche Forschungsgemeinschaft' for the support of these studies (DFG grant: Fr 240-42).

The authors wish to thank the Philippine Bureau of Mines and GeoSciences, particularly G. Bacuta and Dr G. Balce and the geologists of the Acoje Mining Company for their help during field and underground work. Our thanks are also due to Professor H. Wänke for help and hospitality, J. Huth and Dr G. Burchard for carrying out the SEM-photographs and EDS-analyses. Dr K. Traxel kindly enabled us to carry out qualitative proton-microprobe analyses. β-autoradiography investigations were provided by Dr H. M. Prichard. We appreciate the assistance of our students J. Alleweldt, B. Bühn, B. Lünenschloss and E. Rainer during this project. Thanks are due to our colleagues in the Institut für Mineralogie und Lagerstättenlehre for their support and discussions. We are grateful to Professor G. C. Ulmer and Dr M. Hock for their help to initiate this project and to Professor E. F. Stumpfl, who critically reviewed the manuscript and made numerous suggestions.

REFERENCES

Abrajano, T. A. (1984). The petrology and low-temperature geochemistry of the sulphide bearing mafic and ultramafic units of the Acoje Massif, Zambales Ophiolite, Philippines: Characterization of mineral and fluid equilibria. Ph.D. thesis, Washington University, 260 pp.

Augé, T. (1985). Platinum-group mineral inclusions in ophiolitic chromitite from the Vourinos complex, Greece. *Can. Mineral.*, **23**, 163–71.

Bacuta, G. C. (1978). Geology of some alpine-type chromite deposits in the Philippines. *J. Geol. Soc. Philippines*, **33**(2), 44–80.

Bacuta, G. C., Lipin, B. R., Gibbs, A. K. & Kay, R. W. (1988). Platinum-group element abundance in chromite deposits of the Acoje ophiolite block, Zambales ophiolite Complex, Philippines (Abstr.). This volume, pp. 381–2.

Barnes, S.-J., Naldrett, A. J. & Gorton, M. P. (1985). The origin of the fractionation of platinum-group elements in terrestrial magmas. *Chem. Geol.*, **53**, 303–23.

Burgath, K.-P. & Mohr, M. (1986). Chromitites and platinum-group minerals in the Meratus-Bobaris Ophiolite zone, southeast Borneo. In *Metallogeny of Basic and Ultrabasic Rocks*, eds M. J. Gallagher, R. A. Ixer, C. R. Neary & H. M. Prichard. Institution of Mining and Metallurgy, pp. 333–49.

Cabri, L. C., Blank, H., Goresy, A. E., Laflamme, J. H. G., Nobiling, R., Sizogoric, M. B. & Traxel, K. (1984). Quantitative trace-element analysis of sulphides from Sudbury and Stillwater by proton-microprobe. *Can. Mineral.*, **22**, 521–42.

Constantinidis, C. C., Kingston, G. A. & Fisher, R. C. (1980). The occurrence of platinum-group minerals in the chromitites of the Kokkinorotsos chrome mine, Cyprus. *Ophiolites. Proceedings to the International Ophiolite Symposium, Cyprus*, ed. A. Parayioutou, Geological Survey Department, Nicosia, pp. 93–101.

Fuller, M., Williams, T. S., McCabe, R., Encino, J., Almasco, J. & Wolfe, J. A. (1983). Paleomagnetism of Luzon. In *The Tectonic and Geologic Evolution of the Southeast Asian Seas and Islands*, ed. D. E. Hayes. Geophysical Monograph Series, Vol. 27, American Geophysical Union, Washington D.C., pp. 79–94.

Gijbels, R. A., Millard, H. F., Jr, Desborough, G. A. & Bartel, A. J. (1974). Osmium, ruthenium, iridium and uranium in silicates and chromite from the eastern Bushveld Complex, South Africa. *Geochim. Cosmochim. Acta*, **38**, 319–37.

Hawkins, J. W. & Evans, C. A. (1983). Geology of the Zambales Range, Luzon, Philippine Islands: ophiolite derived from an island arc-backarc basin pair. In *The Tectonic and Geologic Evolution of the Southeast Asian Seas and Islands, part 2*, ed. D. E. Hayes. Geophysical Monograph Series, Vol. 27, American Geophysical Union, Washington, D.C., pp. 95–123.

Hiemstra, S. A. (1979). The role of collectors in the formation of platinum deposits in the Bushveld complex. *Can. Mineral.*, **17**, 469–82.

Hock, M. (1983). Podiforme Chromitvorkommen in der Zambales Range, Luzon, Philippines. Dissertation, RWTH Aachen, p. 233.

Hock, M. & Friedrich, G. (1986). Structural features of ophiolitic chromitites in the Zambales Range, Luzon. *Philippines Mineralium Deposita*, **20**, 290–301.

Hulin, C. S. (1950). Results of study of Ni–Pt ores and concentrates: Acoje Mining Company, Philippine Island. *Philippine Geologist*, **4**, 11–23.

Ignacio, F. C. (1979). Composite underground plane of Acoje Mine, Acoje Mining Company Inc., Sta. Cruz, Zambales.

Karig, D. E. (1983). Accreted terranes in the northern part of the Philippine Archipelago. *Tectonics*, **2**(2), 211–36.

Legèndre, O. (1982). Mineralogie et géochimie des platinoides dans les chromitites ophiolitiques. Thèse Doc. 3 ème cycle. L'université Pierre et Marie Curie, Paris VI. p. 280.

Legèndre, O. & Augé, T. (1986). Mineralogy of platinum-group mineral inclusions in chromitites from different ophiolitic complexes. In *Metallogeny of Basic and Ultrabasic Rocks*, eds M. J. Gallagher, R. A. Ixer, C. R. Neary & H. M. Prichard. Institution of Mining and Metallurgy, London, pp. 361–72.

Legèndre, O. & Johan, Z. (1981). Mineralogie des platinoides dans les chromitites ophiolithiques. Rapport annuel d activite CNRS Juillet 1980–Juillet 1981, pp. 32–33.

Mackowicky, M., Mackowicky, E. & Rose-Hansen, I. (1986). Experimental studies on the solubility and distribution of platinum-group elements in base-metal-sulphides in platinum deposits. In *Metallogeny of Basic and Ultrabasic Rocks*, eds M. J. Gallagher, R.

A. Ixer, C. R. Neary & H. M. Prichard. Institution of Mining and Metallurgy, London, pp. 415–25.

Mountain, B. W. & Wood, S. A. (1988). Solubility and transport of platinum-group elements in hydrothermal solutions: Thermodynamic and physical chemical constraints. This volume, pp. 57–82.

Orberger, B., Hock, M., Woermann, E. & Friedrich, G. (1985). Preliminary estimates of oxygen fugacities and Zn-contents in Cr-spinels in the ultramafic and mafic sequence of the Zambales ophiolite complex, Philippines (abstr.). Symposium on Mineral Deposit Modelling, Manila, UNESCO.

Orberger, B., Friedrich, G. & Woermann, E. (1987). Halogen-concentration in the ultramafic sequence of the Acoje ophiolite block, Zambales, Philippines (in press).

Page, N. J, v. Gruenewaldt, G., Haffty, J. & Aruscavage, R. J. (1982). Comparison of platinum, palladium and rhodium distributions in some layered intrusions with special reference to the late differentiates (Upper zone). *Econ. Geol.*, **77**, 1405–18.

Potts, P. J. (1984). Neutron activation-induced beta-autoradiography as a technique for locating minor phases in thin section: application to rare earth element and platinum-group mineral analysis. *Econ. Geol.*, **79**, 738–48.

Prichard, H. M. & Neary, C. R. (1981). Chromite in the Shetland Islands ophiolite complex. UNESCO International Symposium on the metallogeny of mafic and ultramafic complexes, Athens. IGCP Project 160, p. 3.

Prichard, H. M., Neary, C. R. & Potts, P. J. (1986). Platinum-group minerals in the Shetland ophiolite. In *Metallogeny of Basic and Ultrabasic Rocks*, eds M. J. Gallagher, R. A. Ixer, C. R. Neary & H. M. Prichard. Institution of Mining and Metallurgy, London, pp. 395–414.

Stockmann, H. W. & Hlava, P. F. (1984). Platinum-group minerals in the Alpine chromitites from Southwestern Oregon. *Econ. Geol.*, **79**, 491–508.

Talkington, R. W., Watkinson, D. H., Whittaker, P. J. & Jones, P. C. (1983). Platinum-group mineral inclusions in chromite from the Bird River Sill, Manitoba. *Mineralium Deposita*, **18**, 245–55.

Talkington, R. W., Watkinson, D. H., Whittaker, P. J. & Jones, P. C. (1984). Platinum-group minerals and other solid inclusions in chromite of ophiolitic complexes: Occurrence and petrological significance. *Tschermaks Mineralogische und Petrographische Mitteilungen* (TMPM), **32**, 285–300.

Talkington, R. W., Watkinson, D. H., Whittaker, P. J. & Jones, P. C. (1986). Platinum-group element bearing minerals and other solid inclusions in chromite of mafic and ultramafic complexes: chemical compositions and comparisons (in press), IGCP.

Violette, F. (1980). Structure des ophiolites des Philippines Zambales et Palawan et de Chypre-Ecoulement asthenosphérique sous les zones d'expansion oceaniques. Thèse de Docteur 3 ème cycle, Université Nantes. 153 pp.

Volborth, A., Tarkian, M., Stumpfl, E. F. & Housley, R. M. (1986). A survey of the Pt–Pd-mineralization along the 35 km strike of the J-M Reef, Stillwater Complex, Montana. *Can. Mineral.*, **24**, 329–46.

39

Platinum-Group Element Abundance in Chromite Deposits of the Acoje Ophiolite Block, Zambales Ophiolite Complex, Philippines

GEORGE C. BACUTA JR.,[a] BRUCE R. LIPIN,[b] the late ALLAN K. GIBBS[a] & ROBERT W. KAY[a]

[a] Department of Geological Sciences, Cornell University, Ithaca, New York, 14853, USA
[b] Branch of Eastern Mineral Resources, US Geological Survey, Reston, Virginia, 22092, USA

Whole rock platinum-group element (PGE) analyses of 36 stratigraphically controlled chromitite samples from the peridotite unit of the Acoje ophiolite sequence reveal the following: (a) Chromitites occurring within harzburgite of the mantle section (Group I chromitites) have abundances ranging from <20 to 120 Ir, <100 to 150 Ru, 0·7 to 17·8 Rh, <1·0 to 66·0 Pt and <0·5 to 25·0 Pd (all in ppb). (b) Those located within the basal cumulate dunite-wehrlite of the crustal section (Group II chromitites) show concentrations ranging from 2·6 to 759 Rh, 2·8 to 5958 Pt and 2·3 to 8351 Pd; two samples have 30 and 460 Ir, 830 and 1100 Ru (all in ppb).

PGE chondrite normalized (CN) ratios and patterns show that Group I chromitites and some Group II chromitites fall within range and field previously defined for ophiolitic chromitites (e.g. Page & Talkington, 1984; Talkington & Watkinson, 1986). Their Ir and Ru CN ratios are around 0·1 × chondrite while their Pt and Pd CN ratios are till 0·001 × chondrite, resulting in convex upward to negatively sloping CN patterns. Several Group II chromitites with high Rh, Pt and Pd CN-ratios display shallower to positively sloping CN patterns. The above data indicate that chromitites with high PGE abundances, particularly Rh, Pt and Pd, and positively sloping PGE-CN patterns may occur in ophiolitic chromitites and are not exclusive features of stratiform-type chromitites. As such, these deposits may also be economically potential as PGE sources.

The Rh, Pt and Pd-rich Group II chromitites are associated with visible base metal sulphide mineralization; they contain higher sulphur (50–1100 ppm) compared to Rh, Pt and Pd-poor chromitites. Preliminary mineralogical investigations indicate that the PGE abundances may be accounted to PGE alloys (Fe–Pt, Cu–Pd–Fe), sulphides (laurite, erlichmanite(?), vysotskite(?)), telluride (moncheite) and possible solid substitution in base metal sulphides. Petrographic and chemical data of sulphide mineralized chromitites and their associated dunitic host rocks suggest a primary magmatic origin of base metal sulphides superposed by later postmagmatic alteration (Abrajano & Bacuta, 1982). The diverse distribution of the PGE in the Acoje chromitites are examined in terms of PGE fractionation, sulphur saturation and chromite crystallization.

REFERENCES

Abrajano, T. A. & Bacuta, G. C. (1982). Platiniferous Fe–Ni–Cu sulfides in an Alpine terrane, Zambales, Republic of the Philippines. Geol. Soc. Am. Abstract with Meeting, p. 429.

Page, N. J & Talkington, R. W. (1984). Palladium, platinum, rhodium, ruthenium and iridium in peridotites and chromitites from ophiolite complexes in New Foundland. *Can. Mineral.*, **22**(1), 137–49.

Talkington, R. W. & Watkinson, D. H. (1986). Whole rock platinum-group element trends in chromite-rich rocks in ophiolitic and stratiform igneous complexes. In *Metallogeny of Basic and Ultrabasic Rocks*, eds M. J. Gallagher, R. A. Ixer, C. R. Neary & H. M. Prichard. Institution of Mining and Metallurgy, London, pp. 427–440.

40

Platinum-Group Minerals in Ophiolitic Chromitites and Alluvial Placer Deposits, Meratus-Bobaris Area, Southeast Kalimantan

KLAUS-PETER BURGATH

Bundesanstalt für Geowissenschaften und Rohstoffe (BGR), Stilleweg 2, Postfach 510153, D-3000 Hannover 51, Federal Republic of Germany

ABSTRACT

Platinum-group mineral (PGM) from alpine-type chromitites and placer deposits in southeast Kalimantan were studied by reflected-light microscopy, electron microprobe and a few SEM examinations. The PGM found in five chromitites of the Meratus-Bobaris ophiolite are small ($<10 \mu m$). They occur as Ru-rich laurites, Os- and Ir-rich laurites and, rarely, as Os–Ir alloys. Euhedral PGM inclusions in chromite are interpreted to have formed before chromite crystallization at $T > 900°C$ and $\log fS_2$ above -2.3. For the subhedral and anhedral PGM at chromite-serpentine contacts and within serpentine interstitial to chromite a decrease in temperature and sulphur activity is proposed. Os- and Ir-enrichment in some of these grains are assumed to have occurred during serpentinization.

The PGM obtained from the placers are Ag- and Cu-bearing Pt–Fe alloys (sizes up to >1 mm), rutheniridosmine containing Au and Cu, and Cu-bearing osmiridium. The Pt–Fe alloy grains vary in morphology from rounded or angular to embayed or 'spongy', the Ru–Ir–Os alloys are subangular to angular.

The Ru–Ir–Os phases in the placers could have come from the Meratus-Bobaris ophiolite chromitites by erosion alone, but this explanation is not conformable with the large size of the placer grains ($>200 \mu m$) compared to the small PGM found in the chromitites. An origin from non-ophiolitic ultramafic-mafic intrusions is not supported by the vague and disputed existence of such (Pre-Cretaceous) complexes in East Kalimantan. But there is evidence that the Ru–Ir–Os alloys have formed via reworking and reconcentration of PGE during transport from Meratus-Bobaris ophiolite source rocks after serpentinization and partial lateritization. Furthermore, the close association of the Pt-Fe PGM with the Ru–Ir–Os alloys in the placers suggests a common source and similar formation of both varieties of PGM. The individual placers are obviously characterized by a rather limited compositional variation of the Pt–Fe alloys and a separated grouping in the ternary diagram $Pt - \Sigma Fe + Ni + Cu - \Sigma Pd + Rh + Ru + Ir + Os$.

INTRODUCTION

The occurrence of PGM in the alluvial gold-diamond placers of the Riam Kanan, east of Martapura (Southeast Kalimantan) has been known for a long time. Initial mineralogical studies by Stumpfl & Clark (1966) and Stumpfl & Tarkian (1973) revealed the following minerals: Pt–Fe alloy (including 'native platinum'), native osmium, iridosmine, osmiridium, laurite and a Pd-As mineral. Studies by the author of three typical placers in the area demonstrated that the Pt(-Pd-Rh) compounds are dominant. The relationship between these placer minerals, including native Au and diamonds and the geology of the catchment area of the Riam Kanan have not yet been clarified in detail. Primary sources for gold and diamond are discussed in Koolhoven (1935) and Eckhardt *et al.* (1987). Considering the siderophilic behaviour of the PGE, it is the working hypothesis that the PGM in Kalimantan are derived from the ultrabasic to basic rocks.

The oldest rocks known from Southeast Borneo are deep-sea sediments and basic volcanics which belong to the Upper Lower Cretaceous Alino Formation. A spatially associated ophiolite, dismembered into the two parts Meratus and Bobaris, is probably of the same age. Both formations are correlated with a Middle Cretaceous rifting in the Palaeo-Sulawesi Sea and subsequent westerly directed subduction below the 'Sundaland plate'. The subduction is dated at $114 \pm 1 \cdot 4\,Ma$ (H. Kreuzer, BGR). Probable trench sediments are represented by the Paniungan Beds. A further widespread formation includes the calcalkaline magmatic rocks and molasse-type sediments. The igneous rocks are assigned to a magmatic arc above the subduction zone and the sediments are classified as deposits in the former foreland basin. This complex of Turonian to Senonian age is called the Manunggal Formation (Katili, 1981*a,b*; Burgath & Mohr, 1986; Eckhardt *et al.*, 1987). The deeper parts of the Manunggal Formation consist of thick conglomerate beds which are reported to contain diamonds, native gold and platinum besides the detritus of all older formations (van Bemmelen, 1970).

The investigation of the Meratus-Bobaris ophiolite was part of a joint mineral survey of Indonesian ophiolites by BGR and the Directorate of Mineral Resources, Bandung (1981–5). The complex was classified as a MORB-type ophiolite on the basis of field data, petrology and geochemistry of the ultramafic-mafic rock sequence. In the course of the work five small chromitites *in situ* and numerous eluvial chromitite occurrences were discovered. In a few of these chromitites PGM were found (Burgath & Mohr, 1986; Eckhardt *et al.*, 1987). A simplified geological map of the central Meratus-Bobaris zone including PGM-bearing chromitites and some investigated PGM-bearing placers is shown in Fig. 1. This paper deals with PGM found in the chromitites and the gold-diamond placers.

ANALYTICAL METHODS

The PGM (laurites) found in chromitite sections were analysed using the Siemens ELMISONDE microprobe at BGR. Pure elements of Ru, Os, Ir and $(Fe,Ir)_9S_{10}$

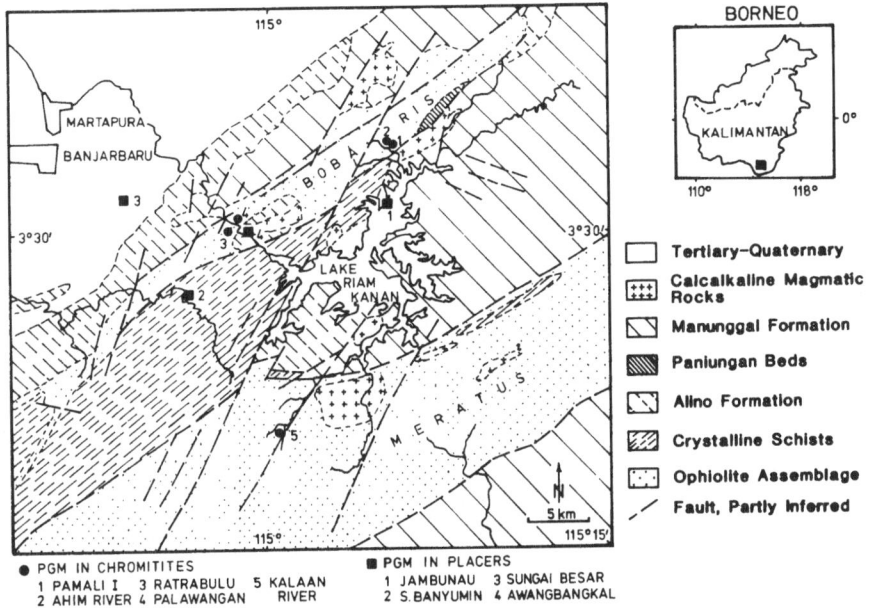

FIG. 1. Geological sketch map of the Central Meratus-Bobaris area and the locations of investigated PGM occurrences. The rock units are explained in the text.

(for S) were used as standards at an operating voltage of 24 kV and a specimen current of 0·7 nA. The results were corrected for matrix effects using the program MAGIC V (Frazer *et al.*, 1966). The results are listed in Table 1.

In principle the analyses show only the Ru, Ir and Os contents. Pt, Pd and Rh could not be detected by analysis, which might be due to a combination of the very low concentrations of these elements and the overlapping of the lines RhL_α by $RuL_{\beta1}$ and RhL_β, or PdL_α by $RuL_{\beta2}$. Moreover, all analyses yielded a small excess of sulphur, which is possibly due to the overlapping of RuL_α by $SK_{\alpha,\beta}$.

The PGM in the concentrates from alluvial placers were analysed with a CAMEBAX microprobe at BGR (20 kV/15 nA; standards for Rh, Ru, Ir, Os, Au, Ag, Cu, Ni: pure elements; for Pt: Pt–Fe alloy; for Pd, Sb: sulphides; for As: Ga arsenide; for Te: $Pd_{35}Te_{42}Hg_{22}$; for S: pyrite; corrections by J. Lodziak/BGR.

The results are listed in the Tables 5 and 6. Preliminary examinations to identify the PGM in a placer concentrate from Sungai Banyumin were carried out with the Etec Autoscan TN 200 (SEM) at BGR used in the backscattered mode and an energy dispersive Si (Li) detector.

Pt–Pd–Rh AND Ru–Ir–Os MINERALS IN OPHIOLITES: A REVIEW

Page (1971) and Page *et al.* (1982, 1984) demonstrated that (1) chromitites have in general higher PGE concentrations than the surrounding ultramafic rocks and that

TABLE 1

Electron microprobe analyses (wt.% and atomic proportions) of PGM from chromitites of the Bobaris-Meratus area

Laurite grain (analysis)		1	2a	2b	3a	3b[b]	4	5	6
Chromitite occurrence			Pamali Ia				Pamali Ib		Kalaan
wt.%[a]	Ru	27·60	44·59	28·82	42·26	20·57	35·62	21·97	35·17
	Os	20·63	11·28	6·42	11·57	1·30	15·42	25·17	20·91
	Ir	5·29	5·93	28·83	5·64	44·48	10·28	16·83	6·16
	Fe	5·51	1·08	1·25	1·59	1·93	2·11	2·07	0·92
	Cr	6·84	0·58	0·82	1·08	1·70	0·36	0·79	1·54
	S	29·17	37·50	30·89	35·35	21·23	36·47	31·92	35·80
Total		95·04	100·96	97·03	97·49	91·21	100·26	98·75	100·50
Atomic prop.	Ru	0·60	0·75	0·59	0·76	0·62	0·62	0·44	0·62
	Os	0·24	0·10	0·07	0·11	0·02	0·02	0·27	0·20
	Ir	0·06	0·05	0·31	0·05	0·70	0·35	0·18	0·06
	Fe	—	0·03	0·02	0·02	0·01	0·01	0·06	—
	S	2·00	2·00	2·00	2·00	2·00	2·00	2·00	2·00

[a]Ni not determined. Cu not detected. The significant Fe and Cr concentrations in the small PGM are attributed either to the fluorescence of the host chromite (samples 1,6) and/or attached iron oxides (samples 2a–5: PGM at chromite-serpentine contacts or in serpentine) or they are due to their direct X-ray emission. The atomic proportions are calculated after subtraction of Cr and Fe using the Cr/Fe ratio of chromite spot analyses very close to the measured PGM. The influence of iron oxides is difficult to ascertain. Because the atomic concentration of Fe is mostly less than 2% after chromite subtraction, the atomic proportions are not further corrected.

[b]Calculation of the atomic proportions to 3 atoms/formula unit in analysis 3b leads to a strong positive deviation in the Ru + Ir + Os sum (1·35). Thus additional masked Ir is assumed, which was taken into account for the calculation of the Ir percentage in Fig. 4.

(2) chromitites in post-Paleozoic ophiolites, in contrast to Precambrian stratiform chromitites, are characterized by a relative enrichment in Ru,Os and Ir and a paucity of Pt,Pd and Rh. Accordingly, almost exclusively PGM of the Ir,Os and Ru group have been so far found in ophiolitic chromitites, in which they occur in the form of sulphides, sulpharsenides, and alloys (Talkington *et al.*, 1982, 1984; compilations in Prichard *et al.*, 1986; and in Legèndre & Augé, 1986).

In some chromitites in the Shetland Complex, which, being Caledonian in age, falls between the two (extreme) ages mentioned above, Ru-Os-Ir phases are infrequently accompanied by minerals of the Pt-Pd-Rh group, and Pt-Fe alloys have not been found yet (Prichard *et al.*, 1981, 1986). Another exception is the occurrence of rare, minute grains of Pt-Ir alloys or of $PtAs_2$ and Pt–Fe alloys adhering to laurite grains (enclosed in chromite) in alpine-type chromitites in southwest Oregon, USA. At least part of the Pt–Fe alloys must be interpreted as having been formed under reducing conditions during the alteration of the peridotitic host rock of the chromitites (Stockmann & Hlava, 1984).

The Zambales ophiolite in the Acoje area, Northern Luzon (Philippines), represents another exception; in this case, too, discrete PGM of the Pt-Pd-Rh group occur. They are not, however, necessarily always associated with chromitites; they form along cracks or in, or on, alteration products after chalcopyrite and pentlandite and are rather associated with serpentinization of dunite in the 'transition zone' between mantle and crust (Orberger, in prep.). Pt–Fe alloys have not been encountered here, either.

In principle, therefore, it should be assumed that the PGM in the Riam Kanan placers originate from two different sources. It seems likely that the chromitites in the adjacent Meratus-Bobaris ophiolite might have been the source of the PGM in which Ir, Os, Ru are predominant.

PGM IN CHROMITITES FROM MERATUS-BOBARIS

The microscopic analysis of the Meratus and Bobaris chromitites demonstrated the presence of PGM in at least five occurrences:

Pamali I, Bobaris	(8 PGM grains)
Ahim river near Pamali I	(3 PGM grains)
Ratrabulu, Bobaris	(4 PGM grains)
Palawangan, Bobaris	(1 PGM grain)
Kalaan river, Meratus	(2 PGM grains)

These PGM are $<10\,\mu$m in diameter and they occur in various habits:

— as idiomorphic to hypidiomorphic inclusions in chromite, locally in association with a silicate grain or with (Cu, Ni, Fe) sulphides (L(6) in Fig. 2);
— as isolated xenomorphic splinter-like fragments in deformed primary silicates (replaced by serpentine) between chromite that shows a slight cataclastic overprint (L(4) in Fig. 3)
— in weakly deformed or cataclastic chromitite in the form of xenomorphic grains intergrown with chromite, at the boundaries between chromite and silicates (serpentine) (e.g. L(5) in Fig. 3).

The position of the various grains indicates that the xenomorphic and splintery shape is probably due to deformation and/or serpentinization (see also Prichard *et al.* (1986)). The idiomorphic shape suggests that the enclosed PGM were enveloped by chromite after they had crystallized (Stockmann & Hlava, 1984; Talkington *et al.*, 1984; Burgath & Mohr, 1986).

The frequent occurrence of PGM and (Ni, Cu, Fe) sulphides within a single polished section or even together within the same chromite grain is of special interest. An example of this is a polyphase inclusion with an octahedral grain shape from the Kalaan, Meratus occurrence consisting of a core of PGM (Os-laurite) with almost idiomorphic grain shape and a margin of chalcopyrite + pentlandite with

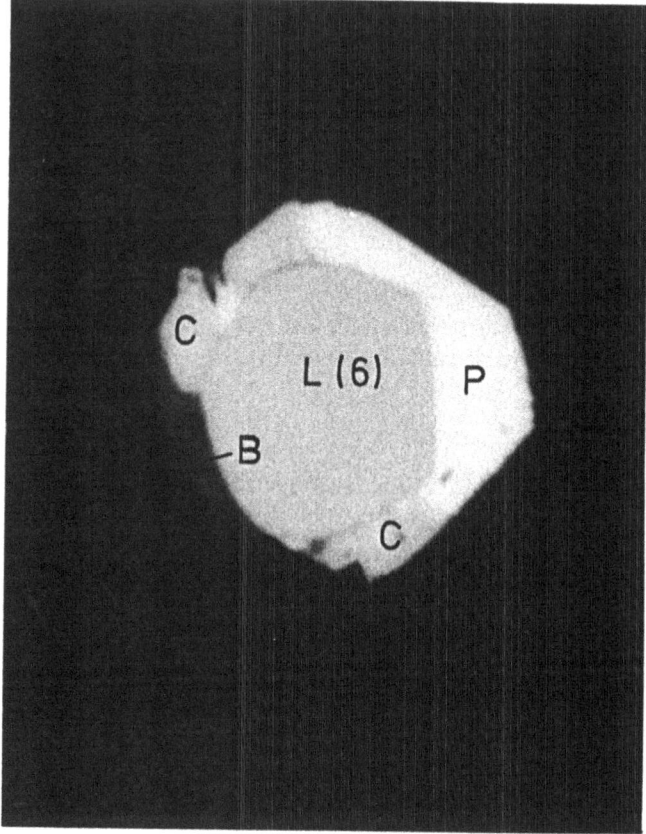

FIG. 2. Octahedral sulphide inclusion in chromite (black). L(6): laurite; P: pentlandite; C: chalcopyrite; B: bornite. P, C and B probably have been produced through decomposition of high-T chalcopentlandite. Chromitite occurrence Kalaan, Meratus. Field of view 21 × 26 μm.

excess NiS + bornite. Inclusions showing a similar paragenesis (pentlandite-bornite-PGM) were also found in chromitites from the Barru ophiolite zone in southwest Sulawesi (Eckhardt *et al.*, 1987). The distinct boundaries between the core and margin as shown in Fig. 2 indicate a stable association. The large PGM grain was apparently enclosed by a Cu-Ni-Fe sulphide droplet before being enveloped by chromite.

A similar complex inclusion comprising laurite + millerite + Ni-Ir-Fe-Cu sulphide was observed by Stockmann & Hlava (1984) in an alpine-type chromitite in southwest Oregon. Unlike the Kalaan idiomorph this one is roundish and therefore was presumably not solid when it was enveloped by the crystallizing chromite, but was in the form of a molten droplet. In contrast the octahedral shape of the Kalaan inclusion could indicate that the grain was already crystallized before being enclosed by the chromite. This would suggest that originally there was a cubic high-temperature phase of the 'chalcopentlandite'—NiS + chalcopyrite + bornite

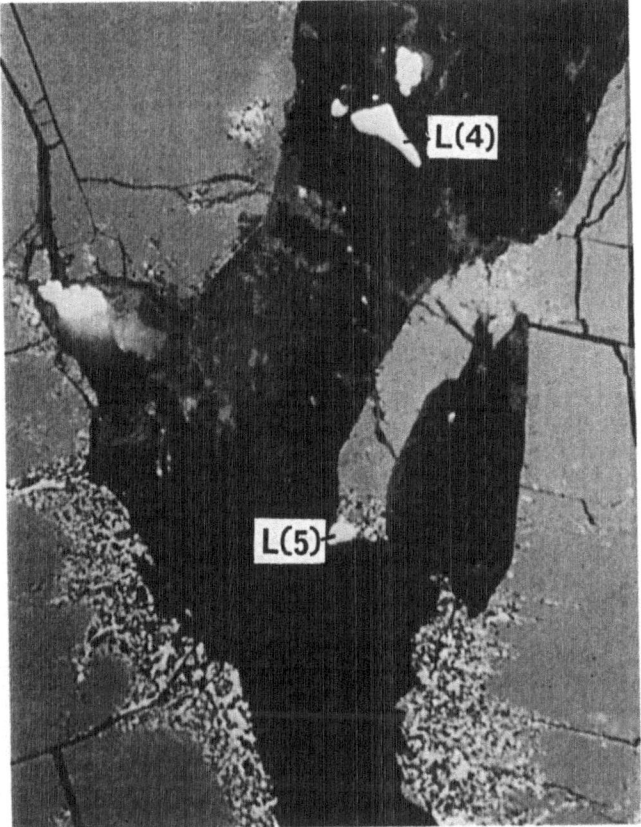

Fig. 3. Two PGM grains in serpentine (black), which is interstitial between chromite (grey). L(4) is Ru-dominated laurite, L(5) is Os-enriched laurite. The chromite is partly replaced by ferrite-chromite in the peripheral zone. Chromitite occurrence Pamali I, Bobaris. Field of view 136 × 170 μm.

on cooling. According to Pauly (1958), a temperature of about 850°C must be reached for crystallization of this kind of 'chalcopentlandite'.

But this assumption of a solid phase incorporation in chromite is opposed by the low upper stability of chalcopentlandite compared with the crystallization temperature of chromite (estimated at >900°C). The 'octahedral' outline may also be a function of the negative crystal shape of the enclosing chromite. Thus it is probable that the aggregate has crystallized from a sulphide droplet (with the earlier crystallized laurite 'core') which was captured during chromite formation.

The purely textural relationship between the laurite inclusion and the narrow sulphide rim corresponds to experimental evidence for the limited solubility of ruthenium in pyrrhotite (up to 3·6 wt.% at 900°C), in pentlandite (up to 12 wt.% at 500°C) while it is not soluble at all in chalcopyrite, bornite and 'intermediate Cu-Fe-S solid solution' (Makovicky *et al.*, 1986). Another indication of lower temperature rim formation is the very low Fe content and absence of Cu in the

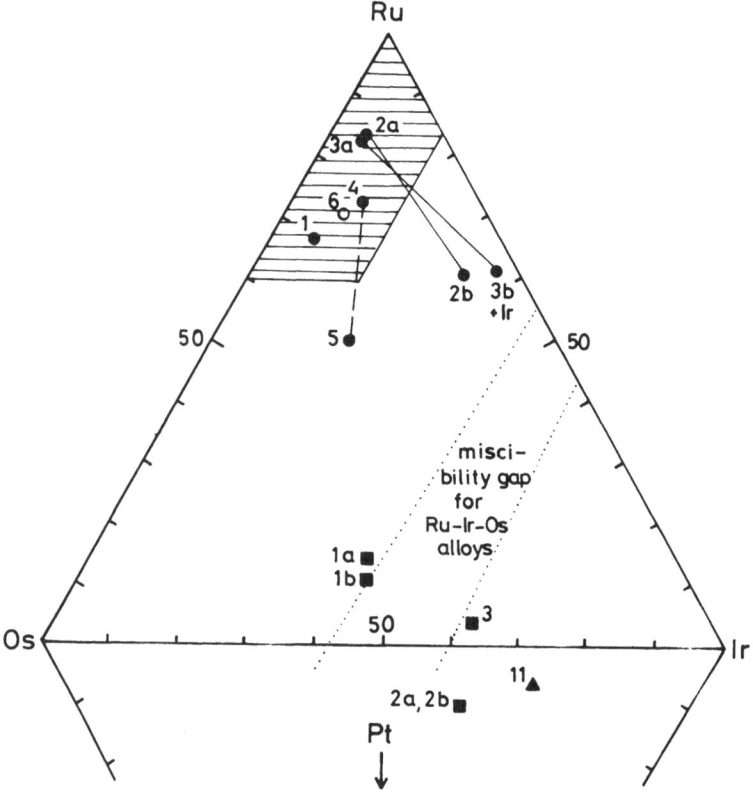

Fɪɢ. 4. Compositions of PGM in Meratus-Bobaris chromitites and in alluvial placers in the Meratus-Bobaris region. Dots: PGM (laurite) in Bobaris chromitites, circles: PGM (laurite) in Meratus chromitite. Numbers as sample nos. in Table 1. The lines connect associated PGM in the same polished section (4–5) or different compositions within the same PGM grain (2a–2b, 3a–3b). The compositions of 2b, 3b and 5 probably resulted from Ru mobilization during serpentinization. Shaded zone: range of laurite inclusions in chromite from other ophiolitic chromitites (data from Constantinides *et al.* (1980), Talkington *et al.* (1984), Augé (1986). Squares: Os–Ir–Ru alloys in the Sungai Besar placer. Triangles: Os–Ir–Ru alloy in the Jambunau placer. The numbers correspond to the grain nos. in Tables 3 and 4.

laurite core (see Table 1); laurite in coexistence with high-temperature Cu-Fe phases or chalcopyrite is able to dissolve up to 5·8 wt.% Cu and up to 6·3 wt.% Fe (Makovicky *et al.*, 1986).

Optical examination confirmed that all PGM in the Meratus-Bobaris area belong to the laurite group ((Ru, Ir, Os)S_2). Only in Pamali I was laurite found once in association with another PGM phase (probably osmiridium).

Six laurites were analysed with the electron probe and their compositions plotted in the Ru-Os-Ir triangle (Fig. 4). Most specimens plot near the Ru apex; five lie within the field in which laurites from other ophiolite chromitites plot. However, if the grains are examined separately, they show distinct differences in composition, which probably depend on whether they occur within, at the margin or outside the host chromite.

Analyses 1 and 6 represent two homogeneous laurite grains which are included in unfractured chromite. From their mode of occurrence it must be assumed that they crystallized at a high temperature and were protected while serpentinization of the silicate minerals in the chromitite took place. Taking their Os contents into account and assuming a crystallization temperature of 900°C (approximate lowest limit of chromite crystallization), the S_2 activities (in log units) during the formation of these laurites can be estimated at values above $-2 \cdot 3$ (Fig. 6; Stockmann & Hlava, 1984).

Analysis 4 (Table 1) refers to a sharp angular laurite fragment of homogeneous composition in the serpentine matrix. In the Ru-Os-Ir triangle (Fig. 4) this fragment plots near the compositions of grains 1 and 6 and consequently, it could represent a fragmented former inclusion. But the fragment has a slightly higher Ir content and its position in Fig. 4 is between the group 1, 6 (probably the grains with the highest crystallization temperature in the system) and a group closer to the Ru apex (analyses 2a and 3a). 2a and 3a represent the major parts of two inhomogeneous laurite grains which occur on the boundary between chromites and adjacent silicate minerals. Thus these grains with the highest Ru contents probably crystallized at lower temperatures than the included laurites 1 and 6. Because the Ru-RuS$_2$ curve is shifted towards lower fS_2 values in the T–fS_2 system (Fig. 6), it is concluded that the sulphur pressure decreased with decreasing temperature during the formation of the laurites 1, 6, then 4, and finally 2a and 3a.

In the same polished section, grain no. 5 occurs near grain no. 4 at the margin of a spongy ferrite-chromite alteration zone around a homogeneous chromite (Fig. 3). However, both grains can be clearly distinguished by their Ru and Os contents. Grain no. 5 with a high percentage of Os may have been formed at a high (magmatic) temperature and increasing fS_2 or at decreasing T with fS_2 remaining the same (see Fig. 5). Due to the very close association of the two grains within a few mm of each other, however, such changes are rather unrealistic. Enrichment of Os coupled with a lower sulphur fugacity than is the case of Ru-rich laurite, however, can also be attained during serpentinization, since this takes place under a low partial pressure of oxygen. Evidence for this includes the following: decrease of $Fe^{2+}/(Fe^{2+} + Fe^{3+})$ from chromite cores to the altered rims of ferrite-chromite as found in chromite ores with serpentinized silicate matrix (Burgath, 1983), and the frequent occurrence of grains of Ni_3Fe in serpentine. With decreasing fO_2, however, the solubility of sulphur increases (MacLean, 1969). The abnormal position of grain no. 5 in the Ru-Os-Ir triangle indicates that enrichment of Os was associated with mobilization of Ru.

Analyses 2b and 3b represent patchy zones within large laurite grains of composition like 2a or 3a; these grains are intergrown with chromite at the contact with serpentine. The patches could be interpreted as due to exsolution, since there is a miscibility gap between Ru and Ir (at least in alloys). At 1300°C this gap is between 45 and 57 at.% Ir and widens with decreasing temperature (Raub, 1964; Harris & Cabri, 1973). The percentage of Ir in the patches varies from 32 to 36 at.%, i.e. the corresponding points in the Ru-Os-Ir triangle are distinctly above, not below the miscibility gap, and for lower temperatures, both would plot within it. Therefore,

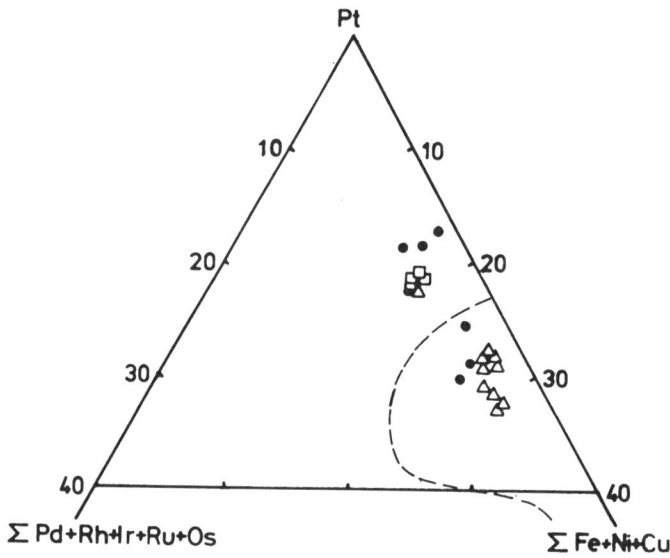

FIG. 5. Pt–Fe alloys from the Sungai Besar and Jambunau placers in the ternary diagram Pt–(Pd + Rh + Ir + Ru + Os) – (Fe + Ni + Cu). Squares: samples from Sungai Besar. Triangles: samples from Jambunau. Dots: Pt–Fe alloys from 'Riam Kanan, SE Borneo' (analyses reported in Stumpfl & Tarkian (1973)). The broken line encloses the approximate field of Pt–Fe alloy compositions from layered and concentric complexes and from Pre-Cambrian placers modified after Tarkian (1987), Fig. 4.

exsolution is unlikely, and these Ir enrichments (probably associated with loss of Ru) are also assumed to be due to serpentinization. The positions of all examined PGM in the T–fS_2 system are shown in Fig. 6.

Finally in the Meratus-Bobaris area, there is apparently no relationship between the occurrence of PGM and the composition of chromitite hosts. The Cr/Fe and Cr/Al ratios in chromite from chromitites containing PGM vary between 1 and 3 or 1·5 and 4·5 (Eckhardt *et al.*, 1987). It is, however, conspicuous that two of the PGM-bearing chromitites, Pamali I and Ahim river, are assigned to the mantle-crust transition zone (or probably the lowest part of the Ultramafic Layered Sequence) and by far the majority of the PGM grains are found in the Pamali I occurrence. This may indicate that the chromitites of the transition zone represent the most favourable zone for enrichment of Ru-Os-Ir compounds. Further investigations are planned.

PGM FROM ALLUVIAL PLACERS IN THE MERATUS-BOBARIS REGION

Samples were taken of PGM from the alluvial placers Jambunau, Sungai Besar, and Sungai Banyumin in the lowlands between the Meratus and Bobaris ranges and to north-west of the Bobaris complex. The PGM grains belong to three principal

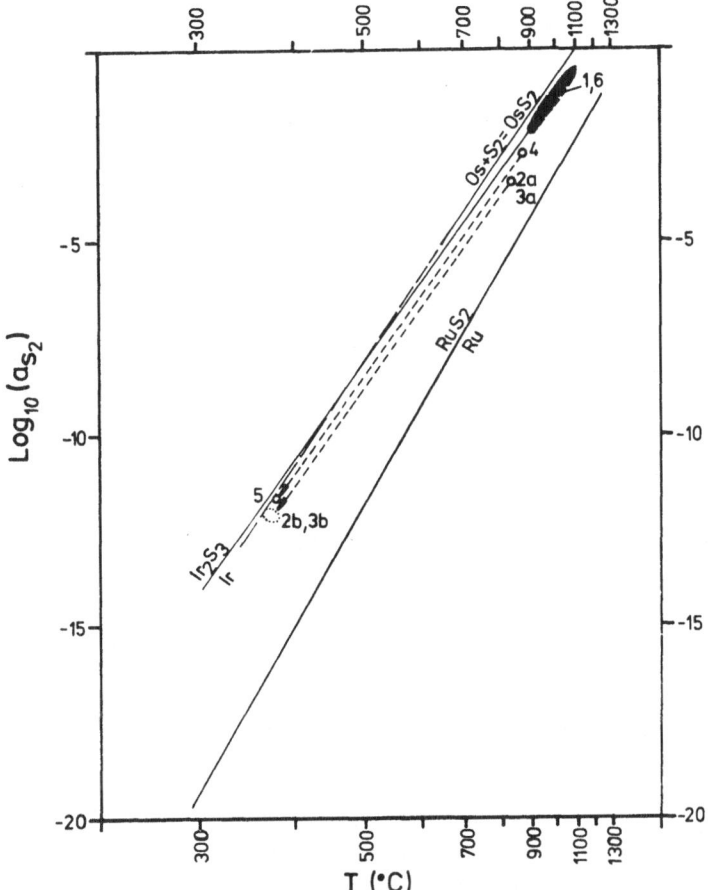

FIG. 6. Temperature and sulphur fugacity supposed during formation of PGM in Meratus-Bobaris chromitites. Numbers as sample nos. in Table 1. Element/sulphide systems from Stockmann and Hlava (1984).

morphological types: rounded flakes, embayed grains, and non-water-worn angular to subangular grains (some of them with indistinct crystal forms). The phases found in Jambunau and Sungai Besar are listed in Tables 2 and 3, and their compositions are given in Tables 4 and 5. The mineral names follow the nomenclatures of Harris & Cabri (1973) and Cabri & Feather (1975).

The compositions of the grains 1–8 in Table 4 are shown in Fig. 5 and grain 9 is plotted in Fig. 4. 1a and 1b are parts of a diffuse accretion (?) rim around grain 1 (see Fig. 7). There is no significant difference in composition between 1 and 1a, 1b except a slight enrichment of Ir in the rim. Grain 5 (Fig. 8) gives the impression that it is an aggregate of several grains which may have become welded together. Morphologically, it has a certain similarity with the 'knobby' nuggets described by Cabri et al. (1981).

TABLE 2

Platinum-group minerals from the alluvial Jambunau placer, at the margin of the Riam Kanan reservoir near Bunglei

	PGM				Associated minerals in concentrate
No. of grain	Composition	Corresp. analyses in Table 4	Morphology	Remarks	Gold diamond silicates (*not yet examined*)
1	Pt–Fe alloy	1	Subrounded		
1a	Pt–Fe alloy	2	Thin coating on grain 1		
1b	Pt–Fe alloy	3	Thin coating on grain 1		
2	Pt–Fe alloy	4	Angular, with projections and deep embayments		
3	Pt–Fe alloy	5	Subrounded, with embayments and projections		
4	Pt–Fe alloy	6	Subrounded		
5	Pt–Fe alloy	7	Embayed, spongy	With pores filled with silicate (Fe-Al-silicate ± Ca, Ti,V)	
6	Pt–Fe alloy	8	Splinter, angular to subrounded		
7	Pt–Fe alloy	9	Splinter, angular to subrounded		
8	Pt–Fe alloy	10	Rounded with embayments		
9	Pt-osmiridium	11	Subrounded, with embayments	With inclusion-like zone (composition: Fe >> Si,Ti,Cr, Al,Ca)	

The compositions of the grains 1–3 in Table 5 are shown in Fig. 4 and the grains 4 and 5 are plotted in Fig. 5. For the latter, the at.% PGE are 82·8–83·7, and the atomic proportions Σ PGE/ΣFe, etc., are 3·3:0·6. Because no X-ray analyses are obtained, the Pt–Fe alloys could be either native platinum or isoferroplatinum (a grain of comparable composition is reported in Cabri & Feather (1975), p. 125).

It is evident from Fig. 5 that the Pt–Fe grains from Jambunau and Sungai Besar roughly correspond to the alloys from 'Riam Kanan, Southeast Borneo' which were reported by Stumpfl & Tarkian (1973). But with one exception the Jambunau and

TABLE 3

Platinum-group minerals from the alluvial Sungai Besar placer, east of Banjarbaru (south of Martapura)

		PGM			Associated minerals in concentrate
No. of grain	Composition	Corresp. analyses in Table 5	Morphology	Remarks	*Gold diamond pink garnet zircon other silicates (not yet examined)*
1	Ruthenirid-osmine	1a,1b	Small sub-angular, compact	Inhomogeneous distribution of minor Pd and Au in grain	
2	Pt-osmiridium	2a,2b	Small sub-angular to partly rounded, compact		
3	Osmiridium	3	Small angular to partly rounded, compact		
4	Pt–Fe alloy	4a,4b	Large rounded grains with a partially lobate marginal zone and few indistinct pores		
5	Pt–Fe alloy	5a,5b	Large rounded grains with a partially lobate marginal zone and few indistinct pores		

Sungai Besar grains are clearly arranged in two separate groups and the Jambunau group is inside the field covering the compositions of Pt–Fe alloys from layered and concentric intrusions and from the Witwatersrand (compiled in Tarkian, 1987).

Finally it is noted that Ru-free iridosmine and Ru-free osmiridium from southeast Kalimantan, which were reported by Stumpfl & Clark (1966) and shown in Harris & Cabri (1973, Fig. 4) have not yet been found in the Jambunau and Sungai Besar concentrates. The same goes also for Ru-poor iridosmine which was found by Stumpfl & Tarkian (1973).

In addition, an initial study was carried out on a PGM-bearing placer concentrate from Sungai Banyumin (a drainage system north of Gunung Damargusang southeast of Banjarbaru). The concentrate was handed over by a gold panner who assured the place of its origin in this area.

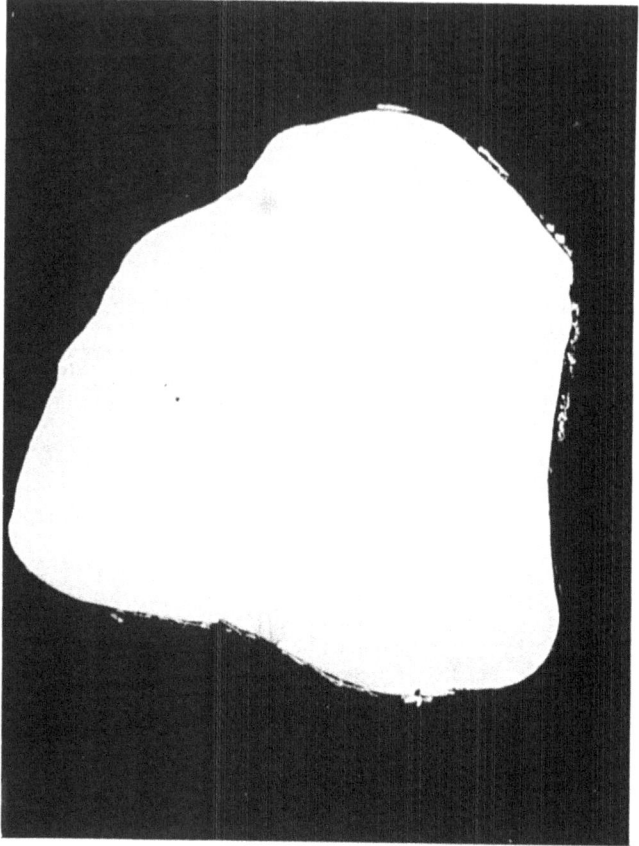

FIG. 7. Pt–Fe alloy with diffuse overgrowth? of the same composition. Jambunau placer deposit. Field of view $1 \cdot 1 \times 1 \cdot 4$ mm.

The grain size of the PGM in this sample varies from $<200 \, \mu$m to $1750 \, \mu$m. Their morphology is almost exclusively angular to subrounded with signs of mechanical wear. Some grains have cubic (octahedral) outlines, other grains show irregular projections. Associated minerals in the concentrate are gold, magnetite, ilmenite, chromite, pink garnet (rounded), whitish brown cloudy garnet (euhedral), bluish green amphibole and other not yet identified silicates.

Initial SEM examinations of three PGM grains yielded Pt–Fe alloy compositions. In one grain an embayment was found to contain $Si > Al \gg Ti, Fe, Cr$.

CONCLUSIONS

A first hypothesis is that the rutheniridosmine and osmiridium grains in the placers might have originated from the Meratus-Bobaris ophiolite-chromitites. However

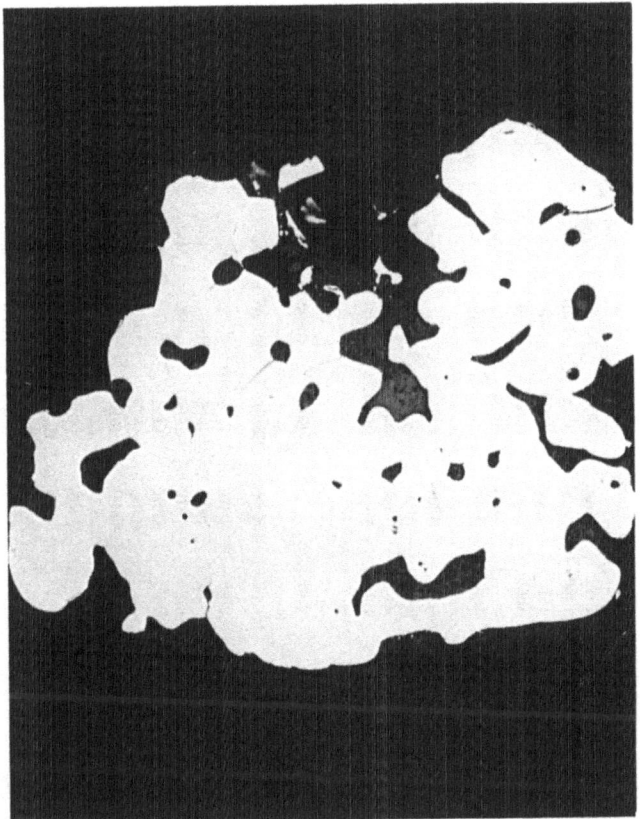

FIG. 8. Spongy Pt–Fe alloy with interstitial silicates. Jambunau placer deposit. Field of view
$1 \cdot 1 \times 1 \cdot 4$ mm.

osmiridium was found only once in close association with laurite in a Bobaris
chromitite and the alloys in the placers contain considerable amounts of Cu and
especially Pt. These have not yet been found in the laurites in chromitite. Difficulties also arise when comparing the size of the PGM; in chromitite they are $<10\,\mu$m
whereas in the Os-Ir-Ru minerals from the placers they are $>200\,\mu$m in size. It is
rather unlikely that PGM grains of this large size might have been overlooked in a
primary environment in the Meratus-Bobaris ophiolite sequence which, at least in
the relevant parts, is well investigated. In addition, the assumption of a pure detrital
origin by erosion of ophiolite source rocks is not supported by the fact that as far as
is known, such large grains of Os–Ir–Ru alloys have also not been reported from
primary sites in other well-exposed and investigated ophiolites (the only exception
known to the author is a $200\,\mu$m large laurite in a Shetland chromite (H. Prichard,
unpublished note)).

Hence a different origin is assumed for the Os-Ir-Ru placer grains. Cabri &
Harris (1975) introduced the bulk $Pt \times 100/(Pt + Ir + Os)$ ratio for placers as an
index of the geological and tectonic environment of the source rocks (ratio >86:

TABLE 4

Electron microprobe analyses (wt.% and atomic proportions) of PGM from the Jambunau placer deposit

PGM grain		1	1a	1b	2	3	4	5	6	7	8	9
Spot analysis		1	2	3	4	5	6	7	8	9	10	11
wt.%	Pt	88·99	88·44	88·19	86·89	86·85	88·91	90·06	86·75	86·56	88·88	6·45
	Pd	0·32	0·62	0·36	1·23	0·94	0·66	0·44	0·92	1·20	0·80	0·07
	Rh	0·81	0·72	0·66	1·02	1·20	0·83	1·53	0·99	1·01	1·04	0·97
	Ru	0·05	—	0·12	—	—	0·09	0·21	0·06	0·08	0·02	0·27
	Ir	0·29	0·77	0·64	0·50	0·44	—	1·60	0·53	0·57	0·19	69·15
	Os	0·44	0·51	0·39	0·42	1·32	0·38	0·95	0·26	0·38	0·47	23·98
	Au	—	—	—	—	—	—	—	—	—	—	—
	Ag	0·15	0·36	0·25	0·25	—	—	0·36	0·02	—	0·52	—
	Cu	0·35	0·52	0·04	1·51	1·21	0·70	0·30	2·17	2·33	0·06	0·64
	Ni	—	—	—	0·03	—	—	0·08	0·03	—	—	—
	Fe	8·49	8·35	8·69	8·25	8·07	8·48	5·05	8·24	8·07	8·51	0·31
	As	0·01	—	0·01	—	0·02	—	—	—	—	—	0·02
	Sb	—	—	—	—	—	—	—	—	—	—	0·03
	Te	—	—	—	—	—	—	—	—	—	—	0·03
	S	0·03	0·18	—	—	0·11	0·06	—	0·08	0·06	0·03	0·03
Total		99·93	100·47	99·35	100·10	100·16	100·11	100·58	100·05	100·26	100·52	101·92
Atomic prop.	Pt	2·89	2·85	2·88	2·76	2·76	2·86	3·10	2·73	2·72	2·85	0·06
	Pd	0·02	0·04	0·02	0·07	0·06	0·04	0·03	0·05	0·07	0·05	<0·01
	Rh	0·05	0·04	0·04	0·06	0·07	0·05	0·10	0·06	0·06	0·06	0·02
	Ru	<0·01	—	0·01	—	—	0·01	0·01	<0·01	<0·01	<0·01	<0·01
	Ir	0·01	0·03	0·02	0·02	0·01	—	0·06	0·02	0·02	0·01	0·66
	Os	0·02	0·02	0·01	0·01	0·04	0·01	0·03	0·01	0·01	0·02	0·23
	Au	—	—	—	—	—	—	—	—	—	—	—
	Ag	0·01	0·02	0·02	0·01	0·15	—	0·02	<0·01	—	—	—
	Cu	0·04	0·05	0·01	0·15	—	0·07	0·03	0·21	0·23	0·05	0·02
	Ni	—	—	—	<0·01	—	—	0·01	<0·01	—	0·01	—
	Fe	0·97	0·94	0·99	0·92	0·90	0·96	0·61	0·91	0·89	0·95	0·01
	As	<0·01	—	<0·01	—	<0·01	—	—	—	—	—	<0·01
	Sb	—	—	—	—	—	—	—	—	—	—	<0·01
	Te	<0·01	0·01	—	—	0·01	<0·01	—	0·01	<0·01	<0·01	<0·01
	S	—	—	—	—	—	—	—	—	—	—	—

—: not detected.

TABLE 5
Electron microprobe analyses (wt. % and atomic proportions) of PGM from the Sungai Besar placer deposit

PGM grain		1		2		3	4		5	
Spot analysis		1a	1b	2a	2b	3	4a	4b	5a	5b
wt.%	Pt	0·40	0·39	9·79	9·92	1·40	92·10	90·61	90·69	91·16
	Pd	0·23	—	0·04	—	—	0·36	0·38	—	0·34
	Rh	0·34	0·41	1·63	1·58	0·59	1·57	1·43	1·69	1·97
	Ru	7·93	6·17	1·69	1·62	2·04	0·11	0·22	0·22	0·10
	Ir	43·77	44·65	54·62	55·03	61·60	0·51	0·54	0·92	0·84
	Os	47·53	49·96	33·34	33·07	35·22	0·93	1·08	0·71	0·72
	Au	0·32	—	—	0·05	—	—	—	—	—
	Ag	—	—	—	0·24	0·03	0·37	0·16	0·51	0·41
	Cu	0·43	0·41	0·48	0·54	0·62	0·53	0·43	0·21	0·21
	Ni	0·03	0·05	0·02	0·10	0·01	0·10	0·14	0·12	0·08
	Fe	0·19	0·16	0·12	0·22	0·23	4·74	4·70	5·10	5·10
	As	0·05	—	—	—	—	—	—	0·04	0·06
	Sb	—	—	0·01	—	0·07	—	—	—	—
	Te	—	—	0·01	0·07	0·08	0·12	0·10	0·02	0·06
	S	—	—	0·02	—	—	—	—	—	—
Total		101·22	102·20	101·75	102·44	101·89	101·44	99·79	100·23	101·05
Atomic prop.	Pt	<0·01	<0·01	0·09	0·09	0·01	3·15	3·16	3·13	3·11
	Pd	<0·01	—	<0·01	—	—	0·02	0·02	—	0·02
	Rh	0·01	0·01	0·03	0·03	0·01	0·10	0·10	0·11	0·13
	Ru	0·14	0·11	0·03	0·03	0·04	0·01	0·10	0·02	0·01
	Ir	0·40	0·41	0·51	0·51	0·58	0·02	0·02	0·03	0·03
	Os	0·43	0·46	0·32	0·31	0·33	0·03	0·04	0·03	0·02
	Au	<0·01	—	—	<0·01	—	—	—	—	—
	Ag	—	—	—	<0·01	<0·01	0·02	0·01	0·03	0·03
	Cu	0·01	0·01	0·01	0·02	0·02	0·06	0·05	0·02	0·02
	Ni	<0·01	<0·01	<0·01	<0·01	<0·01	0·01	0·02	0·01	0·01
	Fe	0·01	<0·01	0·01	0·01	0·01	0·57	0·57	0·62	0·61
	As	<0·01	—	—	—	—	—	—	<0·01	0·01
	Sb	—	—	<0·01	—	<0·01	—	—	—	—
	Te	—	—	<0·01	<0·01	<0·01	0·01	0·01	<0·01	<0·01
	S	—	—	<0·01	—	—	—	—	—	—

—: not detected.

concentrically zoned ultramafic-mafic intrusions, ratio <36: alpine-type perido-tites). A bulk PGE analysis of the investigated placers in southeast Kalimantan has not yet been obtained, but the grain ratio of the Pt–Fe alloys and Os-Ir-Ru phases found in Jambunau and Sungai Besar indicates a clear predominance of the Pt–Fe alloy. It may be supposed therefore that the $Pt \times 100/(Pt + Ir + Os)$ is fairly high. This would imply that the Os–Ir–Ru alloys originated independently of the Meratus-Bobaris ophiolite. They could be pure relics of 'now eroded or hidden Pre-Cretaceous ultramafic-mafic intrusions' when there was a cratonic environ-ment in East Kalimantan. Such a thick crustal environment is indicated by dia-monds associated with the PGM in the Upper Cretaceous Manunggal Formation in the Martapura area. But the presence of cratonic portions in the eastern part of Kalimantan is still disputed (Katili, 1984, 1986; Burret & Slait, 1984; Gatinsky & Hutchison, 1984).

There is also strong evidence for a 'weak oxidation' of Ru-rich laurites in the Meratus-Bobaris chromitites with formation of lower-sulphur phases (Os- and Ir-enriched laurites) during serpentinization or alteration of the peridotite–chromitite association (Fig. 4). It is conceivable that Os–Ir alloys with low Ru content could have formed in the final stage of this process. Such a formation at moderate or low temperatures is called into question by Cabri & Harris (1975) but this explanation would not require the assumption of 'unknown intrusions'. The origin of the Os–Ir–Ru alloys is therefore attributed to reworking of the Meratus-Bobaris ophiolite source rocks after serpentinization and partial lateritization and reconcentration of PGE during transport to the placers.

This explanation is supported by new insights into the mobilization and re-precipitation of the PGE. They indicate a strong dependency of the PGE transport on the Eh–pH of the fluid systems (Mountain & Wood, 1988; Plimer & Williams, 1988; Bowles, 1988).

It has not been possible to identify the equivalent of the Pt–Fe alloys of the placers in the Meratus-Bobaris chromitites and the general rarity of PGM-rich in Pt in alpine-type chromitites places doubt on a primary source in ophiolitic rocks. Additionally, these grains are often characterized by high Ni contents which, like Fe, increases during serpentinization or alteration of the peridotitic chromitite host (Table 4 in Stockmann & Hlava, 1984; Prichard *et al.*, 1986; Orberger, in press). Even more difficulties arise with respect to the comparison of size, since the Pt-Fe PGM in the placers attain sizes of more than 1 mm (size of the Pt-rich PGM in chromitites as described above: maximum 10 μm). But a possible source in 'now eroded or hidden Pre-Cretaceous ultramafic-mafic intrusions' (layered or con-centric type) as discussed above for the Os–Ir–Ru alloys is also rather unlikely. One part only of the Jambunau Pt–Fe alloys is inside the compositional field of alloys from layered or concentric complexes but all investigated grains from Sungai Besar including a grain from Jambunau are enriched in Pt and plot far outside this field (Fig. 5). On the other hand, the close association of the Pt–Fe alloys with Os–Ir–Ru alloys in SE Kalimantan suggests a common source for both varieties of PGM. This is strongly supported by grains composed of Pt–Fe alloy intergrown with osmiri-dium, which have been described by Stumpfl & Clark (1966).

As to the origin of the Pt-Fe PGM in southeast Kalimantan, the 'grains with a growth rim' and the 'embayed, spongy grains' with suggested silicate inclusions from the Jambunau placer are possibly of importance. They indicate grain growth or grain enlargement in the placer. In addition, the separated grouping and limited compositional variation of the Pt–Fe alloys from the individual placers in the ternary diagram Pt–Σ(Fe,Ni,Cu)–Σ(Pd,Rh,Ru,Ir,Os) could point to different conditions in each placer environment.

ACKNOWLEDGEMENTS

The author is grateful to H. Mauer for making the polished sections, to Dr Th. Weiser, J. Lodziak and D. Laszinski for carrying out the microprobe analyses, and to Eng. E. Knickrehm for some reconnaissance SEM investigations. Dr M. Mohr and M. Klimainsky are thanked for their contribution to the poster presented at the Geo-Platinum 87 symposium. H. M. Prichard, J. F. W. Bowles and an unknown reviewer are thanked for critically reading the manuscript and making a number of helpful suggestions. This project was in part funded by the Federal Ministry of Research and Technology, Bonn.

REFERENCES

Augé, T. (1986). Platinum-group-mineral inclusions in chromitites from the Oman ophiolite. *Bull. Minéral.*, **109**, 301–4.

Bowles, J. F. W. (1988). Further studies of the development of platinum-group minerals in the laterites of the Freetown Layered Complex, Sierra Leone. This volume, pp. 273–80.

Burgath, K.-P. (1983). Untersuchungen griechischer Chromitvorkommen. *Geol. Jb.*, **D60**, 67–175.

Burgath, K.-P. & Mohr, M. (1986). Chromitites and platinum-group minerals in the Meratus-Bobaris Ophiolite Zone, Southeast Borneo. In *Proc. Int. Symp. Metallogeny of Basic and Ultrabasic Rocks, Edinburgh*, eds M. J. Gallagher *et al.*, The Institution of Mining and Metallurgy, London, pp. 333–49.

Burret, C. & Slait, B. (1984). Southeast Asia is a part of an Early Paleozoic Gondwanaland. Abstr. *5th Reg. Conf. on Geology and Mineral Resources of SE Asia*, Geological Society of Malaysia, Kuala Lumpur, p. 3.

Cabri, L. J., Criddle, A. J., Laflamme, J. H. Gilles, Bearne, G. S. & Harris, D. C. (1981). Mineralogical study of complex Pt-Fe nuggets from Ethiopia. *Bull. Minéral.*, **104**, 508–25.

Cabri, L. J. & Feather, C. E. (1975). Platinum–iron alloys: a nomenclature based on a study of natural and synthetic alloys. *Can. Mineral.*, **13**, 117–26.

Cabri, L. J. & Harris, D. C. (1975). Zoning in Os–Ir alloys and the relation of the geological and tectonic environment of the source rocks to the bulk Pt:Pt + Ir + Os ratio for placers. *Can. Mineral.*, **13**, 266–74.

Constantinides, C. C., Kingston, G. A. & Fisher, P. C. (1980). The occurrence of platinum-group minerals in the chromitites of the Kokkinorotsos chrome mine, Cyprus. In *Ophiolites, Proc. Int. Ophiolite Symp., Nicosia*, ed. A. Panayiotou, Geological Survey Department, Nicosia, pp. 93–101.

Eckhardt, F. J., Burgath, K. P. & Mohr, M. (1987). Anwendung von Prospektionsmethoden in Ophiolithen Südost-Asiens—Abschlußbericht Indonesien. Abschlußbericht zum

BMFT-Förderungsvorhaben NTS 3016.3, Band 1, Bundesanstalt für Geowissenschaften und Rohstoffe, Hannover, 132 pp.

Frazer, J. Z., Fitzgerald, R. W. & Reid, A. M. (1966). Computer programs EMx and EMx2 for electron-microprobe data processing. Unpublished report of the University of California, Scripps Institute of Oceanography and Institute for the Study of Matter, La Jolla, 67 pp.

Gatinsky, Y. G. & Hutchison, Ch.S. (1984). Cathaysia. Gondwanaland and the Paleotethys in the evolution of Continental Asia. *Abstr. 5th Reg. Conf. on Geology and Mineral Resources of SE Asia*, Geological Survey of Malaysia, Kuala Lumpur, pp. 11–12.

Harris, D. C. & Cabri, L. J. (1973). The nomenclature of the natural alloys of osmium, iridium and ruthenium based on new compositional data of alloys from world-wide occurrences. *Can. Mineral.*, **12**, 104–12.

Katili, J. A. (1981*a*). Contradicting views on the plate tectonic of Indonesia and their bearing on heatflow research. *Bull. Geol. Res. Devel. Centre, Indonesia*, pp. 21–9.

Katili, J. A. (1981*b*). Geology of Southeast Asia with particular reference to the South China Sea. *Energy 6*, **11**, 1077–91.

Katili, J. A. (1984). Evolution of plate tectonic concepts and its implication for the exploration of hydrocarbon and mineral deposits in Southeast Asia. *Pangea*, **3**, 5–18.

Katili, J. A. (1986). On charting new paths to mineral exploration. Conf. Indonesian Mining Industry, General Review, 7.—9.5.1986, Jakarta.

Koolhoven, W. C. B. (1935). The primary occurrence of diamond in South Borneo. *Verh. geol.-mijnb. Genoot. Nederland, Geol. Serie*, pp. 189–232.

Legèndre, O. & Augé, T. (1986). Mineralogy of platinum-group mineral inclusions in chromitites from different ophiolitic complexes. In *Proc. Int. Symp. Metallogeny of Basic and Ultrabasic Rocks, Edinburgh*, eds M. J. Gallagher *et al.*, The Institution of Mining and Metallurgy, London, pp. 361–72.

MacLean, W. H. (1969). Liquidus phase relations in the FeS–FeO–Fe_2O_3–SiO_2 system and their application in geology. *Econ. Geol.*, **64**, 865–84.

Makovicky, M., Makovicky, E. & Rose-Hansen, J. (1986). Experimental studies on the solubility and distribution of platinum-group elements in base-metal sulphides in platinum deposits. In *Proc. Int. Symp. Metallogeny of Basic and Ultrabasic Rocks, Edinburgh*, eds M. J. Gallagher *et al.*, The Institution of Mining and Metallurgy, London, pp. 415–25.

Mountain, B. W. & Wood, S. A. (1988). Solubility and transport of platinum-group elements in hydrothermal solutions: thermodynamic and physical chemical constraints. This volume, pp. 57–82.

Orberger, B. (in press). Chromite- and Ni-Pt-mineralization in the Acoje ophiolite block, Zambales (Philippines)—the influence of oxygen fugacity on the formation of the ultramafic and mafic sequence and their containing chromite and Ni-Cu-sulphide concentration. (See also this volume, pp. 361–80.)

Page, N. J (1971). Sulfide minerals in the G and H chromitite zones of the Stillwater Complex, Montana. US Geol. Survey Prof. Paper 694, 20 pp.

Page, N. J, Pallister, J. S., Brown, M. A., Smewing, J. D. & Haffty, J. (1982). Palladium, platinum, rhodium, iridium and ruthenium in chromite-rich rocks from the Samail ophiolite, Oman. *Can. Mineral.*, **20**, 537–48.

Pauly, H. (1958). Igdlukunguaq nickeliferous pyrrhotite. Texture and composition. A contribution to the genesis of the ore type. *Meddelser om Groenland*, **157**(3), 1–169.

Page, N. J, Engin, T., Singer, D. A. & Haffty, J. (1984). Distribution of platinum-group elements in the Bati Kef chromite deposit, Guleman-Elazig area, eastern Turkey. *Econ. Geol.*, **79**, 177–84.

Plimer, I. R. & Williams, P. A. (1988). New mechanisms for the mobilization of the platinum-group elements in the supergene zone. This volume, pp. 83–92.

Prichard, H. M., Potts, P. J. & Neary, C. R. (1981). Platinum-group element minerals in the Unst chromite, Shetland Isles. *Trans. Inst. Min. Met.*, **90B**, 186–8.

Prichard, H. M., Neary, C. R. & Potts, P. J. (1986). Platinum-group minerals in the Shetland ophiolite. In *Proc. Int. Symp. Metallogeny of Basic and Ultrabasic Rocks, Edinburgh*, eds M. J. Gallagher *et al.*, The Institution of Mining and Metallurgy, London, pp. 395–414.

Raub, E. (1964). Die Ruthenium-Iridium-Legierungen. *Z. Metallkunde*, **55**, 316–19.

Stockmann, H. W. & Hlava, P. F. (1984). Platinum-group minerals in alpine chromitites from southwestern Oregon. *Econ. Geol.*, **79**, 491–508.

Stumpfl, E. F. & Clark, A. M. (1966). Electron-probe microanalysis of gold platinoid concentrates from southeast Borneo. *Trans. Inst. Min. Met.*, **74**, 933–46.

Stumpfl, E. F. & Tarkian, M. (1973). Natural osmium-iridium alloys, iron-bearing platinum and a Pd–As mineral from S.E. Borneo. Reprint, 15th Congr. Geol. Soc. S. Africa, Abstracts, pp. 82–3.

Talkington, R. W., Watkinson, D. H. & Jones, P. (1982). Platinum-group minerals and other solid inclusions in chromite of mafic-ultramafic complexes. Geol. Assoc. Canada program with abstracts, Vol. 7, p. 83.

Talkington, R. W., Watkinson, D. H., Whittaker, P. J. & Jones, P. (1984). Platinum-group minerals and other solid inclusions in chromite of ophiolitic complexes: occurrence and petrological significance. *TMPM Tschermaks Min. Petr. Mitt.*, **32**, 285–301.

Tarkian, M. (1987). Compositional variations and reflectance of the common platinum-group minerals. *Mineral. Petrol.*, **36**, 169–90.

van Bemmelen, R. W. (1970). *The Geology of Indonesia*, 2nd edn. Martinus Nijhoff, The Hague, Vol. IA, 732 pp., Vol. II, 267 pp.

41

Platinum-Group Minerals in the Tiébaghi (New-Caledonia) and Vourinos (Greece) Ophiolites

Thierry Augé

GIS, BRGM-CNRS, 1A rue de la Férollerie, 45071 Orléans Cedex 2, France

Platinum-group minerals (PGM) have been found included in chromite from dunite and chromitite in the Vourinos and Tiébaghi ophiolites but appear to be absent in disseminated chromite from harzburgite: PGM consist of laurite, erlichmanite, xingzhongite, an Ir-Cu sulphide, prassoite and Ir–Os alloys in the Tiébaghi massif and laurite, Ir–Os alloys and sulpharsenides in the Vourinos massif. In both complexes, the same PGM were found in dunite and in chromitite, revealing a genetic relationship between these two rock types. Base-metal sulphides (BMS) have also been found as inclusions in chromite from dunite and chromitite. They are mainly Ni-rich pentlandite and chalcopyrite. Some pentlandites contain platinum-group elements (PGE) (mainly Rh, Ru and Pd) in solid solution (up to 43·7 wt.%) suggesting that in this context, sulphides can act as a collector of the PGE dissolved in the magma.

PGM and BMS formed early in the magma were trapped in chromite crystals. Depending on the further evolution of the magma, this chromite is now found in dunite or, if chromite crystals were concentrated, in chromitite.

42

Platinum-Group Element Chemistry in Relation to the Behaviour of Individual PGEs in Primary and Secondary Geological Processes

L. HAYNES*

Nature Conservancy Council, Northminster House, Peterborough, Cambridgeshire PE1 1UA, UK

The physio-chemical behaviour of PGE as transition metals is outlined and compared to that of the commoner light transition metals with which they are geologically associated. The ability of PGE to form both compounds and alloys is discussed, and the status of some of these unions as alloys or compounds is questioned.

These principles are then applied to explain the behaviour of PGEs during magmatic processes (including evidence obtained during smelting of PGE ores), hydrothermal alteration, serpentinization, carbonatization, weathering and erosion. An interpretation of chemical and mineralogical data from the Unst Ophiolite Complex is used to show how these principles can be of aid in mineral exploration and how the evidence of PGE concentration obtained so far from this intrusion, conforms to what can be expected from an established sequence of geological events.

*Present address: Department of Geology, Imperial College, Prince Consort Road, London SW7 2BP, UK.

43

Gold and Iridium in Sulphides from Submarine Basalt Glasses

C. L. Peach

School of Oceanography, University of Washington, Seattle, Washington 98195, USA

&

E. A. Mathez

Department of Mineral Sciences, American Museum of Natural History, New York, NY 10024, USA

Immiscible sulphide-oxide melts in submarine basalt magmas are preserved as globules in glasses. Their small size (typically $<30\,\mu$m) and even distribution within glass suggest that most had limited residence times in the host magmas and that the two melts did not equilibrate with respect to the chalcophile trace elements. Several MORB glasses from the FAMOUS area have been found for which the sulphides are of sufficient size for separation. For one, chemical and petrographic data also indicate that equilibrium between the two melts may have been approached.

Bulk suphide separates from six glasses were analysed by INAA; globules from one sample, 526-1, were also analysed individually (Table 1). Concentrations of Au and Ir in the sulphides vary sympathetically among samples. Sulphides in 526-1 have significantly higher Au and Ir contents (12·9 ppm and 373 ppb, respectively) than those of other samples, which contain 1–4 ppm Au and 44–210 ppb Ir. Because 526-1 glass is of primitive MORB composition and contains >30% modal olivine, it

TABLE 1
Representative concentrations of selected elements in 526-1 sulphides

Fe (%)	Ni (%)	Co (ppm)	Cr (ppm)	Au (ppm)	Ir (ppb)	Mass (μg)
38·7	11·6	1 624	386	12·0	383	48
39·9	15·3	1 776	408	8·9	309	73
40·2	13·4	1 782	366	12·9	666	91
39·8	14·1	1 757	372	16·3	285	67
35·3	13·8	1 628	252	10·7	346	84
39·8	12·7	1 648	393	12·6	443	25
37·4	11·4	1 651	374	10·8	423	241
41·2	13·5	1 860	404	12·7	441	136

is believed to represent a magma in which early-formed phases accumulated. In addition, the sulphides are numerous and large, have coarse internal textures, exhibit clear evidence of partial resorption and have compositions that are similar and unrelated to their masses. These features indicate that the 526-1 sulphides had remained suspended in the magma for some time, in which case their Au and Ir contents should approach those dictated by sulphide-silicate equilibrium. Assuming that Ir and Au contents of 526-1 glass are similar to those of the few other MORB glasses for which data exist, distribution coefficients for Au and Ir between sulphide and silicate melts are both of the order of 10^4.

PGE and other Trace Element Geochemistry at some Distinct Phanerozoic Stratigraphic Horizons

M. Tredoux,[a] M. J. de Wit,[b] R. J. Hart,[a]
N. M. Lindsay[a] & J. P. F. Sellschop[a]

[a]Wits-CSIR Schonland Research Centre for Nuclear Sciences, [b]BPI-Geophysical Research, University of the Witwatersrand, Wits 2050, Johannesburg, South Africa

PGE and other siderophile anomalies at the Cretaceous-Tertiary (K-T) boundary have been interpreted as evidence for catastrophic extraterrestrial influx. We list some key points from the literature which are difficult to resolve in terms of this model: (i) The relative concentrations of siderophile elements cannot be explained in terms of any known meteorite composition; (ii) The concentration of Re, Ba, As, Sb, Mo, Zn and Cd is high, relative to chrondrite (and usually interpreted to be of terrestrial origin) and Cr, As, Sb and the REE are high relative to iron meteorites; (iii) Stable isotopes of the noble gases suggest a terrestrial source; (iv) osmium isotopic studies cannot distinguish between a meteoritic or mantle component.

Arguments which have been used to support a volcanic origin for the chemostratigraphy at the K-T boundary include: (i) dust from eruptions at Kilauea (Hawaii) has been shown to be extremely enriched in Ir, Re and Au; (ii) microspherule-like particles and shock features in quartz and feldspars (both commonly interpreted as evidence for an impact) are not unique to the K-T boundary and have been documented in terrestrial eruptives.

We have conducted PGE and other trace element chemostratigraphic studies at: (i) the K-T boundary, from globally dispersed sites associated with major extinctions; (ii) near the Devonian-Carboniferous boundary in Ireland, well above the Frasnian-Famennian extinction event.

At the Stevns Klint K-T boundary (fish-clay) we have found *both* quasichondritic PGE patterns and patterns with a distinct negative Pt anomaly. A pyrite separate from the fish-clay has a positive Pt anomaly. Samples below the boundary have positive Pt anomalies and samples above have negative Pt anomalies. The highest PGE concentrations are within the fish-clay, but the other siderophiles (Ni, Co, Sb, As) peak just *below* this marker horizon. We believe that it is difficult to explain the two PGE patterns in terms of secondary alteration of an originally chondritic source, because Pd is much more mobile than Pt in diagenetic environments. The patterns may therefore be primary signatures. We have found similar patterns in Ni-Fe oxide pods in Archean ultramafic rocks. We believe that such pods are embedded in Archean sub-continental lithosphere and are sampled during major volcanic events (e.g. the Cretaceous kimberlite emplacement and major volcanism at 65 Ma, such as the Deccan plateau basalts). Volcanic dust and gas from Kilauea are much more enriched in PGE and other siderophiles than lavas—thus widespread distribution of these elements might be expected.

The widespread and homogeneous Tournasian Balleyvergan shale in Ireland shows an enrichment in PGE relative to average shales but the pattern is different from the K-T components; it resembles continental tholeiites. No major extinctions are recorded across this shale horizon.

Several other boundaries (e.g. the Eocene-Oligocene) have been shown to have Ir anomalies without attendant extinctions. Conversely, several major extinctions (e.g. the Ordovician-Silurian and the Permian-Triassic) do not coincide with any PGE (Ir) anomaly. Thus it is not yet clear how PGE enrichment in terrestrial sediments can be related to extra-terrestrial influx and/or faunal crises in the geological record. The authors would like to stress that a mantle-derived origin for the PGE enhancement at K-T cannot be excluded.

Index of Contributors

413

Subject Index

415